FATS IN ANIMAL NUTRITION

Proceedings of Previous Easter Schools in Agricultural Science, published by Butterworths, London

*SOIL ZOOLOGY   Edited by D. K. McL. Kevan (1955)
*THE GROWTH OF LEAVES   edited by F. L. Milthorpe (1956)
*CONTROL OF THE PLANT ENVIRONMENT   edited by J. P. Hudson (1957)
*NUTRITION OF THE LEGUMES   edited by E. G. Hallsworth (1958)
*THE MEASUREMENT OF GRASSLAND PRODUCTIVITY   Edited by J. D. Ivins (1959)
*DIGESTIVE PHYSIOLOGY AND NUTRITION OF THE RUMINANT   Edited by D. Lewis (1960)
*NUTRITION OF PIGS AND POULTRY   Edited by J. T. Morgan and D. Lewis (1961)
*ANTIBIOTICS IN AGRICULTURE   Edited by A. M. Woodbine (1962)
*THE GROWTH OF THE POTATO   Edited by J. D. Ivins and F. L. Milthorpe (1963)
*EXPERIMENTAL PEDOLOGY   Edited by E. G. Hallsworth and D. V. Crawford (1964)
*THE GROWTH OF CEREALS AND GRASSES   Edited by F. L. Milthorpe and J. D. Ivins (1965)
*REPRODUCTION IN THE FEMALE MAMMAL   Edited by G. E. Lamming and E. C. Amoroso (1967)
*GROWTH AND DEVELOPMENT OF MAMMALS   Edited by G. A Lodge and G. E. Lamming (1968)
*ROOT GROWTH   Edited by W. J. Whittington (1968)
*PROTEINS AS HUMAN FOOD   Edited by R. A. Lawrie (1970)
*LACTATION   Edited by I. R. Falconer (1971)
*PIG PRODUCTION   Edited by D. J. A. Cole (1972)
*SEED ECOLOGY   Edited by W. Heydecker (1973)
 HEAT LOSS FROM ANIMALS AND MAN: ASSESSMENT AND CONTROL   Edited by J. L. Monteith and L. E. Mount (1974)
*MEAT   Edited by D. J. A. Cole and R. A. Lawrie (1975)
*PRINCIPLES OF CATTLE PRODUCTION   Edited by Henry Swan and W. H. Broster (1976)
*LIGHT AND PLANT DEVELOPMENT   Edited by H. Smith (1976)
 PLANT PROTEINS   Edited by G. Norton (1977)
 ANTIBIOTICS AND ANTIBIOSIS IN AGRICULTURE   Edited by M. Woodbine (1977)
 CONTROL OF OVULATION   Edited by D. B. Crighton, N. B. Haynes, G. R. Foxcroft and G. E. Lamming (1978)
 POLYSACCHARIDES IN FOOD   Edited by J. M. V. Blanshard and J. R. Mitchell (1979)
 SEED PRODUCTION   Edited by P. D. Hebblethwaite (1980)
 PROTEIN DEPOSITION IN ANIMALS   Edited by P. J. Buttery and D. B. Lindsay (1981)
 PHYSIOLOGICAL PROCESSES LIMITING PLANT PRODUCTIVITY   Edited by C. Johnson (1981)
 ENVIRONMENTAL ASPECTS OF HOUSING FOR ANIMAL PRODUCTION   Edited by J. A. Clark (1981)
 EFFECTS OF GASEOUS AIR POLLUTION IN AGRICULTURE AND HORTICULTURE   Edited by M.H. Unsworth and D.P. Ormrod (1982)
 CHEMICAL MANIPULATION OF CROP GROWTH AND DEVELOPMENT   Edited by J. S. McLaren (1982)
 CONTROL OF PIG REPRODUCTION   Edited by D.J.A. Cole and G.R. Foxcroft (1982)
 SHEEP PRODUCTION   Edited by W. Haresign (1983)
 UPGRADING WASTE FOR FEEDS AND FOOD   Edited by D.A. Ledward, A.J. Taylor and R.A. Lawrie (1983)

These titles are now out of print but are available in microfiche editions

Fats in Animal Nutrition

J. WISEMAN, PhD
University of Nottingham School of Agriculture

BUTTERWORTHS
London   Boston   Durban   Singapore   Sydney   Toronto   Wellington

All rights reserved. No part of this publication may be reproduced
or transmitted in any form or by any means, including
photocopying and recording, without the written permission of
the copyright holder, application for which should be addressed to
the Publishers. Such written permission must also be obtained
before any part of this publication is stored in a retrieval system of
any nature.

This book is sold subject to the Standard Conditions of Sale of
Net Books and may not be re-sold in the UK below the net price
given by the Publishers in their current price list.

First published 1984

© The several contributors named in the list of contents 1984

British Library Cataloguing in Publication Data

Fats in animal nutrition.
 1. Fats in animal nutrition
 2. Fat—Metabolism
 I. Wiseman, J.
 636.089′2397   SF98.F3

 ISBN 0–408–10864–9

Library of Congress Cataloging in Publication Data
Main entry under title:

Fats in animal nutrition.

   Proceedings of the 37th Nottingham Easter School.
   Includes bibliographies and index.
   1. Oils and fats in animal nutrition—Congresses.
 I. Wiseman, J. (Julian)   II. Easter School in Agricultural
 Science (37th : 1983? : University of Nottingham?)
 SF98.O34F38   1984   636.08′52   84–352
 ISBN 0–408–10864–9

Typeset by Scribe Design, Gillingham, Kent
Printed and bound in Great Britain by Anchor Brendon Ltd,
Tiptree, Essex

PREFACE

The incorporation of fats and oils into diets for farmed animals is receiving an increasing amount of attention and the objective of the 37th Easter School in Agricultural Science was a thorough appraisal of their nutritional role. Traditionally, they have been regarded merely as suppliers of energy-yielding ingredients and accordingly have been assigned a somewhat arbitrary dietary energy value. An appreciation of the complexities of lipid digestion, absorption and transport has revealed that fats and oils are extremely variable commodities, and that their chemical structure has a marked influence upon their nutritive value. Lipids are also suppliers of fat-soluble vitamins and essential fatty acids. Adequate health status and optimum levels of performance of animals may be achieved only if there is a thorough understanding of the mode of action of these nutrients. The utilization of lipids in animal feeding systems was considered subsequently, and this was followed by a review of the importance of carcase fat. Finally, practical problems associated with the blending of fats and oils in feed mills were discussed.

Whilst it is accepted that investigations are far from complete, it is hoped that this book, in gathering together information from many diverse disciplines, will provide a useful text for all those interested in the nutritional utilization of lipids.

UNITS

SI units have been used throughout the text. Where necessary the following conversion factors were used:

1 Cal (kcal) = 0.004184 MJ
1 psi   = 6.9 kPa

The following conventions were used for fatty acid nomenclature. For brevity, common names were used in the text and, in the absence of these, the systematic name. Shorthand notations were adopted for tables and figures. Unsaturated fatty acids assume the *cis* configuration unless otherwise stated. For some texts, it is useful to refer to families of unsaturated fatty acids. The n system is used, where n is the length of the carbon chain and is followed by a number which indicates the position of the first double bond (the carbon atom in the terminal methyl group being number 1). Branched-chain fatty acids are referred to in *Table 1.2* (Chapter 1).

	Systematic name	Common name	Shorthand notation
1. Saturated	Ethanoic	Acetic	2:0
	Propanoic	Propionic	3:0
	Butanoic	Butyric	4:0
	Hexanoic	Caproic	6:0
	Octanoic	Caprylic	8:0
	Decanoic	Capric	10:0
	Dodecanoic	Lauric	12:0
	Tetradecanoic	Myristic	14:0
	Hexadecanoic	Palmitic	16:0
	Octadecanoic	Stearic	18:0
	Eicosanoic	Arachidic	20:0
	Docosanoic	Behenic	22:0
	Tetracosanoic	Lignoceric	24:0
2. Monoethenoic	Δ^9 Tetradecenoic	Myristoleic	9–14:1
	Δ^9 Hexadecenoic	Palmitoleic	9–16:1
	Δ^9 Octadecenoic	Oleic	9–18:1
	Δ^{9t} Octadecenoic	Elaidic	9t–18:1
	Δ^{11t} Octadecenoic	Vaccenic	11t–18:1
	Δ^{11} Eicosenoic	Gondoic	11–20:1

	Systematic name	Common name	Shorthand notation
	Δ^{11} Docosenoic	Cetoleic	11–22:1
	Δ^{13} Docosenoic	Erucic	13–22:1
3. Dienoic	$\Delta^{9,12}$ Octadecadienoic	α linoleic	9,12–18:2
4. Trienoic	$\Delta^{6,9,12}$ Octadecatrienoic	γ linolenic	6,9,12–18:3
	$\Delta^{9,12,15}$ Octadecatrienoic	α linolenic	9,12,15–18:3
	$\Delta^{8,11,14}$ Eicosatrienoic	dihomo-γ-linolenic	8,11,14–20:3
5. Tetraenoic	$\Delta^{5,8,11,14}$ Eicosatetraenoic	Arachidonic	5,8,11,14–20:4
6. Pentaenoic	$\Delta^{5,8,11,14,17}$ Eicosapentaenoic	—	5,8,11,14,17–20:5
	$\Delta^{4,8,12,15,19}$ Docosapentaenoic	Clupanodonic	4,8,12,15,19–22:5

ACKNOWLEDGEMENTS

The contributions of those who presented papers at the conference and their assistance with the preparation of the proceedings is gratefully acknowledged. The meeting was opened by Professor J.D. Ivins, CBE and the sessions chaired by Dr A.G. Garton, FRS, Prof. J.M. Hawthorne, Dr K.N. Boorman, Prof. D. Lewis, Dr T. Walker, Mr G. Harrington and Mr A. de Mulder

The following organizations provided financial assistance as an invaluable contribution towards the expenses of speakers:

  Colborn-Dawes Nutrition Ltd
  Dalgety Spillers Agriculture Ltd
  Favor Parker Ltd
  Imperial Chemical Industries PLC
  Insta-Pro (UK) Ltd
  Vitafoods Northern Ltd
  International Association of Fish Meal Manufacturers
  Kemin (UK) Ltd
  National Renderers Association, Inc
  Nitrovit Ltd
  Nutec Ltd
  W.J. Oldacre Ltd
  Procter and Gamble Ltd
  Prosper de Mulder Ltd
  RHM Agriculture Ltd
  Roche Products Ltd
  Rumenco Ltd
  Smith Kline Animal Health Ltd
  Sun Valley Poultry Ltd
  Vitafoods Northern Ltd

Finally, the secretarial skills of Mrs Jose Newcombe and Mrs Shirley Bruce, together with the administrative and catering staff and the students who helped, ensured the smooth running of the conference.

CONTENTS

I	**Chemistry, Biochemistry and Nutritional Importance of Fats**	1
1	**THE CHEMISTRY AND BIOCHEMISTRY OF PLANT FATS AND THEIR NUTRITIONAL IMPORTANCE** M.I. Gurr, *Nutrition Department, National Institute for Research in Dairying, Shinfield, Reading, UK*	3
2	**THE CHEMISTRY, BIOCHEMISTRY AND NUTRITIONAL IMPORTANCE OF ANIMAL FATS** M. Enser, *Lipids Section, ARC Meat Research Institute, Langford, Bristol, UK*	23
3	**FISH FATS** J. Opstvedt, *Norwegian Herring Oil and Meal Industry Research Institute, N-5033 Fyllingsdalen, Bergen, Norway*	53
II	**Digestion, Absorption and Transport of Fats**	83
4	**DIGESTION, ABSORPTION AND TRANSPORT OF FATS: GENERAL PRINCIPLES** David N. Brindley, *University of Nottingham Medical School, Queen's Medical Centre, Nottingham, UK*	85
5	**THE DIGESTION, ABSORPTION AND TRANSPORT OF FATS—NON-RUMINANTS** C.P. Freeman, *Unilever Research Laboratory, Colworth House, Sharnbrook, Bedford, UK*	105
6	**DIGESTION, ABSORPTION AND TRANSPORT OF FATS IN RUMINANT ANIMALS** J.H. Moore, *Wye College, Ashford, Kent, UK* and W.W. Christie, *Hannah Research Institute, Ayr, Scotland, UK*	123

III Role of Essential Fats   151

7 ESSENTIAL FATTY ACIDS IN POULTRY NUTRITION 153
C.C. Whitehead, *Agricultural Research Council's Poultry Research Centre, Roslin, Midlothian, Scotland, UK*

8 ESSENTIAL FATTY-ACID/MINERAL INTERACTIONS WITH REFERENCE TO THE PIG 167
Stephen C. Cunnane, *Efamol Research Institute, Kentville, Nova Scotia, Canada*

9 ESSENTIAL FATTY ACIDS IN THE RUMINANT 185
R.C. Noble, *Hannah Research Institute, Ayr, Scotland, UK*

10 PROTECTIVE FUNCTIONS OF FAT-SOLUBLE VITAMINS 201
Alfred W. Kormann and Harald Weiser, *F. Hoffmann-La Roche & Co. Ltd, Central Research Units, 4002 Basle, Switzerland*

IV Fats as Energy-Yielding Compounds   223

11 FATS AS ENERGY SOURCES IN ANIMAL TISSUES 225
D.W. Pethick, *School of Veterinary Studies, Murdoch University, Western Australia*, A.W. Bell, *School of Agriculture, La Trobe University, Bundoora, Victoria* and E.F. Annison, *Department of Animal Husbandry, University of Sydney, Camden, NSW, Australia*

12 FATS AS ENERGY-YIELDING COMPOUNDS IN THE RUMINANT DIET 249
J.W. Czerkawski and J.L. Clapperton, *Hannah Research Institute, Ayr, Scotland, UK*

13 THE EXTRA CALORIC VALUE OF FATS IN POULTRY DIETS 265
John D. Summers, *Department of Animal and Poultry Science, University of Guelph, Ontario, Canada*

14 ASSESSMENT OF THE DIGESTIBLE AND METABOLIZABLE ENERGY OF FATS FOR NON-RUMINANTS 277
Julian Wiseman, *University of Nottingham School of Agriculture, Sutton Bonington, Loughborough, Leicestershire, UK*

| V | **Fats in Animal Feeding Systems** | 299 |

15 THE NUTRIENT DENSITY OF PIG   301
DIETS—ALLOWANCES AND APPETITE
D.J.A. Cole, *University of Nottingham School of Agriculture, Sutton Bonington, Loughborough, Leicestershire, UK*

16 USE OF FATS IN DIETS FOR GROWING PIGS   313
Tim S. Stahly, *Department of Animal Sciences, University of Kentucky, Lexington, Kentucky, USA*

17 THE USE OF FAT IN SOW DIETS   333
R.W. Seerley, *Department of Animal and Dairy Science, University of Georgia, Athens, Georgia 30602, USA*
AND
DISCUSSION—SUPPLEMENTAL FATS AND ENERGY   353
DENSITY IN PIG DIETS
Robert H. Wilson, *Wandalup Farms, Mandurah, Western Australia,* and James E. Pettigrew, Jr, *Department of Animal Science and the Swine Center, University of Minnesota, St Paul, Minnesota, USA*

18 USE OF FATS IN DIETS FOR LACTATING DAIRY   357
COWS
Donald L. Palmquist, *Ohio Agricultural Research and Development Center, The Ohio State University, Wooster 44691, USA*

19 THE USE OF FAT IN DOG AND CAT DIETS   383
P.T. Kendall, *Animal Studies Centre, Freeby Lane, Waltham-on-the-Wolds, Leicestershire, UK*

| VI | **Carcase Considerations** | 405 |

20 FAT DEPOSITION AND THE QUALITY OF FAT TISSUE IN   407
MEAT ANIMALS
J.D. Wood, *Animal Physiology Division, ARC Meat Research Institute, Langford, Bristol, UK*

21 FAT DEPOSITION IN BROILERS   437
C. Fisher, *Agricultural and Food Research Council's Poultry Research Centre, Roslin, Midlothian, Scotland*

22 NUTRIENT PARTITIONING IN DOMESTICATED AND   471
NON-DOMESTICATED ANIMALS
M.A. Crawford, W.R. Hare and D.B. Whitehouse, *Institute of Zoology, Regent's Park, London, UK*

VII Practicalities of Fat Utilization   **481**

23 BLEND SOURCES AND QUALITY CONTROL   **483**
A.J. Howard, *Procter & Gamble Limited, Newcastle upon Tyne, UK*

24 APPLICATION OF FATS IN THE MILL   **495**
R.E. Atkinson, *NRA Consultant, 10 Gwentlands Close, Chepstow, Gwent, UK*

LIST OF PARTICIPANTS   **505**

INDEX   **513**

I

Chemistry, Biochemistry and Nutritional Importance of Fats

1

THE CHEMISTRY AND BIOCHEMISTRY OF PLANT FATS AND THEIR NUTRITIONAL IMPORTANCE

M.I. GURR
Nutrition Department, National Institute for Research in Dairying, Shinfield, Reading RG2 9AT, UK

The commercial importance of plant fats

The total world production of seed oils in 1981/82 was 43 million metric tonnes. Most of this was for human consumption but the animal feeds business in the UK currently uses about 75 000 tonnes per annum. Of the several hundred varieties of plants known to have oil-bearing seeds, only a dozen are significant commercially and three of these are used entirely for industrial purposes other than as edible oils (*Table 1.1*). The value of plant fats in nutritional terms cannot be judged entirely by the figures for edible oil production, since leaf crops, although having a low fat content, provide significant quantities of nutritionally important fats to grazing animals.

Types of plant lipids

Structural lipids

The leaves of higher plants contain up to 7% of their dry weight as lipids, some of which are present as surface lipids, the others as components of leaf cells, especially the chloroplast membranes (Hitchcock and Nichols, 1971). The surface lipids are often referred to as waxes, a term that strictly should be reserved for the esters of long-chain alcohols with fatty acids which form a major fraction of the surface lipid mixture. However, the surface lipids also comprise long-chain (C29) hydrocarbons, free fatty acids, and alcohols and ketones. The cuticle, or thick outer part of some plant epidermal cells, contains in addition to the waxes, a lipid fraction, cutin, that is not extractable with organic solvent and which comprises cross-linked polymers of hydroxy and normal fatty acids (Kolattukudy, 1975).

The major leaf lipids are those associated with cellular membranes. The lipids of plasma membranes, mitochondria and endoplasmic reticulum are predominantly phospholipids that are not fundamentally different from those located in animal membranes: phosphatidyl choline, phosphatidyl ethanolamine, phosphatidyl serine, phosphatidyl inositol and phosphatidyl

Table 1.1 SOME COMMERCIALLY IMPORTANT OIL SEED CROPS

Seed	Oil content (%)	Major fatty acid(s)	World total of oil production ('000 metric tons) 1976	1982/3	Chief producing areas	Major uses
Soya bean (*Glycine hispida*)	13–20	9,12-18:2	10250	14700	USA, Brazil, China	Margarine, cooking oil, salad oil, ice cream, paints, soap
Groundnut (peanut) (*Arachis hypogaea*)	45	9,12-18:2	3160	2900	India, China, Africa, USA	Margarine, cooking oil, salad oil, ice cream
Coconut (*Cocos nucifera*)	63	12:0	3130	3200	Philippines, Indonesia	Margarine, cooking oil, soap, lubricants
Sunflower (*Helianthus annuus*)	40	9,12-18:2	2809	5800	USSR, Argentina	Margarine, cooking oil, salad oil, soaps, paints
Oil palm: Palm oil Palm kernel oil (*Elaeis guineensis*)	50(a) 50(b)	16:0, 9-18:1 12:0	2660 527	5000 900	W. Africa, Malaysia, Indonesia	Margarine, shortenings, biscuit fats, frying fat, biscuit and confectionery fats, ice cream, soap
Rape (*Brassica napus*)	35–40	13–22:1 9-18:1 in zero erucic varieties	2520	4600	India, China, Canada, Poland, France, Sweden	Margarine, cooking oils, salad oils, lubricants
Cotton (*Gossypium hirsutum*)	15–23	9,12-18:2, 9-18:1	2500	3300	USSR, China, USA	Margarine, cooking oils, salad oils
Olive (*Olea sativa*)	15	9-18:1	1370	2000	Italy, Spain, Greece	Salad oils, preserving oils, soaps
Linseed (*Linum usitatissimum*)	30–40	9,12,15-18:3	630	760	Canada, USA, Argentina	Paints, varnishes and other industrial uses
Sesame (*Sesame indicum*)	50	9-18:1, 9,12-18:2	610	660	India, China, Mexico	Table oils
Castor (*Ricinus communis*)	45	OH-18:1	280	370	Brazil, India	Paints, lubricants, plastics
Tung (*Aleurites fordii*)		9,11t,13t-18:3	115	100	Argentina	Paints, varnishes

(a): % of mesocarp
(b): % of kernel
(c): 1982/3 are forecast figures

Storage

$$CH_3(CH_2)_7CH{=}CH(CH_2)_7CH_2{\cdot}O{\cdot}CO{\cdot}CH_2(CH_2)_7CH{=}CH(CH_2)_7CH_3$$

Wax esters

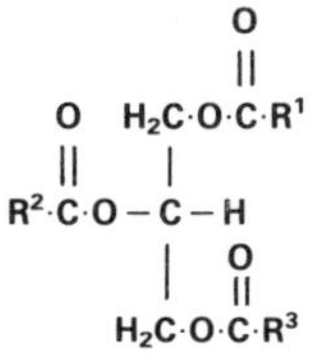

$$\begin{array}{c} \quad\quad\quad\quad O \\ \quad\quad\quad\quad \| \\ O \quad H_2C{\cdot}O{\cdot}C{\cdot}R^1 \\ \| \quad | \\ R^2{\cdot}C{\cdot}O - C - H \\ | \quad O \\ | \quad \| \\ H_2C{\cdot}O{\cdot}C{\cdot}R^3 \end{array}$$

Triacylglycerols

Structural

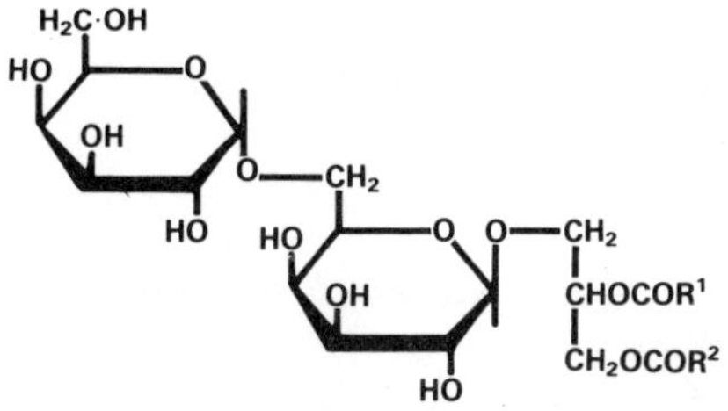

1,2-diacyl-[α-D-galactopyranosyl-(1′ → 6′)-β-D-galactopyranosyl(1′ → 3)]-sn-glycerol

(Digalactosyl diacylglycerol)

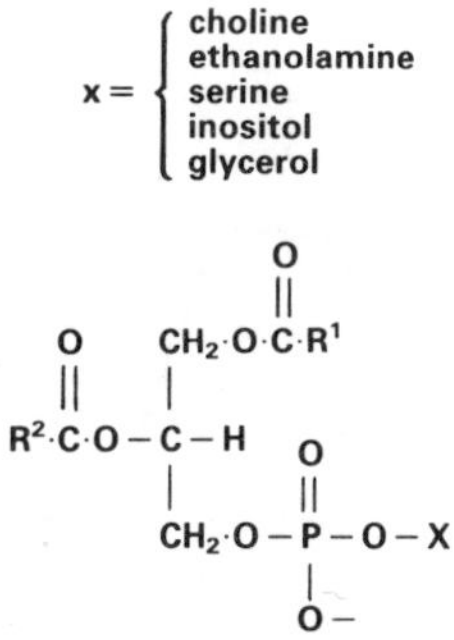

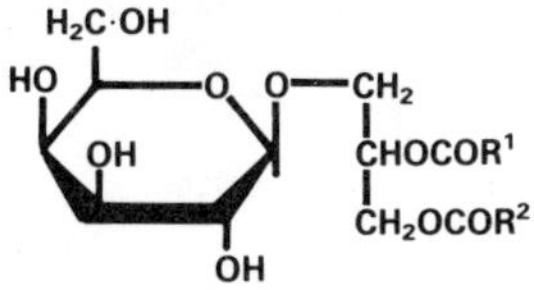

1,2-diacyl-[β-D-galactopyranosyl(1′ → 3)]-sn-glycerol

(Monogalactosyl diacylglycerol)

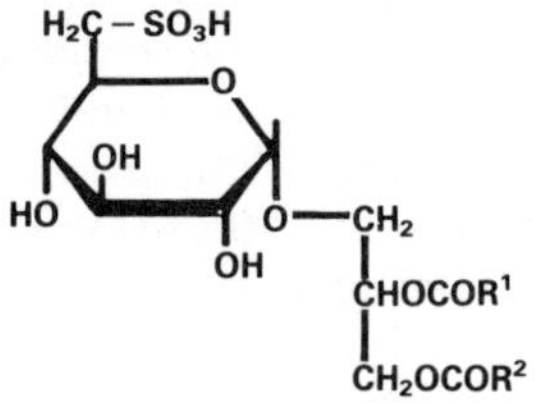

D-quinovose is 6-deoxy-D-glucose. Note the carbon-sulphur bond.

1,2-diacyl-[6-sulpho-α-D-quinovopyranosyl-(1′ → 3)]-sn-glycerol

(Plant sulpholipid (sulphoquinovosyl-diacylglycerol))

$$x = \begin{cases} \text{choline} \\ \text{ethanolamine} \\ \text{serine} \\ \text{inositol} \\ \text{glycerol} \end{cases}$$

$$\begin{array}{c} \quad\quad\quad\quad O \\ \quad\quad\quad\quad \| \\ O \quad CH_2{\cdot}O{\cdot}C{\cdot}R^1 \\ \| \quad | \\ R^2{\cdot}C{\cdot}O - C - H \quad O \\ | \quad \| \\ CH_2{\cdot}O - P - O - X \\ | \\ O^- \end{array}$$

Phosphatidyl-x

Figure 1.1 Structures of some important plant lipids

6   The chemistry and biochemistry of plant fats

glycerol (*Figure 1.1*) (Gurr and James, 1980). The latter lipid is the major phospholipid of the chloroplast membranes but quantitatively more important are the glycolipids: mono- and digalactosyl diacylglycerols which together make up 40–50% of the membrane lipid and the chlorophylls which make up about 20%. A characteristic plant leaf lipid is sulphoquinovosyl diacylglycerol which accounts on average for about 5% of chloroplast membrane lipids but is not found in animal membranes. It is unusual in possessing a carbon–sulphur bond (Harwood and Nicholls, 1979). Given the wide distribution of green plants across the surface of the earth, these plant lipids are among the world's most abundant organic compounds. Other minor lipids present in leaves and flowers are the carotenoids, sterols and acylated sterol glycosides (Hitchcock and Nicols, 1971).

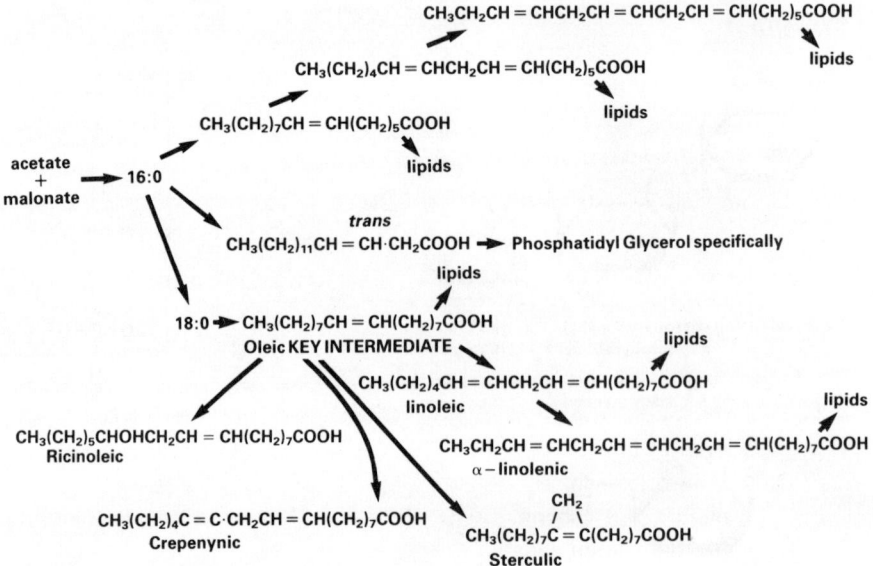

Figure 1.2   Major fatty acid pathways in plants

The fatty acid composition of plant membrane lipids is very simple, six fatty acids generally accounting for over 90% of the total: palmitic, Δ^{3trans} hexadecenoic, oleic, linoleic and α-linolenic (*Figure 1.2*). Of these, α-linolenic acid is quantitatively the most important. The hexadecenoic acid is unusual in having a *trans* double bond between carbon atoms 3 and 4 and is esterified exclusively in phosphatidyl glycerol. The surface lipids contain a wider spectrum of fatty acids with chain lengths from C10 to C30 while the cutins contain a high proportion of C18 hydroxy acids.

Storage lipids

STRUCTURE

Many plants store their reserve energy as carbohydrate in the seed cotyledons (e.g. peas) or the fruit exocarp (e.g. apples). The lipids of these

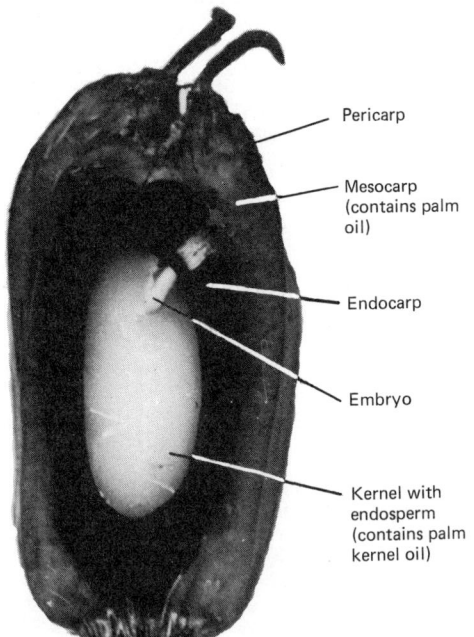

Figure 1.3 Structure of palm fruit indicating lipid storage. Reproduced from 'The biosynthesis of triacylglycerols', Chapter 8 in Volume 4 of *The Biochemistry of Plants* (P.K. Stumpf and E. Conn, Eds). New York, Academic Press, 1980. With permission of the publishers and Dr L.H. Jones

tissues are predominantly the structural lipids (phospholipids and glycolipids) described above, although droplets of triacylglycerols have been identified in small concentrations in the seeds of pea and bean (Mollenhauer and Totten, 1971), wheat (Morrison, Kuo and O'Brien, 1975) and barley (Jones, 1969).

Other plants store large quantities of lipids, mostly triacylglycerols, as an energy reserve in the fleshy fruit exocarp (e.g. avocado) or in the seed endosperm or kernel (e.g. rapeseed). Some, like the palm (*Figure 1.3*), store triacylglycerols in both the exocarp ('palm oil') and the endosperm ('palm kernel oil').

The fatty acids of seed-oil triacylglycerols are more varied than those of the structural lipids and are generally characteristic of a particular plant family. One fatty acid often predominates and is frequently of unusual structure (*Table 1.2*) although the commercially important edible oils listed in *Table 1.1* are mostly those in which the predominant fatty acids are the common ones. An exception is coconut oil which is unusual in containing predominantly saturated medium-chain fatty acids (*Table 1.1*). These characteristic seed-oil fatty acids do not occur in the structural lipids. Marked differences in composition among lipid classes are maintained even within the same organelles, so that, for example, the phospholipids of

The chemistry and biochemistry of plant fats

Table 1.2 SOME 'UNUSUAL' FATTY ACIDS CHARACTERISTIC OF THE SEED OILS OF CERTAIN PLANT FAMILIES

Name	Carbon atoms	Structure[a]	Typical source	Family
Capric	10	n-decanoic	*Cuphea procumbens*	Lythraceae
γ-Linolenic	18	6c,9c,12c trienoic	*Oenothera biennis* (Evening primrose)	Onagraceae
Crepenynic	18	9c,12a acetylenic	*Crepis* spp.	Compositae
α-Eleostearic	18	9c,11t,13t conjugated ethylenic	*Aleurites fordii* (Tung)	Euphorbiaceae
Ximenynic	18	9a,11t, conjugated acetylenic	*Ximenia* spp.	Olacaceae
Malvalic	18	$\underset{CH_2}{-C\overset{9\ \ 8C}{=}C-}$ cyclopropenyl	*Malva* spp.	Malvaceae
Sterculic	19	$\underset{CH_2}{-C\overset{10\ \ 9C}{=}C-}$ cyclopropenyl	*Sterculia foetida*	Sterculiaceae
Vernolic	18	9c,12,13 cis-epoxy	*Vernonia anthelmintica*	Compositae
Ricinoleic	18	12-OH, 9c	*Ricinus communis* (Castor)	Euphorbiaceae
Erucic	22	13c monoenoic	*Brassica napus* (Rape)	Cruciferae

[a] c = *cis*; t = *trans*; a = *acetylenic*; OH = *hydroxy*.

the oil body membrane (see below) have a fatty acid composition quite distinct from that of the oil they encapsulate, as illustrated in *Table 1.3*. In plants that store triacylglycerols in both the exocarp and the endosperm, there is usually a marked difference in fatty acid composition of the triacylglycerols from the two locations, as illustrated for palm oil in *Table 1.4*.

Natural glycerides possess a stereospecific distribution of fatty acids on the three positions of glycerol rather than a completely random or 'restricted random' distribution as once proposed (see Gurr and James, 1980). Analysis of vegetable oils has indicated a marked specificity for saturated acids to occupy position 1 with unsaturated acids at position 2 (Mattson and Volpenhein, 1963; Hitchcock and Nichols, 1971; Weber, De La Roche and Alexander, 1971). In the seed oil of *Crambe abyssinica* (Gurr, Blades and Appleby, 1972), erucic acid is found exclusively at positions 1 and 3 with a marked preference for position 3 (*Table 1.3*). Palmitic and stearic acids are found exclusively at position 1 while the longer-chain arachidic saturated acid is evenly distributed between positions 1 and 3. Oleic, linoleic and linolenic acids are found predominantly in position 2.

SEED-OIL DEVELOPMENT

The presence of a characteristic seed-oil fatty acid, such as erucic, serves as a useful 'marker' in studying the lipid changes taking place during seed

Table 1.3 FATTY ACID COMPOSITION OF *CRAMBE ABYSSINICA* SEED PHOSPHOLIPIDS AND TRIACYLGLYCEROLS

Lipid	Fatty acid (moles/100 moles total fatty acid)										
	16:0	18:0	20:0	22:0	9-16:1	9-18:1	11-20:1	13-22:1	9,12-18:2	9,12,15-18:3
Phospholipids	16.1	2.1	—	—	1.0	32.2	2.0	4.1	31.9	8.3
Triacylglycerols										
total	3.9	1.6	1.6	2.0	0.6	25.7	5.5	41.4	10.5	7.2
position 1	3.3	0.8	1.0	1.1	0.6	6.2	2.6	16.0	1.7	0.0
position 2	0.0	0.0	0.0	0.0	0.0	17.1	0.5	1.0	8.0	6.8
position 3	0.6	0.8	0.6	0.9	0.0	2.4	2.4	24.4	0.8	0.4

After Gurr *et al.*, 1972

Table 1.4 DIFFERENT FATTY ACID COMPOSITIONS OF OIL PALM MESOCARP (PALM OIL) AND KERNEL (PALM KERNEL OIL)

Fatty acid	Fatty acid composition (%)	
	Palm oil	Palm kernel oil
6:0	0	0.1
8:0	0	3.6
10:0	0	4.2
12:0	0.1	44.6
14:0	1.0	17.6
16:0	45.5	9.5
9-16:1	0.1	0
18:0	5.9	2.8
9-18:1	34.6	15.2
9,12-18:2	11.8	2.5
9,12,15-18:3	0.3	0
20:0	0.4	0

Table 1.5 CHANGES IN PROPORTIONS OF LIPID CLASSES IN DEVELOPING SOYA BEANS

Days after flowering	Neutral lipids		Glycolipids		Phospholipids	
	(% wt)	(mg/bean)	(% wt)	(mg/bean)	(% wt)	(mg/bean)
9	17.0	0.01	29.2	0.02	49.0	0.03
18	46.4	0.42	14.7	0.13	36.5	0.33
40	78.1	6.0	5.3	0.4	16.5	1.3
60	90.0	36.6	2.3	0.9	5.5	2.2
80	92.4	32.8	1.6	0.6	3.8	1.3

After Privett et al., 1973

development. During the initial stages, the cotyledons contain little or no triacylglycerol and no detectable amounts of the characteristic seed-oil fatty acids. Membrane lipids predominate and the fatty acid composition reflects this, being characterized by a relatively high proportion of α-linolenic acid. In the second phase, there is an abrupt appearance of the seed-oil fatty acid(s) and gradual accumulation of triacylglycerols indicating the 'switching on' of the enzymes of seed-oil triacylglycerol biosynthesis. During this phase there are marked changes in the proportions of the lipid classes, with decreases in the proportions of structural lipids, although the absolute amounts present in the seed may actually rise, as illustrated for soya bean lipids in *Table 1.5*.

ENVIRONMENTAL FACTORS INFLUENCING COMPOSITION

The characteristic fatty acid patterns of plant triacylglycerols are to some extent under genetic control. In addition, environmental factors may modify the basic patterns, the extent of modification depending on the species. Thus the seed oils of plants grown in cool climates tend to be more unsaturated than those grown in warm climates (Hitchcock and Nichols,

1971). The chief influence seems to be on the characteristic fatty acid of the seed so that, for example, in flax seed oil there is a marked decline in the proportion of linolenic acid between 10°C and 30°C and a corresponding increase in the proportion of its precursor, oleic acid (Canvin, 1965). Similarly, the proportion of linoleic acid in sunflower seed oil steadily declines between 10°C and 30°C, to be replaced by oleic acid. Yet the linoleic acid content of safflower and the ricinoleic acid of castor are unaffected by the same variation in temperature (Canvin, 1965). Appelqvist (1975) showed that different lines of zero-erucic rape could respond differently to the same climatic variations, one producing a greater proportion of linoleic acid as the growth temperature increased. Thus the explanation for the decrease in unsaturation with rising temperature as a reduced concentration of molecular oxygen at the site of the desaturase (Harris and James, 1969; Dompert and Beringer, 1970) may need modification. It is clear that temperature need not necessarily affect total unsaturation, since in rapeseed there is a steady decline in the proportion of erucic acid between 15°C and 30°C and a corresponding increase in its precursor, oleic acid (Canvin, 1965). Hence the total unsaturation is unchanged and the climatic influence seems to be on the activity of the elongating enzymes, not on the desaturase. Light intensity is another variable that may affect the fatty acid composition of seed-oil triacylglycerols. In Cruciferae, Leguminosae and Linaceae, which have a translucent pericarp and testa and in which the major sites of oil storage are the cotyledons, α-linolenic acid is a major seed-oil component. This may be due in part to the presence of chloroplasts in the cotyledons, which may influence the amount and composition of the stored oil. It may also be mentioned that the presence of functional chloroplasts ensures that a constant supply of oxygen is available to surrounding tissues so that the effects of light, temperature and oxygen availability may not be entirely independent.

BIOSYNTHESIS

Some of the more important pathways for plant fatty acid biosynthesis are illustrated in *Figure 1.2*. Fatty acid synthesis in plants proceeds in four distinct stages: (1) the carboxylation of acetyl-CoA to malonyl-CoA; (2) the repeated condensation of malonyl-CoA (with a growing acyl chain bound to an acyl-carrier protein (ACP) component of the fatty acid synthetase to produce palmitoyl-ACP; (3) the elongation of 16:0-ACP to 18:0-ACP and (4) the desaturation of stearate to oleate, linoleate and linolenate. Fatty acid biosynthesis is confined to the plastids of non-photosynthetic tissue and the chloroplasts of photosynthetic tissue (Stumpf, 1980; Roughan and Slack, 1982). The substrate for the desaturation of stearic acid in plants is 18:0-ACP but studies with algae (Gurr, Robinson and James, 1969) and with higher plants (Slack, Roughan and Browse, 1979) have shown that the substrate for the desaturation of oleate is oleoyl-phosphatidylcholine (for reviews see Roughan and Slack, 1982; Stumpf, 1980). Evidence has accumulated for the biosynthesis of seed-oil triacylglycerols via the glycerol phosphate pathway in the seeds of many plants (Gurr, 1980). However, more recent evidence suggests that polyunsaturated diacylglycerols for seed-oil biosynthesis may be derived from phosphatidyl cholines, possibly by a reversal of the enzyme CDP-choline: 1,2-sn-diacylglycerol transferase, normally involved in phosphatidyl choline biosynthesis (Roughan and Slack, 1982).

OIL STORAGE

Microscopic examination of a mature seed or one in the active phase of oil accumulation reveals a cytoplasm packed with spherical organelles (oil bodies) that react strongly with lipid stains and are without doubt the sites of oil storage within the cell (*Figure 1.4*). The controversy about the

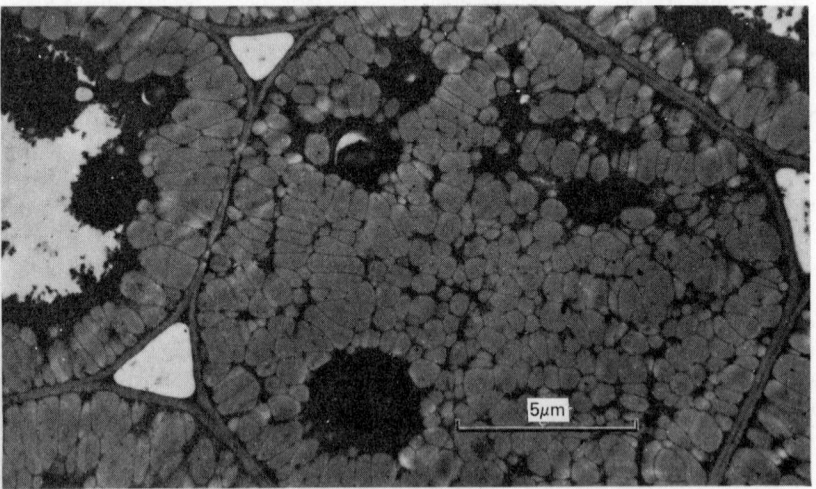

Figure 1.4 Photomicrograph of cytoplasm of mature oil seed showing spherical oil-bearing organelles

naming, origin, ultrastructure and biochemical capacity of these organelles has been reviewed by Gurr (1980). In many seeds the oil bodies are surrounded by a half-unit membrane containing protein and polar lipids. In *Crambe abyssinica* (Gurr *et al.*, 1974) and jojoba (Stumpf, 1980) seeds, the oil-body membrane contains enzymes catalysing several steps in triacylglycerol biosynthesis. In castor beans these enzymes are present not in the membrane but in vacuoles within the oil bodies (Harwood *et al.*, 1971). Thus, oil bodies are not simply inert storage organelles for reserve lipids but contain the enzymic apparatus for the synthesis of their own oil.

Nutritional role of plant fats in animal diets

Digestion and absorption

To have nutritive value, fats must be digested and absorbed from the gastrointestinal tract. Virtually all the published work on fat absorption has been concerned with the absorption of triacylglycerols (see Johnston, 1970; Gurr and James, 1980). Although the fatty acids in the plant phospholipids and glycolipids are presumably as well digested and absorbed as those in triacylglycerols, there has been little detailed study of the mechanisms by which this occurs. In contrast, the waxes are poorly digested and for this reason, the seed oil of jojoba (*Simmodsin chimensis*), a crop of increasing commercial importance which is unusual in storing fatty acids as wax esters rather than as triacylglycerols, has poor nutritive value (Haumann, 1983; Yaron, Samoiloff and Benzioni, 1982).

Energy value

Triacylglycerols provide more than twice the energy per gram eaten as carbohydrates or proteins. The gross energy value of a fatty acid varies slightly with chain length and degree of unsaturation. There are reports that the consumption of unsaturated plant fats results in smaller increases in body energy than saturated fats, presumably because their metabolism is less efficient and more metabolic energy is lost as heat (Kasper, Thiel and Ehl, 1973), but these results need confirmation. The energy value of medium-chain fatty acids present in coconut oil is considerably less than that of long-chain fatty acids because of their pathways of metabolism. Medium-chain acids are absorbed as free fatty acids into the portal bloodstream and carried to the liver where they are oxidized. They are not deposited in adipose tissue. Because of their mode of absorption, these acids in the form of MCT oils refined from coconut, are employed in diets for patients with malabsorption syndromes (Leyland *et al.*, 1969).

Fat-soluble vitamins

Plant lipids also provide important dietary sources of vitamins A and E. Leaf and seed-oil lipids contain the provitamin A, β-carotene, which is converted into retinol in the body (Simpson, 1983). It has been suggested (Lotthammer, Ahlswede and Meyer, 1976) that β-carotene is important in the maintenance of fertility in cattle by a mechanism that does not depend on its conversion into vitamin A. Recent research at NIRD has so far not provided evidence for such a role (Ducker *et al.*, 1982).

Vitamin E activity is shared by several related tocopherols, the most active being α-tocopherol. Because one of its roles is thought to be the inhibition of free-radical oxidation of unsaturated fatty acids, it is generally held that intake should be considered in relation to the polyunsaturated content of the diet rather than in absolute amounts. A ratio of vitamin E to linoleic acid of 0.6 mg/g is generally recommended (Witting, 1970). In general, those vegetable oils containing high concentrations of polyunsaturated acids are also sufficiently rich in vitamin E to give adequate protection.

Essential fatty acids

Certain polyunsaturated fatty acids synthesized only by plants are required in animal diets, for without them the animal will die. These are the essential fatty acids, quantitatively the most important of which is linoleic acid. Animal tissues contain desaturases that introduce double bonds into saturated fatty acids normally at position 9. The insertion of further double bonds to produce polyunsaturated fatty acids also occurs in animals but only between the first double bond and the carboxyl group (*Figure 1.5*). A dietary source of essential fatty acids is required because animals, during the course of evolution, have lost the ability to introduce a double bond into position 12 of the carboxylic acid chain.

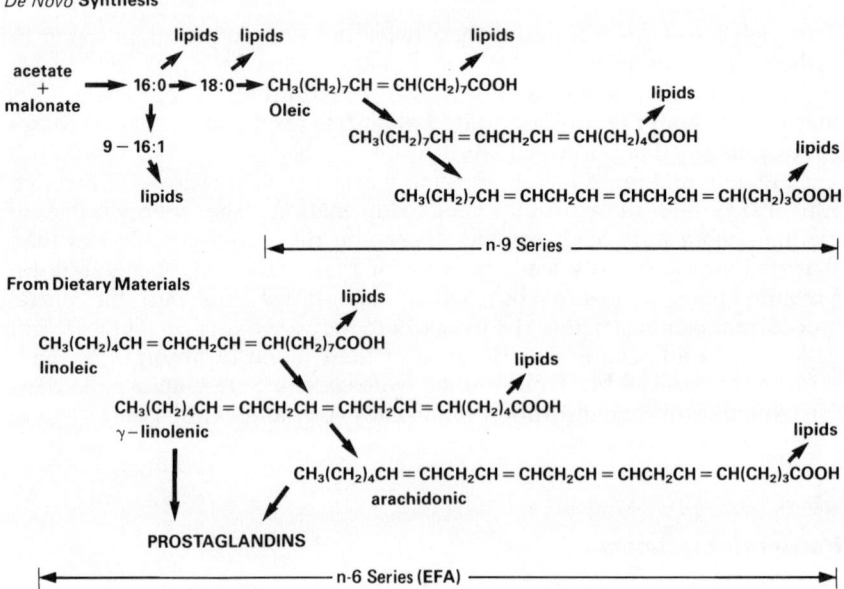

Figure 1.5 Major pathways of fatty acid biosynthesis in animals

A key metabolite of linoleic acid in animal tissues is arachidonic acid. This and related polyunsaturated fatty acids are converted into prostaglandins, oxygenated fatty acids with potent biological activities including the control of haemostasis (Gibney, 1982). The rate-limiting step in the conversion of linoleic to arachidonic acid is the first (Δ^6) desaturation step to form γ-linolenic acid. Several disorders are now recognized in which the conversion of linoleic acid into prostaglandin precursors is limiting. A possible therapeutic measure is the administration of a dietary source of γ-linolenic acid and the seed oil of the evening primrose, an unusually rich source of this acid, is being advocated for this purpose (see *Table 1.2*).

Many, but not all, functions of linoleic acid can be fulfilled by α-linolenic acid, but there is no clear evidence that there is a vital function of α-linolenic acid that cannot be satisfied by linoleic acid, except perhaps in the biosynthesis of long-chain n-3 fatty acids present in brain and in retinal rod outer segments (Leat, 1981).

Polyunsaturated fatty acids may have a role in the regulation of immunity (Meade and Mertin, 1978; Gurr, 1983). Diets containing polyunsaturated seed oils have been employed as adjuncts to conventional immunosuppressive therapy to reduce rejection of kidney grafts (Uldall *et al.*, 1974). Such diets prolong the survival time of skin grafts in mice (Mertin, 1976). A diet containing sunflower seed oil appeared to be beneficial in treating patients with multiple sclerosis (Millar *et al.*, 1973), perhaps by suppressing an abnormal immune response to one of the body's own proteins. Diets containing corn oil reduced the immune responsiveness of guinea pigs *in vivo* and serum from these animals suppressed the responses of lymphocytes to mitogens *in vitro*, a reaction normally used as an index of immune responsiveness (Friend *et al.*, 1980). Essential fatty acid deficiency is also associated with suppressed immunity and it is probable that there is an optimal dietary supply of polyunsaturated fatty acids for the functioning of the immune system above and below which the response is impaired (Mertin and Smith, 1983).

Influence on body lipids

Inclusion of vegetable oils in the diet of monogastric animals influences the composition of fats stored in adipose tissue or milk. The same is not true of ruminant animals, since the micro-organisms in the rumen hydrogenate dietary unsaturated acids so that absorbed fatty acids are either saturated or *cis* and *trans* mono-unsaturated (Moore, 1978). Unsaturated fatty acids can be protected from rumen hydrogenation by encapsulating the oil in proteins that have been cross-linked by formaldehyde treatment (McDonald and Scott, 1977).

Toxicity of plant lipids

Several plant lipids used in animal diets are potentially toxic, as shown by experiments with small animals. Some of the most common are discussed briefly here, but for more detail the reader is referred to reviews by Mattson (1973) and Dhopeshwarkar (1981).

Cyclopropene fatty acids

The most important oil containing these fatty acids (of which sterculic acid is an example, *Table 1.2*) is cottonseed, in which the concentration ranges from 0.6 to 1.2%. After processing, the oil as actually eaten, however, contains only 0.1–0.5%. Sterculic acid inhibits the desaturation of stearic to oleic acid, the effects of which are to alter the permeability of membranes or to increase the saturation and melting point of fats. If cyclopropene fatty acids are present in the diet of laying hens, the permeability of the yolk sac membrane is increased, allowing the release of substances, including pigments, into the white ('Pink-White' disease) (Phelps et al., 1965). In dietary experiments with animals, the source of cyclopropenes is *Sterculia foetida* seed oil which may have up to 70% of these acids. Rats die within a few weeks when fed diets containing 5% of dietary energy as sterculic acid and the reproductive performance of females is completely inhibited with levels as low as 3% (Rascop, Sheehan and Vavich, 1966). Desaturation in the mammary gland of goats is inhibited by feeding sterculic acid, producing a high level of stearic acid in the milk (Bickerstaffe and Johnson, 1972) and in the adipose tissue of pigs, producing a lard with high melting point (Mattson, 1973).

Branched-chain fatty acids

A common dietary branched-chain fatty acid is phytanic acid (3,7,11,15-tetramethylhexadecanoic), formed from phytol, a universal constituent of green plants. The presence of methyl groups in these positions blocks the normal process of β-oxidation, and an α-oxidation is needed at the branched points. Patients with a rare genetic disease (Refsum's disease) lack the alternative α-oxidation pathway. Such people accumulate the branched-chain metabolic product in their tissues, resulting in severe neurological complications. This disease is normally fatal and a phytanic acid-free diet is the only therapeutic measure (Lough, 1973; Gurr and James, 1980).

Long-chain monoenoic acids

When young rats were fed diets containing more than 5% of the energy as rapeseed oil (45% erucic acid) their heart muscles became infiltrated with fat (Abdellatif and Vles, 1973). After about a week the hearts contained three to four times as much fat as normal hearts and although, with continued feeding, the size of the lipid deposits gradually decreased, other pathological changes were noticeable, such as the formation of fibrous tissue in the heart muscle. The biochemistry of the heart muscle was also adversely affected in that the rate of mitochondrial oxidation of substrates was slower, the rate of ATP synthesis impaired and the activity of lipases was slower with triacylglycerols containing erucic acid than with fatty acids of more normal chain length (Vles, 1975). Despite the lack of evidence for harmful effects in man, it has been thought prudent to replace older varieties of rape, having a high erucic acid content, with new varieties of zero-erucic rapes.

Oxidized fats

The storage of fats containing appreciable concentrations of polyunsaturated fatty acids in the presence of oxygen at room temperature can result in the formation of hydroperoxides (Frankel, 1980). When these are ingested they are rapidly degraded in the mucosal cells of the gut to various oxyacids that are further oxidized to CO_2. There is no evidence for the absorption of unchanged hydroperoxides nor for their incorporation into tissue lipids (Bergan and Draper, 1970). The growth of rats fed a fat with a peroxide value of 100 was reported to be normal (Andrews *et al.*, 1960). However, other workers have provided evidence for the potentiation by linoleic acid hydroperoxide of tumour growth in female rats (Cutler and Schneider, 1973).

Fats with even quite low peroxide values are organoleptically unacceptable. While the autoxidation of stored foods and animal feeds may not be important from a toxicological point of view, food spoilage and the reduction of acceptability are important considerations from an animal production and economic point of view.

Improvement of oil composition

Of the many hundreds of species of oil-bearing plants the seed-oil composition of which has been analysed (Hilditch and Williams, 1964; Hitchcock and Nichols, 1971), only a handful are commercially important (*Table 1.1*). Given the increasing cost of oils for human foods and animal feeds, and the increasing knowledge of the role of dietary fats, can the range and availability of edible oils with desirable characteristics be improved? Four means of achieving this objective may be suggested.

Agricultural use of little-used species

A search through the available data (e.g. Hilditch and Williams, 1964; Hitchcock and Nichols, 1971) may reveal seed oils with particularly useful fatty acid composition in species that have hitherto not been used as crops. Recent understanding of the biochemistry of polyunsaturated acids has, for example, highlighted the possible usefulness of γ-linolenic acid, leading to the development of evening primrose as a potential agricultural crop. Examination of the literature reveals that coconut is not the only seed oil containing medium-chain glycerides. *Cuphea* is another provider of these fatty acids that could have agricultural importance (Slabas *et al.*, 1982; Wolf, Graham and Kleiman, 1983). There are undoubtedly many more 'weeds' whose status can be elevated to the benefit of agriculture.

Plant breeding

Within the species *Brassica napus*, selection for plants with a low erucic acid content has resulted in a strain that contains no erucic acid in its seed oil. Genetic analysis of the F_2 and F_3 generations and backcrosses between these plants and normal high-erucic acid types supported the hypothesis that the erucic acid content of the seed oil was controlled by two genes that displayed no dominance and acted in an additive manner. The genes controlled the chain-lengthening pathway from oleic to eicosenoic and erucic acids in such a way that, as the capacity for erucic acid biosynthesis decreased, there was a corresponding increase in the proportion of oleic acid, maintaining a constant oil content of the seed (Downey and Craig, 1964). The work of Knowles (1968) has identified three alleles at a single chromosome locus in safflower (*Carthamus tinctorius*) that govern the relative amounts of oleic and linoleic acids in the mature seed.

The work of the plant breeder, however, is extremely slow and there is always the possibility that breeding in desired characteristics or breeding out undesired characteristics is accompanied by the elimination of other desirable features, such as cold or disease resistance or high yield. Attempts to breed out the high linolenic acid content of soyabean oil to improve flavour and oxidative stability have been disappointing (Anonymous, 1982). Newer, faster methods for improving oil quality will increasingly be sought.

Plant tissue culture

Techniques for growing a variety of plant types in tissue culture have been developed in recent years. Briefly, sterilized pieces of excised tissue, when incubated on a solid medium containing a full range of nutrients and growth substances, give rise to a disorganized mass of largely undifferentiated cells called a callus (see Rhodes and Kirsop, 1982 for a general review). Callus cultures can be maintained in solid media for extended periods, or the cells dispersed and grown in liquid culture. Such cultures can be exploited in two main ways: to regenerate clones of plants for use in conventional agriculture, or as a source of low-volume, high-cost products such as drugs, food colourings or flavours. During the last decade, the ability to propagate clones of genetically identical plantlets in such commercially important crops as the oil palm have become a reality. This approach is especially desirable in view of the problems associated with oil palms for which there are no conventional horticultural methods for making clones. Propagation by tissue culture techniques promises to reduce variability in quality and yield and should bring plants into production more quickly, though at a slightly greater cost than conventional methods (Corley *et al.*, 1982).

Genetic manipulation

Much of the attention in this field has been devoted to increasing the photosynthetic efficiency of green plants and to developing cereals capable of directly fixing atmospheric nitrogen. Several research groups are now directing their attention to the possibility of inserting into specified plant cells the genes for a desired pathway of lipid metabolism. Such an approach needs not only improved techniques for the genetic manipulation of higher plant cells, which is lagging behind the microbial technology, but also more detailed knowledge of the genetic control of the pathways of lipid metabolism. This is a slowly developing area, partly because plant lipid biochemistry has not been a fashionable area for research, but also because the location of many of the enzymes of lipid biosynthesis on biological membranes has not lent itself to an easy understanding of the control of lipid metabolism.

Perhaps the commercial goals of cheaper oils and improved oil quality will provide the spur to more interest by young researchers in plant lipid biochemistry.

References

ABDELLATIF, A.M.M. and VLES, R.O. (1973). Short-term and long-term pathological effects of glycerol trierucate and of increasing levels of dietary rapeseed oil in rats. *Nutrition and Metabolism* **15**, 219–231

ANDREWS, J.S., GRIFFITH, W.H., MEAD, J.F. and STEIN, R.A. (1960). Toxicity of air oxidized soyabean oil. *Journal of Nutrition* **70**, 199–210

ANONYMOUS (1982). Researchers report gains in hunt for low linolenic soybeans. *Journal of the American Oil Chemists' Society* **59**, 882A–884A

APPELQVIST, L-Å. (1975). Biochemical and structural aspects of storage and membrane lipids in developing oil seeds. In *Recent Advances in the Chemistry and Biochemistry of Plant Lipids* (T. Galliard and E.I. Mercer, Eds), pp. 287–299. London, Academic Press

BERGAN, J.G. and DRAPER, H.H. (1970). Absorption and metabolism of I-^{14}C-methyl linoleate hydroperoxide. *Lipids* **5**, 976–982

BICKERSTAFFE, R. and JOHNSON, A.R. (1972). The effect of intravenous infusions of sterculic acid on milk fat synthesis. *British Journal of Nutrition* **27**, 561–570

CANVIN, D.T. (1965). The effect of temperature on the oil content and fatty acid composition of the oils of several oil crops. *Canadian Journal of Botany* **43**, 63–69

CORLEY, R.H.V., WONG, C.Y., WOOI, K.C. and JONES, L.H. (1982). Early results from the first oil palm clone trials. In *Oil Palm in Agriculture in the Eighties*, Vol. 1 (E. Pufparajah and P.F. Chew, Eds), pp. 173–196. Kuala Lumpur, Incorporated Society of Planters

CUTLER, M.G. and SCHNEIDER, R. (1973). Sensitivity of feeding tests in detecting carcinogenic properties in chemicals: examination of 7,12-dimethylbenzanthracene and oxidized linoleate. *Food and Cosmetics Toxicology* **11**, 443–457

DHOPESHWARKAR, G.A. (1981). Naturally occurring food toxicants: toxic lipids. *Progress in Lipid Research* **19**, 107–118

DOMPERT, W. and BERINGER, H. (1970). Fettesynthese in Sonnenblumen-Früchten bei unterschiedlicher Sauerstoff-Konzentration. *Naturwissenschaften* **57**, 40

DOWNEY, R.K. and CRAIG, B.M. (1964). Genetic control of fatty acid biosynthesis in rapeseed (*Brassica napus* L.). *Journal of the American Oil Chemists' Society* **41**, 475–478

DUCKER, M.J., HAGGETT, R.A., BLOOMFIELD, G.A., MORANT, S.V. and GURR, M.I. (1983). The effect of level of feeding and β-carotene on dairy cow fertility. *Animal Production* **34**, 369 (Abst)

FRANKEL, E.N. (1980). Lipid oxidation. *Progress in Lipid Research* **19**, 1–22

FRIEND, J.V., LOCK, S.O., GURR, M.I. and PARISH, W.E. (1980). The effect of different dietary lipids on the immune response of Hartley strain guinea pigs. *International Archives of Allergy and Applied Immunology* **62**, 292–301

GIBNEY, M.J. (1982). The effect of n-3 and n-6 polyunsaturated fatty acids on platelet function in arterial disease. *Biochemical Society Transactions* **10**, 161–163

GURR, M.I. (1980). The biosynthesis of triacylglycerols. In *The Biochemistry of Plants, Vol. 4, Lipids: Structure and Function*, (P.K. Stumpf and E.E. Conn, Eds). New York, Academic Press

GURR, M.I. (1983). The role of lipids in the regulation of the immune system. *Progress in Lipid Research* **22**, 257–287

GURR, M.I. and JAMES, A.T. (1980). *Lipid Biochemistry: An Introduction.* London, Chapman and Hall

GURR, M.I., BLADES, J. and APPLEBY, R.S. (1972). Studies in seed-oil triglycerides: the composition of *Crambe abyssinica* triglyceride during seed maturation. *European Journal of Biochemistry* **29**, 362–368

GURR, M.I., ROBINSON, M.P. and JAMES, A.T. (1969). The mechanism of formation of polyunsaturated fatty acids by photosynthetic tissue. *European Journal of Biochemistry* **9**, 70–78

GURR, M.I., BLADES, J., APPLEBY, R.S., SMITH, C.G., ROBINSON, M.P. and NICHOLS, B.W. (1974). Studies on seed-oil triglycerides: triglyceride biosynthesis and storage in whole seeds and oil bodies of *Crambe abyssinica*. *European Journal of Biochemistry* **43**, 281–290

HARRIS, P. and JAMES, A.T. (1969). Effect of low temperature on fatty acid biosynthesis in seeds. *Biochimica et biophysica acta* **187**, 13–18

HARWOOD, J.L. and NICHOLLS, R.G. (1979). The plant sulpholipid—a major component of the sulphur cycle. *Biochemical Society Transactions* **7**, 440–447

HARWOOD, J.L., SODJA, A., STUMPF, P.K. and SPURR, A.R. (1971). On the origin of oil droplets in maturing castor bean seeds, *Ricinus communis*. *Lipids* **6**, 851–854

HAUMANN, B.A. (1983). Jojoba, desert shrub to commercial crop. *Journal of the American Oil Chemists' Society* **60**, 44A–57A

HILDITCH, T.P. and WILLIAMS, P.N. (1964). *The Chemical Constitution of Natural Fats.* London, Chapman and Hall

HITCHCOCK, C. and NICHOLS, B.W. (1971). *Plant Lipid Biochemistry.* New York, Academic Press

JOHNSTON, J.M. (1970). Intestinal absorption of fats. In *Comprehensive Biochemistry, Vol. 18, Lipid Metabolism*, Chapter 1A (M. Florkin and E.H. Stotz, Eds). Amsterdam, Elsevier

JONES, R.L. (1969). The fine structure of barley aleurone cells. *Planta* **85**, 359–375

KASPER, H., THIEL, H. and EHL, M. (1973). Response of body weight to a low carbohydrate high fat diet in normal and obese subjects. *American Journal of Clinical Nutrition* **26**, 197–204

KNOWLES, P.F. (1968). Modification of quantity and quality of safflower oil through plant breeding. *Journal of the American Oil Chemists' Society* **46**, 130–132

KOLATTUKUDY, P.E. (1975). Biochemistry of cutin, suberin and waxes, the lipid barriers in plants. In *Recent Advances in the Chemistry and*

Biochemistry of Plant Lipids (T. Galliard and E. Mercer, Eds), pp. 203-246. London, Academic Press
LEAT, W.M.F. (1981). Man's requirements for essential fatty acids. *Trends in Biochemical Sciences* **6**, IX-X
LEYLAND, F.C., FOSBROOKE, A.S., LLOYD, J.K., SEGALL, M.M., TAMIR, I., TOMKINS, R. and WOLFF, O.H. (1969). Use of medium chain triglyceride diets in children with malabsorption. *Archives of Disease in Childhood* **44**, 170-179
LOTTHAMMER, K.H., AHLSWEDE, L. and MEYER, H. (1976). Untersuchungen über eine spezifische Vitamin A unabhangige Wirkung des β-Carotins auf die Fertilität des Rindes. *Deutsche Tierärztliche Wochenschrift* **83**, 353-358
LOUGH, A.K. (1973). The chemistry and biochemistry of phytanic, pristanic and related acids. *Progress in the Chemistry of Fats and Other Lipids* **14**, 1-50
McDONALD, I.W. and SCOTT, T.W. (1977). Foods of ruminant origin with elevated content of polyunsaturated fatty acids. *World Review of Nutrition and Dietetics* **26**, 144-207
MATTSON, F.H. (1973). Potential toxicity of food lipids. In *Toxicants Occurring Naturally in Foods*, pp. 189-209. Washington, DC, National Academy of Sciences
MATTSON, F.H. and VOLPENHEIN, R.A. (1963). The specific distribution of unsaturated fatty acids in the triglycerides of plants. *Journal of Lipid Research* **4**, 392-396
MEADE, C.J. and MERTIN, J. (1978). Fatty acids and immunity. *Advances in Lipid Research* **16**, 127-165
MERTIN, J. (1976). The effect of polyunsaturated fatty acids (PUFA) on skin allograft survival and primary and secondary cytotoxic response in mice. *Transplantation* **21**, 1-4
MERTIN, J. and SMITH, A.D. (1983). Immune modulation by prostaglandins and their precursors. In *Handbook of Prostaglandins and Related Lipids* (A.L. Willis, B.H. Vickery and C.P. Asciak, Eds). Cleveland, Ohio, CRC Press Inc
MILLAR, J.H.D., ZILKHA, K.J., LANGMAN, M.J.S., PAYLING-WRIGHT, H., SMITH, A.D., BELIN, J. and THOMPSON, R.H.S. (1973). A double-blind trial of linoleate supplementation of the diet in multiple sclerosis. *British Medical Journal* **1**, 765-768
MOLLENHAUER, H.H. and TOTTEN, C. (1971). Studies on seeds. II. Origin and degradation of lipid vesicles in pea and bean cotyledons. *Journal of Cell Biology* **48**, 395-405
MOORE, J.H. (1978). Cow's milk fat and human nutrition. *Proceedings of the Nutrition Society* **37**, 231-240
MORRISON, I.N., KUO, J. and O'BRIEN, T.P. (1975). Histochemistry and fine structure of developing wheat aleurone cells. *Planta* **123**, 105-116
PHELPS, R.A., SHENSTONE, F.S., KEMMERER, A.R. and EVANS, R.J. (1965). A review of cyclopropenoid compounds: Biological effects of some derivatives. *Poultry Science* **44**, 358-390
PRIVETT, O.S., DOUGHERTY, K.A., ERDAHL, W.L. and STOLYHWO, A. (1973). Studies on the lipid composition of developing soybeans. *Journal of the American Oil Chemists' Society* **50**, 516-520
RASCOP, A.M., SHEEHAN, E.T. and VAVICH, M.G. (1966). Histomorphological

changes in reproductive organs of rats fed cyclopropenoid fatty acids. *Proceedings of the Society for Experimental Biology and Medicine* **122**, 142–145

RHODES, M.J.C. and KIRSOP, B.H. (1982). Plant cell cultures as sources of valuable secondary products. *Biologist* **29**, 134–140

ROUGHAN, P.G. and SLACK, C.R. (1982). Cellular organization of glycerolipid metabolism. *Annual Reviews of Plant Physiology* **33**, 97–132

SIMPSON, K.L. (1983). Relative value of carotenoids as precursors of vitamin A. *Proceedings of the Nutrition Society* **42**, 7–17

SLABAS, A.R., ROBERTS, P.A., ORMESHER, J. and HAMMOND, E.W. (1982). *Cuphea procumbens*: a model for studying the mechanism of medium chain fatty acid biosynthesis in plants. *Biochimica et biophysica acta* **711**, 411–420

SLACK, C.R., ROUGHAN, P.G. and BROWSE, J. (1979). Evidence for an oleoyl-phosphatidylcholine desaturase in microsomal preparations from cotyledons of safflower seed. *Biochemical Journal* **179**, 649–656

STUMPF, P.K. (1980). Biosynthesis of saturated and unsaturated fatty acids. In *The Biochemistry of Plants, Vol. 4, Lipids: Structure and Function* (P.K. Stumpf and E.E. Conn, Eds). New York, Academic Press

ULDALL, P.R., WILKINSON, R., McHUGH, M.I., FIELD, E.J., SHENTON, B.K., TAYLOR, R.M.R. and SWINNEY, J. (1974). Unsaturated fatty acids and renal transplantation. *Lancet* **ii**, 514

VLES, R.O. (1975). Nutritional aspects of rapeseed oil. In *The Role of Fats in Human Nutrition* (A.J. Vergroesen, Ed.), pp. 433–477. London, Academic Press

WEBER, E.J., DE LA ROCHE, I.A. and ALEXANDER, D.E. (1971). Stereospecific analysis of maize triglycerides. *Lipids* **6**, 523–530

WITTING, L.A. (1970). The interrelationships of polyunsaturated fatty acids and antioxidants *in vivo*. *Progress in the Chemistry of Fats and Other Lipids* **9**, 519–556

WOLF, R.B., GRAHAM, S.A. and KLEIMAN, R. (1983). Fatty acid composition of *Cuphea* seed oils. *Journal of the American Oil Chemists' Society* **60**, 103–104

YARON, A., SAMOILOFF, V. and BENZIONI, A. (1982). Absorption and distribution of orally administered Jojoba wax in mice. *Lipids* **17**, 169–171

2

THE CHEMISTRY, BIOCHEMISTRY AND NUTRITIONAL IMPORTANCE OF ANIMAL FATS

M. ENSER
Lipids Section, ARC Meat Research Institute, Langford, Bristol BS18 7DY, UK

Introduction

The aim of this chapter is to concentrate on those aspects of the chemistry, biochemistry and nutritional characteristics of animal fats which relate to animal nutrition and to set the scene for more specific chapters which follow. As a result, and because of limitation of space, many areas of lipid chemistry and metabolism will be mentioned in passing only, or dealt with only briefly. Further information about the biochemistry of lipids may be obtained from the book by Gurr and James (1980) and about their chemistry from the older book by Gunstone (1967) and the encyclopaedic work by Deuel (1951).

In nature, only carnivores and omnivores consume animal fats and in most cases the quantity and fatty acid composition of the fat so consumed differs markedly from that now included in the diets given to domesticated species. Indeed, the feeding of diets with added fat to herbivores such as ruminants is only satisfactory if the fat is supplied in a form not available to the micro-organisms in the rumen. Nevertheless, animal fats have become an important component in the formulation of feeds for farm animals. In many cases the use of animal fats has little to do with their specific properties but to their price relative to other fats or energy sources. On the other hand, their fatty acid composition may be superior to that of plant fats in producing a product, either meat or milk, with the desired physical qualities.

Chemistry

Fatty acids

Fatty acids are major components of many complex lipids. Those containing 12–24 carbon atoms are most common in animal tissues (*Table 2.1*), although shorter-chain fatty acids, C4 to C10, occur in milk. The main fatty acids can be classified as saturated straight- or branched-chain with an even

Table 2.1 SOME FATTY ACIDS PRESENT IN ANIMAL TISSUES

Number of carbon atoms and double bonds	Systematic name	Common name	Melting pt (°C)	Source
14:0	n-Tetradecanoic	Myristic	54.4	Food or synthesis
15:0	n-Pentadecanoic	—	52.3	Food or synthesis
16:0	n-Hexadecanoic	Palmitic	62.9	Food or synthesis
-16:1	cis-9-Hexadecanoic	Palmitoleic	0.0	Food or synthesis from palmitic acid
17:0	n-Heptadecanoic	Margaric	61.3	Food or synthesis
17:0 br[a]	14-Methyl hexadecanoic	—[b]	39.5	Food or synthesis
18:0	n-Octadecanoic	Stearic	69.6	Food or synthesis
-18:1	cis 9-Octadecenoic	Oleic	13.4	Food or synthesis from stearic acid
-18:1	trans-11-Octadecenoic	Vaccenic	39.0	Rumen hydrogenation
-18:2	cis-9,12-Octadecadienoic	Linoleic	−5.0	Food
18:3	cis-9,12,15-Octadecatrienoic	α-Linolenic	−11.0	Food
20:0	n-Eicosanoic	Arachidic	75.4	Food or synthesis
20:4	cis-5,8,11,14-Eicosatetraenoic	Arachidonic	−49.5	Food or synthesis from linoleic acid
22:0	n-Docosanoic	Behenic	80.0	Food or synthesis
22:1	cis-13-Docosenoic	Erucic	33.5	Food
22:5	cis-7,10,13,16,19-Docosapentaenoic	Clupanodonic	−78.0	Synthesis from α-linolenic acid
22:6	cis-4,7,10,13,16,19-Docosahexaenoic	—[b]	—	Synthesis from α-linolenic acid

[a] Branched chain
[b] Not known

or an odd number of carbon atoms, and their unsaturated derivatives, which may contain between one and six double bonds, usually *cis*, with a methylene group between neighbouring double bonds. *Trans-* unsaturated fatty acids are found in ruminants, as a result of fatty acid modification in the rumen, and in other animals fed partly hydrogenated plant oils. A very large number of fatty acids, some with structures different from those in *Table 2.1*, have been described, for example as many as 437 in milks (Patton and Jensen, 1975). These fatty acids are either synthesized *de novo*, obtained directly from the diet, or are synthesized by modification of dietary fatty acids.

The degree of fluidity of the lipids and lipid-containing structures in the animal are of major importance and are determined to a large degree by the fatty acids they contain. The melting point of the saturated straight-chain fatty acids increases with increasing chain-length (*Table 2.1*) and those with 12 or more carbon atoms are solid at body temperature and their presence hardens fat. Fatty acids with a double bond have lower melting points than the saturated straight-chain fatty acids of the same chain-length and the presence of each additional double bond decreases the melting point further. The melting point also depends upon the geometry of the double bond, with *trans* fatty acids melting at a higher temperature than their *cis* isomers, and on the position of the double bond within the chain. For example, for C18 fatty acids, oleic acid (*cis* double bond at position 9) melts at 13.4°C, elaidic acid (the *trans* isomer) melts at 43.7°C and vaccenic acid (*trans* double bond at position 11) melts at 39.0°C.

The branched-chain fatty acids consist of two main types: those with an *iso* structure containing a methyl substituent on the penultimate carbon distal to the carboxyl group, and the *anteiso* series with the methyl substituent on the prepenultimate carbon. Multiple methyl branches may also occur with the methyl group being present on the even-numbered carbon atoms counting from the carboxyl group. The branched-chain fatty acids have lower melting points than straight-chain fatty acids with the same number of carbon atoms (*Table 2.1*). In complex lipids of animals (*Tables 2.2, 2.3*) there is a mixture of fatty acids such that the fat is maintained at a fluid consistency at body temperature.

Table 2.2 MAJOR FATTY ACIDS OF THE TRIGLYCERIDES FROM THE SUBCUTANEOUS ADIPOSE TISSUE OF CATTLE, SHEEP AND PIGS

Fatty acid—number of carbon atoms and double bonds	Composition (% by weight)[a]		
	Cattle	Sheep	Pigs (outer layer)
14:0	3.7	2.9	1.5
16:0	29.8	23.7	27.6
16:1	4.7	3.5	3.2
18:0	17.1	18.3	12.2
18:1	42.3	43.2	45.1
18:2	2.3	3.8	10.4

[a] Minor components to 100%. Cattle: loin fat from 400-day-old Friesian bulls; sheep: inguinal fat from 224-day-old Hampshire lambs; pigs: outer subcutaneous loin fat from Large White pigs 87 kg in weight.

Table 2.3 MAJOR FATTY ACIDS OF THE PHOSPHOLIPIDS FROM THE LONGISSIMUS DORSI MUSCLE OF CATTLE, SHEEP AND PIGS

Fatty acid[a]	Composition (% by weight)[b]		
	Cattle	Sheep	Pigs
14:0	0.4	2.1	0.2
16:0	22.6	22.0	18.9
16:1	2.5	2.3	1.6
18:0	7.8	13.2	12.0
18:1	24.3	30.3	18.8
18:2	23.0	18.0	25.5
18:3	2.0	3.9	0.2
20:4	12.5	NR[c]	7.7

[a] Designated by the number of carbon atoms in the chain followed by the number of double bonds.
[b] Minor components to 100%. Cattle from O'Keefe *et al.*, 1968; sheep from Crouse *et al.*, 1972; pigs from Wood and Lister, 1973.
[c] NR, not reported.

Fatty acids occur in the non-esterified form in significant quantity at only two sites within the animal: in the intestine where, as soaps, they form micellar solutions with bile salts and monoglycerides, and in the blood plasma where they are bound to the protein albumin. The detergent action of the fatty acid salts, the basis of soap, inhibits metabolic processes and the intracellular concentration of fatty acids is kept low. In the rumen the presence of calcium soaps may be an advantage in decreasing the availability of fatty acids to the rumen micro-organisms. It is, however, a disadvantage in the small intestine where the insoluble calcium soaps of the long-chain, saturated fatty acids are poorly absorbed.

Long-chain fatty acids are virtually odourless, but as the fatty acids become shorter there is an increase in unpleasant odour which is particularly marked with C4 to C6 fatty acids. Esters of long-chain fatty acids, formed by reaction of the carboxyl group with an alcoholic hydroxyl group, are odourless. It is the hydrolysis of the butyric esters in the triacylglycerols of milk which produces the characteristic off-odour. The development of off-odours through the hydrolysis of esters of short-chain fatty acids is sometimes termed rancidity, but it should be distinguished clearly from oxidative rancidity, which is the major cause of development of off-odours in non-dairy fats and fat-containing products.

Oxidative rancidity

Fatty acids react chemically with oxygen to give hydroperoxides (*Figure 2.1*). Hydroperoxides are themselves odourless but their subsequent breakdown results in cleavage of long-chain fatty acids into shorter-chain products which may have a strong odour. This process occurs in oil-based paints as they cure and harden and the rancid odour of fat is often described as 'painty'. The breakdown of hydroperoxides yields free radicals which are extremely reactive and attack other fatty acids more readily than the initial attack by oxygen. The reaction then proceeds

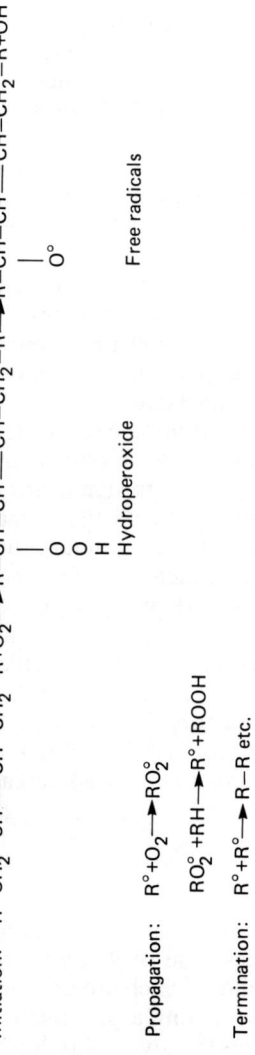

Figure 2.1 Autocatalytic oxidation of fatty acids

through a free-radical chain mechanism and since the products increase the rate of the reaction, it is termed autocatalytic autoxidation. The oxidation process therefore starts slowly but the rate increases almost exponentially until the concentration of free radicals is such that they have a high chance of reacting with each other, thereby terminating the chain. The products of oxidation are shorter fatty acids or fatty acid polymers, aldehydes, ketones, alcohols, epoxides and hydrocarbons. The short-chain unsaturated aldehydes and acids are major odour and flavour components: for example, deca-2,4-dienal can be detected at dilutions in water of 5 in 10^{10}. Many of the aldehydes are reactive molecules and will condense with amino acid residues in proteins, often decreasing the solubility of the protein. The aldehydes are also the products measured in tests for rancidity, such as the Kreis, anisidine and thiobarbituric acid (TBA) tests. Although most aldehydes react to some extent in all the tests, the major reactant in the TBA test is malonic dialdehyde which arises from polyunsaturated fatty acids.

The primary reaction of oxygen with a saturated hydrocarbon chain is slow, but it is increased by the presence of double bonds so that linolenic acid with three double bonds is oxidized 100 times more rapidly than oleic acid with only one double bond. The high concentration of linoleic acid in pork fat compared with beef and lamb (*Table 2.2*) is responsible for its much more rapid oxidative deterioration and shorter life in frozen storage. Antioxidants are present in most fat- or oil-containing natural products, and it is their presence which prevents significant oxidation occurring in feedstuffs or in the animal and prevents the immediate development of oxidation in animal or plant fats. The most important natural antioxidant is vitamin E or α-tocopherol which acts as a free radical acceptor thereby breaking the free radical chain. The presence of oxidized fat in feedstuffs appears to have little effect on the growth of pigs (Dow *et al.*, 1963; L'Estrange *et al.*, 1967; Connolly *et al.*, 1970) or chicken (L'Estrange *et al.*, 1966) or turkeys (Lea *et al.*, 1966) provided that the selenium and vitamin E contents of the diet are satisfactory. Further details about lipid peroxidation may be obtained from the books edited by Schultz, Day and Sinnhuber (1962) and by Emanuel and Lyaskovskaya (1967).

Glycerolipids

The glycerolipids consist of a range of compounds containing glycerol with a fatty acid esterified to at least one of its hydroxyl groups (*Figure 2.2*). They may be subdivided into those which are uncharged, the neutral lipids, and the charged molecules containing a phosphate group, the phospholipids. This division also separates the glycerolipids in terms of their function within animals since neutral glycerides are the major storage form in adipose tissue and phospholipids are mainly structural components of all tissues.

Triglycerides are the predominant neutral lipid present in adipose tissue and may constitute up to 95% of the weight of the tissue in very fat animals. Mono- and diglycerides, on the other hand, are present at only 1% or 2% and are intermediates in the metabolism of other glycerides.

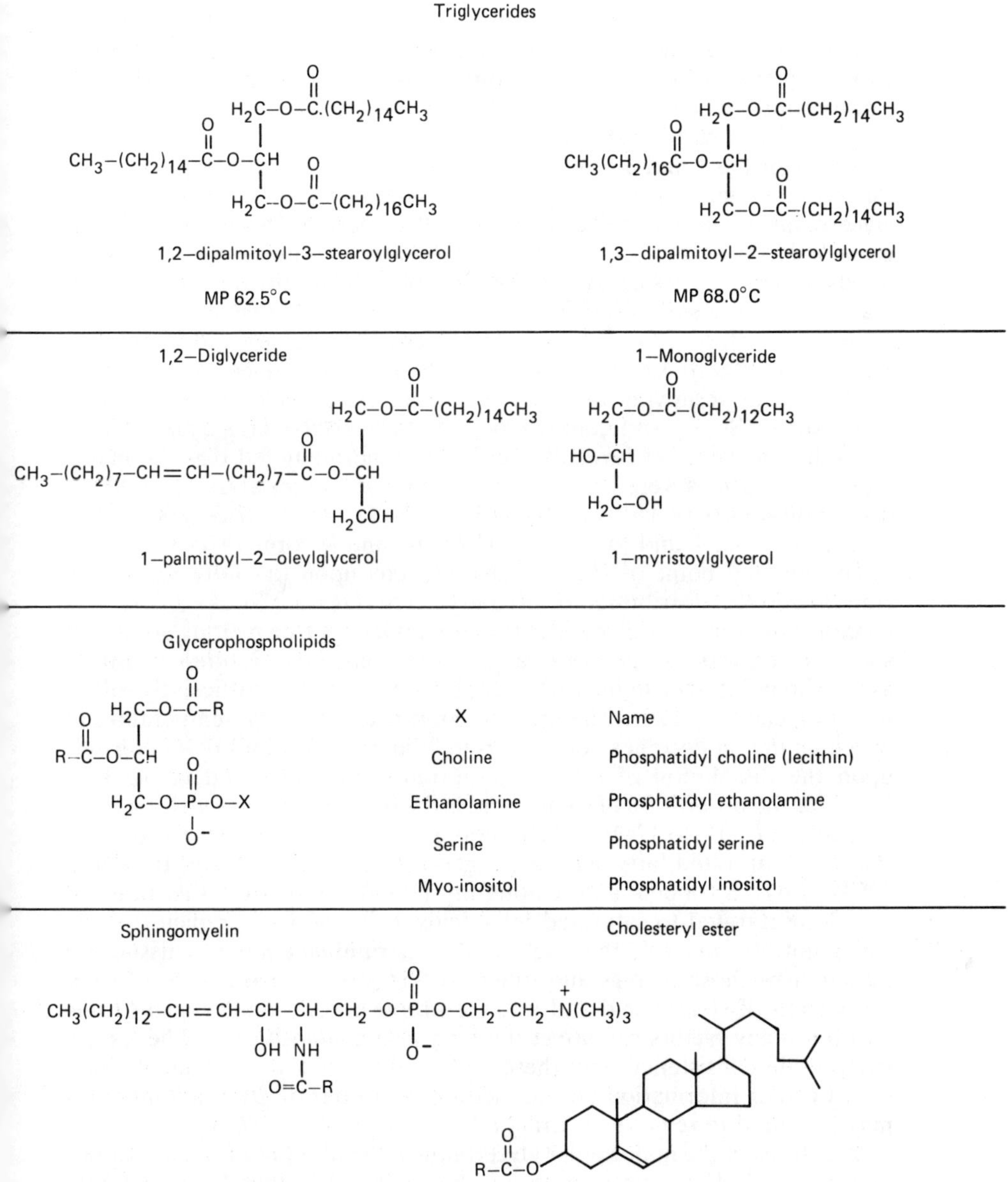

Triglycerides

1,2-dipalmitoyl-3-stearoylglycerol

$H_2C-O-C(=O)-(CH_2)_{14}CH_3$
$CH_3-(CH_2)_{14}-C(=O)-O-CH$
$H_2C-O-C(=O)-(CH_2)_{16}CH_3$

MP 62.5°C

1,3-dipalmitoyl-2-stearoylglycerol

$H_2C-O-C(=O)-(CH_2)_{14}CH_3$
$CH_3(CH_2)_{16}C(=O)-O-CH$
$H_2C-O-C(=O)-(CH_2)_{14}CH_3$

MP 68.0°C

1,2-Diglyceride

1-palmitoyl-2-oleylglycerol

$H_2C-O-C(=O)-(CH_2)_{14}CH_3$
$CH_3-(CH_2)_7-CH=CH-(CH_2)_7-C(=O)-O-CH$
H_2COH

1-Monoglyceride

1-myristoylglycerol

$H_2C-O-C(=O)-(CH_2)_{12}CH_3$
$HO-CH$
H_2C-OH

Glycerophospholipids

$H_2C-O-C(=O)-R$
$R-C(=O)-O-CH$
$H_2C-O-P(=O)(O^-)-O-X$

X	Name
Choline	Phosphatidyl choline (lecithin)
Ethanolamine	Phosphatidyl ethanolamine
Serine	Phosphatidyl serine
Myo-inositol	Phosphatidyl inositol

Sphingomyelin

$CH_3(CH_2)_{12}-CH=CH-CH(OH)-CH(NH-)-CH_2-O-P(=O)(O^-)-O-CH_2-CH_2-N^+(CH_3)_3$

with NH group bonded to $O=C-R$

Cholesteryl ester

$R-C(=O)-O-$ (cholesterol)

Figure 2.2 Structure of major lipids present in animal tissues

Fatty acids are not esterified at random to the glycerol hydroxyl groups in animals. In most species the 2-position in the adipose tissue triglycerides is occupied by an unsaturated fatty acid, whereas in the pig (Hilditch and Stainsby, 1935; Meara, 1945; Quimby, Wille and Lutton, 1953; Mattson, Volpenheim and Lutton, 1964) and human milk (Breckenridge, Marai and Kuksis, 1968) it is occupied by a saturated fatty acid, mainly palmitic acid. Triglycerides in other tissues, for example pig liver, have the more common 2-unsaturated structure.

Although the 1 and 3 positions of glycerol are chemically equivalent in this symmetrical molecule they can be distinguished by enzymes, making glycerol an assymetric molecule in biochemical terms. This allows different groups of fatty acids to be esterified to the 1 and 3 positions. Since the ester bonds of triglycerides cannot be distinguished chemically it is necessary to make use of the stereochemistry of enzymes to determine the fatty acids present at each position (Mattson and Beck, 1956; Savary and Desnuelle, 1956; Brockerhoff, 1965; Lands et al., 1966). Most evidence suggests that the fatty acids occupying any one position are independent of the fatty acid at any other (Slakey and Lands, 1968; Åkesson, 1969; Christie and Moore, 1970a,b). In the inner layer of backfat from normally fed pigs the major types of glycerides were: (S = saturated, M = mono-unsaturated and D = di-unsaturated fatty acids) SSM, 24.1%; MSM, 23.0%; MSS, 7.8%; SSS, 6.8%; SDS, 6.2% and MDS 6.1% (Christie and Moore, 1970a).

The melting point of triglycerides depends upon the fatty acids they contain and the distribution of the fatty acids (*Figure 2.2*). As animal fats contain a mixture of triglycerides they do not have a true melting point, but soften over a wide temperature range. Their consistency is often measured as the slip point—the temperature at which they become sufficiently soft to flow in a capillary tube. The melting point is taken as the temperature at which all the fat becomes liquid. This melting (clarification) point depends upon the dissolution of the more saturated glycerides and in pig triglycerides the proportion of stearic acid is the best predictor of melting point (Wood et al., 1978) (*Figure 2.3*) although other workers have suggested that total saturated fatty acids are a good predictor (Elliot and Bowland, 1969). Lea, Swoboda and Gatherum (1970) considered the ratio of mono-unsaturated to saturated fatty acids to be the best predictor of the slip point. In general, the triglycerides in ruminant adipose tissue are harder than those in pigs and other monogastric animals, with a higher proportion of stearic acid and a lower proportion of linoleic acid (*Table 2.2*) but many factors can affect the fatty acid composition and hence the triglyceride consistency, and these will be discussed in subsequent chapters. Further information on the melting behaviour of glyceride mixtures may be found in reviews by Craig (1957) and Rossell (1967).

Structures of the major polar glycerolipids, the phospholipids, are shown in *Figure 2.2*. The concentration of phospholipids in muscle and adipose tissue is between 0.5% and 1.0% of wet tissue whereas in liver it varies from 2% to 3%. Phosphatidyl choline (lecithin) is the major phospholipid in many tissues, except brain. Phosphatidyl choline is also the major lipid in the erythrocyte membrane of man, rats, rabbits and pigs, but in cattle and sheep it is displaced by sphingomyelin, a non-glyceride phospholipid (De Gier and Van Deenen, 1961).

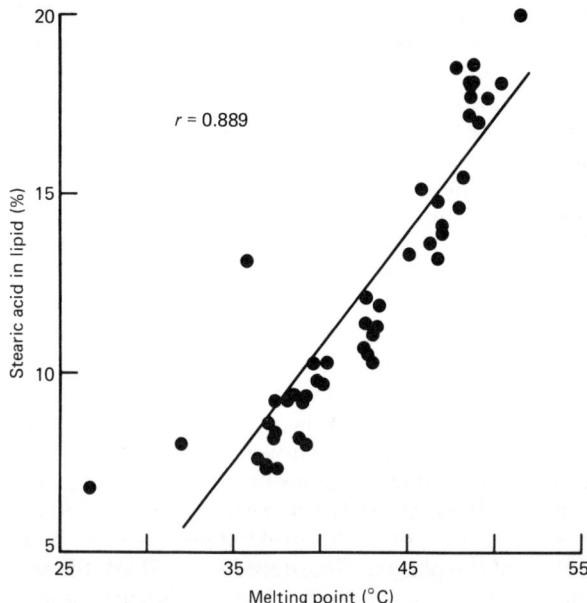

Figure 2.3 Relationship between proportion of stearic acid in pig backfat lipids and the melting point of the lipid

The fatty acid composition of the total phospholipids from muscle of cattle, sheep and pigs is shown in *Table 2.3*. The composition is similar between the three species in contrast to the differences between their adipose tissue triglycerides (*Table 2.2*). Different phospholipids have different fatty acid compositions. In beef muscle, arachidonic acid is the major polyunsaturated fatty acid in phosphatidyl ethanolamine whereas linoleic acid is the major polyunsaturated constituent of phosphatidyl choline (Hornstein, Crowe and Heimberg, 1961). The fatty acid distribution on the glycerol of phospholipids is much more specific than that in the triglycerides. Saturated fatty acids predominate on the 1 position and unsaturated fatty acids on the 2 position and there are also preferred combinations of fatty acids in the molecules. The preponderance of particular pairs varies with the tissue and species: dipalmitoyl phosphatidyl choline is the major type present in the lungs of rats, rabbits, pigs and sheep but 1-palmitoyl 2-oleyl-phosphatidylcholine is the major component in lungs of cattle (Montfoort, Van Golde and Van Deenen, 1971). Dipalmitoyl phosphatidyl choline is believed to be particularly important in decreasing the surface tension in the lungs so that they inflate readily. Other favoured fatty acid combinations are palmitic/linoleic, stearic/linoleic, palmitic/arachidonic and stearic/arachidonic. Phospholipids containing vinyl ether bonds, called plasmalogens, are much more common than the neutral lipids containing vinyl ethers. In ox heart approximately 50% of the phosphatidyl choline and phosphatidyl ethanolamine fractions consist of plasmalogens (Gray and MacFarlane, 1958). The aldehydes obtained by acidic hydrolysis of the vinyl ethers are mainly palmitaldehyde and stearaldehyde (Grigor, Moehl and Snyder, 1972). Sphingomyelin (*Figure 2.2*) is an important phospholipid which does not contain glycerol. This and related compounds are present in membranes, forming between 10% and 25% of the phospholipids of red cell membranes in different species (De Gier and Van Deenen, 1961). They constitute similar proportions of the brain phospholipids but are only approximately 5% of the muscle phospholipids. The long-chain fatty acid in sphingomyelin is attached to the sphingosine by an amide linkage rather than the ester linkage in other lipids. In pig muscle the sphingomyelin fatty acids resemble those of phosphatidyl choline in composition (Kuchmak and Dugan, 1965).

Cholesterol and its fatty acid esters (*Figure 2.2*) are the most important non-glyceride neutral lipids. Cholesterol occurs in membranes where the hydrophilic hydroxyl group is associated with the polar end of the phospholipids. The properties of the membranes appear to be related to the cholesterol:phospholipid ratio. On a fresh-weight basis, muscle and adipose tissue contain similar quantities of cholesterol, ranging from 600 to 900 mg/kg (Feeley, Criner and Watt, 1972; Reiser, 1975). Cholesteryl-esters are important in the transport of cholesterol between tissues, by the blood. Approximately 80% of the plasma cholesterol is esterified, the main fatty acids involved being linoleic acid or oleic acid. Animals can synthesize cholesterol or obtain it from the diet although the normal ruminant diet does not contain cholesterol. The relationship between the content of cholesterol and saturated fatty acids in the diet and the concentration of plasma cholesterol has been extensively investigated because hypercholesterolaemia is a contributory factor in atherosclerosis. Atherosclerosis can be induced in pigs fed diets to which cholesterol has been added (Scott, Daoud and Florentin, 1971).

Metabolism

Fatty acid synthesis

Fatty acids are synthesized from any body component which yields a two-carbon acetyl unit during its metabolism. The overall equation for the process is:

$$CH_3CO-S-CoA + 7\,HOOC-CH_2-CO-S-CoA + 14\,NADPH + 14H^+ \rightarrow C_{15}H_{31}COOH + 7CO_2 + 6H_2O + 8CoASH + 14NADP^+$$

Two enzyme complexes are involved—acetyl-CoA carboxylase and fatty acid synthetase (*Figure 2.4*). Acetyl-CoA carboxylase adds carbon dioxide to acetyl-CoA to yield malonyl-CoA. The malonyl group and an acetyl group are then transferred from CoA to the fatty acid synthetase complex and condensed to give acetoacetyl-S-enzyme with the release of the carbon dioxide. The enzyme system then carries out sequentially the reduction of the ketoacyl group, dehydration of the hydroxyacyl group and reduction of the enoyl double bond to yield a saturated fatty acid two carbons longer (*Figure 2.4*). The cycle is repeated with more malonyl-CoA until the fatty acid is released from the enzyme. The length of the fatty acid synthesized

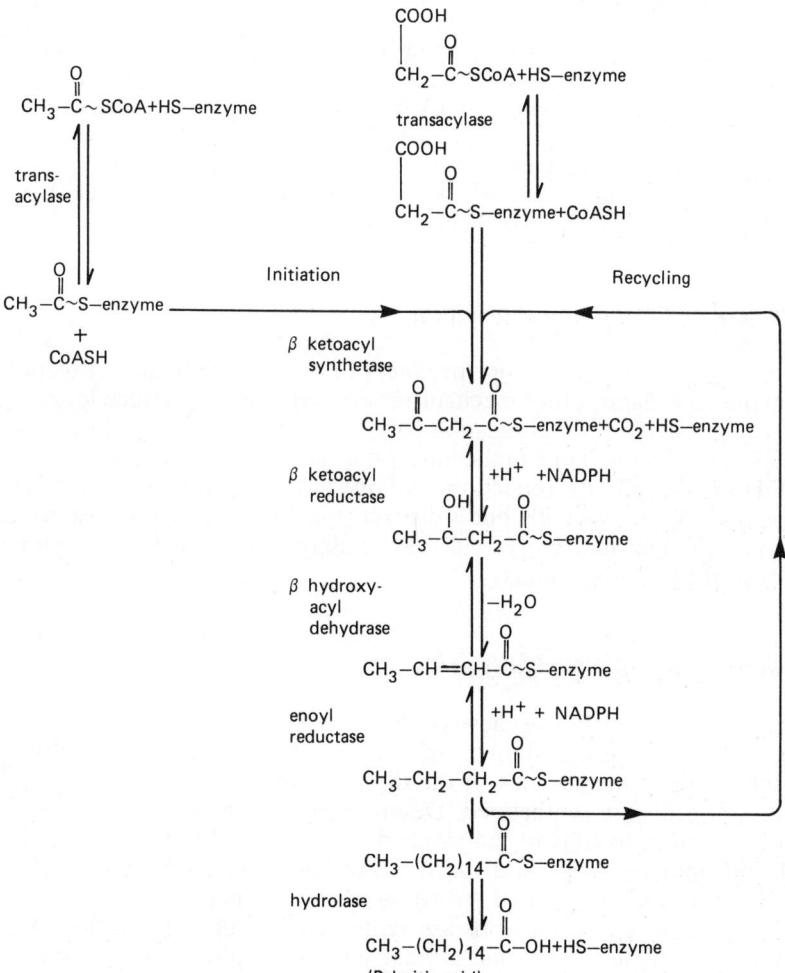

Figure 2.4 Pathway of fatty acid synthesis *de novo* in animals

depends upon the tissue. In liver and adipose tissue the major product is palmitic acid. In the mammary gland, shorter-chain fatty acids are produced in the presence of a cytosolic protein which has acylthioester hydrolase activity (Knudsen, Clark and Dils, 1975).

The addition of two carbon units to a two-carbon acetyl primer results in the formation of the common even-chain fatty acids. If the primer is a three-carbon propionyl group, odd-chain fatty acids result. Propionic acid is a common product of rumen metabolism which is converted to methylmalonyl-CoA by the animal. An isomerase, which contains vitamin B_{12}, then converts it to succinyl CoA which is oxidized. Ruminants produce increased amounts of propionic acid when they are fed large quantities of cereals and in the case of the lamb this causes the deposition of increased quantities of odd-chain fatty acids. Large quantities of *anteiso* branched-chain fatty acids and fatty acids with methyl groups on even carbons, from the carboxyl group, are also synthesized and deposited so that the fat becomes soft (Garton, Hovell and Duncan, 1972). These fatty acids arise by the replacement of malonyl-CoA by methylmalonyl-CoA for fatty acid synthesis. *Anteiso* fatty acids may also be synthesized when 2-methylbutyryl-CoA, derived from the amino acid isoleucine, is the primer. The amino acids valine and leucine are degraded to isobutyryl-CoA and isovaleryl-CoA which are the primers for isoacid synthesis.

Fatty acid extension and shortening

Since palmitic acid is the major product of fatty acid synthetase, except in the mammary gland, other mechanisms are required to produce longer or shorter fatty acids. Two elongation pathways exist which extend the chain by a 2C unit at a time: one in the mitochondria, which uses acetyl-CoA and NADH or NADPH for reduction, and one in the microsomes, which uses malonyl-CoA and NADPH but is distinct from fatty acid synthetase which is a cytosolic enzyme. Fatty acids also undergo shortening by sequential removal of two carbon units.

Synthesis of unsaturated fatty acids

Unsaturated fatty acids, because of their low melting points, are essential to maintain the gross fluidity of adipose tissue and the fluidity of phospholipids in membranes. If the dietary supply is insufficient, unsaturated fatty acids are synthesized. Desaturation is also essential, in concert with elongation, to convert the major dietary essential fatty acids, linoleic acid and linolenic acid, into more unsaturated longer-chain derivatives which are needed for normal growth and maintenance.

Mono-unsaturated fatty acids are synthesized by the fatty acyl-CoA Δ^9 desaturase complex. This microsomal enzyme requires oxygen and reduced pyridine nucleotide and shows maximum velocity with stearic acid. The major product in animal tissues is oleic acid, with smaller quantities of palmitoleic acid. Desaturase activity is present in sheep intestinal mucosa but absent from the intestine of pig and chicken (Bickerstaffe and Annison, 1969) thus indicating the necessity for the ruminant to soften the saturated fat produced by the rumen. The desaturase activity of sheep adipose tissue was greater than that of the liver but in rats and chicken the liver activity was greater (Wahle, 1974). The enzyme also occurs in bovine mammary gland (McDonald and Kinsella, 1973).

Animals do not possess enzymes capable of inserting double bonds between the 9-carbon and the terminal methyl group. Fatty acids with

$CH_3-CH_2-CH_2-CH_2-CH_2-CH_2-CH_2-CH_2-CH=CH-(CH_2)_7-COOH$ n9 octadecenoic acid (oleic acid)

$CH_3-CH_2-CH_2-CH_2-CH_2-CH=CH-CH_2-CH=CH-(CH_2)_7-COOH$ n6 octadecadienoic acid (linoleic acid)

$CH_3-CH_2-CH=CH-CH_2-CH=CH-CH_2-CH=CH-(CH_2)_7-COOH$ n3 octadecatrienoic acid (linolenic acid)

Figure 2.5 Relationships between the structures of unsaturated fatty acids with 18 carbon atoms

Oleic acid

9–18:1 $\xrightarrow{-2H}$ 6,9–18:2 $\xrightarrow{+2C}$ 8,11–20:2 $\xrightarrow{-2H}$ 5,8,11–20:3

Linoleic acid

9,12–18:2 $\xrightarrow{-2H}$ 6,9,12–18:3 $\xrightarrow{+2C}$ 8,11,14–20:3 $\xrightarrow{-2H}$ 5,8,11,14–20:4

Linolenic acid

9,12,15–18:3 $\xrightarrow{-2H}$ 6,9,12,15–18:4 $\xrightarrow{+2C}$ 8,11,14,17–20:4 $\xrightarrow{-2H}$ 5,8,11,14,17–20:5

$\xrightarrow{+2C}$ 7,10,13,16,19–22:5 $\xrightarrow{-2H}$ 4,7,10,13,16,19–22:6

Figure 2.6 Metabolic pathways of essential fatty acids and oleic acid in animals

methylene-interrupted double bonds in this region are required in the diet and are termed essential fatty acids. The two essential fatty acids most common in the diet are linoleic acid and linolenic acid. These two fatty acids and oleic acid give rise to series of polyunsaturated fatty acids designated n-9, n-6 and n-3 according to the position of the double bond nearest the methyl group (*Figure 2.5*). The sequence of further desaturation and elongation of these fatty acids is shown in *Figure 2.6*. The final product formed depends upon the affinity of the Δ^4, Δ^5 and Δ^6 desaturases for their substrates and competition or inhibition by substrates from the different series. For instance oleic acid does not undergo desaturation and extension unless linoleic and linolenic acid are virtually absent and the ratio of $\Delta^{5,8,11}20:3$ eicosatrienoic to arachidonic ($\Delta^{5,8,11,14}20:4$ eicosatetraenoic) is taken to indicate the essential fatty acid status of the animal. Because of the extensive loss of linoleic acid and linolenic acid in the rumen, the products of oleic acid metabolism are increased in ruminants than in other meat animals.

Regulation of fatty acid synthesis

The regulation of fatty acid synthesis has been reviewed by Volpe and Vagelos (1976). Acetyl-CoA carboxylase is under short-term metabolic control and longer-term hormonal and dietary regulation. The enzyme is activated by increased concentrations of citric acid, which also acts as a source of acetyl-CoA, and is inhibited by high concentrations of long-chain acyl-CoA. It is also inactivated by a cyclic AMP-dependent protein kinase (Brownsey, Hughes and Denton, 1979) so that fatty acids are not synthesized at times when they are being mobilized for degradation. Acetyl-CoA carboxylase is generally believed to be the rate-limiting enzyme in fatty acid synthesis but both it and fatty acid synthetase decrease when the animal is starved and increase during refeeding. As the ratio of fatty acid synthetase to acetyl-CoA carboxylase is less than 1.0 in chicken liver, it is possible that the synthetase may be rate-limiting in this species (Donaldson, 1979). In ruminants, however, the rates of fatty acid synthesis and acetyl-CoA carboxylase activity are closely correlated (Ingle *et al.*, 1973; Pothoven and Beitz, 1975).

The effect of dietary fat on the lipogenic activity and enzymes of fatty acid synthesis has been little studied in the adipose tissue of meat animals, where most fatty acid synthesis occurs. Two factors have complicated the interpretation of many studies. High concentrations of dietary fat may decrease fatty acid synthesis, not through inhibition of the enzymes but because the diets contain insufficient substrates for fatty acid synthesis. For instance, in chicken liver, fatty acid synthesis and the enzymes of synthesis were decreased by dietary fat, but if the dietary carbohydrate remained constant, fat had no effect (Hillard, Lundin and Clarke, 1980). The other factor is the type of fat in the diet. Saturated fatty acids and their glycerides added to diets may be poorly absorbed and may have no effect on fatty acid synthesis, whereas vegetable oils which are well absorbed may be active. However, in rats, 3% methyl linoleate in the diet decreased fatty acid synthesis in the liver by 50%, although 8% methylstearate (which was absorbed to the same extent as the linoleate) had no effect. Neither fatty acid affected fatty acid synthesis in adipose tissue (Clarke, Romsos and Leveille, 1977). In pigs, fats low in linoleic acid consistently depress fatty acid synthesis in adipose tissue but fats containing high proportions of linoleic acid have produced variable results (Allee *et al.*, 1972; Waterman *et al.*, 1975; Steffen *et al.*, 1978). Fatty acid synthesis in the adipose tissue of lambs was inhibited by dietary tallow (Vernon, 1976), However, Hood *et al.* (1980) observed inhibition of fatty acid synthesis in the adipose tissue of lambs, *in vivo*, only when they were fed protected safflower oil, which contained 59.8% linoleic acid, and not when the protected lipid was beef tallow with 2.4% linoleic acid or palm oil with 9.1% linoleic acid. In cows, dietary fat decreased the concentration of short-chain fatty acids in milk by decreasing their synthesis (Christie, 1979). Both saturated and unsaturated fatty acids were effective, whether protected from rumen hydrogenation or not.

The synthesis of mono-unsaturated fatty acids is also under metabolic, hormonal and dietary control. Dietary saturated fatty acids, or feeding a fat-free diet (which results in increased synthesis of saturated fatty acids *de novo*), increases the Δ^9 acyl-CoA desaturase activity in rat liver (Inkpen, Harris and Quackenbush, 1969). Polyunsaturated fatty acids exert the opposite effect: they repress the enzyme and, in mouse liver, the activity is inversely proportional to the concentration of linoleic acid in the liver lipids (Enser, 1979). However, repression by linoleic acid can be overcome by saturated fatty acids, either exogenous or synthesized *de novo* (Enser and Roberts, 1982). Although the linoleic acid content of lamb's liver increases between birth and 8 days of age, this fails to prevent a simultaneous increase in desaturase activity (Shand, Noble and Moore, 1978). The effect of dietary fatty acids on desaturase activity of adipose tissue does not appear to have been studied in cattle, sheep and pigs.

Taylor and Thomke (1964) reported that dietary copper produced soft fat in pigs and the increased concentrations of oleic acid (Elliot and Bowland, 1968; Moore *et al.*, 1968) were demonstrated to be associated with increased Δ^9 acyl-CoA desaturase activity (Ho and Elliot, 1973; Thompson, Allen and Meade, 1973). The mechanism of action of copper is not understood since the desaturase, although it reacts with oxygen, does not contain copper (Strittmatter *et al.*, 1974) but the importance of copper

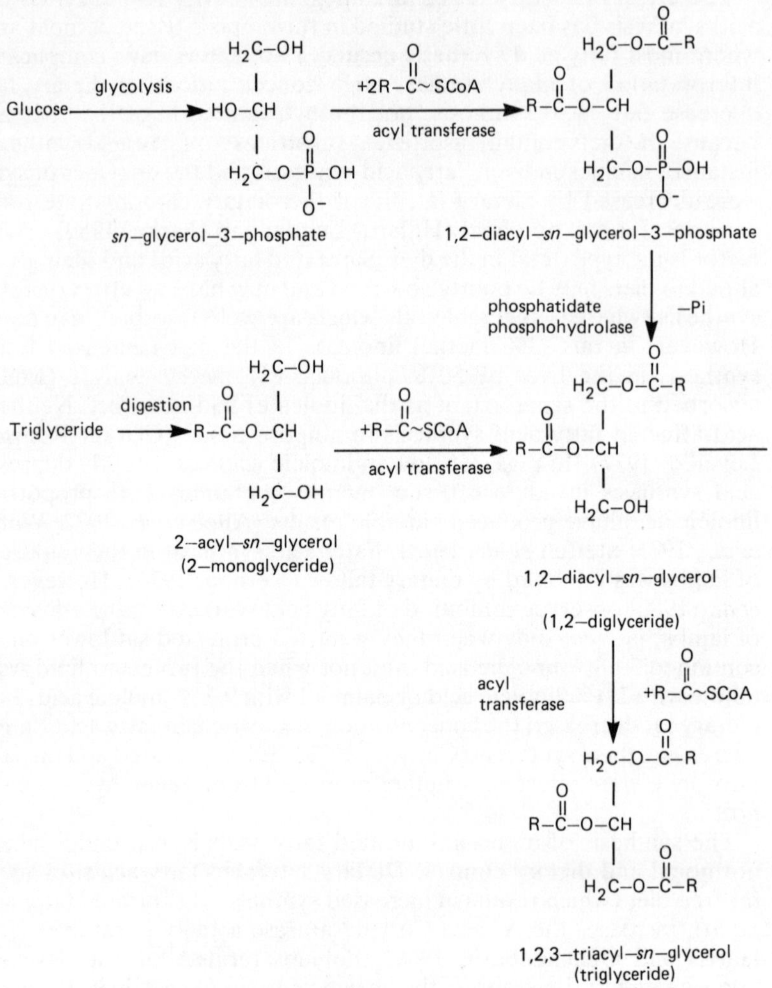

Figure 2.7 Pathways of glycerolipid synthesis

has been confirmed in copper-deficient rats in which the enzyme activity was decreased by 65% (Wahle and Davies, 1974).

Glycerolipid metabolism

The pathways of glycerolipid synthesis are shown in *Figure 2.7*. In adipose tissue the glycerol moiety is derived from glucose metabolism via L-α-glycerophosphate, but in the intestinal mucosa the starting point may be absorbed monoglycerides. The specific fatty acid distribution may be introduced either at the stage of acylation of the L-α-glycerophosphate or 2-monoglyceride or through partial hydrolysis and reacylation of the glycerides. The short-chain fatty acids in milk are esterified to the sn3-hydroxyl of 1,2-diglycerides by a transacylase which is specific for short-chain acyl-CoA. Changes in fatty acid availability may affect the pattern of triglycerides synthesized. Pigs fed a fat-free diet (Anderson, Bottino and Reiser, 1970) deposit similar types of triglyceride to that in pigs fed a normal grain diet (Christie and Moore, 1970a) with oleic acid replacing dietary linoleic acid. However, in copper-fed pigs there were larger changes in triglyceride types, although the total change in fatty acid

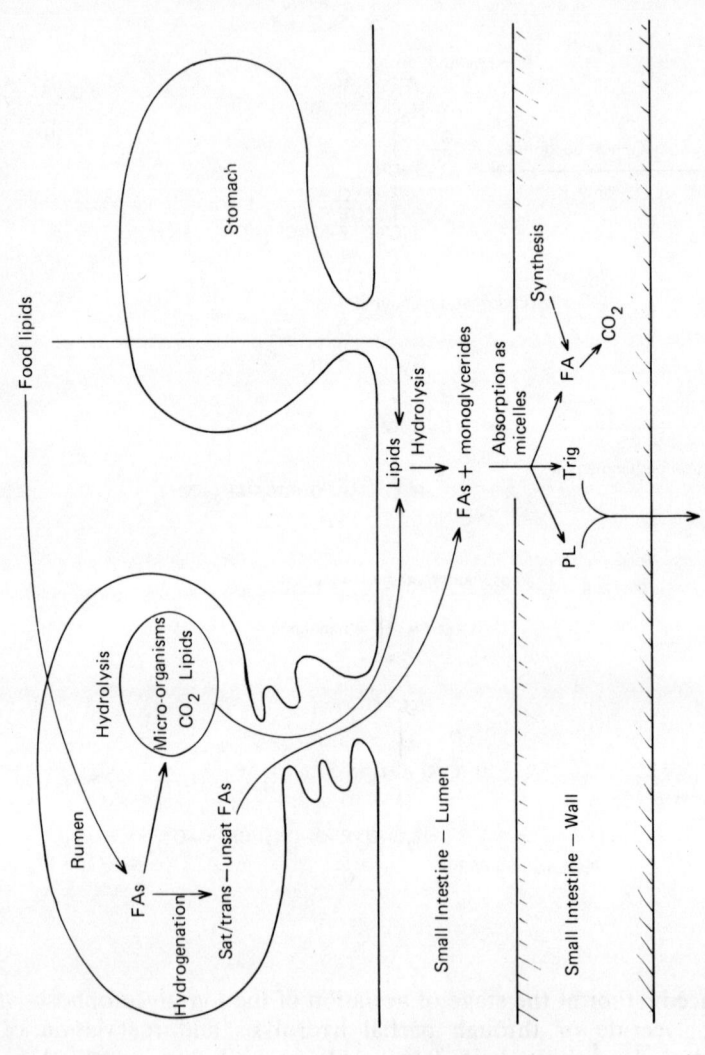

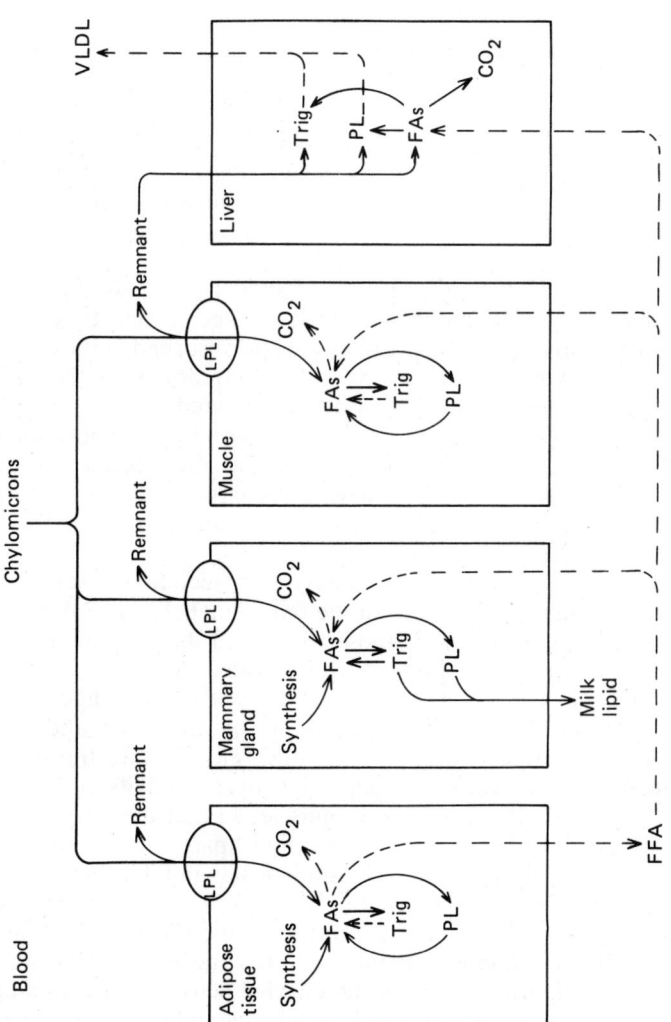

Figure 2.8 Lipid metabolism in meat animals in the fed state. FAs, fatty acids; FFA, free fatty acids; LPL, lipoprotein lipases; PL, plasma lipids; Trig, triglycerides; VLDL, very low density lipoproteins

composition was small (Christie and Moore, 1969, 1970b; Amer and Elliot, 1973). The large quantities of linoleic acid made available to cows through the use of sunflower seed protected from hydrogenation in the rumen produced considerable changes in the pattern of milk triglycerides (Morrison and Hawke, 1977). Transacylase enzymes are decreased in the adipose tissue of starved pigs, but increase in the liver where the synthesis of very low density lipoproteins (VLDL) is increased during starvation (Steffen *et al.*, 1978).

Triglycerides do not pass intact through the intestinal wall but must be hydrolysed. In the rumen, microbial lipases release free fatty acids for the organisms to metabolize. The metabolism of the fatty acids in the rumen is discussed in detail in later chapters. In general, unsaturated fatty acids are hydrogenated to give saturated fatty acids. Partial hydrogenation of polyunsaturated fatty acids also results in the formation of *trans* isomers and positional isomers. Only small quantities of dietary polyunsaturated fatty acids, some as constituents of microbial lipids, pass unchanged through the rumen. In the monogastric animal, lipids pass through the stomach and into the duodenum with little change, unless they contain short-chain fatty acids which may be released and absorbed in the stomach. In the duodenum, pancreatic lipase hydrolyses glycerides. After resynthesis in the epithelial cells, the triglyceride is secreted into the lymphatic system as chylomicrons in mammals or into the portal vein as portamicrons in birds (Bensadoun and Rothfeld, 1972). The triglycerides in these particles are hydrolysed by the enzyme lipoprotein lipase, which is attached to the capillary endothelium (Enser *et al.*, 1967) and the fatty acids enter the tissues.

Differences in the lipoprotein lipase activity between tissues regulate the distribution of plasma triglycerides. In the fed animal (*Figure 2.8*) the flow of lipid is from the intestine to adipose tissue and the enzyme activity is high in adipose tissue. In the postabsorptive state, nutrients are required by other tissues and there is a release of fatty acids from adipose tissue. One-third of these fatty acids is taken up by the liver and a high proportion are converted to triglycerides and released into the blood as VLDL. Under these conditions, lipoprotein lipase activity decreases in adipose tissue (Enser, 1973) but not in muscle, so that the fatty acids are supplied to muscle. The activity of lipoprotein lipase in adipose tissue is proportional to the concentration of plasma insulin and decreases with the starvation-induced decrease in insulin. In chicken, however, starvation does not decrease plasma insulin concentration and this may explain why lipoprotein lipase in adipose tissue does not decrease (Benson and Bensadoun, 1977). During lactation the enzyme activity is low in adipose tissue and high in the mammary glands—a change induced by prolactin in the rat (Zinder *et al.*, 1974)—whereas in the laying hen the lipoprotein lipase activity of the ovarian follicles increases rapidly as the rapid accumulation of yolk lipid takes place (Bensadoun and Kompiang, 1979).

Fatty acids entering tissues are either oxidized to supply energy, incorporated into structural lipids or stored as triglycerides. Liver, intestine and mammary gland may subsequently secrete the triglycerides either into the blood or milk ducts, as particulate lipoprotein, but in adipose tissue the triglycerides are first hydrolysed and then the fatty acids are

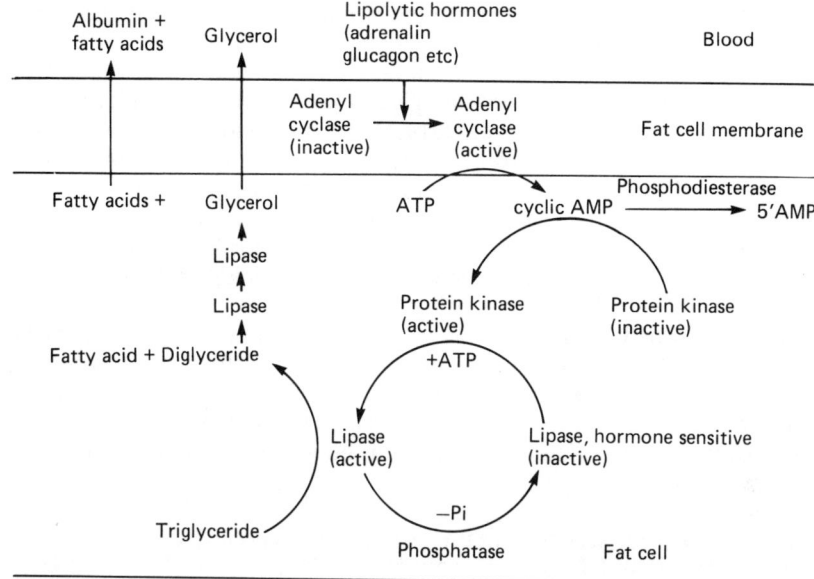

Figure 2.9 Lipolysis in adipose tissue

released into the plasma. The hydrolytic enzyme in adipose tissue is termed hormone-sensitive lipase because it is activated by hormones such as adrenalin or glucagon, as illustrated in *Figure 2.9*. Glucagon is an important lipolytic hormone in birds but probably not in meat animals. Insulin regulates release of fatty acids from adipose tissue by stimulating glucose uptake, thereby increasing the supply of L-glycerophosphate for fatty acid esterification. It also may act to inhibit adenyl cyclase and stimulate cyclic-AMP phosphodiesterase.

The turnover of fatty acids in adipose tissue depends on the energy status of the animal. Growing pigs fed *ad libitum* are probably always in positive energy balance and hence do not have to rely on stored fat to supply their energy needs. Even a slightly restricted commercial regimen of twice-daily feeding will probably not result in extensive release of fatty acids from adipose tissue. There is a small basal release of fatty acids from adipose tissue but these may not all come from the triglycerides of the fat globule, since there is a delay between the time fatty acids enter the tissue or are synthesized and their incorporation into the lipid globule. When lipolysis in rat adipose tissue is stimulated by adrenaline, it is the recently acquired fatty acids which are released (Ekstedt and Olivecrona, 1970). Whether the adipose tissue of pigs and ruminants behaves similarly has not been established. During starvation of pigs and sheep for 2–4 days the composition of the mobilized fatty acids resembles the fatty acid composition of the adipose tissue triglycerides, but prolonged starvation of sheep results in selective mobilization (Wood *et al.*, 1977; Jackson and Winkler, 1970). The half-life of the triglyceride fatty acids is approximately 180 days in pigs (Cunningham, 1968). A fatty acid deposited in adipose tissue therefore has a 50% or better chance of being present when the animal is slaughtered at bacon weight. The half-life of fatty acids in the growing ruminant has not been determined by labelling the adipose tissue lipids. Values based on plasma fatty acid flux and estimated glyceride stores indicate a half-life of 88 days in the non-pregnant ewe. During starvation and lactation, turnover is increased and this value drops to one-half and one-quarter respectively (Emery, 1980). These results may underestimate the actual half-life and higher values probably occur in growing lambs and steers, since the rumen results in a more even supply of nutrient than in monogastric animals. Only in starvation or in the lactating female are the fatty acids likely to turn over more rapidly.

In view of the long half-life of fatty acids in adipose tissue and the apparent lack of selectivity during mobilization, the fatty acid composition of the tissue will be determined by the type of fatty acids available for storage throughout the life of the animal. The high lipoprotein lipase activity of the adipose tissue in animals in the fed state results in the rapid

Table 2.4 COMPARISON OF THE INCORPORATION OF DIETARY LINOLEIC ACID INTO THE SUBCUTANEOUS ADIPOSE TISSUE OF CHICKEN, PIGS AND SHEEP

Animal species[a]	Supplementary fat in diet (%)	Linoleic acid in supplement (%)	Linoleic acid in adipose tissue fatty acids (%)
Chicken	10	10.3	14.1
Chicken	10	28.7	44.1
Pig (normal diet)	(3)[b]	40	9.3
Pig	10	5.6	4.1
Pig	10	≥ 60[c]	27.9
Sheep (normal diet)	(3.4)[b]	39.2	5.6
Lamb (protected lipid)	8.9	75	19.3

[a] Chicken data from Isaaks et al., 1964; pig data for normal diet from Wood et al., 1978; other pig diets from Mason and Sewell, 1967; sheep data for normal diet from Crouse et al., 1972; protected lipid from Faichney, Scott and Cook, 1973.
[b] Total dietary fat, no fat supplement.
[c] Approximate value for corn oil used; data not reported.

deposition of dietary fat in this tissue. In conditions of caloric excess these fatty acids will remain in the tissue and, if the diet is low in fat, they will be joined by fatty acids synthesized in the animal. As discussed earlier, the type of fatty acids synthesized is regulated, in part, by the type and quantity of fatty acid absorbed from the intestine, so that the latter are the major regulators of the composition of adipose tissue. In the monogastric animal the dietary fatty acids are absorbed unchanged: hence, changes in the type and amount of fat in the diet will be reflected in the composition of adipose tissue fatty acids (*Table 2.4*). Diets containing a high concentration of linoleic acid result in high concentrations of linoleic acid in the adipose tissue of chicken or pigs, but the high linoleic acid content of ruminant diets is destroyed in the rumen, resulting in the low linoleic acid content of ruminant adipose tissue (*Table 2.2*).

Nutritional importance of lipids

In terms of energy-yielding potential, fat is not an essential dietary ingredient and may be replaced by carbohydrate. However, in the nutrition of modern farm animals the high energy density of fat is advantageous. The low appetite of young pigs, a side-effect of selection for leanness, can be mitigated by increasing the fat content of the diet. With high-yielding dairy cows the use of protected tallow in the diet can maintain the energy status of the animal and increase the fat content of the milk (Storry, Brumby and Cheeseman, 1974).

Essential fatty acids (EFA) are required by all animals. Their absence eventually results in abnormalities such as reduced growth rates, poor reproductive performance, skin lesions and many other changes. The functions of EFA in animals appear to be twofold: they act as components of membrane phospholipids and they are the substrates for the synthesis of at least four families of metabolic regulators—the prostaglandins, prostacyclins, thromboxanes and leukotrienes. Among other activities these compounds affect blood pressure, blood clotting and the immune response. Different compounds are formed within the families, depending upon the EFA precursor. Prostaglandins PGE_1, PGE_2 and PGE_3 are formed, respectively from dihomo-γ-linolenic acid, arachidonic acid and $\Delta^{5,8,11,14,17}$-eicosapentaenoic acid.

The recommended linoleic acid content of pig diets is 1.5% of digestible energy for the young pig and 0.7% of digestible energy for the finishing ration to bacon weight (Agricultural Research Council, 1981). However, it is difficult to produce overt signs of EFA deficiency in pigs and normal growth has been reported in pigs taken to bacon weight on 0.1% dietary linoleic acid. The backfat of pigs usually contains approximately 10% linoleic acid in the lipids, double that in humans, and the softening effect of excess linoleic acid on the tissue lipids is likely to be more of a problem than linoleic acid deficiency. The requirements for linolenic acid are not known and the function of its metabolites are less well established than those of linoleic acid. However the high content of its metabolite, n-3 docosahexaenoic acid, in brain phospholipids and the competition between the metabolites of linoleic and linolenic acid suggest a significant role for linolenic acid. Diets supplying sufficient linoleic acid are likely to contain linolenic acid. Ruminants have low supplies of essential fatty acids because of the hydrogenation in the rumen. This, coupled with the poor transfer of essential fatty acids across the placenta, results in increased concentration of fatty acids of the n-9 oleic acid series in their young. However, they do not show other symptoms of deficiency and appear to have evolved an efficient system to utilize the available EFA.

The relationship between dietary saturated fatty acids and cardiovascular disease in man has been much discussed but more specific toxic effects of certain dietary fatty acids are known. Dietary cyclopropene fatty acids, inhibitors of acyl-CoA Δ^9 desaturase, cause defects in eggs and disease in poultry and other animals (Phelps *et al.*, 1965). Long-chain monounsaturated fatty acids are also toxic, if they form a large amount of the dietary fatty acids. Particularly important in this respect are erucic acid and cetoleic acid, from rapeseed and fish oils, both of which can cause fat deposition in, and subsequent permanent damage to, heart muscle (Beare-Rogers, 1977).

In conclusion, it must be remembered that the fat of meat animals is a major source of dietary fat for man. Changes in the deposited fat may result in changes in the consumer acceptability of the product or in its stability. The production of polyunsaturated ruminant fat using protected oil seeds resulted in bland-tasting products, which rapidly oxidized. This oxidative instability certainly contributed to the failure of such meat and milk to become established as components in low-atherogenic diets. Even when the change is desired by the customer, such as very lean pigs, the associated changes in the colour and softness of the fat may be unacceptable.

References

AGRICULTURAL RESEARCH COUNCIL (1981). *The Nutrient Requirements of Pigs.* Slough, England, Commonwealth Agricultural Bureaux

ÅKESSON, B. (1969). Composition of rat liver triacylglycerols and diacylglycerols. *European Journal of Biochemistry* **9**, 463–477

ALLEE, G.L., ROMSOS, D.R., LEVEILLE, G.A. and BAKER, D.H. (1972). Lipogenesis and enzymatic activity in pig adipose tissue as influenced by source of dietary fat. *Journal of Animal Science* **35**, 41–47

AMER, M.A. and ELLIOT, G.I. (1973). Effects of supplemental dietary copper on glyceride distribution in the backfat of pigs. *Canadian Journal of Animal Science* **53**, 147–152

ANDERSON, R.E., BOTTINO, N.R. and REISER, R. (1970). Animal endogenous triglycerides: 1. Swine adipose tissue. *Lipids* **5**, 161–164

BEARE-ROGERS, J.L. (1977). Docosenoic acids in dietary fats. *Progress in the Chemistry of Fats and Other Lipids* **15**, 29–56

BENSADOUN, A. and KOMPIANG, I.P. (1979). Role of lipoprotein lipase in plasma triglyceride removal. *Federation Proceedings* **38**, 2622–2626

BENSADOUN, A. and ROTHFELD, A. (1972). The form of absorption of lipids in the chicken. *Proceedings of the Society for Experimental Biology and Medicine* **141**, 814–817

BENSON, J.D. and BENSADOUN, A. (1977). Response of adipose tissue lipoprotein lipase to fasting in the chicken and the rat—a species difference. *Journal of Nutrition* **107**, 990–997

BICKERSTAFFE, R. and ANNISON, E.F. (1969). Glycerokinase and desaturase activity in pig, chicken and sheep intestinal epithelium. *Comparative Biochemistry and Physiology* **31**, 47–54

BRECKENRIDGE, W.C., MARAI, L. and KUKSIS, A. (1968). Triglyceride structure of human milk fat. *Canadian Journal of Biochemistry* **47**, 761–769

BROCKERHOFF, H. (1965). A stereospecific analysis of triglycerides. *Journal of Lipid Research* **6**, 10–15

BROWNSEY, R.W., HUGHES, W.A. and DENTON, R.M. (1969). Adrenaline and the regulation of acetyl-Coenzyme A carboxylase in rat epididymal adipose tissue. *Biochemical Journal* **184**, 23–32

CHRISTIE, W.W. (1979). The effects of diet and other factors on the lipid

composition of ruminant tissues and milk. *Progress in Lipid Research* **17**, 245–277
CHRISTIE, W.W. and MOORE, J.H. (1969). The effect of dietary copper on the structure and physical properties of adipose tissue triglycerides in pigs. *Lipids* **4**, 345–349
CHRISTIE, W.W. and MOORE, J.H. (1970a). A comparison of the structure of triglycerides from various pig tissues. *Biochimica et biophysica Acta* **210**, 46–56
CHRISTIE, W.W. and MOORE, J.H. (1970b). The variation of triglyceride structure with fatty acid composition in pig adipose tissue. *Lipids* **5**, 921–928
CLARKE, S.D., ROMSOS, D.R. and LEVEILLE, G.A. (1977). Influence of dietary fatty acids on liver and adipose tissue lipogenesis and on liver metabolites in meal-fed rats. *Journal of Nutrition* **107**, 1277–1287
CONNOLLY, J.F., SPILLANE, T.A., POOLE, D.R.B. and McALEESE, D.M. (1970). Nutritional effects of oxidized lipids in fresh and stored pig diets. *Irish Journal of Agricultural Research* **9**, 39–58
CRAIG, B.M. (1957). Dilatometry. *Progress in the Chemistry of Fats and Other Lipids* **4**, 198–226
CROUSE, J.D., KEMP, J.D., FOX, J.D., ELY, D.G. and MOODY, W.G. (1972). Effects of castration on ovine neutral and phospholipid deposition. *Journal of Animal Science* **34**, 388–392
CUNNINGHAM, H.M. (1968). Effect of caffeine on nitrogen retention, carcass composition, fat mobilization and the oxidation of C^{14} labelled body fat in pigs. *Journal of Animal Science* **27**, 424–430
DE GIER, J. and VAN DEENEN, L.L.M. (1961). Some lipid characteristics of red cell membranes of various animal species. *Biochimica et biophysica acta* **49**, 288–296
DEUEL, H.J. (1951). *The Lipids: their Chemistry and Biochemistry, Vol. 1.* New York, Chemistry Interscience Publishers Inc.
DONALDSON, W.E. (1979). Regulation of fatty acid synthesis. *Federation Proceedings* **38**, 3617–3621
DOW, C., LAWSON, G.H.K., McFERRAN, J.B. and TODD, J.R. (1963). Mulberry heart disease. *Veterinary Record* **75**, 76–77
EKSTEDT, B. and OLIVECRONA, T. (1970). Uptake and release of fatty acids by rat adipose tissue: Last in—first out? *Lipids* **5**, 858–859
ELLIOT, J.I. and BOWLAND, J.P. (1968). Effects of dietary copper on the fatty acid composition of porcine depot fats. *Journal of Animal Science* **27**, 956–960
ELLIOT, J.I. and BOWLAND, J.P. (1969). Correlation of melting point with the sum of the unsaturated fatty acids in samples of porcine depot fat. *Canadian Journal of Animal Science* **49**, 397–398
EMANUEL, N.M. and LYASKOVSKAYA, Y.N. (1967). *The Inhibition of Fat Oxidation Processes.* Oxford, UK, Pergamon Press
EMERY, R.S. (1980). Mobilization, turnover and disposition of adipose tissue lipids. In *Digestive Physiology and Metabolism in Ruminants* (Y. Ruckebusch and P. Thivend, Eds), pp. 541–558. Lancaster, England, M.T.P. Press Ltd
ENSER, M. (1973). Clearing factor lipase in muscle and adipose tissue of pigs. *Biochemical Journal* **136**, 381–385

ENSER, M. (1979). The role of insulin in the regulation of stearic acid desaturase activity in liver and adipose tissue from obese-hyperglycaemic (ob/ob) and lean mice. *Biochemical Journal* **180**, 551–558

ENSER, M. and ROBERTS, J.L. (1982). The regulation of hepatic stearoyl-coenzyme A desaturase in obese-hyperglycaemic (ob/ob) mice by food intake and the fatty acid composition of the diet. *Biochemical Journal* **206**, 561–570

ENSER, M.B., KUNZ, F., BORENSZTAJN, J., OPIE, L.H. and ROBINSON, D.S. (1967). Metabolism of triglyceride fatty acid by the perfused rat heart. *Biochemical Journal* **104**, 306–317

FAICHNEY, G.J., SCOTT, T.W. and COOK, L.J. (1973). The utilization by growing lambs of casein-safflower oil supplement treated with formaldehyde. *Australian Journal of Biological Science* **26**, 1179–1188

FEELEY, R.M., CRINER, P.E. and WATT, B.K. (1972). Cholesterol content of foods. *Journal of the American Dietetic Association* **61**, 134–139

GARTON, G.A., HOVELL, F.D.D. and DUNCAN, W.R.H. (1972). Influence of dietary fatty acids on the fatty acid composition of lamb triglycerides with special reference to the effect of propionate on the presence of branched-chain components. *British Journal of Nutrition* **28**, 409–416

GRAY, G.M. and MacFARLANE, M.G. (1958). Separation and composition of the phospholipids of ox heart. *Biochemical Journal* **70**, 409–425

GRIGOR, M.R., MOEHL, A. and SNYDER, F. (1972). Occurrence of ethanolamine and choline-containing plasmalogens in adipose tissue. *Lipids* **7**, 766–768

GUNSTONE, F.D. (1967). *An Introduction to the Chemistry and Biochemistry of Fatty Acids and their Glycerides*, 2nd edn. London, Chapman and Hall

GURR, M.I. and JAMES, A.T. (1980). *Lipid Biochemistry: An Introduction*, 3rd edn. London, Chapman and Hall

HILDITCH, T.P. and STAINSBY, W.J. (1935). The body fats of the pig. IV. Progressive hydrogenation as an aid in the study of glyceride structure. *Biochemical Journal* **29**, 90–99

HILLARD, B.L., LUNDIN, P. and CLARKE, S.D. (1980). Essentiality of dietary carbohydrate for maintenance of liver lipogenesis in the chick. *Journal of Nutrition* **110**, 1533–1542

HO, S.K. and ELLIOT, J.I. (1973). Supplemental dietary copper and the desaturation of 1-^{14}C stearoyl-coenzyme A by porcine hepatic and adipose microsomes. *Canadian Journal of Animal Science* **53**, 537–545

HOOD, R.L., COOK, L.J., MILLS, S.C. and SCOTT, T.W. (1980). Effect of feeding protected lipids on fatty acid synthesis in ovine tissues. *Lipids* **15**, 644–650

HORNSTEIN, I., CROWE, P.F. and HEIMBERG, M.J. (1961). Fatty acid composition of meat tissue lipids. *Journal of Food Science* **26**, 581–586

INGLE, D.L., BAUMAN, D.E., MELLENBERGER, R.W. and JOHNSON, D.E. (1973). Lipogenesis in the ruminant: Effect of fasting and refeeding on fatty acid synthesis and enzymatic activity of sheep adipose tissue. *Journal of Nutrition* **103**, 1479–1488

INKPEN, C.A., HARRIS, R.A. and QUACKENBUSH, F.W. (1969). Differential responses to fasting and subsequent feeding by microsomal systems of rat liver: 6- and 9-desaturation of fatty acids. *Journal of Lipid Research* **10**, 277–282

ISAAKS, R.E., DAVIES, R.E., FERGUSON, T.M., REISER, R. and COUCH, J.R. (1964). Studies on ovian fat composition. *Poultry Science* **43**, 105–113

JACKSON, H.D. and WINKLER, V.W. (1970). Effects of starvation on the fatty acid composition of adipose tissue and plasma lipids of sheep. *Journal of Nutrition* **100**, 201–207

KNUDSEN, J., CLARK, S. and DILS, R. (1975). Acyl-CoA hydrolase(s) in rabbit mammary gland which control the chain length of fatty acids synthesized. *Biochemical and Biophysical Research Communications* **65**, 921–926

KUCHMAK, M. and DUGAN, L.R. (1965). Composition and positional distribution of fatty acids in phospholipids isolated from pork muscle tissues. *Journal of the American Oil Chemists Society* **42**, 45–48

LANDS, W.E.M., PIERINGER, R.A., SLAKEY, S.P.M. and ZSCHOCKE, A. (1966). A micromethod for the stereospecific determination of triglyceride structure. *Lipids* **1**, 444–448

LEA, C.H., SWOBODA, P.A.T. and GATHERUM, D.P. (1970). A chemical study of soft fat in cross-bred pigs. *Journal of Agricultural Science* **74**, 279–284

LEA, C.H., PARR, L.J., L'ESTRANGE, J.L. and CARPENTER, K.J. (1966). Nutritional effects of autoxidized fats in animal diets. 3. The growth of turkeys on diets containing oxidized fish oil. *British Journal of Nutrition* **20**, 123–133

L'ESTRANGE, J.L., CARPENTER, K.J., LEA, C.H. and PARR, L.J. (1966). Nutritional effects of autoxidized fats in animal diets. 2. Beef fat in the diet of broiler chicks. *British Journal of Nutrition* **20**, 113–122

L'ESTRANGE, J.L., CARPENTER, K.J., LEA, C.H. and PARR, L.J. (1967). Nutritional effects of autoxidized fats in animal diets. 4. Performance of young pigs on diets containing meat meals of high peroxide value. *British Journal of Nutrition* **21**, 377–390

McDONALD, T.M. and KINSELLA, J.E. (1973). Stearyl-CoA desaturase of bovine mammary microsomes. *Archives of Biochemistry and Biophysics* **156**, 223–231

MASON, J.V. and SEWELL, R.F. (1967). Influence of diet on the fatty acid composition of swine tissues. *Journal of Animal Science* **26**, 1342–1347

MATTSON, F.H. and BECK, L.W. (1956). The specificity of pancreatic lipase for the primary hydroxyl groups of glycerides. *Journal of Biological Chemistry* **219**, 735–740

MATTSON, F.H., VOLPENHEIM, R.A. and LUTTON, E.S. (1964). The distribution of fatty acids in the triglycerides of the Artiodactyla (even-toed animals). *Journal of Lipid Research* **5**, 363–365

MEARA, M.L. (1945). The configuration of naturally occurring mixed glycerides. Part II. The configuration of some monopalmito-glycerides from various natural sources. *Journal of the Chemical Society* 23–24

MONTFOORT, A., VAN GOLDE, L.M.G. and VAN DEENEN, L.L.M. (1971). Molecular species of lecithins from various animal tissues. *Biochimica et biophysica acta* **231**, 335–342

MOORE, J.H., CHRISTIE, W.W., BRAUDE, R. and MITCHELL, K.J. (1968). The effect of 250 ppm of copper in the diet of growing pigs on the fatty acid composition of the adipose tissue lipids. *Proceedings of the Nutrition Society* **27**, 45A–46A

MORRISON, I.M. and HAWKE, J.C. (1977). Triglyceride composition of bovine milk fat with elevated levels of linoleic acid. *Lipids* **12**, 994–1004

O'KEEFE, P.W., WELLINGTON, G.H., MATTICK, L.R. and STOUFFER, J.R.

(1968). Composition of bovine muscle lipids at various carcass locations. *Journal of Food Science* **33**, 188–192

PATTON, S. and JENSEN, R.G. (1975). Lipid metabolism and membrane functions of the mammary gland. *Progress in the Chemistry of Fats and Other Lipids* **14**, 167–277

PHELPS, R.A., SHENSTONE, F.S., KEMMERER, A.R. and EVANS, R.J. (1965). A review of cyclopropenoid compounds: Biological effects of some derivatives. *Poultry Science* **44**, 358–394

POTHOVEN, M.A. and BEITZ, D.C. (1975). Changes in fatty acid synthesis and lipogenic enzymes in adipose tissue from fasted and fasted refed steers. *Journal of Nutrition* **105**, 1055–1061

QUIMBY, O.T., WILLE, R.L. and LUTTON, E.S. (1953). On the glyceride composition of animal fats. *Journal of the American Oil Chemists Society* **30**, 186–190

REISER, R. (1975). Fat has less cholesterol than lean. *Journal of Nutrition* **105**, 15–16

ROSSELL, J.B. (1967). Phase diagrams of triglyceride systems. *Advances in Lipid Research* **5**, 353–426

SAVARY, D. and DESNUELLE, P. (1956). Sur quelques éléments de spécificité pendant l'hydrolise enzymatique des triglycerides. *Biochimica et biophysica acta* **21**, 349–360

SCHULTZ, H.W., DAY, E.A. and SINNHUBER, R.O. (Eds) (1962). *Symposium on Foods: Lipids and their Oxidation.* Westport, Connecticut, USA, The Avi Publishing Company, Incorporated

SCOTT, R.F., DAOUD, A.S. and FLORENTIN, R.A. (1971). Animal models in atherosclerosis. In *The Pathogenesis of Atherosclerosis* (R.W. Wissler and J.C. Geer, Eds), pp. 120–146. Baltimore, MD, USA, Williams & Wilkins

SHAND, J.H., NOBLE, R.C. and MOORE, J.H. (1978). Dietary influences on fatty acid metabolism in the liver of the neonatal lamb. *Biology of the Neonate* **34**, 217–224

SLAKEY, S.P.M. and LANDS, W.E.M. (1968). The structure of rat liver triglycerides. *Lipids* **3**, 30–36

STEFFEN, D.G., CHAI, E.Y., BROWN, L.J. and MERSMANN, H.J. (1978). Effects of diet on swine glyceride lipid metabolism. *Journal of Nutrition* **108**, 911–918

STORRY, J.E., BRUMBY, P.E. and CHEESEMAN, G.C. (1974). Protected lipids in ruminant nutrition. *Agricultural Development Advisory Service Quarterly Review* **15**, 96–106

STRITTMATTER, P., SPATZ, L., CORCORAN, D., ROGERS, M.J., SETLOW, B. and REDLINE, R. (1974). Purification and properties of rat liver microsomal stearoyl coenzyme A desaturase. *Proceedings of the National Academy of Sciences of the United States of America* **71**, 4565–4569

TAYLOR, M. and THOMKE, S. (1964). Effect of high level copper on the depot fat of bacon pigs. *Nature* **201**, 1246

THOMPSON, E.H., ALLEN, C.E. and MEADE, R.J. (1973). Influence of copper on stearic acid desaturation and fatty acid composition in the pig. *Journal of Animal Science* **36**, 868–873

VERNON, R.G. (1976). Effect of dietary fats on ovine adipose tissue metabolism. *Lipids* **11**, 662–669

VOLPE, J.J. and VAGELOS, P.R. (1976). Mechanisms and regulation of biosynthesis of saturated fatty acids. *Physiological Review* **56**, 339–417

WAHLE, K.W.J. (1974). Desaturation of long-chain fatty acids by tissue preparations of the sheep, rat and chicken. *Comparative Biochemistry and Physiology* **48B**, 87–105

WAHLE, K.W.J. and DAVIES, N.T. (1974). Effect of dietary copper deficiency on the desaturation of stearic acid by rat liver microsomal fractions. *Biochemical Society Transactions* **2**, 1283–1285

WATERMAN, R.A., ROMSOS, D.R., TSAI, A.C., MILLER, E.R. and LEVEILLE, G.A. (1975). Influence of dietary safflower oil and tallow on growth, plasma lipids and lipogenesis in rats, pigs and chicks. *Proceedings of the Society for Experimental Biology and Medicine* **150**, 347–351

WOOD, J.D. and LISTER, D. (1973). The fatty acid and phospholipid composition of *Longissimus dorsi* muscle from Pietrain and Large White pigs. *Journal of the Science of Food and Agriculture* **24**, 1449–1456

WOOD, J.D., GREGORY, N.G., HALL, G.M. and LISTER, D. (1977). Fat mobilization in Pietrain and Large White pigs. *British Journal of Nutrition* **37**, 167–185

WOOD, J.D., ENSER, M.B., MacFIE, H.J.H., SMITH, W.C., CHADWICK, J.P., ELLIS, M. and LAIRD, R. (1978). Fatty acid composition of backfat in Large White pigs selected for low backfat thickness. *Meat Science* **2**, 289–300

ZINDER, O., HAMOSH, M., FLECK, T.R.C. and SCOW, R.O. (1974). Effect of prolactin on lipoprotein lipase in mammary gland and adipose tissue of rats. *American Journal of Physiology* **226**, 744–748

3

FISH FATS

J. OPSTVEDT
Norwegian Herring Oil and Meal Industry Research Institute, N-5033 Fyllingsdalen, Bergen, Norway

Introduction

Fish has been an important part of the diet for humans and animals in many parts of the world since ancient times. Although prized mainly for its content of valuable protein, the importance of the fish lipids in supplying the requirements for calories and essential fatty acids has become increasingly realized. In their natural form, fish lipids comprise various lipid classes, each having their characteristic fatty acid composition. Furthermore, the fatty acid profiles of fish lipids may vary according to species and to feeding patterns. However, although variable in their fatty acid composition, fish lipids have some features in common which separate them from other types of lipids. These are:

1. A high content of long-chain (more than 18 carbon atoms) fatty acids;
2. A high content of long-chain polyethylenic fatty acids of the linolenic (n-3) family.
3. In partially hydrogenated fish oil (PHFO), a high content of long-chain isomeric fatty acids.

The peculiarities of fish lipids are closely linked to these characteristics.
Fish lipids are nowadays mainly consumed by man and animals in

1. Fish fillets and fabricated fish food;
2. Fish meal (fish protein concentrate = FPC);
3. Fish silage;
4. Fish oil;
5. PHFO (partially hydrogenated fish oil);
6. PHFAO (partially hydrogenated fish acid oil).

These various sources provide different types of lipid and are used to a varying degree. This chapter reviews the chemistry and biochemistry of fish lipids with the objective of evaluating their nutritional value, with particular emphasis on the feeding of animals.

Chemistry

The chemistry of fish lipids has been reviewed by Reichwald (1976), El-Shattory (1979), Ackman (1980) and Ackman (1982). This review will particularly focus attention on fatty acids of nutritional importance. Fish lipids contain all lipid classes found in living organisms (El-Shattory, 1979). For convenience it is usual to divide them into phospholipids (PL) and

Table 3.1 TOTAL LIPID (TL), PHOSPHOLIPID (PL) AND NEUTRAL LIPID (NL) CONTENT (%) OF TISSUES FROM VARIOUS SPECIES OF FISH

Species/tissue	TL	PL	NL	References
Cod (*Gadus morhua*)				
white muscle	0.59	0.52	0.07	Addison, Ackman and Hingley (1968)
flesh	0.70	0.55	0.10	Jangaard *et al.* (1967)
fillet	0.75	0.65	0.06	Bligh and Scott (1966)
flesh	0.77	0.57	0.12	Hardy, McGill and Gunstone (1979)
Hake (*Merluccius capensis*)				
muscle	1.55	0.46	1.00	de Koning (1966)
Rockfish (*Morone saxatilis*)				
fillets	2.14	0.67	1.36	Wood and Hintz (1971)
Capelin (*Mallotus villosus*)				
whole	2.89	0.56	1.99	Ackman *et al.* (1969)
Pilchard (*Sardinops ocellata*)				
whole	5.00	0.91	3.90	de Koning and McMullan (1966)
Menhaden (*Brevortia tyrannus*)				
flesh	8.20	1.23	6.97	Ackman, Eaton and Hingley (1976)
Herring, Baltic (*Clupea harengus*)				
fillet	4.60	1.10	3.31	Linko (1965)
Herring, Atlantic (*Clupea harengus*)				
whole	16.40	1.13	15.20	Drozdowski and Ackman (1969)
Mackerel (*Scomber scombrus*)				
dorsal muscle	2.10	0.85	1.23	Viviani *et al.* (1967)
flesh June	9.10	0.88	7.93	Hardy and Keay (1972)
flesh Dec.	24.10	0.84	22.70	Hardy and Keay (1972)
light muscle	10.20	0.50	9.10	Ackman and Eaton (1971)
dark muscle	14.40	1.60	10.70	Ackman and Eaton (1971)
Mackerel (*Scomber japanicus*)				
flesh Aug.	10.80	1.10	9.40	Ueda (1976)
flesh Jan.	15.50	0.99	14.40	Ueda (1976)
Horse mackerel (*Tracharus japanicus*)				
dorsal muscle	1.10	0.39	0.18	Toyomizu, Nakamura and Shono (1976)
dorsal muscle	11.20	0.35	7.81	Toyomizu, Nakamura and Shono (1976)
Scad, Black Sea (*Trachurus mediterraneus ponticus*)				
light muscle	4.90	1.15	2.20	Shchepkin *et al.* (1974) (cited by Ackman, 1980)
dark muscle	10.50	2.07	6.10	Shchepkin *et al.* (1974) (cited by Ackman, 1980)
Mackerel scads (*Decapterus pinnulatus*)				
dorsal muscle	9.30	0.36	6.27	Toyomizu, Nakamura and Shono (1976)
Trout (*Salvelinus iridis*)				
muscle	5.30	0.63	4.68	Gray and MacFarlane (1961)
Carp (*Cyprinus carpio*)				
whole	4.83	0.44	3.36	Takeuchi and Watanabe (1982)
whole	7.30	1.10	6.20	Takeuchi and Watanabe (1982)
Average	7.09	0.83	5.63	
CV[a] (%)	84	49	99	

[a] CV = coefficient of variation = (average/standard deviation) × 100

neutral lipids (NL). Of the NL the triglycerides (TG) constitute the main part. The phospholipids serve structural and metabolic purposes while the triglycerides are used mainly for storage of fat. *Table 3.1* shows the content of total lipids (TL) and of PL and NL in various tissues in different species of fish. Certain features are evident from this Table. Firstly, the content of TL varies widely between different species of fish, from a low of less than 1% in cod flesh to a high of 24% in mackerel, resulting in a CV (coefficient of variation) of 84%. Secondly, the content of NL shows even greater variation than the TL (CV = 99%), although following a fairly similar pattern. The content of PL is relatively constant (CV = 49%) compared with that of TL and NL, and seems only to a limited extent to be related to the content of TL. For further elaboration the regressions of PL and NL with TL are shown in *Figure 3.1*. Between NL and TL there is a highly significant ($P<0.001$) correlation. For each percentage unit increase in TL, there is an increase of 0.9% percentage units in NL. The content of PL, on the other hand, is not significantly ($P>0.05$) correlated with the content of TL. These findings mirror some basic biological facts concerning fish. Some fish, like cod, store fat in separate organs (i.e. the liver) and the lipids in the flesh are mainly PL. Other fish, like mackerel, store fat throughout the body, resulting in high, but somewhat variable contents of TL and NL, dependent on nutritional status.

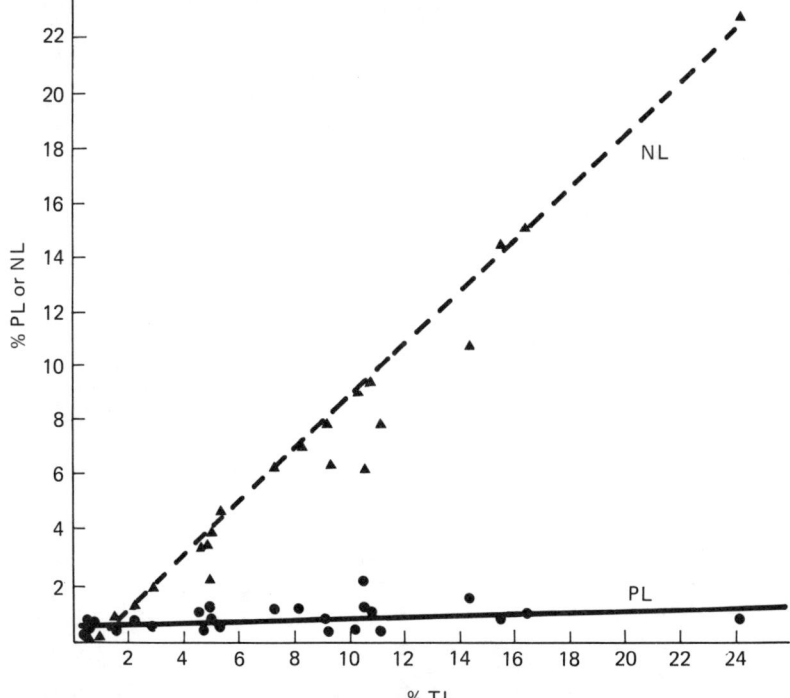

Figure 3.1 Correlation between % total lipids (TL) and percentages of phospholipids (PL; ●—●) and of neutral lipids (NL; ▲---▲) in various tissues from different species of fish. PL = 0.634 + 0.028 TL ($r^2 = 0.1624$; $P > 0.05$): NL = 0.915 TL − 0.914 ($r^2 = 0.9702$; $P < 0.001$)

Table 3.2 FATTY ACID COMPOSITION (w/w%) OF NEUTRAL LIPIDS (TRIGLYCERIDES) IN DIFFERENT FISH SPECIES

Fish species and tissue	14:0	16:0	18:0	Fatty acid[a] 9-16:1	9-18:1	11-20:1	11-22:1	Total n-6	Total n-3	References[b]
Anchovy (*Engraulis ringens*)	11.2	20.4	6.8	7.9	12.2	2.8	2.3	3.3	22.8	1
Menhaden (*Brevoortia tyrannus*) eviscerated	8.1	25.5	4.3	8.4	17.9	1.8	—	2.2	25.0	2
Horse mackerel (*Trachurus japanicus*) muscle	4.7	21.4	7.7	8.6	22.4	0.5	1.0	4.5	24.6	3
Mackerel scads (*Decapterus pinnulatus*) fillets	9.3	24.2	7.4	8.4	25.7	1.2	trace	3.8	22.7	3
Herring (*Clupea harengus*)	5.6	12.5	1.1	13.6	15.5	13.7	19.4	1.4	11.6	4
Capelin (*Mallotus villosus*) eviscerated	8.0	9.4	1.1	16.1	8.2	17.4	15.6	2.6	13.9	5
Rockfish (*Morone saxatilis*) fillets	2.9	18.0	3.2	9.5	31.2	4.1	—	2.3	27.4	6
Redfish (*Sebastus marinus*) fillets	6.4	14.3	1.8	7.0	21.5	10.8	9.3	1.9	21.5	7
liver	2.3	12.7	3.2	9.4	40.2	6.5	2.5	1.5	15.5	7
Cod (*Gadus morhua*) flesh	3.4	12.3	2.7	9.3	18.8	7.7	6.3	2.1	34.0	8
liver	1.8	8.7	1.8	6.5	17.6	4.7	3.0	6.2	45.5	9
Coho salmon (*Oncorhynchus kisutch*)	3.2	10.3	3.5	10.7	21.2	3.4	0.5	6.3	23.5	10
Carp (*Cyprinus carpio*) whole	2.5	20.1	3.6	9.0	32.0	7.8	2.3	16.1	4.3	11
Average	5.3	16.1	3.7	9.6	21.9	6.3	4.8	4.2	22.5	
CV[c] (%)	57	36	61	27	40	80	131	94	46	

[a] Unsaturated fatty acids given are major isomeric forms. Other positional isomers present in minor quantities.
[b] References: 1. Masson and Burgos (1973); 2. Ackman, Eaton and Hingley (1976); 3. Toyomizu, Nakamura and Shono (1976); 4. Addison, Ackman and Hingley (1969); 5. Ackman *et al.* (1969); 6. Wood and Hintz (1971); 7. Lambertsen (1972); 8. (calculated from) Addison, Ackman and Gunstone, Wijesundera and Scrimgeour (1978); 10. Braddock and Dugan (1972); 11. Takeuchi and Watanabe (1982).
[c] As in Table 3.1

Table 3.3 FATTY ACID COMPOSITION (w/w%) OF PHOSPHOLIPIDS IN DIFFERENT FISH SPECIES

Fish species and tissue	14:0	16:0	18:0	Fatty acid[a] 9-16:1	9-18:1	11-20:1	11-22:1	Total n-6	Total n-3	References[b]
Anchovy (*Engraulis ringens*)	4.5	25.2	5.7	7.0	15.0	2.6	2.0	3.2	24.8	1
Menhaden (*Brevortia tyrannus*) eviscerated	1.6	24.6	9.5	3.6	18.5	0.7	0.1	3.9	32.2	2
Horse mackerel (*Trachurus japanicus*) muscle	—	19.4	11.0	1.9	9.6	trace	—	5.0	50.5	3
Mackerel scads (*Decapterus pinnulatus*) fillets	—	18.1	12.4	1.5	9.6	trace	—	6.4	49.8	3
Herring (*Clupea harengus*)	1.8	21.4	3.2	4.6	13.0	2.4	1.6	2.5	46.2	4
Capelin (*Mallotus villosus*) eviscerated	2.0	18.3	2.7	5.7	11.2	3.0	2.6	3.1	44.4	5
Rockfish (*Morone saxatilis*) fillets	0.7	22.1	4.0	2.1	9.4	1.4	—	4.2	55.4	6
Redfish (*Sebastus marinus*) fillets	2.6	24.9	2.5	3.2	11.9	2.7	1.3	4.0	42.8	7
liver	2.9	19.3	3.3	4.9	18.0	3.0	1.3	2.5	38.7	7
Cod (*Gadus morhua*) flesh	1.1	20.8	4.0	3.6	12.7	1.9	0.2	4.1	48.1	8
liver	2.0	9.9	2.5	4.5	13.9	3.7	2.4	8.6	51.1	9
Coho salmon (*Oncorhyncus kisutch*)	3.0	15.6	4.7	8.9	25.3	2.3	0.6	5.5	21.0	10
Carp (*Cyprinus carpio*) whole	3.2	22.2	10.4	5.1	19.9	2.9	0.3	15.0	18.9	11
Average	2.3	20.1	5.8	4.4	14.5	2.0	1.0	5.2	40.3	
CV[c] (%)	47	21	62	47	33	59	96	65	30	

[a,b,c] As in *Table 3.2*

Despite the fact that fish cannot be divided into distinct and separate classes based on the NL content in the flesh (*Table 3.1*), it has become practice to call fish lean (i.e. those which mainly store fat in the liver) or fatty (i.e. those which store fat in the flesh). The latter group comprises the pelagic species used for the production of fish meal, while the former group is the raw material for white fish meal. When whole fish, including the liver, is utilized for meal production, the difference with regard to fat content between white fish meal and fish meal may be small.

It is known that the different lipid classes show characteristic fatty acid patterns. *Tables 3.2* and *3.3* show the fatty acid composition of NL and PL respectively in various tissues from different species of fish. Only the major fatty acids are shown, and they are reported so as to facilitate nutritional evaluation.

Tables 3.2 and *3.3* demonstrate the relatively high content of n-3 and the low content of n-6 fatty acids in fish lipids. The n-3 fatty acids originate from the phytoplankton (Pohl and Wagner, 1972). The high, but somewhat variable, content of C20 and C22 monoethylenic fatty acids in fish lipids are also evident from *Tables 3.2* and *3.3*. These acids are known to originate from ingested copepods (Ackman, 1982). As most of the polyethylenic fatty acids in fish lipids have chain lengths of more than 18 carbon atoms, *Tables 3.2* and *3.3* also demonstrate the high content of long-chain fatty acids in fish lipids.

Comparison of the fatty acid profiles of NL and PL within species reveal that the PL contain about twice as much n-3 polyethylenic and less of palmitoleic and oleic acids than of NL. However, there are relatively large variations in the fatty acid profiles of the PL and NL between species. Thus the relatively low content of n-3 fatty acids in the freshwater carp is noteworthy.

The interspecies variation is generally less in the PL than in the NL. This reflects the well-known fact that the fatty acid patterns of the PL are less affected by dietary lipids than are those of the NL.

Processing of fish with high fat content provides fish oil in addition to fish meal. *Table 3.4* shows an example of the separation of the lipids in

Table 3.4 DISTRIBUTION OF LIPID CLASSES AND FATTY ACID COMPOSITION IN WHOLE CAPELIN AND CAPELIN MEAL AND OIL[a]

	Capelin, whole	Meal	Oil
Total lipids (TL), % of dry matter	32	13	100
Percentage of TL in whole fish	—	40	60
Neutral lipids, % of TL	77	60	97
Phospholipids, % of TL	16	24	1
Fatty acid[b]			
14:0	6.7	4.5	8.1
16:0	11.3	15.9	9.0
18:0	1.3	2.0	1.1
9-16:1	8.2	7.6	8.7
9-18:1	17.3	16.0	17.4
11-20:1	20.5	10.0	24.9
11-22:1	15.6	7.1	19.5
Total n-6	1.5	2.0	1.4
Total n-3	15.3	31.6	6.3

[a] From Urdahl, N. and Nygard, E. (1970) (unpublished data)
[b] See footnote (a), *Table 3.2*

Table 3.5 CONTENT (w/w%) OF NUTRITIONALLY IMPORTANT FATTY ACIDS IN SOME FISH MEALS

Type	Reference[a]	Species[b]	14:0	16:0	18:0	9-16:1	9-18:1	11-20:1	11-22:1	Total n-6	Total n-3
Anchovy	1	N.S.	2.8	12.9	2.9	4.6	20.2	4.7	4.2	6.9	38.5
	2	A	8.7	23.3	6.4	8.7	10.5	—	—	2.5	33.4
	3	P	6.2	20.5	5.5	8.6	11.7	—	0.4	3.6	33.8
	4	A	7.4	22.8	4.2	7.2	13.1	1.3	0.7	3.4	31.5
		Average	6.3	19.9	4.8	7.3	11.4	3.0	1.8	4.1	34.3
		SD	2.5	4.8	1.5	1.9	7.0	—	2.1	1.9	2.3
Herring	1	N.S.	3.2	10.3	0.6	7.0	11.5	10.3	8.7	3.6	41.6
	4	C	4.8	16.6	1.6	6.9	17.6	8.9	7.1	3.5	27.8
	4	M	6.1	15.9	4.1	4.9	13.8	10.1	14.7	4.4	18.6
	4	H	5.5	16.3	2.0	4.2	14.7	12.9	16.9	2.4	20.2
		Average	4.9	14.8	2.1	5.8	14.4	10.6	11.9	3.5	27.1
		SD	1.3	3.0	1.5	1.4	2.5	1.7	4.7	0.8	10.5
White fish	1	N.S.	3.2	11.1	1.7	6.8	16.9	9.7	9.1	3.4	35.5

[a] References: 1. Gunstone and Wijesundera (1978); 2. Bassler and Putzka (1975); 3. Wessels et al. (1971); 4. Opstvedt (1971).
[b] A: anchovera; P: pilchard; C: capelin; M: mackerel; H: herring; N.S.: not specified; SD: standard deviation.
[c] See footnote (a), Table 3.2

capelin into meal and oil. Of the total lipids in the fish, 40% followed the fat-free matter in the meal, while 60% was pressed out as fish oil. The fish oil was almost exclusively TG in line with previous findings (Ackman, Eaton and Hingley, 1976). Of the meal lipids, 24% were PL. Previous studies have found between 20% and 40% of the TL as PL in fish meal from herring (Lea, Parr and Carpenter, 1958) and pilchard (Wessels et al., 1971). The distribution of the fatty acids in whole capelin between the meal and the oil reflects the content of NL and PL. Thus a considerably higher content of polyethylenic fatty acids is found in the meal, compared with that in the oil.

Table 3.5 shows the content of some fatty acids in anchovy and herring-type fish meals compared with those in a sample of white fish meal. The major difference between the anchovy and the herring-type fish meal is a somewhat higher content of n-3 fatty acids in the anchovy meals. All fish meal must, however, be regarded as good sources of n-3 but poor sources of n-6 fatty acids. The oxidation of the lipids in fish meal may cause a substantial reduction in the content of polyethylenic fatty acids (Opstvedt, 1971). This is currently prevented by the addition of an antioxidant.

Table 3.6 CONTENT (w/w%) OF NUTRITIONALLY IMPORTANT FATTY ACIDS IN SOME FISH OILS

Fatty acid[a]	Anchovy[b]		Menhaden[b]	Herring[b]	Capelin	
	Peruvian	South African			Summer	Winter
14:0	7.5	6.9	10.5	6.1	7.0	8.2
16:0	17.5	20.3	21.5	10.8	11.2	11.3
18:0	4.0	3.7	3.4	1.4	1.2	1.2
9-16:1	9.0	9.4	14.2	7.3	8.3	8.2
9-18:1	11.6	13.7	10.3	10.3	12.5	20.2
11-20:1	1.6	3.5	1.2	13.4	15.0	19.8
11-22:1	1.2	2.6	0.1	21.3	16.4	16.7
Total n-6	2.1	1.8	3.7	1.3	2.6	1.8
Total n-3	33.7	30.3	27.8	21.4	21.8	13.8

[a] See footnote (a), Table 3.2
[b] Data from Ackman (1982)

The fatty acid composition of some types of commercial fish oils is shown in Table 3.6. Fish oils contain relatively low levels of n-6 and high levels of n-3 fatty acids. The content of n-3 fatty acids is some 30% higher in anchovy and menhaden oils than in herring-type (i.e. herring and capelin) fish oils. It is of interest that oil from capelin harvested in the summer is considerably higher in n-3 fatty acids than oil from capelin harvested in the winter. Capelin and herring fish oils contain considerably more gondoic and cetoleic fatty acids than do anchovy and menhaden oils.

Table 3.7 shows the fatty acid composition of some PHFO and PHFAO. During hydrogenation the polyethylenic fatty acids with *cis* configurations are converted to a variety of isomeric mono-, di-, tri-, tetra- and pentaethylenic fatty acids (Ackman, 1982). The degree of unsaturation decreases with the increasing m.p. (melting point) of the PHFO. The main difference between the different types of PHFO, hydrogenated to the same m.p., is a somewhat higher content of gondoic and cetoleic in the herring type compared with the menhaden and anchovy type.

Table 3.7 FATTY ACID COMPOSITION (w/w%) OF SOME PARTIALLY HYDROGENATED FISH OILS (PHFO) AND PARTIALLY HYDROGENATED FISH ACID OILS (PHFAO)

Fatty acid[a]	Herring-type[b] (herring, capelin) (melting point 30–40°C)	PHFO[b]	Menhaden[b] (melting point 34–40°C)	PHAFO (melting point 40°C)
14:0	7.6		9.4	7.6
16:0	14.2		22.1	12.9
18:0	5.0		7.6	8.4
20:0	3.5		2.1	4.7
22:0	3.7		0.8	3.9
9-16:1	7.4		13.1	5.6
9-18:1	14.1		15.6	13.3
11-20:1	13.9		7.5	13.5
11-22:1	15.4		2.6	14.6
18:X[c]	2.7		2.0	0.4
20:X[c]	4.9		9.7	2.9
22:X[c]	4.2		4.8	4.0

[a] See footnote (a), *Table 3.2*
[b] From Anonymous (1977).
[c] X denotes different geometrical and positional isomers with 2–4 double bonds.

Nutritive value

Digestibility and energy value

The digestibility of different fish lipids in poultry and pigs is shown in *Table 3.8*, in ruminants in *Table 3.9* and in mink and fish in *Table 3.10*. Only those studies in which the fish lipids were the sole or main lipid in the diet are shown in the tables. The data show no consistent difference between the digestibilities of fish meal lipids and fish oil in poultry, pigs and ruminants although there is a tendency that fish oil may be more digestible than meal lipids. However, it appears that both types of fish lipids have true digestibilities of 90% or higher in all types of animals tested, including mink and fish. The digestibility of PHFO with an m.p. of 31°C or higher is lower than that of the corresponding fish oil in all animals tested. Further, the digestibility of PHFO decreases with increasing m.p. in all animals tested, but the decrease appears to be greater in fish than in warm-blooded animals. It is, however, likely that the addition of unsaturated fats to PHFO will increase its digestibility compared to that found when PHFO is fed as the sole fat. The digestibility of fatty acids in different fish lipids for various animals is shown in *Table 3.11*. The data show that the digestibility decreases with increasing chain length and increases with increasing unsaturation. Because the different animals have been tested in different studies using different techniques, interspecies variation cannot be evaluated, although it appears that the polyethylenic fatty acids ($\Delta^{5,8,11,14,17}$-eicosapentaenoic and $\Delta^{4,7,10,13,16,19}$-docosahexaenoic) are highly digestible in all species. Further, gondoic and cetoleic acids are generally better digested than palmitic and stearic acids. The higher digestibility of gondoic and cetoleic in mink and trout than in chicks has been attributed to a dietary adaptation in these animals (Austreng, Skrede and Eldegard, 1979).

Table 3.8 DIGESTIBILITY (%) OF DIFFERENT FISH LIPIDS IN POULTRY AND PIGS

Animal	Fish meal residual lipids		Fish oil		Partially hydrogenated fish oil, Melting point (°C)									Partially hydrogenated fish acid oil (melting point 40°C)	Reference[d]
					32		38		44		50		53		
	D_A[a]	TD[b]	D_A	TD	TD	D_A	TD	D_A	TD	D_A	TD	D_A	D_A		
Poultry	80	—	—	—	—	—	—	—	—	—	—	—	—	—	1
	—	88	—	94	85	—	—	—	70	—	50	—	—	—	2
	93	—	91	—	—	—	—	—	—	—	—	—	—	—	3
	83	92	84	91	—	—	—	—	—	—	—	—	—	—	4
	—	—	—	—	—	—	74	—	—	—	—	—	—	—	5
	—	—	—	98	87	—	75	—	74	—	55	—	—	—	6
	—	—	—	94	—	—	—	—	—	—	—	—	—	—	7
	—	—	—	—	—	—	—	—	—	—	—	—	—	—	8
Average (poultry)	85	90	88	94	86	—	75	—	72	—	53	—	—	—	
Pigs	87	—	—	—	—	—	—	—	—	—	—	—	—	—	9
	92	—	—	—	—	—	81	—	—	—	—	—	—	—	10
	—	—	—	—	78	72	—	—	—	61	72	53	—	—	11
	—	—	—	—	—	—	—	—	—	—	—	—	64	54–75[c]	12–13
Average (pigs)	90	—	—	—	78	72	81	—	—	61	72	53	64	54–75	

[a] D_A = apparent digestibility.
[b] TD = true digestibility, e.g. corrected for metabolic faecal fat.
[c] Highest value with added lecithin and monoglycerides.
[d] References: 1. Potter et al. (1962); 2. Laksesvela (1966); 3. Hoffmann and Schiemann (1971); 4. Cuppett and Soares (1972); 5. Opstvedt (1973b); 6. Veen, Grimbergen and Stappers (1974); 7. Herstad (1975); 8. Ackman (1980); 9. Homb (1962); 10. Schiemann, Jentsch and Hoffmann (1969); 11. Sundstøl (1974a); 12. Lysø (1980); 13. Lysø (1983). Personal communication.

Table 3.9 DIGESTIBILITY (%) OF DIFFERENT FISH LIPIDS IN RUMINANTS

Lipid	Animal	Oil melting point (°C)	$D_A^{(a)}$	$TD^{(b)}$	Reference
Fish meal residual lipids	Sheep		93	—	Breirem and Homb (1970)
Fish oil	Sheep		77	84	Andrews and Lewis (1970)
Partially hydrogenated fish oil (PHFO)	Sheep	31–33	74	88	
		38–40	72	86	Sundstøl (1974b)
		43–45	73	80	
		48–50	68	82	
	Preruminant calves	31–33	89	—	Flatlandsmo (1972)
		35	92	—	Bjørnstad and Hansen (1974)
		38–40	87	—	Flatlandsmo (1972)

[a,b] As in *Table 3.8*

Table 3.10 DIGESTIBILITY (%) OF DIFFERENT FISH LIPIDS IN MINK AND RAINBOW TROUT

Oil	Melting point (°C)	Mink $D_A^{(a)}$	$TD^{(b)}$	Trout D_A	References
Cod-liver oil		94	—	89	Austreng, Skrede and Eldegard (1979)
Capelin fish oil		94	—	84	Austreng, Skrede and Eldegard (1979)
Capelin fish oil		—	—	86	Austreng and Gjefsen (1981)
Partially hydrogenated fish oil	21	92	—	75	Austreng, Skrede and Eldegard (1979)
	32	81	91	—	Rimeslåtten (1971)
	33	84	—	69	Austreng, Skrede and Eldegard (1979)
	40	75	80	—	Rimeslåtten (1971)
	41	67	—	48	Austreng, Skrede and Eldegard (1979)
	46	24	37	—	Rimeslåtten (1971)

[a,b] As in *Table 3.8*.

The high digestibility of the fish lipids indicates a high dietary energy value. This is generally confirmed in the literature although data on values for fish lipids are scarce (*Table 3.12*). Despite the fact that the studies have been conducted at different centres over a considerable period and using different techniques, the agreement between the individual figures is relatively good. It appears that the dietary energy value of fish oil is slightly higher (approx. 15%) than that of fish meal lipids. Further, the energy value for pigs may be slightly higher than that for poultry. In line with what was observed for digestibility, the energy value of PHFO seems to be lower than that for the corresponding unhydrogenated oil, and to decrease as the melting point increases.

Metabolism

The absorbed fatty acids may be deposited or catabolized. Comparison of deposited fatty acids with those absorbed does not usually take into account chain shortening and chain elongation, or saturation or desaturation in the organism. Furthermore, fatty acids which may arise from the *de*

Table 3.11 APPARENT DIGESTIBILITY (%) OF FATTY ACIDS IN VARIOUS FISH LIPIDS FOR DIFFERENT ANIMALS

Fish lipid	M.p[a] (°C)	Species	Fatty acid[b]								References[c]
			14:0	16:0	18:0	9-16:1	9-18:1	11-20:1	11-22:1	5,8,11,14,17-20:5	4,7,10,13,16,19-22:6	
Fish meal lipids		Chicks	93	83	32	93	81	88	82	95	95	1
Whole sprat lipids		Cod	68	62	56	86	80	67	58	—	—	2
Fish oil		Sheep	88	82	—	—	94	91	96	96	100	3
		Chicks	94	86	78	93	85	85	78	92	86	1
		Mink	96	89	77	91	94	99	99	100	100	4
		Trout	89	79	59	86	81	92	95	100	100	4
Partially	37	Chicks	84	70	49	88	78	—	—	—	—	5
hydrogenated	32	Calves	95	90	72	97	93	—	83	—	—	6
fish oil	35	Calves	98	94	88	98	97	95	94	—	—	7
	33	Mink	93	76	70	82	78	92	96	—	—	4
	33	Trout	59	49	43	81	73	76	82	—	—	4

[a] Melting point
[b] See footnote (a), *Table 3.2*
[c] References: 1. Opstvedt (1973b); 2. Lied and Lambertsen (1982); 3. Andrews and Lewis (1970); 4. Austreng, Skrede and Eldegard (1979); 5. Veen, Grimbergen and Stappers (1974); 6. Flatlandsmo (1972); 7. Bjørnstad and Hansen (1974)

Table 3.12 METABOLIZABLE ENERGY (ME) VALUE OF DIFFERENT FISH LIPIDS FOR POULTRY AND PIGS

Lipid source		Animal	ME (kJ/g)	Reference
Fish meal,	menhaden	Poultry	29.3	Potter et al. (1962)
	unspecified	Poultry	31.0	Hoffman and Schiemann (1971)
	herring	Poultry	28.0	Opstvedt (1973a)
	anchovy	Poultry	32.2[a]	Rojas and Arana (1981)
Fish meal,	average	Poultry	30.1 ± 1.9	
Fish meal,	unspecified	Pigs	33.7	Schiemann, Jentsch and Hoffmann (1969)
Fish oil,	menhaden	Poultry	33.7	Hill (1964)
		Poultry	38.7	Artman (1964)
		Poultry	35.4	Cuppet and Soares (1972)
	herring	Poultry	32.5	Opstvedt (1973a)
	capelin	Poultry	36.9	Herstad (1975)
	anchovy	Poultry	35.8	Rojas and Arana (1981)
Fish oil,	average	Poultry	35.5 ± 2.2	
Partially hydrogenated fish oil, melting pt.				
31–35°C		Poultry	33.1	Herstad (1975)
37°C		Poultry	28.5	Veen, Grimbergen and Stappers (1974)
38–40°C		Poultry	30.5	Herstad (1975)
38–40°C		Poultry	34.5	Rojas and Arana (1981)
43–45°C		Poultry	27.9	Herstad (1975)

[a] Figure calculated by author

novo synthesis will dilute the absorbed fatty acids. Despite these limitations, data for fatty acid deposition may give valuable indications of the extent to which fish fatty acids are catabolized or deposited. *Table 3.13* compares figures for deposition in a warm-blooded animal, the chick (Opstvedt, 1973b) with that of a cold-blooded animal, the trout (Yu, Sinnhuber and Putnam, 1977). In general, the deposition was greater in the trout than in the chick, and this was particularly the case for the polyethylenic n-3 fatty acids. However, for both types of animals the fraction of the long-chain fatty acids that was catabolized or converted to other fatty acids in the body was larger than that directly deposited.

The discovery of the extramitochondrial peroxisomal β-oxidation system by Lazarow and colleagues (Lazarow and de Duve, 1976; Lazarow, 1978) enlarged the classical scheme for fatty acid oxidation. The peroxisomal β-oxidation involves to a large extent the same chemical reactions as those

Table 3.13 DEPOSITION (% OF CONSUMED) OF FATTY ACIDS IN HERRING OIL IN THE CHICK AND THE RAINBOW TROUT

Fatty acid[a]	Chick[b]	Trout[c]
9-18:1	31	50
11-20:1	21	29
11-22:1	11	18
9,12-18:2	16	40
9,12,15-18:3	14	60
5,8,11,14,17-20:5	5	25
4,7,10,13,16,19-22:6	9	48

[a] See footnote (a), *Table 3.2*
[b] Data from Opstvedt (1973b).
[c] Data calculated from Yu, Sinnhuber and Putnam (1977).

66   Fish fats

found in the mitochondria, but is distinctly different from that of the mitochondria with regard to the enzyme proteins and the end-products. Thus, while the mitochondria oxidize the fatty acids to acetyl-CoA units, the end-products of the peroxisomal oxidation are, in addition to acetyl-CoA, a fatty acid with reduced chain length compared with that of the substrate fatty acid. Further, since the peroxisomes are devoid of an electron transport chain, peroxisomal flavoproteins react directly with molecular oxygen to produce H_2O_2. Thus, for each turn of the fatty acid oxidation cycle, the peroxisomes produce two molecules of ATP less than the mitochondria. The peroxisomal fatty acid oxidation is, therefore, 40% less energy-efficient than that of the mitochondrial fatty acid oxidation.

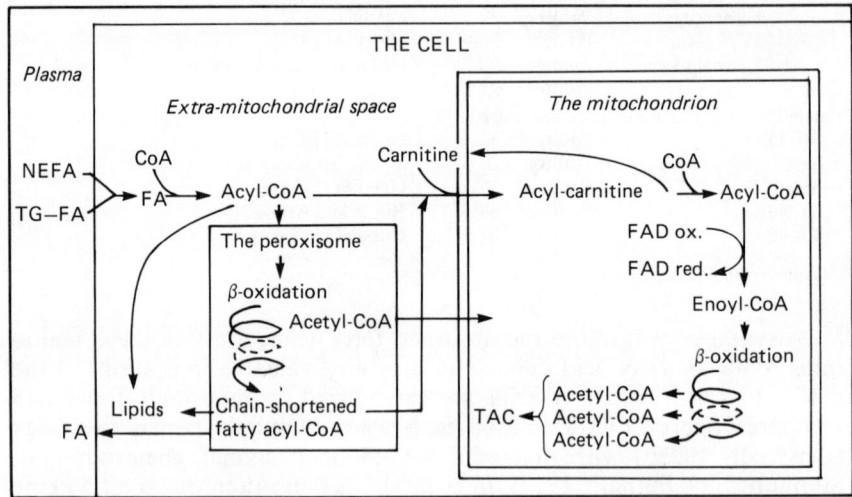

Figure 3.2   Modified scheme of β-oxidation

The knowledge of the peroxisomal fatty acid oxidation and its particular role in the catabolism of the long-chain fatty acids in fish lipids has been greatly extended through recent work at the University of Oslo (Bremer and Norum, 1982; Christophersen et al., 1982). Fatty acids with 20 and 22 carbon atoms are not readily oxidized by the mitochondria. It therefore appears that the peroxisomes function mainly by chain shortening of fatty acids with 20 or 22 carbon atoms to fatty acids with 16 and 18 carbon atoms, which are subsequently oxidized in the mitochondria. The role of the peroxisomes in the total oxidation of long-chain fatty acids is shown in Figure 3.2.

Essentiality of n-3 fatty acids in fish lipids

The role of the n-3 fatty acids in fish lipids in the nutrition of different species of animals has recently been reviewed by Tinoco et al. (1979), Dyerberg and Anker Jørgensen (1982), and Lands (1982). For the trout it has been clearly demonstrated that n-3 fatty acids are nutritionally essential (Lee *et al.*, 1967; Castell *et al.*, 1972; Yu and Sinnhuber, 1972, 1975) and that satisfactory growth and reproduction can be achieved without a dietary supply of n-6 fatty acids (Yu, Sinnhuber and Hendriks, 1979). Further, the long-chain n-3 fatty acids in fish lipids are more effective in promoting growth than is linolenic acid, possibly because of limiting capacity for chain elongation (Yu and Sinnhuber, 1976; Watanabe and Takeuchi, 1976). Results similar to those obtained with trout have also been found with turbot (Cowey, 1976; Cowey *et al.*, 1976), carp (Watanabe *et al.*, 1975; Watanabe, Takeuchi and Ogino, 1975) and red sea bream (Yone and Fujii, 1975; Fujii and Yone, 1976). Recently, a close correlation between dietary content of long-chain, polyethylenic n-3 fatty acids and growth rate in yellow-tail was demonstrated (Deshimaru, Kuroki and Yone, 1982). Finally, available data indicate that the shellfish prawn has a requirement for n-3 fatty acids (Kanazawa *et al.*, 1977), which is better met by the long-chain polyethylenic acids in fish lipids than by linolenic acid (Kanazawa, Teshima and Tokiwa, 1977).

The essentiality of n-3 fatty acids for warm-blooded animals is still debated (Tinoco *et al.*, 1979; Holman and Johnson, 1982). However, the high concentration of n-3 fatty acids in certain organs and organelles (i.e. the brain, cerebral cortex, retina, testis and spermatozoa), which is built up against a great concentration gradient from the blood plasma, has been taken as indirect evidence of an essentiality of n-3 fatty acids also in warm-blooded animals. The observations that n-3 fatty acids appear to have a role in prostaglandin synthesis (Dyerberg and Anker Jørgensen, 1982) support this conclusion. Further, recent experimental results in humans (Holman and Johnson, 1982) strongly support an n-3 fatty acid requirement in man.

It is commonly accepted that dietary intake of n-6 fatty acids higher than that needed to avoid clinical signs of deficiency promotes growth in animals, but opinions vary as to the optimal level for different animal functions. Nor is the question settled as to what extent n-3 fatty acids may substitute for n-6 fatty acids with regard to promoting production in farm animals. Edwards and colleagues (Edwards, Marion and Driggers, 1962; Edwards and Marion, 1963) showed that menhaden fish oil stimulated growth in chicks. This has since been confirmed by Engster, Carew and Foss (1975) although results from studies by Opstvedt (1981) do not entirely support these conclusions. Thus the replacement of partially hydrogenated cocoa oil with fish meal lipids did not increase growth, whereas a comparable replacement with sunflower seed oil gave a growth improvement. However, in the study by Opstvedt (1981) fish meal lipids did improve dietary energy utilization.

Menge and colleagues (Menge, Calvert and Denton, 1965) found that menhaden fish oil improved fertility in hens. However, a positive effect of n-3 fatty acids in fish oil compared with that of n-6 fatty acids in plant oil on the development of gonadal organs was not fully confirmed in experiments by Engster, Carew and Foss (1975).

Mundheim and Bergsrønning (Opstvedt, 1981) studied the effect of an addition of soyabean oil (SBO) to an all-vegetable diet and to diets containing fish meal from capelin or blue whiting. The addition of SBO increased the dietary content of linoleic acid from about 1.1% to 1.7% and caused a significant increase in egg production and feed utilization on the all-vegetable diet. The effect of the SBO addition to the fish meal diets was different for capelin and blue whiting meals. Thus, the addition of SBO affected egg production and feed utilization in hens fed capelin meal, in a similar manner to that found for the all-vegetable diet. Hens fed blue whiting meal without SBO on the other hand had similar production to those fed SBO. It thus appears that a dietary supply of n-3 fatty acids in fish lipids up to a certain level enhance production in poultry, but the effect is somewhat more variable than that seen for linoleic acid in plant oils.

Effects of polyethylenic fatty acids in fish lipids on milk-fat secretion in ruminants

Polyethylenic fatty acids in fish lipids have been used as model substrates for studying milk-fat secretion and the 'low milk-fat syndrome' in ruminants. Recent interest in the use of fish meal as a source of undegradable protein in lactating cows has focused attention on the effect of fish lipids as a natural ingredient in dairy cow rations. The metabolism of the polyethylenic fatty acids in fish lipids in ruminants and their effect on milk-fat secretion have been extensively reviewed elsewhere (Van Soest, 1963; Christie, 1981; Storry, 1981). It is established that feeding or infusion at high levels of fish lipids containing polyethylenic fatty acids causes a depression of milk-fat secretion. The reduction in milk-fat secretion is accompanied by physiological changes similar to those found when milk-fat secretion is reduced due to high levels of concentrate feeding (Opstvedt and Ronning, 1967; Opstvedt, Baldwin and Ronning, 1967; Brumby, Storry and Sutton, 1972) (i.e. reduced concentration of acetic acid and

Table 3.14 EFFECT OF DIETARY COD-LIVER OIL ON THE SECRETION OF MILK FAT IN COWS

References	Fish lipids (g/cow/d)	Yield		Changes compared with control animals	
		Milk (kg/d)	Fat (%)	Fat (%)	Milk fat (g/d)
Hvidsten, Mehlum and Simonsen (1952)	{ 147 { 130	8.2 13.5	4.18 3.88	−0.05 −0.68	−16 −55
Varman, Schultz and Nichols (1968)	250	17.0	3.17	−0.60	—
Tanaka (1970a)	350	14.0	1.9	−1.24	−199
Tanaka (1970b)	{ 50 { 150 { 300 { 450	19.9 19.0 18.4 15.9	3.14 2.95 2.73 2.44	0.13 −0.02 −0.13 −0.45	27 −10 −42 −129
Nicholson and Sutton (1971)	{ 150 { 300 { 450	16.6 16.0 15.2	3.4 2.9 2.5	−0.4 −0.9 −1.3	−67 −167 −251
Brumby, Storry and Sutton (1972)	300	16.5	3.3	−0.8	−174
Storry et al. (1974)	300	21.9	2.99	−0.82	−162
Pennington and Davis (1975)	225	23.0	3.05	−0.58	−90

Table 3.15 EFFECTS OF LIPIDS IN WHALE-MEAT MEAL (WM), COD-LIVER MEAL (CLM), FISH MEAL (FM), WHITE FISH MEAL (WFM) AND FISH SILAGE (FS) WHEN INCLUDED IN DIETS, ON THE SECRETION OF MILK FAT IN COWS

References	Source	Fish lipid (g/cow/d)	Yield Milk (kg/d)	Yield Milk fat (%)	Changes compared with control animals Milk fat (%)	Changes compared with control animals Milk fat (g/day)
Sebelien (1894)	WM	360	9.3	3.47	−0.16	6
Isaachsen et al. (1915)	CLM	327	10.5	3.05	−0.35	−36
Isaachsen, Aashamar and Bang-Sandmo (1919)	CLM	100	10.2	3.38	−0.05	−9
Isaachsen and Ulvesli (1926)	FM	128	10.7	3.48	−0.11	8
Isaachsen and Ulvesli (1927)	WFM	16	10.9	3.40	0.01	−20
Hvidsten, Mehlum and Simonsen (1952)	FM	approx. 50	12.9	3.90	−0.11	82
	FM	approx. 100	8.4	4.34	0.11	5
Ekern (1961)	FM	50	14.5	4.07	0.01	−8
Christiansen, Klausen and Andersen (1970)	FM	approx. 80	22.2	3.39	−0.44	−87
Miller et al. (1981)	FM	49	32.4	3.92	±0	106
	WFM	approx. 60	20.9	4.4	0.11	96
	WFM	approx. 45	24.9	4.8	±0	−5
	WFM	approx. 60	28.5	3.9	−0.10	−20
Ørskov, Reid and MacDonald (1981)	FM	approx. 12	22.7	3.92		Control
L. Vik-Mo and E. Thuen (personal communication)	FM	approx. 39	24.0	3.99	0.07	73
	FM	approx. 68	24.8	4.17	0.25	137
Thuen (1982)	FM	66	20.3	4.07	0.01	12
Bergsrønning, Opstvedt and Rodt (unpublished data)	FM	45	27.2	3.70	−0.04	−15
	FS	42	19.7	4.03	−0.03	−42
Johnson (1981)	FS	81	16.6	3.53	−0.18	(a)
	FS	152	16.3	3.44	−0.24	(a)

[a] Cannot be calculated from the data available

70   Fish fats

increased concentration of propionic acid in the rumen, increased fat synthesis in adipose tissue and reduced fat synthesis in the mammary gland). It is commonly accepted that the effect of the polyethylenic fatty acids on milk-fat secretion is mediated mainly via their effects on rumen metabolism, although additional effects at a tissue level cannot be ruled out.

The effect of level of fish-lipid consumption on milk-fat secretion is summarized in *Table 3.14* which presents the results of experiments in which cod-liver oil has been fed to dairy cows. When the level of cod-liver oil was 200 g or more per day, the percentage milk-fat and total milk-fat output was consistently and substantially reduced. When the consumption was 100–150 g/day, the effects were variable in the different experiments, and below 100 g no effects were found.

Table 3.15 summarizes results of studies on feeding fish meal and similar products to dairy cows. The early studies by Sebelien (1894) and Isaachsen *et al.* (1915) used whalemeat meal and cod-liver meal which supplied 327–360 g lipids per day and which caused a decrease in milk-fat secretion. In the experiments with fish meal the levels of fish lipids consumed were in most instances well below 100 g per cow per day. The effects on milk-fat secretion were generally negligible, and varied from -0.11% to 0.25% and from -42 to 137 g milk fat per cow per day. When fish-meal feeding caused a reduction in percentage fat, this was usually compensated by increased milk production leaving milk-fat production practically unchanged. The

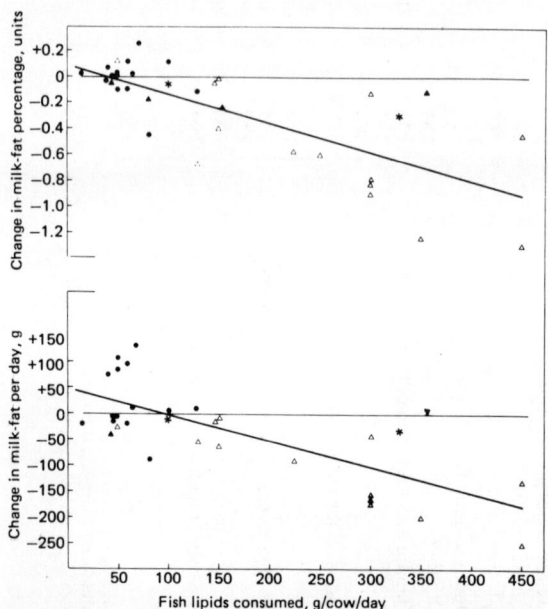

Figure 3.3   Covariation between amount of fish lipid consumed (x = g/cow/day) and changes in milk-fat percentage (y_1) and milk-fat production (y_2 = g/cow/day). $y_1 = 0.085 - 0.002238x$ ($r^2 = 0.5625; 0.01 > P > 0.001$): $y_2 = 46.9 - 0.5023x$ ($r^2 = 0.5373; 0.01 > P > 0.01$). Key: ● fish meal and white fish meal; ▲ fish silage; * cod-liver meal; ▼ whale-meat meal; △ cod-liver oil

exception, for which there is no apparent explanation, is the Danish study by Christiansen, Klausen and Andersen (1970) in which a consumption of 80 g fish lipids per cow per day caused a reduction in milk-fat.

In contrast to what has been found with fish meal, the feeding of fish silage appears to cause a reduction in milk fat, even when it supplies less than 100 g of fish lipids per cow per day. The reason for this difference between fish meal and fish silage is not clear, but it has been argued to be due to the high content of free fatty acids in fish silage (Johnson, 1981).

In order to elucidate the effect of amount of fish lipid ingested on milk-fat secretion, the overall regressions between fish-lipid consumption and changes in milk-fat percentage and total milk-fat production, respectively, in the different experiments have been calculated. The results are presented in *Figure 3.3*. The correlations between amount of fish lipid consumed and changes in milk-fat percentage and milk-fat production were highly significant ($0.01 > P > 0.001$). The regression line crosses the zero-effect line at 38 g and 93 g of fish lipids consumption per cow per day for milk-fat percentage and milk-fat production, respectively. However, the individual data points reveal that when the consumption of fish lipids was less than 100 g/cow per day, no consistent reduction was seen in either of the two criteria.

Breirem (1949) concluded that, although feeding of marine lipids tends to reduce the percentage of fat in the milk, the effect will be insignificant if the daily supply is below 100–150 g/cow per day. Later studies have confirmed this conclusion. However, although there are no signs, either from experiments or from practical experience, of its existence, the possible interaction between low roughage intake and fish lipid consumption in precipitating the 'low milk-fat syndrome' warrants further studies.

Factors that affect the quality of fish lipids and possible undesirable effects

Toxicological effects, in warm-blooded animals, of oxidized lipids in fish meal have been reviewed by Barlow and Pike (1977) who concluded that 'the range of peroxide values generally found in commercial fish meals will not cause any adverse effect on mortality, growth or feed conversion of poultry and pigs'. Because of the practice of using high levels of fish oil in fish feeding, the question of oxidation is apparently more important in fish feeding than in the feeding of warm-blooded animals. Murai and Andrews (1974) fed oxidized menhaden oil (peroxide value (PV) = 60 mEq/kg) at levels of 1% and 10% of the diet for channel catfish with and without the addition of vitamin E or ethoxyquin. The fish fed the diet without vitamin E or ethoxyquin showed reduced growth, increased mortality and gross symptoms of vitamin E deficiency which were partly counteracted by the addition of ethoxyquin and fully prevented by the addition of vitamin E. Studies with rainbow trout using 7.5% of oxidized herring oil (PV = 50–314 mEq/kg) in the diets have not resulted in reduced growth rate, but have increased mortality and caused symptoms of vitamin E deficiency which could be counteracted by the addition of α-tocopheryl acetate

(Hung, Cho and Slinger, 1980; Hung and Slinger, 1980; Silas et al., 1981). It has also been shown that oxidized fish oil may cause a reduction in the dietary content of vitamin C (Hung and Slinger, 1980). It thus appears that fish oil for fish feeding should be protected from oxidation, e.g. by adding antioxidant and by proper packing. The content of FFA (free fatty acids) has been used as a quality criteria for fats and oils without much experimental evidence. Gjefsen and Lysø (1979) compared PHFO with less than 1% and 8% FFA in diets for growing, finishing pigs. The diets contained 8% PHFO. The level of FFA in the PHFO had no effect on growth and feed conversion or on carcass quality. These results have been confirmed and extended by Lysø (1980) who did not observe negative effects from the feeding of 4% PHFAO to growing finishing pigs. Similar results have also been obtained by Austreng and Gjefsen (1981) feeding fish oil with graded levels of FFA to rainbow trout and salmon parr. In that study, increasing levels of FFA from 0.1% to 11% showed no consistent effects on either type of fish when the oil was fed at levels of up to 15% of the diet. FFA thus appears to be an unreliable criterion of the quality of fish lipids.

It has long been recognized that high intakes of fish lipids may cause off-flavours in meat and eggs. The subject has been extensively reviewed by Lineweaver (1970), Opstvedt (1971), Hartfiel and Tuschy (1973) and Barlow and Pike (1977). It appears to be commonly accepted that the off-flavour (fishy flavour) originates from the n-3 fatty acids in the fish lipids which, when deposited in carcass and egg fats, oxidize when the products are stored. It is further commonly accepted that the degree of off-flavour is related to the level of intake of fish lipids, and that intakes below a certain level compatible with practical feeding of fish meal to poultry in most situations, do not create off-flavour. On the basis of experimental data it was postulated that the risk of off-flavour is correlated with the level of n-3 fatty acids in the total dietary fat (Opstvedt, 1974). Thus the risk of fishy off-flavour may be reduced by the addition to the diet of fat that does not contain n-3 fatty acids.

The question of fishy off-flavours in milk and dairy products as a consequence of fish-meal feeding to dairy cows has been raised. Experimental data on this problem are extremely meagre. Hvidsten, Mehlum and Simonsen (1952) found no effect on the oxidative stability (as judged by taste and smell) of milk caused by feeding 0.5–1.1 kg fish meal (50–108 g fish lipids) per day, whereas the addition of 120–150 g cod-liver oil per day led to increased susceptibility to oxidation. This finding is in agreement with recent results from dairy goats (Skjevdahl and Opstvedt, unpublished) in which fish-meal feeding did not have any detrimental effect on the flavour quality of the milk or the cheese made therefrom. Contrary to these results, the feeding of fish meal at a level to provide 70 g of fish lipids per cow per day in Danish experiments (Christiansen et al., 1970) resulted in a tendency to increased off-flavour in the milk and in the fresh but not in the stored (1 month) butter made therefrom, compared with milk and butter from cows on diets without fish meal. Butter made from milk from cows fed on fish meal had better consistency than butter from cows without fish meal in their diets. The latter finding is in agreement with the increase in the content of polyethylenic fatty acids in the milk fat from cows fed 300 g or more per day of cod-liver oil, observed in Canadian experiments (Nicholson and Sutton, 1971). In experiments with silages containing fish entrails (including the liver) fed at levels to provide fish lipids at 41.7, 73.5 and 147 g/cow per day. Johnson (1981) found a tendency to increased off-flavour in the milk and increased content of polyethylenic fatty acids in the milk fat at the lowest level of fish silage fed, compared with control diets with no fish silage. The effect increased with increasing level of silage fed. It is, however, doubtful how the negative effects which are observed in some experiments of the feeding of some fish products on the organoleptic quality of the milk, apply to practical conditions. Fish meal has, in recent years, been used to an increasing extent in the feeding of dairy cows in many countries, and this has not been associated with any flavour problem in the milk.

Some preliminary reports indicate that some types of PHFO, when fed in excessive amounts to young rats, may cause transitory cardiac lipidoses and myocardial necroses of the same type as those observed with feeding of conventional rapeseed oil. Data from recent comprehensive studies have, however, not shown cardiotoxic effects of PHFO (Svaar, 1982; Duthie and Barlow, 1982).

Conclusion

Fish lipids are characterized by a high content of long-chain (more than 18 carbon atoms) fatty acids. In the natural form these are mainly polyethylenic of the n-3 type, although lipids of fish from the northern hemisphere may contain appreciable amounts of monoethylenic fatty acids. During hydrogenation to produce PHFO the polyethylenic fatty acids with *cis* configurations are converted to a mixture of geometrical and positional isomers with one to four double bonds.

Unhydrogenated fish lipids have a high digestibility of more than 90% in all animals tested. The digestibility of PHFO is generally lower than that of the corresponding raw oil and decreases on increasing the melting point. The digestibility of the individual fatty acids decreases with increasing chain length and increases with increasing unsaturation. The high digestibility gives fish lipids a high energy value.

Absorbed fish fatty acids are deposited in the animal body, but the majority are catabolized for energy or converted to other fatty acids. The oxidation of the long-chain fatty acids in the body differs from that of the shorter-chain acids by being initiated in the peroxisomes and completed by the classic oxidation process in the mitochondria.

The n-3 fish fatty acids are essential for fish, and recent findings indicate that they serve important structural and metabolic functions in warm-blooded animals. Intakes higher than those required to prevent clinical deficiency symptoms promote growth in fish and warm-blooded animals.

The polyethylenic fish fatty acids may have some adverse effects when consumed in excessive amounts. Thus they may cause a reduction in the milk fat in ruminants and impart off-flavours to animal products. However, this is not a problem under modern practical feeding conditions. No adverse effects from oxidized fish lipids have hitherto been reported with regard to normal feeding situations of warm-blooded animals. For fish feeding, the oxidation of fish oil should be prevented. The content of FFA commonly used as a quality criterion of fat has not been found to affect the quality of fish lipids in pigs and fish.

References

ACKMAN, R.G. (1980). Fish lipids. Part 1. In *Advances in Fish Science and Technology* (J. Connell and J.J. Farnham, Eds), pp. 86–103. Surrey, UK, Fishing News Books Ltd

ACKMAN, R.G. (1982). Fatty acid composition of fish oils. In *Nutritional Evaluation of Long-Chain Fatty Acids in Fish Oil* (S.M. Barlow and M.E. Stansby, Eds), pp. 25–139. London, Academic Press

ACKMAN, R.G. and EATON, C.A. (1971). Mackerel lipids and fatty acids. *Canadian Institute of Food Science and Technology Journal* **4**, 169–174

ACKMAN, R.G., EATON, C.A. and HINGLEY, J.H. (1976). Menhaden body lipids: details of fatty acids in lipids from an untapped food resource. *Journal of the Science of Food and Agriculture* **27**, 1132–1136

ACKMAN, R.G., KE, P.J., MacCALLUM, W.A. and ADAMS, D.R. (1969). Newfoundland capelin lipids: fatty acid composition and alterations during frozen storage. *Journal of the Fisheries Research Board of Canada* **26**, 2037–2060

ADDISON, R.F., ACKMAN, R.G. and HINGLEY, J. (1968). Distribution of fatty acids in cod flesh lipids. *Journal of the Fisheries Research Board of Canada* **25**, 2083–2090

ADDISON, R.F., ACKMAN, R.G. and HINGLEY, J. (1969). Free fatty acids of herring oil: Possible derivation from both phospholipids and triglycerides in fresh herring. *Journal of the Fisheries Research Board of Canada* **26**, 1577–1583

ANDREWS, R.J. and LEWIS, D. (1970). The utilization of dietary fats by ruminants. I. Digestibility of some commercially available fats. *Journal of Agricultural Science* **75**, 47–52

ANONYMOUS (1977). *Dietary Fats and Oils in Human Nutrition. FAO Food and Nutrition Paper 3*. Rome, FAO

ARTMAN, N.R. (1964). Interactions of fats and fatty acids as energy sources for the chick. *Poultry Science* **43**, 994–1004

AUSTRENG, E. and GJEFSEN, T. (1981). Fish oils with different contents of free fatty acids in diets for rainbow trout fingerlings and salmon parr. *Aquaculture* **25**, 173–183

AUSTRENG, E., SKREDE, A. and ELDEGARD, Å. (1979). Effect of dietary fat source on the digestibility of fat and fatty acids in rainbow trout and mink. *Acta agriculturae scandinavica* **29**, 119–126

BARLOW, S.M. and PIKE, I.H. (1977). *The Role of Fat in Fish Meal in Pig and Poultry Nutrition. I.A.F.M.M. Technical Bulletin No. 4*. International Association of Fish Meal Manufacturers, Hoval House, Orchard Parade, Mutton Lane, Potters Bar, Hertfordshire, SN6 3AR, UK. 37 pp

BASSLER, R. and PUTZKA, H.-A. (1975). Der Einfluss verschiedener Fettbestimmungsmethoden auf die quantitativen Extraktausbeuten und auf deren Fettsaurespektrum bei stabilisiertem und unbehandeltem peruanischem Fischmehl. [The effect of different methods of fat determination on the quantitative yield of extraction and on their fatty acid spectrum in stabilized and unstabilized Peruvian fishmeal.] *Landwirtschaftliche Forschung* **28**, 56–68

BJØRNSTAD, J. and HANSEN, P.J. (1974). Digestibility of hydrogenated marine fat (HMF) in milk replacers for calves. *Zeitschrift für Tierphysiologie, Tierernährung und Futtermittelkunde* **33**, 126–137

BLIGH, E.G. and SCOTT, M.A. (1966). Lipids of cod muscle and the effect of frozen storage. *Journal of the Fisheries Research Board of Canada* **23**, 1025–1036

BRADDOCK, R.J. and DUGAN, L.R. Jr. (1972). Phospholipid changes in muscle from frozen stored Lake Michigan coho salmon. *Journal of Food Science* **37**, 426–429

BREIREM, K. (1949). Norwegian experiments regarding the effects of the feed on the composition and quality of the milk and milk products. In *Proceedings, XII International Dairy Congress, Stockholm, 15–19 August*, pp. 28–60

BREIREM, K. and HOMB, T. (1970). *Fórmidler of Fórkonservering.* [*Feedstuffs and Feed Preservation.*] Gjøvik, Norge, Forlag Buskap og Avdrått A/S. 459 pp

BREMER, J. and NORUM, K.R. (1982). Metabolism of very long-chain monounsaturated fatty acids (22:1) and the adaptation to their presence in the diet. *Journal of Lipid Research* **23**, 243–256

BRUMBY, P.E., STORRY, J.E. and SUTTON, J.D. (1972). Metabolism of cod-liver oil in relation to milk fat secretion. *Journal of Dairy Research* **39**, 167–182

CASTELL, J.D., SINNHUBER, R.O., WALES, J.H. and LEE, D.J. (1972). Essential fatty acids in the diet of rainbow trout (*Salmo gairdneri*): Growth, feed conversion and some gross deficiency symptoms. *Journal of Nutrition* **102**, 77–86

COWEY, C.B. (1976). Use of synthetic diets and biochemical criteria in the assessment of nutrient requirements of fish. *Journal of the Fisheries Research Board of Canada* **33**, 1040–1045

COWEY, C.B., OWEN, J.M., ADRON, J.W. and MIDDLETON, C. (1976). Studies on the nutrition of marine flatfish. The effect of different dietary fatty acids on the growth and fatty acid composition of turbot (*Scophthalmus maximus*). *British Journal of Nutrition* **36**, 479–486

CHRISTIANSEN, B., KLAUSEN, S. and ANDERSEN, P. (1970). Anvendelse af fiskemel som proteinkilde til malkekøer og dettes indvirkning på melkeydelse og -kvalitet. [The use of fish meal as a source of protein for milking cows and its effect on milk yield and quality.] *Landøkonomisk Forsøgslaboratorium Efterårsmøde. Årbog (Institute of Animal Sciences. Fall Meeting. Yearbook). Copenhagen, 28–30 October*, pp. 300–307

CHRISTIE, W.W. (1981). The effect of diet and other factors on the lipid composition of ruminant tissues and milk. In *Lipid Metabolism in Ruminant Animals* (W.W. Christie, Ed.), pp. 193–226. Oxford, Pergamon Press

CHRISTOPHERSEN, B.O., NORSETH, J., THOMASSEN, M.S., CHRISTIANSEN, E.N., NORUM, K.R., OSMUNDSEN, H. and BREMER, J. (1982). Metabolism and metabolic effects of C22:1 fatty acids with special reference to cardiac lipidosis. In *Nutritional Evaluation of Long-Chain Fatty Acids in Fish Oil* (S.M. Barlow and M.E. Stansby, Eds), pp. 89–139. London, Academic Press

CUPPET, S.L. and SOARES, J.H. Jr. (1972). The metabolizable energy values and digestibilities of menhaden fish meal, fish solubles, and fish oils. *Poultry Science* **51**, 2078–2083

DE KONING, A.J. (1966). Phospholipids of marine origin. I. The hake (*Merluccius capensis*, Castelman). *Journal of the Science of Food and Agriculture* **17**, 112–117

DE KONING, A.J. and McMULLAN, K.B. (1966). Phospholipids of marine origin. III. The pilchard (*Sardina ocellata*, Jenyns) with particular reference to oxidation in pilchard meal manufacture. *Journal of the Science of Food and Agriculture* **17**, 385–388

DESHIMARU, O., KUROKI, K. and YONE, Y. (1982). Nutritive values of various oils for yellowtail. *Bulletin of the Japanese Society of Scientific Fisheries* **48**, 1155–1157

DROZDOWSKI, B. and ACKMAN, R.G. (1969). Isopropyl alcohol extraction of oil and lipids in the production of fish protein concentrate from herring. *Journal of the American Oil Chemists Society* **46**, 371–376

DUTHIE, I.F. and BARLOW, S.M. (1982). A rat life span study comparing partially hydrogenated fish oils, partially hydrogenated soybean oil and rapeseed oil included in the diet at high levels: outline description and interim communication. In *Nutritional Evaluation of Long-Chain Fatty Acids in Fish Oil* (S.M. Barlow and M.E. Stansby, Eds), pp. 185–214. London, Academic Press

DYERBERG, J. and ANKER JØRGENSEN, K. (1982). Marine oils and thrombogenesis. *Progress in Lipid Research* **21**, 255–269

EDWARDS, H.M. Jr. and MARION, J.E. (1963). Influence of dietary menhaden oil on growth rate and tissue fatty acids of the chick. *Journal of Nutrition* **81**, 123–130

EDWARDS, H.H. Jr., MARION, J.E. and DRIGGERS, J.C. (1962). Studies on fat and fatty acid requirements of poultry. In *Proceedings, XIIth World Poultry Congress*, p. 182

EKERN, A. (1961). Sildemel i foret til melkekyr. [Herring meal in the diet of lactating cows.]. Beretn. nr. 107. *Landbrukshøgskolens Fóringsforsøk* 1–68

EL-SHATTORY, Y. (1979). Review on fish phospholipids. *Die Nahrung* **23**, 179–186

ENGSTER, H.M., CAREW, L.B. and FOSS, D.C. (1975). Effect of herring oil on body weight, comb size and gonadal development in the chick. *Poultry Science* **54**, 2118–2121

FLATLANDSMO, K. (1972). Marine fat. Digestibility of its fatty acids in young calves. *Acta veterinaria scandinavica* **13**, 260–262

FUJII, M. and YONE, Y. (1976). Studies on nutrition of red sea bream—XIII Effect of dietary linolenic acid and w3 polyunsaturated fatty acids on growth and feed efficiency. *Bulletin of the Japanese Society of Scientific Fisheries* **42**, 583–588

GJEFSEN, T. and LYSØ, A. (1979). Hydrogenated marine fat with high content of free fatty acids in feed mixtures for growing-finishing pigs. *Acta agriculturae scandinavica* **29**, 65–70

GRAY, G.M. and MacFARLANE, M.G. (1961). Composition of phospholipids of rabbit, pigeon and trout muscle and various pig tissues. *Biochemical Journal* **81**, 480–488

GUNSTONE, F.D. and WIJESUNDERA, R.C. (1978). The component acids of the lipids in four commercial fish meals. *Journal of the Science of Food and Agriculture* **29**, 28–32

GUNSTONE, F.D., WIJESUNDERA, R.C. and SCRIMGEOUR, C.M. (1978). The component acids of lipids from marine and freshwater species with special reference to furan-containing acids. *Journal of the Science of Food and Agriculture* **29**, 539–550

HARDY, R. and KEAY, J.N. (1972). Seasonal variations in the chemical composition of Cornish mackerel, *Scomber scombrus* (L) with detailed references to the lipids. *Journal of Food Technology* **7**, 125–137

HARDY, R., McGILL, A.S. and GUNSTONE, F.D. (1979). Lipid and autoxidative changes in cold stored cod (*Godus morhua*). *Journal of the Science of Food and Agriculture* **30**, 999–1006

HARTFIEL, W. and TUSCHY, D. (1973). Der Fettverderb und sein Einfluss auf den Geschmack von Mastgeflugel. [Fat deterioration and its effect on the flavour of broilers.] *Kraftfutter* **56**, 64–72

HERSTAD, O. (1975). Omsettelig energi og meltegrad av fôrfeitt. [Metabolizable energy and digestibility of feed fat.] *Husdyrforsøksmøtet ved Norges Landbrukshøgskole (Animal Science Research Meeting) The Agricultural University of Norway, Aas, Norway, 4–5 February*, pp. 49–54

HILL, F.W. (1964). The experimental basis of advances in efficiency of poultry nutrition. *Federation Proceedings* **23**, 857–862

HOFFMANN, L. and SCHIEMANN, R. (1971). Verdaulichkeit und Energiekennzahlen von Futterstoffen beim Huhn. [Digestibility and energy value of feedstuffs for hens.] *Archiv für Tierernährung* **21**, 65–81

HOLMAN, L.T. and JOHNSON, S.B. (1982). Linolenic acid deficiency in man. *Nutrition Reviews* **40**, 144–147

HOMB, T. (1962). Proteinkvaliteten i ulike sildemeltyper belyst ved balanseforsøk med voksende svin. [The protein quality of different types of herring meal determined in balance experiments with growing pigs.] *Meldinger fra Norges Landbrukshøgskole (Scientific Report of the Agricultural University of Norway)* **41**, (5), 1–28

HUNG, S.S.O., CHO, C.Y. and SLINGER, S.J. (1981). Measurement of oxidation in fish oil and its effect on vitamin E nutrition of rainbow trout (*Salmo gairdneri*). *Canadian Journal of Fisheries and Aquatic Sciences* **37**, 1248–1253

HUNG, S.S.O. and SLINGER, S.J. (1980). Effect of oxidized fish oil on the ascorbic acid nutrition of rainbow trout (*Salmo gairdneri*). *International Journal for Vitamin and Nutrition Research* **50**, 393–400

HVIDSTEN, H., MEHLUM, J. and SIMONSEN, H. (1952). Undersøkelser over fóringas virkning på oksydasjonsfeil i mjølk. [Studies on the effect of feeding on oxidative deterioration of milk.] *Meieriposten* **41**, 145–148, 170–173, 186–190

ISAACHSEN, H. and ULVESLI, O. (1926). Produksjonsværdien av forstoffer fremstillet av fisk. IV. Sildemel. [The production value of feedstuffs from fish. IV. Herring meal.] 20. beretn. *Landbrukshøgskolens Foringsforsøk* 1–36

ISAACHSEN, H. and ULVESLI, O. (1927). Melkeproduksjonsværdien av forstoffer fremstillet av fisk. V. Fiskemel. [The production value of feedstuffs from fish. V. Fish meal.] 21. beretn. *Landbrukshøgskolens Foringsforsøk* 1–19

ISAACHSEN, H., AASHAMAR, O. and BANG-SANDMO, O. (1919). Forverdien av ekstrahert levermel til melkefæ. [The feeding value of extracted liver-meal for lactating cows.] 12. beretn. *Landbrukshøgskolens Foringsforsøk* 5–25

ISAACHSEN, H., FREDRICKSEN, L.A., LALIM, A. and WOLD, I.K. (1915). Forværdien av levermel til melkefæ. [The feeding value of extracted liver-meal for lactating cows.] 9. beretn. *Landbrukshøgskolens Foringsforsøk* 1–52

JANGAARD, P.M., BROCKERHOFF, H., BURGHER, R.D. and HOYLE, R.J. (1967). Seasonal changes in general condition and lipid content of cod from inshore waters. *Journal of the Fisheries Research Board of Canada* **24**, 607–612

JOHNSON, F. (1981). *Ensilert fiskeslo som for til drøvtyggere.* [Ensiled fish entrails as a feed for ruminants.] PhD thesis, Agricultural University of Norway, Aas

KANASAWA, A., TESHIMA, S. and TOKIWA, S. (1977). Nutritional requirements of prawn—VIII Effect of dietary lipids on growth. *Bulletin of the Japanese Society of Scientific Fisheries* **43**, 849–856

KANAZAWA, A., TOKIWA, S., KAYAMA, M. and HIRATA, M. (1977). Essential fatty acids in the diet of prawn—I Effect of linoleic and linolenic acids on growth.*Bulletin of the Japanese Society of Scientific Fisheries***43**,1111–1114

LAKSESVELA, B. (1966). Sildefett som kyllingfór. [Herring fat as a feed for chicks.] *Medlinger fra Sildolje-og Sildemel-industriens Forskningsinstitutt (Reports from the Norwegian Herring Oil and Meal Industry Research Institute)* (4), 92–102

LANDS, W.E.M. (1982). Biochemical observations on dietary long chain fatty acids from fish oil and their effect on prostaglandin synthesis in animals and humans. In *Nutritional Evaluation of Long-Chain Fatty Acids in Fish Oil* (S.M. Barlow and M.E. Stansby, Eds), pp. 267–282. London, Academic Press

LAMBERTSEN, G. (1972). Lipids in fish fillet and liver—a comparison of fatty acid compositions. *Fiskeridirektoratets Skrifter Serie Teknologiske Undersøkelser (Reports on Technological Research Concerning Norwegian Fish Industry)* **5**, (6), 1–15

LAZAROW, P.B. (1978). Rat liver peroxisomes catalyse the β-oxidation of fatty acids. *Journal of Biological Chemistry* **253**, 1522–1528

LAZAROW, P.B. and de DUVE, C. (1976). A fatty acid-CoA oxidation system in rat liver peroxisomes, enhancement by clofibrate, a hypolipidemic drug. *Proceedings of the National Academy of Sciences of the United States of America* **73**, 2043–2046

LEA, C.H., PARR, L.J. and CARPENTER, K.J. (1958). Chemical and nutritional changes in stored herring meal. *British Journal of Nutrition* **12**, 297–312

LEE, D.J., ROEHM, J.N., YU, T.C. and SINNHUBER, R.O. (1967). Effect of w3 fatty acids on the growth rate of rainbow trout, *Salmo gairdneri. Journal of Nutrition* **92**, 93–98

LIED, E. and LAMBERTSEN, G. (1982). Apparent availability of fat and individual fatty acids in Atlantic cod (*Gadus morhua*). *Fiskeridirektoratets Skrifter Serie Ernæring (Reports on Nutrition from the Directorate of Fisheries, Bergen, Norway)* **2**, 63–75

LINEWEAVER, H. (1970). Effect of feed ingredients on the development of off flavors in turkey meat. *Feedstuffs* **42**, (9) 30–31

LINKO, R.R. (1965). The chemical composition of lipids in the Baltic herring. In *Fat Oil Chemistry, Fourth Scandinavian Symposium, Almquist Wiksell, Stockholm, Sweden*, pp. 34–39

LYSØ, A. (1980). Futterungsversuche an Schweinen mit freien Fettsauren hydrierter Seetierøle in der Rationen. [Feeding experiments with swine on diets with free fatty acids from hydrogenated marine oil.] *Fette-Seifen-Anstrichmittel* **82**, 279–282

MASSON, L. and BURGOS, M.T. (1973). Fatty acid composition of the Chilian anchovy (anchoveta: *Engraulis ringens*) and of its neutral and polar fractions. *Grasas y Aceites* **24**, 327–330

MENGE, H., CALVERT, C.C. and DENTON, C.A. (1965). Influence of dietary oils on reproduction in the hen. *Journal of Nutrition* **87**, 365–370

MILLER, E.L., GALWAY, N.W., NEWMAN, G. and PIKE, I.H. (1981). Report of co-ordinated trials carried out on commercial farms in U.K. *Animal Production* **32**, 131–141

MURAI, T. and ANDREWS, J.N. (1974). Interactions of dietary tocopherol, oxidized menhaden oil and ethoxyquin on channel catfish (*Ictalurus punctatus*). *Journal of Nutrition* **104**, 1416–1431

NICHOLSON, I.W.G. and SUTTON, I.D. (1971). Some effects of unsaturated oils given to dairy cows with rations of different roughage content. *Journal of Dairy Research* **38**, 363–372

OPSTVEDT, J. (1971). Fish taints in eggs and poultry meat. In *University of Nottingham Nutrition Conference for Feed Manufacturers No. 5* (H. Swan and D. Lewis, Eds), pp. 70–93. London, Churchill Livingstone

OPSTVEDT, J. (1973a). Influence of residual lipids on the nutritive value of fish meal. IV. Effects of drying and storage on the energy value of the protein and lipid fractions of herring meal. *Acta agriculturae scandinavica* **23**, 200–208

OPSTVEDT, J. (1973b). Influence of residual lipids on the nutritive value of fish meal. V. Digestion and deposition of marine fatty acids in chickens. *Acta agriculturae scandinavica* **23**, 217–224

OPSTVEDT, J. (1974). Influence of residual lipids on the nutritive value of fish meal. VI. Effects of fat addition to diets high in fish meal on fatty acid composition and flavour quality of broiler meat. *Acta agriculturae scandinavica* **24**, 61–75

OPSTVEDT, J. (1981). The significance of fish meal lipids in animal nutrition and feeding. In *Fats in Feeds and Feeding. Proceedings, LIPIDFORUM Seminar, Gotenborg, Sweden, 9–10 March 1981* (R. Marcuse, Ed.), pp. 13–24. Scandinavian Forum for Lipid Research and Technology, SIK, Box 5401, S-40229, Gøteborg, Sweden

OPSTVEDT, J. and RONNING, M. (1967). Effect upon lipid metabolism of feeding alfalfa hay or concentrate ad libitum as the sole feed for milking cows. *Journal of Dairy Science* **50**, 345–354

OPSTVEDT, J., BALDWIN, R.L. and RONNING, M. (1967). Effect of diet upon activities of several enzymes in abdominal adipose and mammary tissues in the lactating dairy cow. *Journal of Dairy Science* **50**, 108–109

ØRSKOV, E.R., REID, G.W. and McDONALD, I. (1981). The effects of protein degradability and food intake on milk yield and composition in cows in early lactation. *British Journal of Nutrition* **45**, 547–555

PENNINGTON, J.A. and DAVIS, C.L. (1975). Effects of intra-ruminal and intra-abomasal additions of cod-liver oil on milk fat secretion in the cow. *Journal of Dairy Science* **58**, 49–55

POHL, P. and WAGNER, H. (1972). Fettsauren im Pflanzen – und Tierreich (eine Ubersicht) I: Gesattigte und *cis*-ungesattigte Fettsauren. [Fatty acids in the plant and animal kingdom (a review). I Saturated and *cis*-unsaturated fatty acids.] *Fette-Seifen-Anstrichmittel* **74**, 424–435

POTTER, L.M., PUDELKIEWICZ, J.W., WEBSTER, L. and MATTERSON, L.D. (1962). Metabolizable energy and digestibility evaluation of fish meal for chickens. *Poultry Science* **41**, 1745–1752

REICHWALD, I. (1976). Chemie der Fischlipide. [Chemistry of fish lipids.] *Fette-Seifen-Anstrichmittel* **78**, 328–334

RIMESLÅTTEN, H. (1971). Fordøyelsesforsøk med herdet marint fett. [Digestibility trials with hydrogenated marine fat.] *Nordiske Jordbruksforskeres Forening (Association of Nordic Agricultural Scientists) Congress, Uppsala, Sweden, 29 June–2 July*, pp. 1–18

ROJAS, S. and ARANA, C.M. (1981). Metabolizable energy values of anchovy fish meal and oil for chicks. *Poultry Science* **60**, 2274–2277

SCHIEMANN, R., JENTSCH, W. and HOFFMANN, L. (1969). Die energetische Verwertung der Futterstoffe. [The energetical utilization of feedstuffs.] 10, Mitteilung. *Archiv für Tierernährung* **19**, 331–344

SEBELIEN, J. (1894). Fosøg med sild- og kvalkjødmel som foder for melkekjør. [Experiments with herring- and whalemeat-meal as feed for lactating cows.] *Beretning om den høyere Landbruksskole i Aas, 1892–1893* 142–189

SILAS, S., HUNG, O., CHO, C.Y. and SLINGER, S.J. (1981). Effect of oxidized fish oil, DL-α-tocopheryl acetate and ethoxyquin supplementation on the vitamin E nutrition of rainbow trout (*Salmo gairdneri*) fed practical diets. *Journal of Nutrition* **111**, 648–657

STORRY, J.E. (1981). The effect of dietary fat on milk composition. In *Recent Advances in Animal Nutrition* (W. Haresign, Ed.), pp. 3–33. London, Butterworths

STORRY, J.E., BRUMBY, P.E., HALL, A.J. and TUCKLEY, B. (1974). Effects of free and protected forms of codliver oil on milk fat secretion in the dairy cow. *Journal of Dairy Science* **57**, 1046–1049

SUNDSTØL, F. (1974a). Hydrogenated marine fat as feed supplement. I Digestibility of rations containing hydrogenated marine fat in pigs. *Meldinger fra Norges Landbrukshøgskole (Scientific Report of the Agricultural University of Norway)* **53**, (22), 1–24

SUNDSTØL, F. (1974b). Hydrogenated marine fat as feed supplement. III. Digestibility of rations containing hydrogenated marine fat in sheep. *Meldinger fra Norges Landbrukshøgskole (Scientific Report of the Agricultural University of Norway)* **53**, (24), 1–29

SVAAR, H. (1982). The long-term lesion phenomenon in animals and humans. In *Nutritional Evaluation of Long-Chain Fatty Acids in Fish Oils* (S.M. Barlow and M.E. Stansby, Eds), pp. 163–184. London, Academic Press

TAKEUCHI, T. and WATANABE, T. (1982). The effects of starvation and environmental temperature on proximate and fatty acid compositions of carp and rainbow trout. *Bulletin of the Japanese Society of Scientific Fisheries* **48**, 1307–1316

TANAKA, K. (1970a). The effect of the type of dietary fat on milk fat secretion in cows. *Japanese Journal of Zootechnical Science* **41**, 254–261

TANAKA, K. (1970b). The effect of increasing amounts of dietary cod oil on milk fat secretion in cows. *Japanese Journal of Zootechnical Science* **41**, 453–458

THUEN, E. (1982). Forsøk med avfetta, ensilert fiskeslo i sammenligning med sildemjøl og urea som proteinsupplement til mjølkekyr. [Experiments with fat-extracted, ensiled fish entrails in comparison with herring meal and urea as protein supplements for dairy cows.] *Husdyrforsøksmøtet. NLH 1980–81 (Animal Science Research Meeting), The Agricultural University of Norway, Aas, Norway, 26–28 January*, pp. 66–70

TINOCO, J., BABCOCK, R., HINCENBERGS, I., MEDWADOWSKI, B., MILJANICH, P. and WILLIAMS, M.A. (1979). Linoleic acid deficiency. *Lipids* **14**, 166–173

TOYOMIZU, M., NAKAMURA, T. and SHONO, T. (1976). Fatty acid composition of lipid from horse mackerel muscle—discussion of fatty acid composition of fish lipid. *Bulletin of the Japanese Society of Scientific Fisheries* **42**, 101–108

UEDA, T. (1976). Changes in the fatty acid composition of mackerel lipid and probably related factors. I. Influence of the season, body length and lipid content. *Bulletin of the Japanese Society of Scientific Fisheries* **42**, 479–484

VAN SOEST, P.J. (1963). Ruminant fat metabolism with particular reference to factors affecting low milk fat and feed efficiency. A review. *Journal of Dairy Science* **46**, 204–216

VARMAN, P.N., SCHULTZ, L.H. and NICHOLS, R.E. (1968). Effect of unsaturated oils on rumen fermentation, blood components, and milk composition. *Journal of Dairy Science* **51**, 1956–1963

VEEN, A.G., GRIMBERGEN, A.H.M. and STAPPERS, H.P. (1974). The true digestibility and calorific value of various fats used in feeds for broilers. *Archiv für Geflügelkunde* **6**, 213–220

VIVIANI, R., BORGATTI, A.R., MANCINI, L. and CORTESI, P. (1967). Changes in the muscular lipids of the mackerel (*Scomber scombrus* L.) during frozen storage. *Atti della societa italiana di scienze veterinarie* **21**, 706–710

WATANABE, T. and TAKEUCHI, T. (1976). Evaluation of pollock liver oil as a supplement to diets for rainbow trout. *Bulletin of the Japanese Society of Scientific Fisheries* **42**, 893–906

WATANABE, T., TAKEUCHI, T. and OGINO, C. (1975). Effect of dietary methyl linoleate and linolenate on growth of carp—II. *Bulletin of the Japanese Society of Scientific Fisheries* **41**, 263–269

WATANABE, T., UTSUE, O., KOBAYASHI, I. and OGINO, C. (1975). Effect of dietary methyl linoleate and linolenate on growth of carp—I. *Bulletin of the Japanese Society of Scientific Fisheries* **41**, 257–263

WESSELS, J.P.H., ATKINSON, A., MERWE, R.P. and SWART, L.G. (1971). Rations with fish meal and other additives vs. flavour of chickens (I). In *Twenty-fifth Annual Report of the Director, Fishing Industry Research*

Institute, University of Cape Town, Rondebosch, South Africa, pp. 10–18

WOOD, G. and HINTZ, L. (1971). Decomposition in foods. Lipid changes associated with the degradation of fish tissue. *Journal of the Association of Official Analytical Chemists* **54**, 1019–1023

YONE, Y. and FUJII, M. (1975). Studies on nutrition of red sea bream—XI Effect of w3 fatty acid supplement in a corn oil diet on growth rate and feed efficiency. *Bulletin of the Japanese Society of Scientific Fisheries* **41**, 73–77

YU, T.C. and SINNHUBER, R.O. (1972). Effects of dietary linolenic acid and docosahexaenoic acid on growth and fatty acid composition of rainbow trout (*Salmo gairdneri*). *Lipids* **7**, 450–454

YU, T.C. and SINNHUBER, R.O. (1975). Effect of dietary linolenic and linoleic acids upon growth and lipid metabolism of rainbow trout. *Lipids* **10**, 63–66

YU, T.C. and SINNHUBER, R.O. (1976). Growth responses of rainbow trout (*Salmo gairdneri*) to dietary w3 and w6 fatty acids. *Aquaculture* **8**, 309–317

YU, T.C., SINNHUBER, R.O. and HENDRICKS, J.D. (1979). Reproduction and survival of rainbow trout (*Salmo gairdneri*) fed linolenic acid as the only source of essential fatty acids. *Lipids* **14**, 572–575

YU, T.C., SINNHUBER, R.O. and PUTNAM, G.B. (1977). Effect of dietary lipids on fatty acid composition of body lipid in rainbow trout (*Salmo gairdneri*). *Lipids* **12**, 495–499

II

Digestion, Absorption and Transport of Fats

4

DIGESTION, ABSORPTION AND TRANSPORT OF FATS: GENERAL PRINCIPLES

DAVID N. BRINDLEY
Department of Biochemistry, University of Nottingham Medical School, Queen's Medical Centre, Nottingham NG7 2UH, UK

Introduction

The term fat or lipid is used to describe a wide variety of compounds that are insoluble in water, and that dissolve in organic solvents such as chloroform and diethylether. In terms of nutrition the major lipids are triacylglycerols (triglycerides), phospholipids, sterols and the fat-soluble vitamins. The fact that these compounds are insoluble in the aqueous environment of the intestinal lumen, cells and the circulation, presents the body with a number of special metabolic problems. The efficient metabolism and transport of fats requires that they should be dispersed in a relatively stable form.

The purpose of this article is to give a general outline of the processes that are involved in the digestion, absorption and transport of fat in mammals. More detailed accounts can be obtained from the reviews and research papers that are cited, and from the next two chapters.

Digestion

Relatively little digestion of fat is thought to take place in the stomach. However, proteolytic digestion helps to release fat from food particles. In addition, the churning action of the stomach helps to form a coarse emulsion of the fat. The bulk of dietary lipids (triacylglycerols and sterols) have been classified as Type IB non-swelling amphiphiles which are polar (*Table 4.1*). They form insoluble crystals, or oils in which can dissolve the non-polar lipids including hydrocarbons, sterol esters, waxes and the vitamin esters. These are partially stabilized by a coat of amphiphilic proteins and Type IIB swelling amphiphiles (phospholipids, glycolipids and monoacylglycerols). These lipids orientate themselves at oil/water interfaces to form monolayers.

A distinct gastric lipase that has properties which differ from those of pancreatic lipase has been isolated from stomach contents (Cohen, Morgan and Hofmann, 1971). The gastric lipase has been demonstrated in man (Schonheyder and Volqvartz, 1964), rat (Clark, Brause and Holt, 1969;

Table 4.1 CLASSIFICATION OF DIETARY FATS AND THEIR HYDROLYSIS PRODUCTS BASED UPON THEIR INTERACTION WITH WATER AND BILE SALT. TABLE BY COURTESY OF BORGSTRÖM (1974)

Class	Mode of interaction with	
	Water	Bile salt solution
A. *Non-polar*		
Hydrocarbons	Insoluble oil or crystals do not orientate at interfaces	Low micellar solubility. Excess forms oil or crystals
Sterol esters		
Waxes		
Vitamin esters		
B. *Polar*		
Insoluble non-swelling amphiphiles	Insoluble oil or crystals orientate at interfaces, form stable mono-layers	Low micellar solubility. Excess forms oil or crystals
Triacylglycerol		
Diacylglycerol		
Long-chain protonated fatty acids		
Stearol		
Vits A, D, E and K		
II. Insoluble swelling amphiphiles	Solubility very low, swell to form liquid crystals, form stable monolayers at interfaces	Mixed micelles with a salt to bile salt ratio generally >0.5. Excess forms liquid crystals
Phospholipids		
Glycolipids		
Monoacylglycerol		
'Acid soaps'		
III. Soluble amphiphiles		
(a) Soaps of long-chain fatty acids.	Molecular solution → micellar solution → liquid crystalline phase	Mixed micelles
Lyso-phosphatidylcholine (most synthetic detergents)		
(b) Bile salts	Molecular solution → micellar solution → solid phase	

Barrowman and Darnton, 1970; Helander and Olivecrona, 1970) and dog (Engstrom *et al.*, 1968). Although the lipase has a pH optimum of 6–7, it is stable at pH 3.5 where it still has significant activity (Engstrom *et al.*, 1968; Clark, Brause and Holt, 1969). The lipase preferentially hydrolyses the esters of medium- rather than of long-chain fatty acids (Engstrom *et al.*, 1968; Clark, Brause and Holt, 1969). The medium-chain length acids that are liberated can be absorbed directly in the stomach (Clark, Brause and Holt, 1969). The lipase may be particularly important in suckling animals since the milk of many species contains high proportions of medium-chain fatty acids (Helander and Olivecrona, 1970). The native milk fat droplet is not readily hydrolysed by pancreatic lipase. By contrast, the gastric lipase hydrolyses triacylglycerols mainly by liberating the short- or medium-chain fatty acids which are located mainly at position 3. This initial digestion in the stomach provides a modified substrate that is readily attacked by pancreatic lipase (Cohen, Morgan and Hofmann, 1971; Olivecrona *et al.*, 1973).

Apart from the limited hydrolysis of esters containing short- and medium-chain fatty acids, the major part of fat digestion takes place in the proximal part of the small intestine. The physical properties of the fat emulsion that enters the intestine from the stomach become modified after mixing with the bile, pancreatic juice and the secretions from the small intestine. The pH of the proximal part of the small intestine, where most of the fat digestion takes place, rises to 5.8–6.5 (Borgström, 1974). This means that any unesterified fatty acids will be converted to soluble amphiphiles. Bile salts also possess the properties of Type III soluble amphiphiles. They have detergent-like properties since one side of the rigid planar structure of the steroid nucleus is hydrophobic. This surface can either interact with the equivalent face of other bile salt molecules, or it can dissolve at an oil/water interface. The hydrophilic groups of the bile salt are concentrated on the other face of the steroid nucleus, and these interact with the aqueous environment.

The bile salt molecules accumulate on the surface of the lipid droplet, dislodging other surface-active constituents and donating a negative charge to the oil droplet. This attracts to the surface a protein, with a relative molecular mass of 10 000, which is called colipase. Pancreatic lipase in turn is attracted to the surface of the droplet (Hofmann, 1978; Patton et al., 1978). The function of the colipase is to anchor the pancreatic lipase near the surface of the oil droplet despite the presence of bile salts that might otherwise displace it. The bile salts, colipase, and lipase are thought to interact together to form a ternary complex. Calcium ions are also needed to stimulate the action of pancreatic lipase (Borgström, 1974).

This lipase preferentially hydrolyses the fatty acids from the 1- and 3-positions of triacylglycerols to yield 2-monoacylglycerols (Borgström, 1974). There is relatively little hydrolysis of the fatty acid at the 2-position (*Figure 4.1*). In addition the rate of isomerization of 2-monoacylglycerol to 1-monoacylglycerol is slow relative to the rate of fat absorption and therefore little further hydrolysis takes place in non-ruminant animals. In ruminants, microbial lipases are responsible for the complete hydrolysis of triacylglycerols to glycerol.

Pancreatic juice also contains other enzymes that are important in lipid digestion. Phospholipases of the A_1 and A_2 types remove fatty acids from

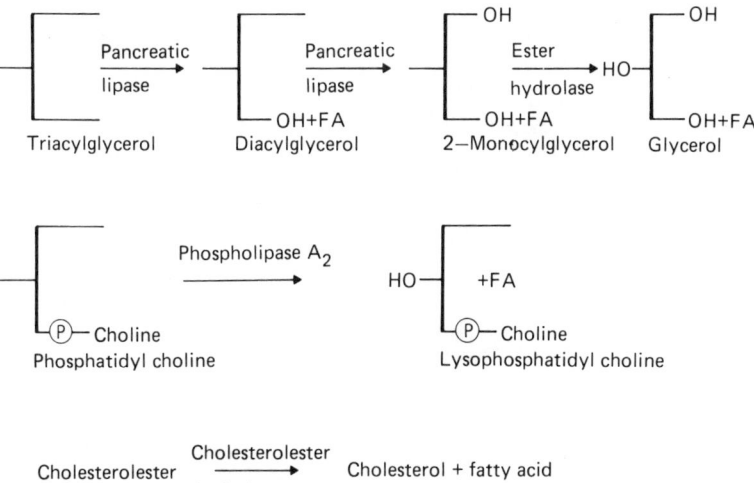

Figure 4.1 Hydrolysis of triacylglycerols, phosphatidylcholine and cholesterol esters

88 *Digestion, absorption and transport of fats: general principles*

the 1- and 2-positions respectively of phospholipids. Phospholipase A_2 has been identified as an inactive proenzyme in pancreatic juice. It can be activated by a trypsin-catalysed removal of a heptapeptide from the N-terminus of the single polypeptide chain (van den Bosch, 1982). The enzyme requires calcium ions and bile salts for activity. The phospholipid substrates may be from the diet itself, or from the mixed micelles with cholesterol that are derived from the bile. Their hydrolysis is not complete and lysophospholipids may accumulate in the intestinal contents.

A carboxylic ester hydrolase that is present in pancreatic juice can act upon a variety of ester bands, and in particular is responsible for the conversion of cholesterol esters into cholesterol. It may also remove some of the fatty acids from the 2-position of acylglycerols and it may be responsible for the hydrolysis of the esters of fat-soluble vitamins (Friedman and Nylund, 1980).

The overall process of fat digestion therefore converts fats into more polar derivatives that are able to interact with water (*Figure 4.1*). Triacylglycerols are transformed mainly to monoacylglycerols (insoluble swelling amphiphiles). Cholesterol esters which are non-polar are converted to polar non-swelling amphiphiles. Phospholipids are hydrolysed to lyso-derivatives which are soluble amphiphiles that readily form micellar solutions (Type III). The fatty acids that are released in these reactions are also Type III amphiphiles when they are ionized (*Table 4.1*).

The lipids in the intestinal lumen exist in equilibrium between the oil phase and a micellar phase that is promoted by the presence of bile salts and lysophospholipids (*Figure 4.2*). As digestion continues the monoacylglycerols and ionized fatty acids leave the surface of the oil droplet to be incorporated into micelles. This partitioning of fatty acids into the micellar phase is favoured by the gradual increase in pH that occurs in the luminal

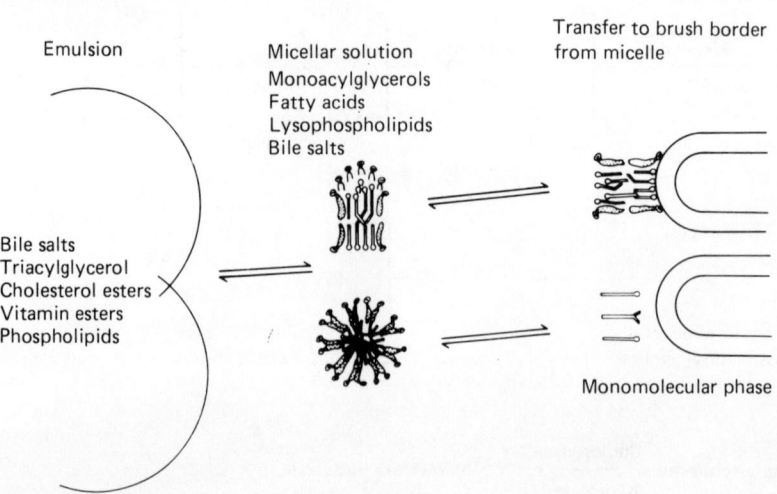

Figure 4.2 The transfer of lipids from the emulsion to micelles and absorption into the enterocyte (after Hofmann, 1968)

contents as they pass to the distal parts of the small intestine. The presence of lysophosphatidylcholine in the micelles promotes the incorporation of cholesterol and stearic acid into this phase (Borgström, 1974).

Some limited digestion can also take place when the micellar lipid comes into contact with the microvillus membranes of the enterocytes. These structures contain phospholipase A_2 and cholesterol ester hydrolase activities. The enterocytes also contain other acylglycerol hydrolases and phospholipases that can degrade adsorbed lipids if they are not efficiently esterified (for details see Brindley, 1974).

Absorption

The digested lipids in the micellar phase are carried to the surface of the enterocytes where they encounter two barriers for their absorption. The first is a layer of unstirred water at the surface of the microvillus membranes. This layer is thought to provide the major rate-limiting factor in the uptake process (Dietschy, 1978). It is believed that absorption across the second barrier, the lipid membrane of the microvillus, could occur when the micelles collide with it. Alternatively, the micelles are in equilibrium with mono-molecular dispersed lipid which could partition into the lipid membrane (*Figure 4.2*). The subsequent transport of the lipids across the membranes is by an energy-independent process. Most of the absorption of micellar lipids takes place in the jejunum, except for bile salts which are taken up in the distal ileum (Borgström, 1974; Brindley, 1974). The bile salts may cross the walls of the jejunal enterocytes, but this is part of a bidirectional transport that results in no net upake (Feldman, 1968; Miskin and Kessler, 1970).

The absorption of lipids into the enterocytes depends upon the continued establishment of an inward diffusion gradient. On entering the cells it is likely that the fatty acids become bound to intracellular proteins. A low molecular weight (M.W. 12 000) binding protein has been characterized from the intestinal mucosa (Ockner, Pittman and Yager, 1972; Ockner *et al.*, 1972). It binds unsaturated acids in preference to saturated acids, and long-chain acids are preferred to short- and medium-chain acids. This property may be part of the mechanism which enables oleate to be absorbed more rapidly than stearate. The other way in which the inward diffusion gradient is maintained is by the re-esterification of the absorbed lipids. This is the energy-dependent part of absorption.

The difference between the energy-dependent and independent phases of fat absorption is demonstrated when pieces of the intestine are incubated with micellar solutions containing monoacylglycerols and fatty acids. The intestinal cells are able to absorb the lipid at 0°C, but triacylglycerol is not formed (Johnston and Borgström, 1964). By using radioactive substrates and electron microscopy, it can be demonstrated that the fatty acids that are absorbed at 0°C are distributed diffusely through the cytoplasm of the enterocytes. Subsequent incubation at 37°C leads to the esterification of the acids to form triacylglycerols which can also be detected as fat droplets (Strauss, 1966).

Resynthesis of absorbed lipids

The metabolism of fatty acids first involves their activation to acyl-CoA esters. The process requires the expenditure of two high-energy bonds from ATP and it is catalysed by acyl-CoA synthetase:

$$\text{Fatty acid} + \text{ATP} \xrightarrow{\text{Mg}^{2+}} \text{Fatty acyl-AMP} + P_1P_1$$
$$\text{Fatty acyl-AMP} + \text{CoA} \longrightarrow \text{Fatty acyl-CoA} + \text{AMP}$$

Although the enterocytes contain a number of enzymes which can activate fatty acids of different chain lengths, the major activity is towards long-chain fatty acids (Brindley, 1974). Once in the form of acyl-CoA esters, these acids can then be used to esterify various acyl-acceptors.

In non-ruminant animals the major acceptor is the 2-monoacylglycerol that is absorbed from the lumen after the partial hydrolysis of triacylglycerol (*Figure 4.3*). This pathway was first described in the small intestine of rabbits (Clark and Hübscher, 1960, 1961), but it has since been reported in the intestines of many other mammals (see Brindley, 1974, 1977; Johnston, 1978; Kuksis, Shaikh and Hoffman, 1979). The monoacylglycerol acyltransferase that initiates this synthesis is found predominantly in the smooth endoplasmic reticulum (Brindley, 1974), at the villus tips (Hoffman and Kuksis, 1982). The diacylglycerol that is formed is then rapidly converted to triacylglycerol.

It has been estimated that 75–85% of dietary triacylglycerol could be metabolized via the monoacylglycerol pathway during its absorption. These experiments employed ether analogues at position 2 of the triacylglycerols, or the use of specifically labelled triacylglycerols (see Brindley, 1974). These results refer to non-ruminants (i.e. rat, hamster and man). As mentioned previously, the 2-monoacylglycerols are broken down by bacterial enzymes in ruminants.

In the latter animals the triacylglycerol that is synthesized is formed *de novo* using either glycerol 3-phosphate, or possibly dihydroxyacetone phosphate as the acylacceptor (*Figure 4.3*). These are derived mainly from glucose, although glycerol can also be phosphorylated in the small intestine. Little is known about the importance of dihydroxyacetone phosphate as a precursor for triacylglycerol synthesis in the small intestine except for the fact that its esterification can be demonstrated (Brindley, 1974; Johnston, 1978). The acyl-dihydroxyacetone phosphate is then reduced, using NADPH, to lysophosphatidate (*Figure 4.3*).

The formation of lysophosphatidate from glycerol phosphate is normally thought to be the major route of synthesis *de novo* in the small intestine. Further esterification of the lysophosphatidate yields phosphatidate which is converted to diacylglycerol by the loss of the phosphate group. This compound can then be converted to triacylglycerol by another esterification reaction (*Figure 4.3*; Brindley and Sturton, 1982). This route of synthesis provides most of the triacylglycerol that is produced in the ruminant intestine. It also produces the remainder of the triacylglycerol that is not synthesized from 2-monoacylglycerol in non-ruminants.

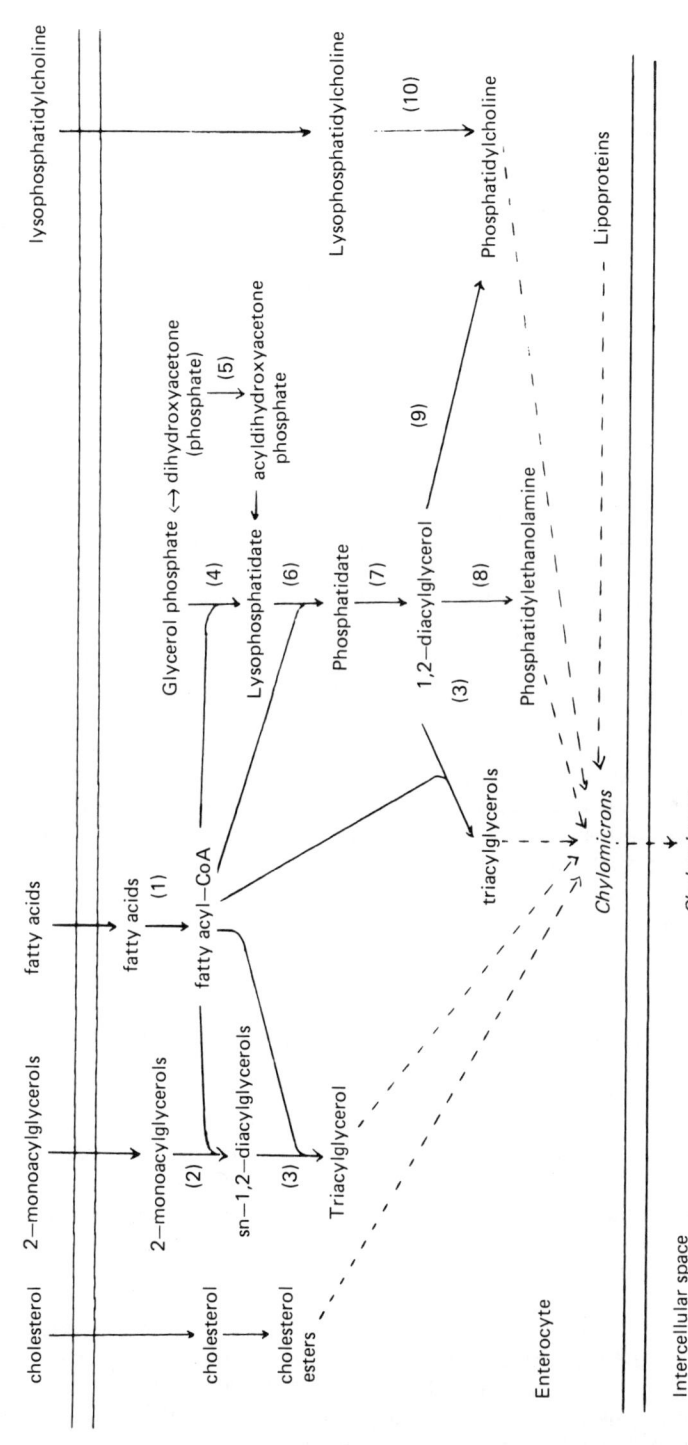

Figure 4.3 The metabolism of lipids in the enterocytes of the small intestine. Enzymes are indicated as follows: (1) acyl-CoA synthetase (EC 6.2.1.3); (2) monoacylglycerol acyltransferase (EC 2.3.1.22) monoglyceride acyltransferase; (3) diacylglycerol acyltransferase (EC 2.3.1.20) diglyceride acyltransferase; (4) glycerol phosphate acyltransferase (EC 2.3.1.15); (5) dihydroxyacetone phosphate acyltransferase (EC 2.3.1.42); (6) lysophosphatidate acyltransferase; (7) phosphatidate phosphohydrolase (EC 3.1.3.4); (8) ethanolamine phosphotransferase (EC 2.7.8.1); (9) choline phosphotransferase (EC 2.7.8.2); (10) lysophosphatidylcholine acyltransferase (EC 2.3.1.23). This figure is reproduced by the courtesy of Excerpta Medica from Brindley (1977)

In addition to this, the diacylglycerol that is formed as an intermediate in these reactions can be converted to phosphatidylcholine (*Figure 4.3*). This lipid is important in stabilizing the chylomicron which is the particle responsible for the transport of fat from intestine. However, the predominant location of choline phosphotransferase in the intestinal mucosa is in the crypts rather than in the villus tips (Mansbach, 1973). It is therefore thought that its major function is to generate phosphatidylcholine for membrane synthesis and cell proliferation. The alternative route for the synthesis of phosphatidylcholine is by the acylation of lysophosphatidylcholine (*Figure 4.3*). The activity of this acyltransferase is concentrated in the villus tips where most of the fat absorption takes place (Mansbach, 1973). It has been estimated that up to 30% of the dietary phosphatidylcholine is metabolized by this route and without total degradation (Scow, Stein and Stein, 1967; Nilsson, 1968). The other source of phosphatidylcholine is from the bile and it has been suggested that an enterohepatic circulation of bile phospholipids may take place (Boucrot, 1972).

Cholesterol is transported through the intestines much more slowly than triacylglycerols and phospholipids because it becomes diluted with the cholesterol of the enterocyte membranes. It is then slowly incorporated into chylomicrons, and its $T_{1/2}$ in the mucosa is about 12 h (Treadwell and Vahouny, 1968). Its absorption from the lumen is incomplete and about 50% of that portion which is absorbed is lost by desquamation. A large proportion of the cholesterol that crosses the enterocyte is re-esterified. It was thought that this occurred by a reversal of the cholesterol esterase reaction since a requirement for ATP and CoA could not be demonstrated (Treadwell and Vahouny, 1968). This type of reaction is not favoured energetically. A requirement for CoA and acyl-CoA has since been demonstrated and it appears that the enterocyte, like many other cells, does contain an acyl-CoA—cholesterol acyltransferase (Haugen and Norum, 1976; Suckling, Strange and Dietschy, 1983). Despite this it was claimed that cholesterol esterase does play an important part in cholesterol transport and absorption (Gallo *et al.*, 1983).

Formation of chylomicrons and very-low-density lipoproteins

The transport of the lipids that have been resynthesized in the enterocytes requires that they are packaged in a stable physical form in order to exist in an aqueous environment. This is achieved by coating the hydrophobic lipids (e.g. triacylglycerol, cholesterol esters) with a surface coat of more amphiphilic compounds. The latter consist of phospholipids (mainly phosphatidylcholine), cholesterol and various proteins referred to as apoproteins. The major apoproteins that are produced by the small intestine are apo-B and apo-A_I and apo-A_{IV}.

Most of the fat that is transported from the small intestines is packaged in particles called chylomicrons. These particles have diameters in the range 50–450 nm. On average these contain 85–95% of triacylglycerol, 4–9% of phospholipid (mainly phosphatidylcholine), 0.2–1.4% of free cholesterol, 0.2–0.7% of esterified cholesterol and about 0.6% of protein

(Zilversmit, 1978). As described later, this composition changes when the chylomicrons enter the circulation.

The size and composition of the chylomicrons that are secreted from the intestine depend upon the relative rates of lipid and apoprotein synthesis, and upon the composition of the dietary fat. Thus, larger chylomicrons are produced after the consumption of high fat loads, at the peak of absorption, and when apoprotein synthesis is inhibited. Feeding unsaturated rather than saturated fat also increases the average size of the chylomicrons (Zilversmit, 1978). The proportion of the surface material (i.e. cholesterol, phospholipid and protein) relative to the core lipids (triacylglycerol and cholesterol esters) consequently decreases as the size of the chylomicron increases.

The intestine can also secrete lipids in another class of lipoproteins, namely very-low-density lipoproteins (VLDL). These contain about 60% of triacylglycerol. During fasting they contain about 47% and 54% respectively of the triacylglycerol and cholesterol of fasting lymph (see Brindley, 1974; Shiau, 1981). In fasting human volunteers VLDL were seen within the cells of the villus tips (Tytgat, Rubin and Saunders, 1971). After a fatty meal the lipoproteins extracted from the cells changed from small particles into larger chylomicrons. It was therefore suggested that VLDL could expand and coalesce to form chylomicrons during fat absorption. It has been assumed that VLDL may be largely responsible for carrying endogenous fat from the intestine, whereas chylomicrons carry dietary fat. However, this has been questioned since it is claimed that VLDL in intestinal lymph may carry more exogenous than endogenous triacylglycerol (Shiau, 1981).

Transport of fat across the enterocytes and the synthesis of chylomicrons

Fat droplets are seen in the apical cytoplasm of the enterocyte within minutes of their exposure to micelles of mixed lipids (Strauss, 1966; Friedman and Nylund, 1980). The droplets are located within the cisternae of the smooth endoplasmic reticulum, often in bulbous expansions (Cardell, Badenhausen and Porter, 1967). This site coincides with the subcellular distribution of the enzymes involved in triacylglycerol synthesis which are almost exclusively found in the microsomal fraction (Brindley, 1974; Hülsmann and Kurpershoef-Davidov, 1976; Négrel and Ailhaud, 1978). Cytochemical techniques have shown that monoacylglycerol acyltransferase is located on the inner surface of the smooth endoplasmic reticulum, whereas glycerol phosphate acyltransferase is in the rough endoplasmic reticulum (Higgins and Barnett, 1971). Although the rough and smooth endoplasmic reticulum are continuous, lipid droplets are rarely seen within the cisternae of the rough endoplasmic reticulum, and this does not appear to be a major route of absorption (Cardell, Badenhausen and Porter, 1967). However, the rough endoplasmic reticulum is important in providing the phospholipid and the apoproteins that coat and stabilize the lipid droplets.

As absorption continues, the lipid droplets increase in size and number, and eventually the surrounding smooth endoplasmic reticulum pinches off

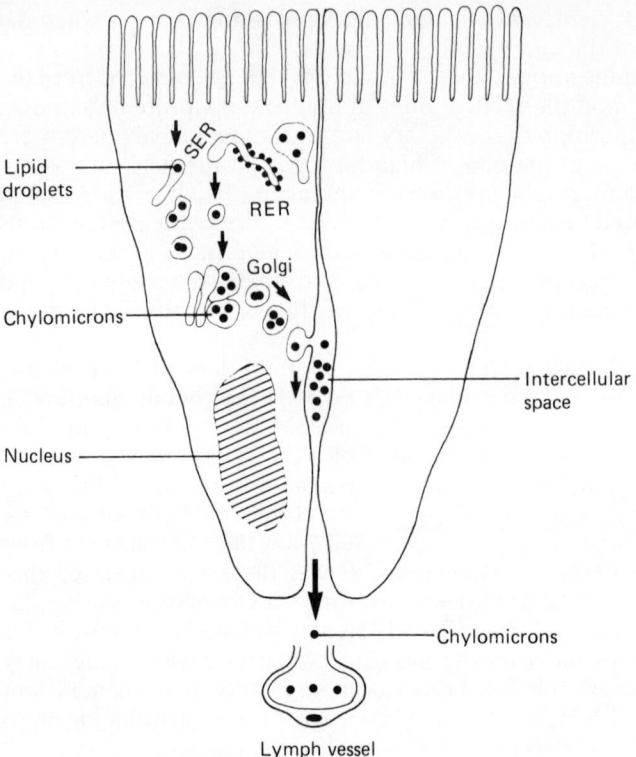

Figure 4.4 Schematic representation of the formation of chylomicrons and their passage through the enterocyte

to form lipid-containing vesicles (Friedman and Nylund, 1980). These vesicles are then believed to fuse with Golgi vacuoles and the lipid droplets are deposited within the Golgi apparatus (*Figure 4.4*). At this stage the droplets resemble chylomicrons. The Golgi apparatus is thought to provide two important functions in the transport of fat: it is the site where carbohydrate groups are added to surface proteins of the chylomicrons; the Golgi membranes also provide a vehicle for the transfer of the chylomicrons to the lateral surfaces of the enterocytes where membrane fusion takes place. The overall process is called exocytosis and the chylomicrons are secreted into the intercellular spaces (*Figure 4.4*).

The transport of fat from the smooth endoplasmic reticulum to the Golgi and the secretion from the Golgi apparatus involves the formation of microtubules. If this is disrupted by colchicine or vinblastine then the appearance of chylomicrons in the lymph is decreased and delayed. The triacylglycerol accumulates in the enterocytes in the Golgi and the endoplasmic reticulum (Glickman, Perrotto and Kirsch, 1976; Reaven and Reaven, 1977).

The importance of the apoproteins in the transport of fat across the enterocyte is best indicated by inherited disorders in which the synthesis of the different apoproteins is impaired. When the synthesis of β-lipoproteins is absent or low, the transport of lipid droplets from the endoplasmic reticulum to the Golgi is defective and triacylglycerols accumulate in the enterocyte (Dobbins, 1966). A deficiency in the production of the α-lipoproteins does not produce the impairment of fat absorption that was seen when the β-lipoprotein synthesis is deficient (Fredrickson, Levy and Lees, 1967). A malabsorption of fat can also be produced by a variety of protein synthesis inhibitors. Part of this undoubtedly results from the inhibition of apoprotein synthesis. However, the interpretation of the results is complicated by the fact that these inhibitors also decrease the synthesis of proteins and phospholipids required for the turnover of membranes that accompanies fat transport. They will also delay gastric emptying and decrease the flow of lymph (see Brindley, 1974; Friedman and Nylund, 1980).

Partitioning of fatty acids between portal blood and chyle

The description so far has considered the absorption of lipids and their incorporation into chylomicrons. These enter the lymphatics and pass via the thoracic duct to the jugular vein. Most of the long-chain fatty acids (chain length greater than C12) are transported in this way in their esterified form. However, acids with chain lengths from C10 to C12 may be transported either in chylomicrons, or as unesterified acids bound to albumin in the portal blood. Short-chain acids are transported almost exclusively by the portal route.

There are a number of reasons for this partitioning of the fatty acids on the basis of chain length. First, short-chain fatty acids are readily hydrolysed from triacylglycerols, which means that they are not likely to be retained in monoacylglycerols. Secondly, the fatty acids partition between the aqueous phase and the lipid emulsions and micelles that are present in the intestinal lumen. The short-chain acids diffuse into the aqueous phase and they are absorbed more rapidly than are the longer-chain acids. Finally, the enzymes that are responsible for the activation and re-esterification of the fatty acid act preferentially on the long-chain acids. This facilitates their incorporation into chylomicrons (see Brindley, 1974).

Despite the preferential partitioning of long-chain acids into the chyle, these acids are also detected in the portal blood. Their quantity can be increased in a number of conditions where the process of digestion and absorption may be slow or impaired, e.g. in neonatal animals, in animals with bile fistulae, in a-β-lipoproteinaemia etc. (see Dawson, 1967; Brindley, 1974). The clinical management of malabsorption syndromes is facilitated by feeding medium-chain triacylglycerols so that the fatty acids can be efficiently absorbed via the portal route rather than by chylomicron formation. This provides a high-energy diet and a supply of readily oxidizable substrate for the liver.

Metabolism of chylomicrons and very-low-density lipoproteins

The chylomicrons and VLDL that are produced in the small intestine enter the circulation in the jugular vein. The nascent chylomicrons then interact with other lipoproteins that are in the blood and they become modified chemically. Phospholipid is lost from the surface coat, and there is a net gain of cholesterol esters. The surface of the chylomicrons also acquires apo-C and apo-E (Zilversmit, 1978; Havel, 1982).

When the chylomicrons pass through the capillaries of muscle, heart, adipose tissue, mammary gland etc., they become trapped on the walls through interaction with the enzyme lipoprotein lipase (*Figure 4.5*). This in turn is anchored to the capillary wall through the chains of glycosaminoglycans. Apo-C_{II} lowers the K_m for the interaction between chylomicrons

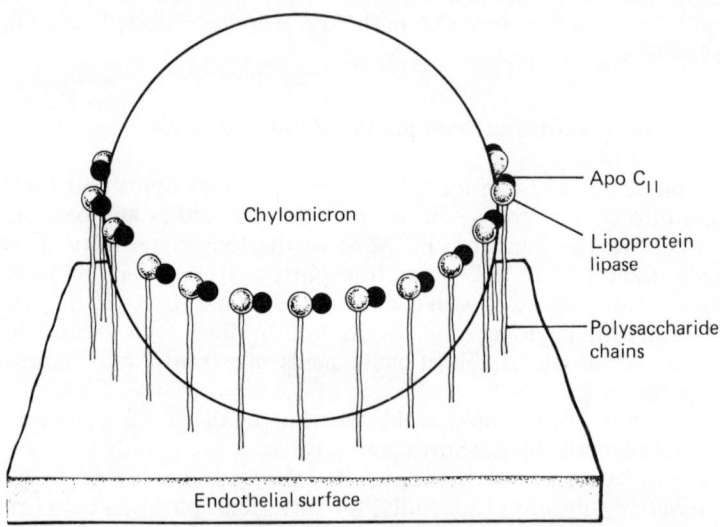

Figure 4.5 Schematic representation of the binding of a chylomicron to the endothelial surface of a capillary

and lipoprotein lipase (Schrecker and Greten, 1979), and it increases the rate of hydrolysis of the triacylglycerol in the core of the lipoprotein (Olivecrona and Bengtsson, 1979; Cryer, 1981). It is calculated that the rate of hydrolysis is nearly maximum when 21–22 molecules (dimers) of lipoprotein lipase interact per chylomicron. About 37 molecules of activator protein are estimated to be present on the chylomicron surface (Olivecrona and Bengtsson, 1979). Under these conditions *in vitro* about 70% of the triacylglycerol can be hydrolysed in about 3 min. This agrees well with the $T_{1/2}$ of about 2–3 min as estimated in rats for the chylomicron triacylglycerol (Harris and Felts, 1973). Work from this laboratory (Brindley, D.N. and Campbell, D.B., unpublished work) confirms this finding, but shows that the disappearance of [^{3}H]labelled triacylglycerol which has been infused in chylomicrons shows more than one decay rate (*Figure 4.6*). It can be resolved to give at least two $T_{1/2}$ values of 30 s–1 min and about 7 min respectively. This may reflect the removal rate catalysed by lipoprotein lipase, and the subsequent metabolism of the chylomicron remnant which will be discussed later.

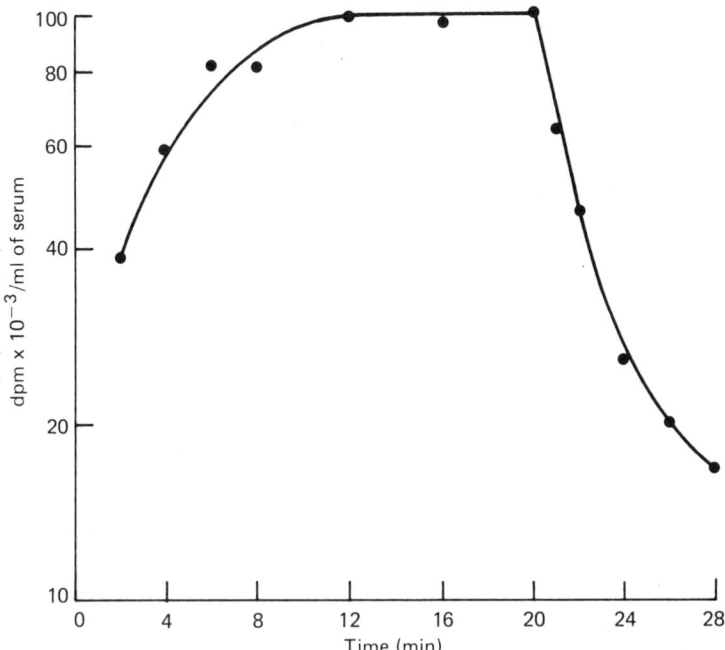

Figure 4.6 Kinetics of removal of chylomicrons containing triacylglycerols labelled with [^{3}H]-palmitate. A rat was fed [^{3}H]-palmitate in evaporated milk and the chyle was collected through a thoracic duct cannula. It was diluted to a concentration of 4.1 mg of lipid and 9.4×10^6 dpm per ml with 0.9% NaCl. Ninety per cent of the ^{3}H was present as triacylglycerol. This solution was infused through an indwelling jugular catheter into a conscious rat at a rate of 50 µl/min for 20 min. Blood samples (0.25 ml) were taken from a catheter in the carotid artery at the times indicated and the plasma was analysed for triacylglycerol. Blood loss was replaced with 0.9% NaCl

The fatty acids and monoacylglycerols that are produced by lipoprotein lipase are rapidly transferred to the cells that underlie the capillary bed. One mechanism that has been proposed is the lateral diffusion of the fatty acids in an interfacial continuum of cell membranes extending from the capillary to a site of metabolism in the recipient cell (Blanchette-Mackie and Scow, 1981). Monoacylglycerols are probably hydrolysed during this process. The fate of the fatty acids depends upon the site of uptake: in adipose tissue they are rapidly esterified to triacylglycerol for storage, whereas in muscle cells the fatty acids may be stored temporarily, or oxidized immediately.

The capacity to remove fatty acids from the chylomicron may not be able to keep pace with the rapid rates of hydrolysis. In this case lipoprotein lipase is subjected to feedback inhibition by the fatty acids through the abolition of the apo-C_{II} activation. The affinity of the lipase for the chylomicron is thus weakened, and it can be released (Olivecrona and Bengtsson, 1979). A chylomicron may undergo a series of attachment –hydrolysis–detachment cycles (Fielding, 1978; Cryer, 1981). At the same time the chylomicron will be losing surface material as the triacylglycerol core decreases in size. This loss results in the transfer of apo-C and apo-A to high-density lipoproteins (Havel, 1982).

These changes decrease the affinity of the particle to such an extent that it cannot compete effectively with undegraded chylomicrons and so the chylomicron remnant leaves the capillary bed. When it reaches the liver it binds to the hepatocytes through receptors that recognize the apo-E on its surface. The remnants are taken up by the cells by endocytosis and transferred to the area of the bile canaliculus. The lipid and protein components are then degraded by lysosomal catabolism (Havel, 1982).

The removal of VLDL from the circulation is similar to that of chylomicrons. The triacylglycerol is hydrolysed by lipoprotein lipase and the remnant particles are converted to low-density lipoproteins. These can then be removed from the circulation by the liver or extrahepatic tissue including macrophages. The latter uptake is by a low-affinity system. By contrast, fibroblasts have a high-affinity uptake system that recognizes apo-B and apo-E (Goldstein and Brown, 1982).

Hormonal and dietary control of the uptake of triacylglycerols by tissues

The rate of uptake of fatty acids is controlled by the activity of lipoprotein lipase, and this activity is under differential control in various tissues. Insulin plays a major part in stimulating lipoprotein lipase activity in adipose tissue, and glucocorticoids can further enhance this effect (Ashby *et al.*, 1979). By contrast catecholamines, corticotropin, glucagon, TSH, dibutyrylcyclic AMP, caffeine and theophylline decrease the activity (Cryer, 1981; Lawson *et al.*, 1981). These latter compounds stimulate hormone-sensitive lipase and thus the mobilization of fatty acids from the adipose tissue stores.

The administration of oestrogens *in vivo* decreases the activity of lipoprotein lipase in adipose tissue (Wilson *et al.*, 1976; Ramirez, 1981). It was proposed that this effect of oestrogens could divert triacylglycerol transport away from adipose tissue and towards muscle. Similarly, prolactin administration decreases lipoprotein lipase activity in adipose tissue and increases it in the mammary gland of lactating animals (Zinder *et al.*, 1974). This action is thought to be important in ensuring the uptake of circulating triacylglycerol and its use for milk production rather than for storage.

High insulin concentrations are not required in order to maintain the activity of lipoprotein lipase in skeletal and cardiac muscle. Glucocorticoids are thought to be the most important hormones in this respect (Friedman, Stein and Stein, 1978; Cryer, 1981). However, some insulin may be required in order to maintain optimum activities of lipoprotein lipase in muscles.

The balance between insulin, glucocorticoids and the other stress hormones is important in the reciprocal regulation of lipoprotein lipase activities in adipose tissue and muscle. In starvation, diabetes and in conditions of stress, the circulating triacylglycerols are diverted away from adipose tissue to cardiac and skeletal muscle. By contrast, after feeding high-carbohydrate diets the insulin concentrations are high, and chylomicrons and VLDL are mainly taken up by adipose tissue. High-fat diets can be associated with decreases in lipoprotein lipase activity in adipose tissue and increases in that in muscle (Weisenberg-Delorme and Harris, 1975; Childs et al., 1981; Lawson et al., 1981). The changes in the adipose tissue activity could be associated with the insulin-insensitivity that often accompanies the feeding of high-fat diets (see Lawson et al., 1981). However, it is difficult to generalize about the effects of dietary constituents because there are so many variables that can affect hormonal balance and metabolic activity.

Summary

The digestion of triacylglycerols, phospholipids and cholesterol esters converts them to compounds that are amphiphilic. These dissociate from the oil phase in the intestinal lumen with the aid of bile salts to form micelles. These interact with the microvillus membranes of the enterocytes. The lipids are absorbed by an energy-independent process. Their subsequent re-esterification represents an energy-dependent phase of absorption which establishes an inward diffusion gradient. Most short- and medium-chain fatty acids are not esterified and they are transported away from the intestine via the portal blood. The triacylglycerols, cholesterol esters and phospholipids contain the majority of the 'long-chain' fatty acids. They are assembled into chylomicrons with the aid of apo-proteins that are also made in the enterocytes. The chylomicrons are secreted into the lymphatic system and they enter the circulation via the jugular vein. On their passage through the capillaries of adipose tissue, muscle, mammary gland etc. the chylomicrons become attached to lipoprotein lipase which is anchored to the endothelial wall. This enzyme hydrolyses the triacylglycerols and the fatty acids are taken up into the tissues. The chylomicron remnant that is formed after hydrolysis of the majority of the triacylglycerol detaches from the capillary wall and travels to the liver where it is taken up and metabolized.

References

ASHBY, P., PARKIN, S., WALKER, K., BENNETT, D.P. and ROBINSON, D.S. (1979). Hormonal control of adipose tissue lipoprotein lipase activity. In *Obesity—Cellular and Molecular Aspects* (G. Ailhaud, Ed.), pp. 149–159. Paris, Inserm

BARROWMAN, J.A. and DARNTON, S.J. (1970). The lipase of rat gastric mucosa. *Gastroenterology* **59**, 13–21

BLANCHETTE-MACKIE, E.J. and SCOW, R.O. (1981). Lipolysis and lamellar structures in white adipose tissue of young rats: lipid movement in membranes. *Journal of Ultrastructural Research* **77**, 295–318

BORGSTRÖM, B. (1974). Fat digestion and absorption. In *Biomembranes*, vol. 4B (D.H. Smyth, Ed.), pp. 555–620. London and New York, Plenum Press

BOUCROT, P. (1972). Is there an enterohepatic circulation of bile phospholipids? *Lipids* **7**, 282–288

BRINDLEY, D.N. (1974). The intracellular phase of fat absorption. In *Biomembranes*, vol. 4B (D.H. Smyth, Ed.), pp. 621–671. London and New York, Plenum Press

BRINDLEY, D.N. (1977). Absorption and transport of lipids in the small intestine. In *Intestinal Permeation* (M. Kramer and F. Lauterbach, Eds), pp. 350–362. Amsterdam, Excerpta Medica

BRINDLEY, D.N. and STURTON, R.G. (1982). Phosphatidate metabolism and its relation to triacylglycerol biosynthesis. In *Phospholipids* (J.N. Hawthorne and G.B. Ansell, Eds), pp. 179–213. Amsterdam, Elsevier Biomedical Press

CARDELL, R.R., BADENHAUSEN, S. and PORTER, K.R. (1967). Intestinal triglyceride absorption in the rat—an electron microscopical study. *Journal of Cell Biology* **34**, 123–156

CHILDS, M.T., TELLEFSON, J., KNOPP, R.H. and BOWDEN, D.A. (1981). Lipid metabolism in pregnancy. VIII. Effects of dietary fat versus carbohydrate on lipoprotein and hepatic lipids and tissue triglyceride lipase. *Metabolism: Clinical and Experimental* **30**, 27–35

CLARK, B. and HÜBSCHER, G. (1960). Biosynthesis of glycerides in the mucosa of the small intestine. *Nature* **185**, 35–37

CLARK, B. and HÜBSCHER, G. (1961). Biosynthesis of glycerides in subcellular fractions of intestinal mucosa. *Biochimica et biophysica acta* **46**, 479–494

CLARK, S.B., BRAUSE, B. and HOLT, P.R. (1969). Lipolysis and absorption of fat in the rat stomach. *Gastroenterology* **56**, 214–222

COHEN, M., MORGAN, R.G.H. and HOFMANN, A.F. (1971). Lipolytic activity of human gastric and duodenal juice against medium and long chain triglycerides. *Gastroenterology* **60**, 1–15

CRYER, A. (1981). Tissue lipoprotein lipase activity and its action in lipoprotein metabolism. *International Journal of Biochemistry* **13**, 525–541

DAWSON, A.M. (1967). Absorption of fats. *British Medical Bulletin* **23**, 247–251

DIETSCHY, J.M. (1978). General principles governing movement of lipids across biological membranes. In *Disturbances of Lipid and Lipoprotein Metabolism* (S.M. Dietschy, A.M. Gotto and J.A. Ontko, Eds), pp. 1–28. Bethesda, American Physiological Society

DOBBINS, W.O. (1966). An ultrastructural study of the intestinal mucosa in congenital β-lipoprotein deficiency with particular emphasis upon the intestinal absorptive cell. *Gastroenterology* **50**, 195–210

ENGSTROM, J.F., RYBAK, J.J., DUBER, M. and GREENBERGER, N.J. (1968). Evidence for a lipase system in canine gastric juice. *American Journal of the Medical Sciences* **256**, 346–351

FEDLMAN, E.B. (1968). Factors modifying cholesterol uptake by intestinal rings. *Biochimica et biophysica acta* **150**, 727–729

FIELDING, C.J. (1978). Origin and properties of remnant lipoproteins. In *Disturbances in Lipid and Lipoprotein Metabolism* (J.M. Dietschy, A.M. Gotto and J.A. Ontko, Eds), pp. 83–98. Bethesda, American Physiological Society

FREDRICKSON, D.S., LEVY, R.I. and LEES, R.S. (1967). Fat transport in lipoproteins—an integrated approach to mechanisms and disorders. *New England Journal of Medicine* **276**, 94–103

FRIEDMAN, G., STEIN, O. and STEIN, Y. (1978). Lipoprotein lipase of cultured mesenchymal rat heart cells. III. Effect of glucocorticoids and insulin on enzyme formation. *Biochimica et biophysica acta* **531**, 222–232

FRIEDMAN, H.I. and NYLUND, B. (1980). Intestinal fat digestion, absorption and transport. A review. *American Journal of Clinical Nutrition* **33**, 1108–1139

GALLO, L.L., CLARK, S.B., MYERS, S. and VAHOUNEY, G.V. (1983). Cholesterol absorption in rat intestine: role of cholesterol esterase and acylcoenzyme A: cholesterol acyltransferase. *Federation Proceedings* **42**, 1256

GLICKMAN, R.M., PERROTTO, J.L. and KIRSCH, K. (1976). Intestinal lipoprotein formation: effect of colchicine. *Gastroenterology* **70**, 347–352

GOLDSTEIN, J.L. and BROWN, M.S. (1982). The LDL receptor defect in familial hypercholesterolaemia. *Medical Clinics of North America* **66**, 335–362

HARRIS, K.L. and FELTS, J.M. (1973). Kinetics of chylomicron triacylglycerol removal from plasma in rats. *Biochimica et biophysica acta* **316**, 288–295

HAUGEN, R. and NORUM, K.R. (1976). Coenzyme-A dependent esterification of cholesterol in rat intestinal mucosa. *Scandinavian Journal of Gastroenterology* **11**, 615–621

HAVEL, R.J. (1982). Approach to a patient with hyperlipidaemia. *Medical Clinics of North America* **66**, 319–333

HELANDER, H.F. and OLIVECRONA, T. (1970). Lipolysis and lipid absorption in the stomach of the suckling rat. *Gastroenterology* **59**, 22–35

HIGGINS, J.A. and BARNETT, R.J. (1971). Fine structural localization of acyltransferases. The monoglyceride and α-glycerophosphate pathways of intestinal absorptive cells. *Journal of Cell Biology* **50**, 102–120

HOFFMAN, A.G.D. and KUKSIS, A. (1982). Relative acylglycerol acyltransferase activities in homogenates of enzymically dispersed rat jejunal villus and crypt cells. *Biochimica et biophysica acta* **710**, 53–62

HOFMANN, A.F. (1978). Lipase, colipase, amphipathic dietary proteins and bile acids: new interactions on an old interface. *Gastroenterology* **75**, 530–532

HOFMANN, A.F. (1968). Functions of bile in the alimentary canal. In *The Alimentary Canal*, Chapter 117 (C.F. Code and W. Heidel, Eds), pp. 2507–2533. Baltimore, Williams and Wilkins Co.

HÜLSMANN, W.C. and KURPERSHOEF-DAVIDOV, R. (1976). Topographic distribution of enzymes involved in glycerolipid synthesis in rat small intestinal epithelium. *Biochimica et biophysica acta* **450**, 288–300

JOHNSTON, J.M. (1978). Esterification reactions in the intestinal mucosa and lipid absorption. In *Disturbances of Lipid and Lipoprotein Metabolism* (S.M. Dietschy, A.M. Gotto and J.A. Ontko, Eds), pp. 57–68. Bethesda, American Physiological Society

JOHNSTON, J.M. and BORGSTRÖM, B. (1964). The intestinal absorption and metabolism of micellar solutions of lipids. *Biochimica et biophysica acta* **84**, 412–423

KUKSIS, A., SHAIKH, N.A. and HOFFMAN, A.G.D. (1979). Lipid absorption and metabolism. *Environmental Health Perspectives* **33**, 45–55

LAWSON, N., POLLARD, A.D., JENNINGS, R.J., GURR, M.I. and BRINDLEY, D.N. (1981). The activities of lipoprotein lipase and of enzymes involved in triacylglycerol synthesis in rat adipose tissue. *Biochemical Journal* **200**, 285–294

MANSBACH, C.M. (1973). Complex lipid synthesis in hamster intestine. *Biochimica et biophysica acta* **296**, 386–402

MISKIN, S. and KESSLER, J.I. (1970). The uptake and release of bile salt and fatty acids by hamster jejunum. *Biochimica et biophysica acta* **202**, 222–224

NÉGREL, R. and AILHAUD, G. (1975). Localization of the monoglyceride pathway enzymes in the villus tips of intestinal cells and their absence from the brush-border. *FEBS Letters* **54**, 183–188

NILSSON, A. (1968). Intestinal absorption of lecithin and lysolecithin by lymph fistula rats. *Biochimica et biophysica acta* **152**, 379–390

OCKNER, R.K., PITTMAN, J.P. and YAGER, J.L. (1972). Differences in the intestinal absorption of saturated and unsaturated long chain fatty acids. *Gastroenterology* **62**, 981–992

OCKNER, R.K., MANNING, I.M., POPPENHAUSEN, R.B. and HO, W.K.L. (1972). A binding protein for fatty acid in cytosol of intestinal mucosa, liver, myocardium and other tissues. *Science* **177**, 56–58

OLIVECRONA, T. and BENGTSSON, G. (1979). Molecular basis for the interaction of lipoprotein lipase with triglyceride rich lipoproteins at the capillary endothelium. In *Obesity—Cellular and Molecular Aspects* (G. Ailhaud, Ed.), pp. 125–135. Paris, Inserm

OLIVECRONA, T., HERNOLL, O., EGELRUD, T., BILLSTROM, A., HELANDER, H., SAMUELSON, G. and FREDRICKSON, B. (1973). Studies on the gastric lipolysis of milk lipids in suckling rats and in human infants. In *Dietary Lipids and Postnatal Development* (C. Galli, G. Jacini and A. Pecili, Eds), pp. 77–89. New York, Raven Press

PATTON, J.S., ALBERTSON, P-A., ERLANSON, C. and BORGSTRÖM, B. (1978). Binding of porcine pancreatic lipase and colipase in the absence of substrate studied by two-phase partition and affinity chromatography. *Journal of Biological Chemistry* **253**, 4195–4202

RAMIREZ, I. (1981). Estradiol-induced changes in lipoprotein lipase, eating, and body weights in rats. *American Journal of Physiology* **240**, E533–E538

REAVEN, E.P. and REAVEN, G.M. (1977). Distribution and content of microtubules in relation to the transport of lipid. *Journal of Cell Biology* **75**, 559–572

SCHONHEYDER, F. and VOLQVARTZ, K. (1946). The gastric lipase in man. *Acta physiologica scandinavica* **11**, 349–380

SCHRECKER, O. and GRETEN, H. (1979). Activation and inhibition of lipoprotein lipase. Studies with artificial lipoproteins. *Biochimica et biophysica acta* **572**, 244–256

SCOW, R.O., STEIN, Y. and STEIN, O. (1967). Incorporation of dietary lecithin and lipolecithin into lymph chylomicrons in the rat. *Journal of Biological Chemistry* **242**, 4919–4924

SHIAU, Y-F. (1981). Mechanism of intestinal fat absorption. *American Journal of Physiology* **240**, G1–G9

STRAUSS, E.W. (1966). Electron microscopic study of intestinal fat absorption in vitro from mixed micelles containing linoleic acid, monoolein and bile salt. *Journal of Lipid Research* **7**, 307–323

SUCKLING, K.E., STRANGE, E.F. and DIETSCHY, J.M. (1983). Dual modulation of hepatic and intestinal acyl-CoA cholesterol acyltransferase activity by (de-)phosphorylation and substrate supply in vitro. *FEBS Letters* **151**, 111–116

TREADWELL, C.R. and VAHOUNY, G.C. (1968). Cholesterol absorption. In *Handbook of Physiology, Section 6, Alimentary Canal, Vol. 3 Intestinal Absorption* (C.F. Code and W. Heidel, Eds), pp. 1407–1438. Baltimore, Williams and Wilkins Co.

TYTGAT, G.N., RUBIN, C.E. and SAUNDERS, D.R. (1971). Synthesis and transport of lipoprotein particles by intestinal absorptive cells in man. *Journal of Clinical Investigation* **50**, 2065–2078

VAN DEN BOSCH, H. (1982). Phospholipases. In *Phospholipids* (J.N. Hawthorne and G.B. Ansell, Eds), pp. 313–357. Amsterdam, Elsevier Biomedical Press

WEISENBERG-DELORME, C.L. and HARRIS, K.L. (1975). Effects of diet on lipoprotein lipase activity in the rat. *Journal of Nutrition* **105**, 447–451

WILSON, D.E., FLOWERS, C.M., CARLILE, S.I. and UDALL, K.S. (1976). Estrogen treatment and gonadal function in the regulation of lipoprotein lipase. *Atherosclerosis* **24**, 491–499

ZILVERSMIT, D.B. (1978). Assembly of chylomicrons in the intestinal cell. In *Disturbances of Lipid and Lipoprotein Metabolism* (J.M. Dietschy, A.M. Gotto and J.A. Ontko, Eds), pp. 69–81. Bethesda, American Physiological Society

ZINDER, O., HAMOSH, M., FLICK, R.G. and SCOW, R.O. (1974). Effect of prolactin on lipoprotein lipase in mammary gland and adipose tissue of rats. *American Journal of Physiology* **226**, 744–748

5

THE DIGESTION, ABSORPTION AND TRANSPORT OF FATS—NON-RUMINANTS

C.P. FREEMAN
Unilever Research Laboratory, Colworth House, Sharnbrook, Bedfordshire, UK

Introduction

Much of our knowledge of the processes of digestion, absorption and transport of fat has been derived from studies in the rat and human. The broad evidence available suggests that the basic physicochemical mechanisms concerned in fat assimilation are similar in monogastric species. Although less extensively studied, particular reference in this chapter is made to these processes in the pig and the chick. As well as their economic importance in animal production, these species provide interesting comparative models for the physiological processes concerned in fat assimilation in mammalian and avian species.

Digestion

Pre-duodenal digestion

Intragastric lipolysis has been demonstrated in a number of non-ruminant species, including the rat and the human (Clark, Brause and Holt, 1969; Hamosh *et al.*, 1975). Although extensive pre-duodenal breakdown of lipid has not been described in the pig or the chick, lipolytic activity has been demonstrated in the oral secretions of the pig, and is an activity which appears to be widespread in mammalia (see Nelson, Jensen and Pitas, 1977). It is frequently difficult to distinguish whether lipolytic activity observed in the stomach is attributable to lipolytic (or esterolytic) enzymes of oral origin ('lingual lipases') or to the presence of a true gastric lipase. Barrowman and Darnton (1970) provide convincing histological and biochemical evidence in the rat for the secretion of a lipase by the fundic region of the glandular stomach, stimulated by the presence of neutral lipid. It is likely that in most mammalian species both oral and gastric lipases operate in the stomach to achieve an initial modification of dietary lipid. The properties of these enzymes are distinguishable in a number of respects from pancreatic lipase. The oral lipase of the rat exhibits a marked specificity for the sn-3 position of the triglyceride molecule (Paltauf, Esfandi and Holasek, 1974) and may also exhibit a specificity for short- and medium-chain fatty acids; the rat gastric lipase described by Barrowman and Darnton (1970) exhibited an almost total specificity for medium- as opposed to long-chain triglycerides.

Pre-duodenal lipolytic activity develops rapidly in the neonatal rat (Liao, Hamosh and Hamosh, 1982) and the activity is important in the newborn when pancreatic lipase activity and bile secretion are limited. As a result of the combined specificities of pre-duodenal lipases the main products of intragastric lipolysis are diglycerides and free fatty acids, of which short- and medium-chain acids predominate; the latter may be particularly important in the neonate since they can be absorbed directly through the stomach wall (Aw and Grigor, 1980) to provide a rapidly available source of energy.

The activities of these enzymes become less significant with age as pancreatic lipase and bile salt secretion develop, although the hydrolysis of 3–12% of ingested triglyceride has been observed in the stomach of adult humans (Hamosh *et al.*, 1975). It may be inferred that intragastric lipolysis in the adult serves as a functional, but not obligatory, preparative activity prior to digestion in the small intestine.

In the suckling animal, the activities of pre-duodenal lipase(s), pancreatic lipase, and lipases, including lipoprotein lipase, present in milk, appear to be adapted to the assimilation of the milk fat globule. Jensen *et al.* (1982) have described in detail the activities and physiological significance of the 'triad' of human lingual, breast milk, and pancreatic lipase in the digestion of fat in the human infant. Human breast milk contains a non-specific, bile-salt-activated lipase (BSAL) which does not appear to be found in the milks of lower mammals (Hernel, Blackberg and Olivecrona, 1981). The ability of BSAL to hydrolyse retinol esters is of particular significance to human milk. The existence of a similar dedicated enzyme system in the pig may be a contributory, but perhaps overlooked factor in the generally disparate performance observed between sow-suckled and artificially reared piglets.

Digestion in the small intestine

Large fat particles entering the duodenum from the gizzard or stomach (depending on species), and which may be partly modified by intragastric activity, are subdivided by the emulsifying action of the conjugated bile salts secreted in bile. In this finely divided state, the fat particles present an appropriate surface to the action of pancreatic lipase, which acts at the oil–water interface of the water-insoluble substrate. The rate of lipolysis is a direct function of the surface area of substrate presented to the enzyme. Although bile salts are essential for the emulsification of lipid, pancreatic lipase itself has no specific requirement for bile salts.

Colipase (a polypeptide cofactor) is required for the attachment and function of lipase at the substrate–water interface. A close kinetic relationship is known to exist between colipase, bile salts and lipase activity, and a number of mechanisms have been proposed for the action of colipase

(Borgström, 1975; Charles et al., 1975; Vandermeers et al., 1976). It is thought that the charge characteristics of colipase enable it to penetrate the adsorbed layer of bile salts that surround the substrate surface under physiological conditions and which effectively prevent the attachment of lipase. Having penetrated this shield, colipase acts as an 'anchor' to bind lipase into the surface in a favourable configuration. At lower bile salt concentrations, colipase and bile salts are competitive inhibitors for binding sites on the substrate: at very high bile salt concentrations colipase is displaced from the surface by bile salts, with a resultant inhibition of pancreatic lipase.

The colipase/lipase interaction has important physiological implications. Colipase and bile salts stabilize pancreatic lipase against inactivation, for example, by proteolytic enzymes at the water–substrate interface (Borgström, 1982). Lipase activity is no longer inhibited in the presence of bile salts above their critical micellar concentration (CMC), and the optimum pH for activity of the complex is shifted from pH 9 to 6, i.e. much closer to the ambient pH of the duodenal lumen.

The specificity of porcine pancreatic lipase for the terminal fatty acid groups of triglycerides has been well established (Schonheyder and Volquartz, 1954; Savary and Desnuelle, 1956) and the properties of the chick enzyme appear to be similar (Laws and Moore, 1963).

Other enzymes secreted by the pancreas and which are of importance in the duodenal digestion of lipids are carboxylic ester hydrolase (cholesterol esterase) and phospholipase. Carboxylic ester hydrolase has a wide optimum pH, requires bile salts for its activation, and exhibits a low specificity for either the sterol moiety or the esterified acyl groups (Erlanson, 1975). The hydrolysis of cholesterol esters appears to be an obligatory step for cholesterol absorption (Shiratori and Goodman, 1965). The enzyme also catalyses the re-synthesis of cholesterol and free fatty acids. Because of competition for intra-micelle space, the accumulation of cholesteryl esters will impair cholesterol absorption. Under normal physiological conditions, bile salt concentrations in the intestinal lumen inhibit esterification and promote hydrolysis, as well as protecting the enzyme from proteolysis (Vahouny, Weersing and Treadwell, 1964). Two major phospholipases are secreted by the pancreas—phospholipase A_1, which is specific for the sn-1 position of the glycerol moiety and phospholipase A_2 which hydrolyses the ester bond at position 2. Both phospholipases have an obligatory requirement for calcium ion as a cofactor, and are effective on solubilized substrates. Bile salts are therefore an important component of this enzyme system.

As a result of the combined activities of the pancreatic enzymes, the primary products of lipolysis are 2-monoglycerides (which isomerize slowly to the 1-isomer) and free fatty acids, with cholesterol (and other sterols), lysophospholipids, 1,2-diglycerides and glycerol as minor components.

Absorption

Intraluminal processes

Little is known of the mechanism of transfer of lipolytic products from the oil–water interface to the bulk micellar phase in the intestinal lumen, but it is thought that the bile acids play an important part both at the interface and in the transitional phase. This aspect has been considered in more detail in the ruminant animal (Lough, 1970), where the oil 'phase' rather than free is strongly adsorbed to the surface of fibrous food particles entering the duodenum from the abomasum; under these conditions and at a pH generally lower than in monogastric species, bile salts, in conjunction with lysolecithin and lecithin, are probably involved in a classic detergent, i.e. soil removal, role. The degree of agitation of the luminal contents, imparted by gut motile activity, is an important factor in the efficiency of the transitional process.

Under normal physiological conditions, bile salts are present in the intestinal lumen at a concentration in excess of their critical micellar concentration (CMC) and at a temperature above their critical micellar temperature (CMT), and spontaneously aggregate to form polymolecular aggregates, i.e. micelles. The essential and physiologically important feature of micellar solutions is their ability to dissolve ('solubilize') water-insoluble materials, by incorporating appropriately shaped and charged molecules either into the core or the outer sheath of bile salt molecules comprising the micellar matrix. In this way the products of lipolysis are solubilized in the bile salt micelle *in vivo*. Monoglycerides, short-chain ($\leq$C12) and unsaturated fatty acids and phospholipids which have low or negligible solubilities in water, behave in bile salt solutions as classic swelling amphiphiles (see Small, 1968), with finite solubilities. Their ability to expand the micelle and its hydrophobic interior has important physiological significance (Freeman, 1969). The behaviour of fatty acids in bile salt solutions is influenced by their degree of ionization: at the luminal pH of the monogastric duodenum (pH 6.0–6.4) fatty acids are only partly ionized; nevertheless, their solubility in bile salt solution is considerably greater than that in water and probably greater than that in solutions of typical ionic detergents for a given micellar concentration (Hofmann and Mekhjian, 1973). The solubility of fatty acids in mixed bile salt micelles increases markedly with increasing pH. Fatty acids, like monoglycerides, lower the CMC of bile salt solutions. Whether bile salts are conjugated with glycine as in the pig, or with taurine as exclusively in the chick, appears to have little effect on their solubilizing properties (Norman, 1960), although deoxy salts are more effective in this respect than the cholate derivatives (Hofmann and Borgström, 1962).

The precise structure of the mixed lipolytic-products/bile-salt micelle has not been elucidated. By analogy with the postulated structure of the bile-acid/lecithin micelle (Small, Penketh and Chapman, 1969), it probably consists of bimolecular leaflets of monoglyceride and amphiphilic fatty acids surrounded by a cylindrical sheath of bile salt molecules orientated with their hydrophilic regions facing the exterior and the hydrophobic 'backs' directed towards the interior of the micelle (*Figure 5.1*). The hydrophilic regions of the monoglyceride and fatty acid molecules occupy the face of the cylinder with their hydrophobic chains extending into the micelle interior. The interior of the micelle consists therefore of a predominantly hydrophobic 'core'. Estimates of the size of the simple bile salt micelle indicate a face radius of 1.5–2.0 nm; such estimates, however, may have limited value since the size of the bile salt micelle is known to

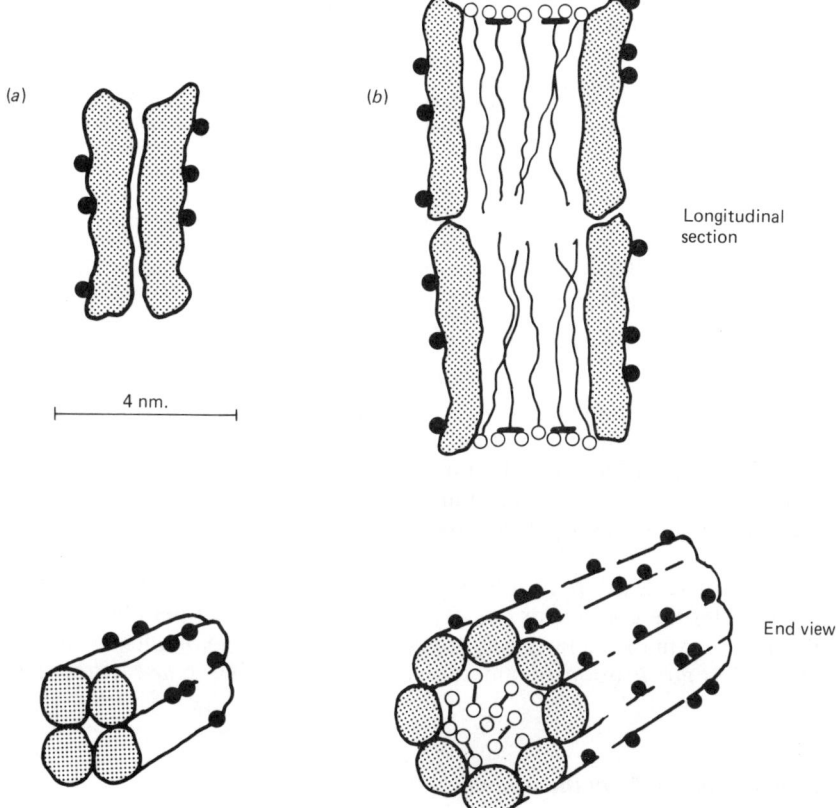

Figure 5.1 Diagrammatic representation of (a) simple bile-salt micelle and (b) lipolytic-product/bile-salt micelle (after Small et al., 1969)

vary with the counterion concentration, pH, temperature and the ratio of dihydroxy to trihydroxy bile salts (Small, 1968), and aggregation has been shown to occur. Moreover, inclusion of lipolytic products increases the dimensions of the mixed micelle. A state of rapid and continuous interchange is considered to exist between the molecular components of the micelle and the bile salt and lipolytic product molecules of the disperse phase surrounding it.

Long-chain saturated fatty acids with non-amphiphilic characteristics, and other non-polar lipids of physiological significance such as sterols and fat-soluble vitamins, are able to be solubilized in the hydrophobic interior of the mixed micelle. In the case of lipids such as cholesterol, bile salts are considered to be obligatory for their absorption since they can reach quantitatively significant concentrations in the aqueous phase only in micellar form. Although lecithin has been shown to increase the absolute solubility of cholesterol in micellar bile salt solution, its presence in the bile salt micelle creates a polar interfacial barrier which has to be crossed in order to gain access to the hydrophobic core, effectively reducing the dissolution rate of cholesterol (Higuchi et al., 1972).

Table 5.1 SUMMARY OF THE PHYSICOCHEMICAL PROPERTIES OF LIPIDS UNDER PHYSIOLOGICAL CONDITIONS

	Classification	Physicochemical characteristics
Triglycerides, diglycerides	Insoluble amphiphiles	V. low solubility in bile salt soln. Remain as oil droplets in disperse systems
Long-chain saturated fatty acids Cholesterol, and other sterols	Non-polar solutes	Low solubility in bile salt soln. High bile salt CMC. Entry into micellar phase controlled by soluble amphiphiles
Medium-chain fatty acids Long-chain unsaturated fatty acids Monoglycerides Phospholipids, lysophospholipids	Polar solutes or 'swelling' amphiphiles	Finite solubility in bile salt solns. Low bile salt CMC. Increase solubility and $K_{m/o}$ of non-polar solutes

Triglycerides and diglycerides, because of their low or negligible solubilities in bile acid solution, remain as oil droplets. In intestinal contents therefore there exists a bulk micellar phase consisting mainly of the products of lipolysis, in equilibrium with a particulate oil phase consisting primarily of intact and partial glycerides, together with other non-polar lipids.

Free fatty acids are partitioned between the two phases in a distribution described by the coefficient ($K_{m/o}$), that is dependent on pH (*Figure 5.2*), the chain length of the fatty acid, the composition of the oil phase, and the type and concentration of amphiphile in the micellar phase (Freeman, 1969).

Mode of uptake from the lumen

The major route of fat absorption in both the pig and the chick has been shown to be by micelles (Freeman *et al.*, 1968b; Freeman, 1976). A bile-independent mechanism of fatty acid uptake, however, has been demonstrated in the chick (Sklan and Hurwitz, 1980), mediated through the formation of a fatty-acid/carrier-protein complex. Vodovar, Flanzy and François (1966), from electron-microscopic studies of the pig intestine, postulated the feasibility of particulate fat absorption provided that the size of the particle was less than the inter-microvilli space, i.e. of the order of ≤40 nm; however, the probability distribution of such particles in the intestinal lumen is low. The uptake of particulate lipid by the pig small intestine has been shown to be small (< 0.1 mg/min) and a fraction of the uptake of micellar lipid (Freeman *et al.*, 1968b). The low but finite water-solubility of some fatty acids and monoglycerides affords the possibility of a limited uptake by simple diffusion from molecular solution, i.e. independent of bile salts.

The uptake of micellar lipid by the intestine is considered to be energy-independent, i.e. a passive absorption takes place. The precise mechanism by which the lipolytic products pass from the mixed micelle into the mucosal cell remains speculative. The observation in mammals that the major intestinal sites of fat and of bile salt absorption are not the same, has been taken as indirect evidence that the mixed bile-salt/

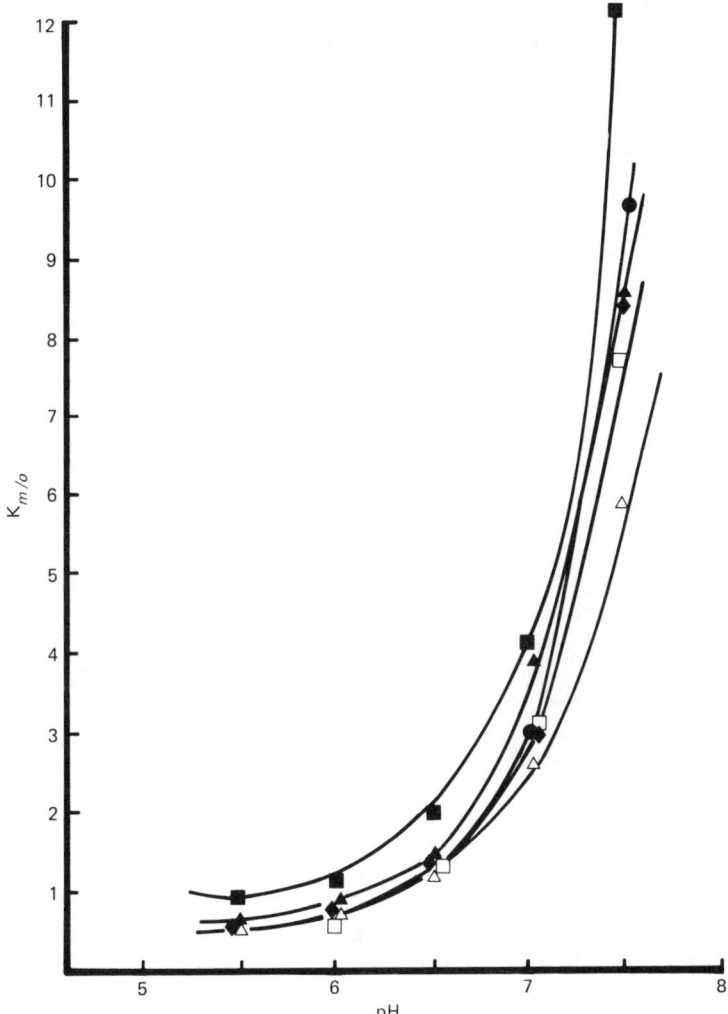

Figure 5.2 Effect of pH on the distribution coefficient, $K_{m/o}$, of different fatty acids. ■, lauric acid; ▲, myristic acid; △, palmitic acid; ◆, stearic acid; □, oleic acid; ●, linoleic acid. From Freeman, 1969, courtesy of Cambridge University Press

lipid micelle is not absorbed intact. Hofmann (1966) suggests that micellar 'disruption' need not be invoked, but that if the rates of exchange of fatty acid and monoglyceride molecules in micelles and those molecularly dispersed in the aqueous phase surrounding the micelle were rapid, then lipid absorption could occur by simple diffusion following random molecular collisions of lipolytic products with the cell membrane. In this sense, the mixed micelle serves to provide a high 'driving' concentration (Hofmann and Mekhjian, 1973) of lipid in the unstirred water layer which lies immediately adjacent to the luminal surface of the brush border. It has been suggested that the unstirred layer offers a marked resistance to the free movement of micelles, impeding their transfer to the mucosal cell surface (Wilson, Salee and Dietschy, 1971); at the same time, however, the layer functions as an effective 'sink' for lipid-rich micelles.

In common with other monogastric species, the major intestinal site of fat absorption in the pig and the chick is the jejunum. In the fowl, uptake of lipid has been shown to occur from the ileum (Hurwitz et al., 1973) and from the duodenum (Noyan et al., 1964; Hurwitz et al., 1973). The duodenum of the chick appears to be associated with an unusual degree of reflux or 'antiperistaltic' activity and from the above and our own observations it may be suggested that the duodenum has a relatively greater significance in the chick, both as a preparative and absorptive site.

The rate of uptake of micellar lipid by the small intestine of the pig has been shown to be directly proportional to its luminal concentration (Freeman et al., 1968b).

Bile salts, on the completion of their role of transporting micellar lipid to the mucosal surface, are, in the pig as in other mammals, almost quantitatively absorbed by an active process in the ileum. A passive absorptive mechanism exists in the jejunum for undissociated bile acids with appropriate solubilities in the lipoid membrane of the mucosal cell. The extent of this process is determined by the pH of the luminal contents and the pKa of the acid (which defines the proportion of molecules in the undissociated state) and the polarity of the bile salt molecule. Taurine-conjugated bile salts and trihydroxy acids conjugated with glycine, by virtue of their low solubility in lipid-like phases and low pKa, do not participate significantly in this process (Lack and Weiner, 1973). Unconjugated bile acids which have appropriate physicochemical properties can be extensively absorbed by this mechanism; however, because under normal physiological conditions bile salts in this region of the intestine are conjugated, little passive absorption of bile salts occurs in the mammalian jejunum. The demonstration of an active ileal transport site, and the absence of a similar process in the mammalian jejunum, offers an explanation for the efficient enterohepatic cycling of bile acids—a physiological process which allows a large circulating pool of multi-functional bile salts despite a relatively small synthesis rate. The site-selective absorptive processes result in a high bile salt concentration in the proximal small intestine and a relatively low concentration in the large intestine.

In contrast to mammalian species, the active absorption mechanism for bile salts appears to be more extensively located throughout the length of the small intestine in the fowl, and similar rates of bile salt absorption occur in both the jejunum and ileum (Hurwitz et al., 1973). Despite the relatively distal location of the bile duct in the chick intestine and substantial jejunal uptake, high bile salt concentrations are maintained in the proximal duodenum, presumably as a result of the considerable antiperistaltic activity in this region. The bile acid concentration in bile also appears to be comparatively high in the fowl (Hurwitz et al., 1973).

The active mucosal absorption of bile salts implies that an enterocyte pathway remote from that of the pasively diffused lipolytic products exists in the intestinal brush border.

The enterohepatic circulation of bile acids is completed by the transport of bile salts via the mesenteric portal system to the liver. In both the fowl and the pig, only a limited hepatic synthesis of bile acids is necessary to maintain the bile salt pool. The formation of the bile-salt/phospholipid complex, the main component of bile, appears to occur at a location close to the bile canaliculus, and the exocytitic secretion of the complex from hepatocytes is dependent on the availability of phospholipid, and on the presence of an associated apoprotein (Gonzalez, Sutherland and Simon, 1979). Release of bile from the gall bladder into the intestine is effected principally by the action of the polypeptide hormone cholecystokinin (CCK), the release of which in turn is mediated by sensors located near the intestinal mucosal surface sensitive to the presence of absorbed fat; gastrin (II) and large gastrointestinal polypeptide (GIP) appear to complement the action of CCK by altering gall-bladder and duct muscle sensitivity (Rehfeld, 1981).

Glycerol, the water-soluble lipolytic product, is absorbed from the gut by both passive and facilitated diffusion processes. Passive diffusion and a saturable, carrier-mediated process of glycerol uptake have been observed in the rabbit intestine (Rubin and Deren, 1974); the former process probably predominates in the proximal small intestine where glycerol concentrations are generally higher.

Mucosal triglyceride resynthesis

In the fowl and the pig, resynthesis of triglyceride from long-chain fatty acids occurs in the small-intestinal epithelium by both the monoglyceride and glycerol-3-phosphate pathways (Bickerstaffe and Annison, 1969a) following uptake. As in other animals, the triglyceride synthetase activity by either pathway is located in the microsomal fraction of the intestinal epithelium. Some specificity in the rate at which monoglycerides were incorporated into triglycerides was observed, mono-oleate proving a better acceptor of activated fatty acids than monopalmitate or monostearate.

Triglyceride resynthesis via the glycerol-3-phosphate pathway was shown to be markedly dependent on the presence of the magnesium cation. That the glycerol liberated in the small intestine during fat digestion may contribute to glycerol-3-phosphate synthesis is suggested by the presence of glycerokinase activity in the intestinal mucosa (Bickerstaffe and Annison, 1969b).

Monoglyceride hydrolase activity has also been detected in pig and chick intestinal mucosa (Bickerstaffe and Annison, 1969a), but significant hydrolysis occurs only if monoglyceride is present in excess of the amount of fatty acid required for its esterification. Moreover, this enzyme has been shown, in the rat, to have a preferential specificity for short- and medium-chain monoglycerides (Senior and Isselbacher, 1963) and this, together with the poor activation of medium-chain fatty acids, may account for the absence of substantial re-esterification of these acids in intestinal tissue.

Transport

In mammals, lipid secreted by the mucosal cells is almost exclusively transported from the intestine as chylomicrons via the lymphatic system. The cellular mechanisms of chylomicron assembly and secretion have been dealt with in the preceding chapter. Short-chain fatty acids are transported directly through the cell into the portal system as albumin-bound free fatty acids. Medium-chain (C8–C12) fatty acids are transported in lymph (as chylomicrons) or in portal blood depending on their chain length. By contrast, in the fowl the lymphatic system is poorly developed, the villus core being uniformly rather than peripherally occupied with a capillary network and containing no central lacteal (Kiyasu, 1955). This structural configuration was considered to favour the absorption of all nutrients via the mesenteric portal system, a view that was confirmed by Noyan et al. (1964) from radio-isotopic studies in the chicken. Examination of the plasma lipids of portal blood indicated that the principal form in which absorbed lipid is transported in the fowl is as triglycerides of the very-low-density lipoprotein (VLDL) fraction. The composition of this fraction has been described by Husbands (1971); these data indicate that chick VLDL are characterized not only by a relatively low content of triglyceride compared with mammalian chylomicrons (consistent with a generalized observation) but by a low content of triglyceride and a significantly higher content of protein than serum VLDL fractions of the pig and man. In the laying hen, where lipoprotein synthesis in the liver is very active, VLDL are present in serum at remarkably high concentrations (Husbands, 1971), and appear to be the major lipid precursor for egg-yolk synthesis.

Some factors affecting fat assimilation

Composition of dietary fat

In contrast to the ruminant, there is a very wide divergence in the digestibilities of different fats and fatty acids in monogastric animals. In the latter, the composition and to some extent the configuration of dietary fat has a significant effect on their assimilation, through both the digestive and absorptive processes.

In digestion, the positional specificity of pancreatic lipase for fatty acids at the terminal positions of triglycerides is well established. Some preference is also shown by this enzyme for short- and medium-chain fatty acids at the terminal positions, i.e. some intramolecular specificity is exhibited (Desnuelle and Savary, 1963). Intermolecular specificity also occurs, although it is not clear whether this is a true intermolecular specificity, i.e. preference for particular triglyceride species, or an indirect consequence of the surface-dependent nature of pancreatic lipase. Undoubtedly, however, the physical properties of the fat, as determined by the fatty acid composition and to some extent configuration of the constituent triglycerides, have an important influence on its rate of digestion.

Studies in the pig (Freeman et al., 1968b) have indicated that the capacity of the small intestine to absorb micellar lipid is substantial, and

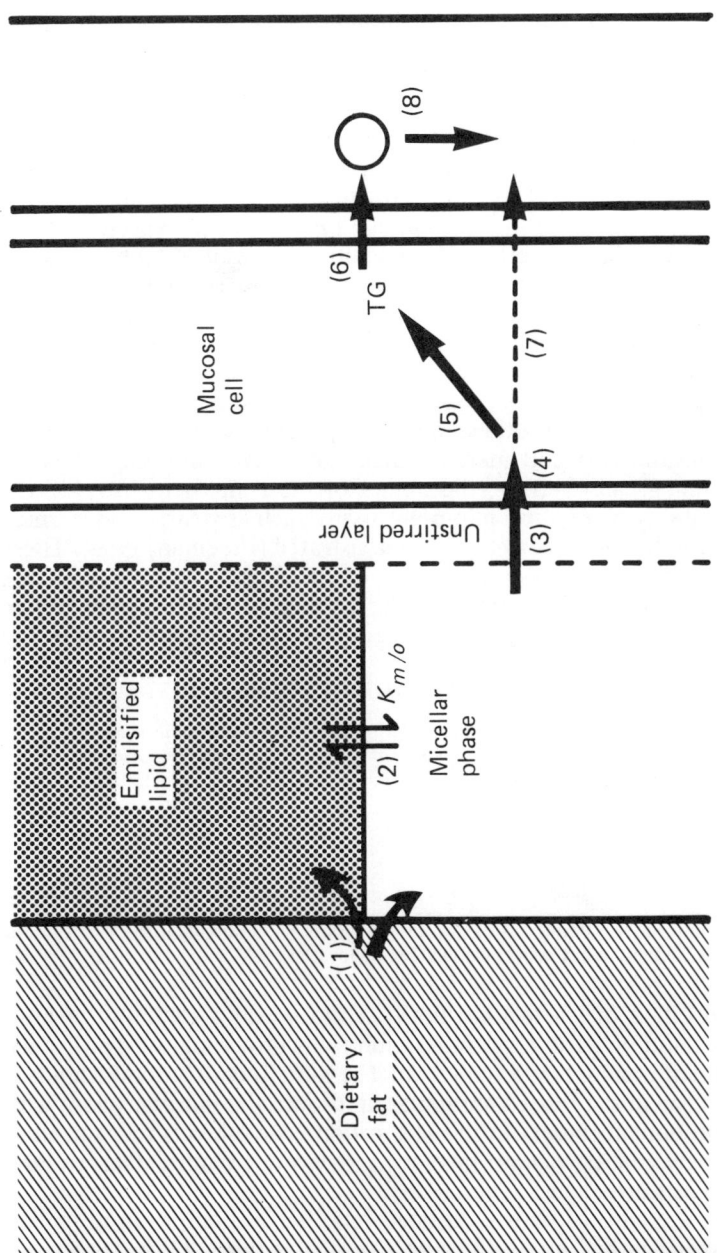

Figure 5.3 Schematic depiction of integrated processes concerned in the digestion, absorption and transport of fat. (1) Emulsification and lipolysis of fat; (2) entry into micellar phase; (3, 4) transfer of micellar lipid through unstirred layer and penetration of mucosal cell membrane; (5) triglyceride re-synthesis; (6) chylomicron (pig) or VLDL (chick) assembly; (7) direct portal transport of short-chain fatty acids; (8) transport of VLDL and chylomicrons in hepatic portal (chick) or lymphatic system respectively

significantly exceeds the estimated influx of lipid to the gut. Moreover, the uptake of fatty acids from mixed micellar solutions was shown to be non-specific. It is likely therefore that *in vivo* the entry of lipid into the micellar phase is both the rate-limiting and discriminatory process in the overall absorption of lipids (see *Figure 5.3*). Entry of lipid into the micellar phase is determined by (1) the solubility of the lipid in bile salt solution and (2) its partition between the particulate oil phase and the micellar phase, described by the coefficient, $K_{m/o}$. Relatively non-polar lipid solutes such as long-chain saturated fatty acids and cholesterol have very low solubilities in bile salt solutions. They are characterized by a high bile salt CMC and a low saturation ratio (Hofmann, 1963), which represent large impediments to their absorption. Polar solutes or swelling amphiphiles, on the other hand, by their ability to infiltrate the cylindrical bile salt micelle (*Figure 5.1*), expand the micelle and in particular greatly increase the capacity of its hydrophobic core. As a result, the solubility of non-polar solutes in the mixed bile salt micelle is enhanced. Monoglycerides are frequently quoted as the classic example of swelling amphiphiles: however, it has been shown that other lipids of physiological significance, in particular unsaturated and medium-chain fatty acids, are capable of a similar, if not greater, effect; the ability of these lipids to increase the micellar solubility of a typical non-polar solute such as stearic acid (termed the amphilic index) has been clearly demonstrated (Freeman, 1969). These same amphiphiles, with the exception of oleic acid, also increase the partition coefficient, $K_{m/o}$, of non-polar solutes in the direction of the micellar phase (*Table 5.2*). It is interesting in this respect to note the

Table 5.2 AMPHIPHILIC PROPERTIES OF SOME POLAR LIPID SOLUTES

Amphiphile	Amphiphilic index[a]	Increase or decrease in $K_{m/o}$ of stearic acid (%)
Oleic acid	0.138	−11.0
1-Mono-olein	0.138	+36.5
Linoleic acid	0.154	—
Lauric acid	0.164	—
Lysolecithin	0.280	+114.6

[a] The amphiphilic index is defined as the increase in stearic acid solubility in bile salt solution per unit increase in amphiphile concentration (Freeman, 1969)

pronounced effect of lysolecithin both on the micellar solubility of non-polar solutes and on their $K_{m/o}$, in relation to its physiological significance in the ruminant animal (see Chapter 6). The net effect of amphilic lipids is thus an increase in the flux of non-polar lipids into the micellar phase with a consequent enhancement of their absorption. The effect is recognized as synergism between particular fats or fatty acids (Young, Garrett and Griffith, 1963; Lewis and Payne, 1966).

Except in the case of poorly digested fats, the configuration of constituent triglycerides has little effect on overall fat digestibility (Freeman, Holme and Annison, 1968a), even though the digestibility of individual, long-chain saturated fatty acids may differ depending on their positional location on the triglyceride molecule. Similar conclusions may be drawn from the limited data available in the chick (Renner, 1960).

Age

In the very young chick (<1 week old), the ability to assimilate dietary fat is impaired by a deficiency in bile production. During this period of chick growth, fats which emulsify readily (e.g. corn oil) are digested to a greater extent than fats with a higher melting point, such as tallow (Carew *et al.*, 1972). Recent experiments of Kussaibati, Guillaume and Leclerq (1982) confirm this view.

Sow's milk contains about 40% of its dry matter as fat, which in turn supplies approximately 60% of the dietary energy requirement of the neonatal pig. The efficiency with which the piglet is able to convert the nutrients of sow's milk into weight gain, indicates that it is capable of digesting and utilizing large amounts of fat. On the transition to dry feeding or the feeding of artificial milk, the inclusion of fat in the diet has led to a poorer response than predicted, implying an impairment in fat digestion (see Kidder and Manners, 1978). Among the factors which may contribute to the disparity in fat assimilation between suckled and artificially reared piglets are the finely divided state of fat particles in sows' milk, the frequency of feeding, the possible existence of an enzyme system dedicated to the digestion of milk fat globules similar to that in the human (see earlier), and the lower luminal pH found in the small intestine of artificially reared piglets. By analogy with other species, pancreatic lipase activity may be limited in the piglet, although other (pre-gastric) lipases may be relatively more active.

Gut environment

The influence of the gut microflora on the absorption of nutrients by the fowl has been reviewed by Jayne-Williams and Fuller (1971) and by Coates (1976). However, little specific information is available on the effect of the intestinal flora on the absorption of fat. Young *et al.* (1963) found that fatty acids were absorbed better by chicks reared under sterile laboratory conditions or fed on diets containing antibiotics than by those in a contaminated environment. Boyd and Edwards (1967) found that the germ-free chick absorbed palmitic and stearic acids more efficiently than conventional chicks; at the same time the absorption of the more readily assimilated oleic and linoleic acids was unaffected by the gut environment.

Gut micro-organisms have been shown to be capable of modifying bile acids intraluminally, but the effect of bile salt deconjugation on the absorption of lipids is unquantified. Kenworthy (1967), in studies in the pig, has attributed the generally impaired absorption of nutrients in the presence of particular micro-organisms to changes in the morphology of the intestinal villus.

Feeding regimen

The feeding pattern of both the *ad libitum* and the restricted meal-fed pig gives rise to intermittent, large fluxes of digesta into the small intestine

(Freeman *et al.*, 1968b). Since it has been demonstrated, *in vivo*, that the rate of flow of micellar lipid through the active absorptive region of the small intestine inversely affects the efficiency of absorption (Freeman *et al.*, 1968b), some impairment of lipid absorption during such surges is probable. In comparison, the eating habit of the fowl, combined probably with the regulatory influence of the crop, produces a relatively more uniform flow of digesta into the small intestine (P.J. Strachan, unpublished observations) which, together with the degree of retrograde flow observed, create conditions favourable to a more efficient uptake of lipid in this species.

Other dietary factors

Of the non-lipid dietary factors affecting the absorption of dietary fat, the concentration of the divalent elements calcium and magnesium is perhaps the most significant. The effects of these cations on the digestibility of high-melting-point fats and saturated fatty acids in the rat has long been recognized (Cheng, Morehouse and Deuel, 1949). A similar interaction has also been observed in the chick (Whitehead, Dewar and Downie, 1971). In the laying hen, the digestibilities of several animal fats were depressed when the calcium content of the diet was increased from 2.87% to 4.34% of the diet (Hakansson, 1974). In all cases, the effect is most severe with long-chain saturated fatty acids, presumably because of the insolubility of their calcium or magnesium soaps. A mechanism for the interaction, as it affects intraluminal micellar solubilization, has been described by Freeman (1969).

Dietary fibre has also been shown to depress fat digestibility in the chick and the laying hen (Scheele, 1981).

References

AW, T.Y. and GRIGOR, M.R. (1980). Digestion and absorption of milk triacylglycerols in 14-day-old suckling rats. *Journal of Nutrition* **110**, 2133–2140

BARROWMAN, J.A. and DARNTON, S.J. (1970). The lipase of rat gastric mucosa. A histochemical demonstration of the enzymatic activity against a medium-chain triglyceride. *Gastroenterology* **59**, 13–21

BICKERSTAFFE, R. and ANNISON, E.F. (1969a). Triglyceride synthesis by the small-intestinal epithelium of the pig, sheep and chicken. *Biochemical Journal* **111**, 419–429

BICKERSTAFFE, R. and ANNISON, E.F. (1969b). Glycerokinase and desaturase activity in pig, chicken and sheep intestinal epithelium. *Comparative Biochemistry and Physiology* **31**, 47–54

BORGSTRÖM, B. (1975). On the interactions between pancreatic lipase and colipase and the substrate, and the importance of bile salts. *Journal of Lipid Research* **16**, 411–417

BORGSTRÖM, B. (1982). The temperature-dependent interfacial inactivation of porcine pancreatic lipase. *Biochimica et biophysica acta* **712**, 490–497

BOYD, F.M. and EDWARDS, H.M. (1967). Fat absorption by germ-free chicks. *Poultry Science* **46**, 1481–1483

CAREW, L.B., MACHEMER, R.H., SHARP, R.W. and FOSS, D.C. (1972). Fat absorption by the very young chick. *Poultry Science* **57**, 738–742

CHARLES, M., SARI, H., ENTRESSANGLES, B. and DESNUELLE, P. (1975). Interaction of pancreatic colipase with a bile salt micelle. *Biochemical and Biophysical Research Communications* **65**, 740–745

CHENG, A.L.S., MOREHOUSE, M.G. and DEUEL, J.J. (1949). The effect of the level of dietary calcium and magnesium on the digestibility of fatty acids, simple triglycerides and some natural and hydrogenated fats. *Journal of Nutrition* **37**, 237–250

CLARK, B.S., BRAUSE, B. and HOLT, P.R. (1969). Lipolysis and absorption of fat in the rat stomach. *Gastroenterology* **56**, 214–222

COATES, M.E. (1976). The influence of the gut microflora on digestion and absorption in the fowl. In *Digestion in the Fowl* (K.N. Boorman and B.M. Freeman, Eds), pp. 179–191. Edinburgh, British Poultry Science Ltd

DESNUELLE, P. and SAVARY, P. (1963). Specificities of lipases. *Journal of Lipid Research* **4**, 369–384

ERLANSON, C. (1975). Purification, properties, and substrate specificity of a carboxylesterase in pancreatic juice. *Scandinavian Journal of Gastroenterology* **10**, 401–408

FREEMAN, C.P. (1969). Properties of fatty acids in dispersions of emulsified lipid and bile salt and the significance of these properties in fat absorption in the pig and the sheep. *British Journal of Nutrition* **23**, 249–263

FREEMAN, C.P. (1976). Digestion and absorption of fat. In *Digestion in the Fowl* (K.N. Boorman and B.M. Freeman, Eds), pp. 117–142. Edinburgh, British Poultry Science Ltd

FREEMAN, C.P., HOLME, D.W. and ANNISON, E.F. (1968a). The determination of the true digestibilities of interesterified fats in young pigs. *British Journal of Nutrition* **22**, 651–660

FREEMAN, C.P., NOAKES, D.E., ANNISON, E.F. and HILL, K.J. (1968b). Quantitative aspects of intestinal fat absorption in young pigs. *British Journal of Nutrition* **22**, 739–749

GONZALEZ, M.C., SUTHERLAND, E. and SIMON, F.R. (1979). Regulation of hepatic transport of bile salts. *Journal of Clinical Investigation* **63**, 684–694

HAKANSSON, J. (1974). Factors affecting the digestibility of fats and fatty acids in chicks and hens. *Swedish Journal of Agricultural Research* **4**, 33–47

HAMOSH, M., KLAEVEMAN, H.L., WOLF, R.O. and SCOW, R.O. (1975). Pharyngeal lipase and digestion of dietary triglyceride in man. *Journal of Clinical Investigation* **55**, 908–913

HERNELL, O., BLACKBERG, L. and OLIVECRONA, T. (1981). Human milk lipases. In *Textbook of Gastroenterology and Nutrition in Infancy* (E. Lebenthal, Ed.), pp. 347–354. New York, Raven Press

HIGUCHI, W.I., PRAKONGPAN, S., SURPURIYA, V. and YOUNG, F. (1972). Cholesterol dissolution rate in micellar bile acid solutions: retarding effect of added lecithin. *Science* **178**, 633–634

HOFMANN, A.F. (1963). The function of bile salts in fat absorption. *Biochemical Journal* **89**, 57–68

HOFMANN, A.F. (1966). A physico-chemical approach to the intraluminal phase of fat absorption. *Gastroenterology* **50**, 56–64

HOFMANN, A.F. and BORGSTRÖM, B. (1962). Physico-chemical state of lipids in intestinal contents during their digestion and absorption. *Federation of American Societies for Experimental Biology* **21**, 43–50

HOFMANN, A.F. and MEKHJIAN, H.S. (1973). Bile acids and the intestinal absorption of fat and electrolytes in health and disease. In *The Bile Acids, Vol. 2. Physiology and Metabolism* (P.P. Nair and D. Kritchevsky, Eds), pp. 103–153. New York, Plenum Press

HURWITZ, S., BAR, A., KATZ, M., SKAN, D. and BUDOWSKI, P. (1973). Absorption and secretion of fatty acids and bile acids in the intestine of the laying fowl. *Journal of Nutrition* **103**, 543–547

HUSBANDS, D.R. (1971). Lipid and acetate metabolism. Cited by E.F. Annison in *Physiology and Biochemistry of the Domestic Fowl, Vol. 1*. (D.J. Bell and B.M. Freeman, Eds), p. 331. London, Academic Press

JAYNE-WILLIAMS, D.J. and FULLER, R. (1971). The influence of the intestinal microflora on nutrition. In *Physiology and Biochemistry of the Domestic Fowl, Vol. 1*. (D.J. Bell and B.M. Freeman, Eds), pp. 73–92. London, Academic Press

JENSEN, R.G., CLARK, R.M., deJONG, F.A., HAMOSH, M., LIAO, T.H. and MEHTA, N.R. (1982). The lipolytic triad: human lingual, breast milk, and pancreatic lipases. *Journal of Pediatric Gastroenterology and Nutrition* **1**, 243–255

KENWORTHY, R. (1967). The influence of bacteria on absorption from the small intestine. *Proceedings of the Nutrition Society* **26**, 19

KIDDER, D.E. and MANNERS, M.J. (1978). *Digestion in the Pig*, pp. 184–186. Bristol, Scientechnica

KIYASU, J.Y. (1955). *Fat Absorption in the Rat and Chicken*. PhD thesis, University of California, Berkeley

KUSSAIBATI, R.K., GUILLAUME, K. and LECLERCQ, B. (1982). The effects of age, dietary fat and bile salts and feeding rate on apparent and true metabolizable energy value in the chick. *British Poultry Science* **23**, 393–403

LACK, L. and WEINER, I.M. (1973). Bile salt transport systems. In *The Bile Acids, Vol. 2. Physiology and Metabolism* (P.D. Nair and D. Kritchevsky, Eds), pp. 33–54. New York, Plenum Press

LAWS, B.M. and MOORE, J.H. (1963). The lipase and esterase activities of the pancreas and small intestine of the chick. *Biochemical Journal* **87**, 632–638

LEWIS, D. and PAYNE, G.C. (1966). Fats and amino acids in broiler rations. 6. Synergistic relationships in fatty acid utilization. *British Poultry Science* **6**, 209–218

LIAO, T.H., HAMOSH, P. and HAMOSH, M. (1982). Development of preduodenal fat digestion in the rat. *Federation Proceedings. Federation of American Societies for Experimental Biology* **41**, 1003

LOUGH, A.K. (1970). Aspects of lipid digestion in the ruminant. In *Proceedings of the 3rd International Symposium on the Physiology of*

Digestion and Metabolism in the Ruminant (A.T. Phillipson, Ed.), pp. 519–522. Newcastle upon Tyne, Oriel Press
NELSON, J.H., JENSEN, R.G. and PITAS, R.E. (1977). Pregastric esterase and other oral lipases—a review. *Journal of Dairy Science* **60**, 327–362
NORMAN, A. (1960). The beginning solubilization of 20-methyl cholanthrene in aqueous solution of conjugated and unconjugated bile acid salts. *Acta chemica scandinavica* **14**, 1295–1299
NOYAN, A., LOSSOW, W.J., BROT, N. and CHAIKOFF, I.L. (1964). Pathways and form of absorption of palmitic acid in the chicken. *Journal of Lipid Research* **5**, 538–541
PALTAUF, F., ESFANDI, F. and HOLASEK, A. (1974). Stereospecificity of lipases. Enzymic hydrolysis of enantiomeric alkyl diacylglycerols by lipoprotein lipase, lingual lipase and pancreatic lipase. *FEBS Letters* **40**, 119–123
REHFELD, J.F. (1981). Four basic characteristics of the gastrin-cholecystokinin system. *American Journal of Physiology* **240**, G255
RENNER, R. (1960). *Site of Fat Absorption in the Chick*. PhD thesis, University of Cornell
RUBIN, A.W. and DEREN, J.J. (1974). Studies of glycerol transport across the rabbit brush border. *Gastroenterology* **66**, 378–383
SAVARY, P. and DESNUELLE, P. (1956). Sur quelques éléments de spécificité pendant l'hydrolyse enzymatique des triglycerides. *Biochimica et biophysica acta* **21**, 349–360
SCHEELE, C.W. (1981). Mengvoeder samenstelling en mengvoeder benutting. *Plumveehouderij* **36**, 20–21
SCHONHEYDER, F. and VOLQUARTZ, K. (1954). Studies on the lipolytic enzyme action. VI. Hydrolysis of trilauryl glycerol by pancreatic lipase. *Biochimica et biophysica acta* **15**, 288–290
SENIOR, J.R. and ISSELBACHER, K.J. (1963). Demonstration of an intestinal monoglyceride lipase: an enzyme with a possible role in the intracellular completion of fat digestion. *Journal of Clinical Investigation* **42**, 187–195
SHIRATORI, T. and GOODMAN, D.S. (1965). Complete hydrolysis of dietary cholesterol esters during intestinal absorption. *Biochimica et biophysica acta* **106**, 625–627
SKLAN, D. and HURWITZ, S. (1980). Intestinal uptake of fatty acids complexed to proteins in the chick intestine. *Journal of Nutrition* **110**, 270–274
SMALL, D.M. (1968). A classification of biological lipids based upon their interaction in aqueous systems. *Journal of the American Oil Chemists Society* **45**, 108–119
SMALL, D.M., PENKETT, S.A. and CHAPMAN, D. (1969). Studies on simple and mixed bile salt micelles by nuclear magnetic resonance spectroscopy. *Biochimica et biophysica acta* **176**, 178–189
VAHOUNY, G.V., WEERSING, S. and TREADWELL, C.R. (1964). Taurocholate protection of cholesterol esterase against proteolytic inactivation. *Biochemical and Biophysical Research Communications* **15**, 224–229
VANDERMEERS, A., VANDERMEERS-PIRET, M-C., RATHÉ, J. and CHRISTOPHE, J. (1976). Competitive inhibitory effect exerted by bile salt

micelles on the hydrolysis of tributyrin by pancreatic lipase. *Biochemical and Biophysical Research Communications* **69**, 790–797

VODOVAR, N., FLANZY, J. and FRANÇOIS, A.C. (1966). Pénétration et acheminement des graisses dans la cellule épithéliale absorbante de l'intestine du porc. *Comptes rendus hebdomadaires des Séances de l'Académie des Sciences, Paris* **262**, 812–815

WHITEHEAD, G.C., DEWAR, W.C. and DOWNIE, J.N. (1971). Effect of dietary fat on mineral retention in the chick. *British Poultry Science* **12**, 249–254

WILSON, F.A., SALEE, V.L. and DIETSCHY, J.M. (1971). Unstirred water layers in intestine: rate determinant of fatty acid absorption from micellar solutions. *Science* **174**, 1031–1033

YOUNG, R.J., GARRETT, R.L. and GRIFFITH, M. (1963). Factors affecting the absorbability of fatty acid mixtures high in saturated fatty acids. *Poultry Science* **42**, 1146–1154

6

DIGESTION, ABSORPTION AND TRANSPORT OF FATS IN RUMINANT ANIMALS

J.H. MOORE
Wye College, Ashford, Kent TN25 5AH, UK
and
W.W. CHRISTIE
Hannah Research Institute, Ayr KA6 5HL, Scotland, UK

Digestion of lipids in the rumen

Dietary lipids

The diet of the grazing ruminant animal usually consists of pasture grasses and legumes which may also be consumed in a preserved form (dried or ensiled) by animals housed indoors. In addition, ruminant animals are often given dietary supplements of cereals, oil cakes or meals, usually referred to as 'concentrates'. The lipid content of forage crops is low (5–10 g lipid per 100 g dry plant tissue), but as a grazing cow of 550 kg live weight consumes about 15 kg of forage dry matter per day, it can ingest between 750 and 1500 g lipid per day. The lipids of forage plants are normally concentrated in the leaf chloroplast which contains about 22 g lipid per 100 g dry tissue (Menke, 1966) and consist mainly of glycosyl-diacylglycerols and phospholipids. Linolenic acid (53%), linoleic acid (13%) and oleic acid (10%) together account for a high proportion of the total fatty acids of forage crops (Shorland, Weenink and Johns, 1955; Garton, 1959, 1960). Although some changes occur during drying (Czerkawski, 1967) and ensiling (Lough and Anderson, 1973), the lipids of preserved forages are broadly similar in composition to those of the fresh grass or clover. A lactating cow maintained indoors on a daily ration of 10 kg of hay and 8 kg of concentrates will ingest about 550 g lipid per day, but an appreciable portion of this will be in the form of triacylglycerols derived from components of the concentrate mixture such as maize or decorticated ground nut meal. Much larger intakes of dietary lipid may be achieved in ruminant animals given diets containing 'protected' lipid supplements, either to increase the energy intake (e.g. 'protected' tallow) or to produce polyunsaturated meat or milk (e.g. 'protected' safflower oil); under these dietary regimens the efficient absorption by lactating cows of up to 1.5 kg fatty acids per day has been reported (Storry, Brumby and Dunkley, 1980).

The composition of lipids in the rumen

In cows given a diet of hay, for example, the concentration of total lipid in the contents of the rumen remains relatively constant during the day and amounts to about 500 mg per 100 g fresh material; approximately 80% of this lipid is associated with food particles, 16% with protozoa and 4% with bacteria (Katz and Keeney, 1966; Keeney, 1970).

LIPIDS ASSOCIATED WITH FOOD PARTICLES

The acyl ester bonds of dietary lipids are rapidly hydrolysed in the rumen and the resulting unesterified acids are adsorbed on to particulate matter (Harfoot, Noble and Moore, 1973; Harfoot, 1981). As the hydrogenation of unesterified, unsaturated fatty acids in the rumen is also a rapid process, the lipid associated with food particles consists largely of unesterified, saturated fatty acids (Harfoot, 1981).

BACTERIAL LIPIDS

The lipid classes in rumen bacteria have been incompletely characterized. For instance, Viviani *et al.* (1968) reported that 30% of the total lipid extracted from mixed rumen bacteria was accounted for by unesterified fatty acids, 20% by phosphatidylethanolamine, 6.5% by phosphatidylserine and 0.4% by phosphatidylcholine; thus, 43% of the total lipid was unaccounted for. Sphingolipids (not determined by Viviani *et al.*, 1968) may constitute a sizeable proportion of this unaccounted lipid, for Kunsman (1973) has shown that sphingolipids comprised about 50% of the total lipids in *Bacteroides ruminicola*. Rumen bacteria have also been reported to contain glycolipids (Keeney, 1970). The composition of the total fatty acids in mixed rumen bacteria determined by Viviani *et al.* (1968) was myristic, 3.9%; pentadecanoic, 8.0%; palmitic, 31.0%; margaric, 1.6%; stearic, 15%; oleic, 6.0%; linoleic, 2.7% and branched-chain fatty acids, 15.8%.

PROTOZOAL LIPIDS

According to Harfoot (1981), the composition of the lipids extracted from mixed rumen protozoa obtained from sheep was as follows: unesterified fatty acids, 10.1%; monoacylglycerols, 1.4%; diacylglycerols, 1.0%; triacylglycerols, 1.0%; phospholipids, 85.5%; sterol esters, waxes, etc., 0.7%. The principal phospholipids present were phosphatidylcholine, phosphatidylethanolamine, diacylglycerol aminoethylphosphonate and ethanolamine plasmalogen, which accounted for 36.3, 18.7, 11.0 and 9.5% respectively of the total (Dawson and Kemp, 1967). Harfoot (1981) also determined the composition of the total fatty acids present in mixed rumen protozoa and obtained the following values: pentadecanoic, 3.4%; palmitic, 43.1%; stearic, 9.3%; oleic, 18.4%; linoleic, 16.1% and branched-chain fatty acids, 4.9%.

Lipolysis

The acylester linkages in triacylglycerols, phospholipids, galactosylglycerides, sterol esters, and methyl and ethyl esters are rapidly hydrolysed in the rumen (Garton, Hobson and Lough, 1958; Dawson, 1959; Garton, Lough and Vioque, 1961; Dawson et al., 1974). The rumen bacterium, Anaerovibrio lipolytica, was shown to secrete a lipase that hydrolysed triacylglycerols containing medium- and long-chain fatty acids; the intermediary formation of di- and monoacylglycerols could not be demonstrated, which was consistent with the observation that the enzyme hydrolysed diacylglycerols more rapidly than it did triacylglycerols (Hobson and Mann, 1961; Hobson and Summers, 1966; Henderson, 1971; Henderson and Hodgkiss, 1973). On the other hand, the lipase of Anaerovibrio lipolytica did not hydrolyse the acyl ester bonds in galactosylglycerides (Henderson, 1968). Dawson and Hemington (1974) showed that the galactosylglycerides of plant chloroplasts were hydrolysed by mixed rumen micro-organisms to fatty acids, galactose and glycerol: the major pathway appeared to involve the initial cleavage of the acyl ester bonds by the action of a 'galactolipase', followed by the liberation of galactose from digalactosylglycerol or monogalactosylglycerol by the action of α-galactosidase or β-galactosidase respectively. Protozoal and α-galactosidases have been shown to hydrolyse digalactosylglycerides to galactose and diacylglycerol (Howard, 1963); the diacylglycerols so formed could then be hydrolysed by bacterial lipases such as that secreted by Anaerovibrio lipolytica. Some rumen bacteria, including Butyrivibrio fibrisolvens, are able to hydrolyse phosphatidylcholine, lysophosphatidylcholine and phosphatidylethanolamine to fatty acids and glycerolphosphorylcholine or glycerolphosphorylethanolamine (Hazlewood and Dawson, 1975). On the other hand, the work of Dawson and Kemp (1969) would seem to indicate that protozoa play little part in the hydrolysis of phospholipids in the rumen. There remains a possibility that lipases and phospholipases derived from dietary herbage might contribute to the hydrolysis of the acyl bonds in the rumen (see Harfoot, 1981).

Biohydrogenation

The ability of rumen micro-organisms to hydrogenate unsaturated fatty acids was first demonstrated by Reiser (1951) from experiments in which linseed oil was incubated with sheep rumen contents. It is now clear that the biohydrogenation of unsaturated fatty acids in the rumen occurs only after these acids have been liberated in unesterified form by the hydrolysis of acyl ester linkages (Hawke and Silcock, 1970; Dawson et al., 1974). Although the important role of bacteria in the hydrogenation of unsaturated fatty acids in the rumen has been established with certainty, the function of protozoa in the process seems likely to be of secondary importance (Viviani, 1970; Harfoot, 1981). Thus, Dawson and Kemp (1969) have shown that the biohydrogenating activity of rumen contents from which the protozoa had been removed was only slightly less than that of rumen contents with a normal complement of protozoa.

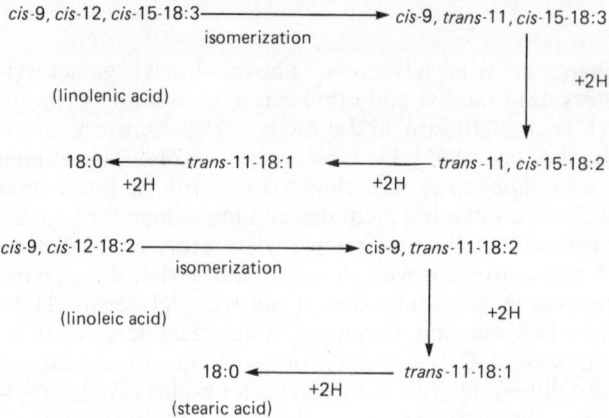

Figure 6.1 Biohydrogenation of linolenic and linoleic acids

The evidence that is available at present (reviewed by Harfoot (1981)) indicates that the major, but probably not the only, metabolic pathways involved in the biohydrogenation of linolenic and linoleic acids in the rumen are as shown in *Figure 6.1*.

According to Harfoot, Noble and Moore (1973, 1975), the site of these metabolic reactions in the rumen is the surfaces of particles of digesta on to which unesterified fatty acids become adsorbed immediately after liberation from acyl ester combination. The extracellular and intracellular enzymes that catalyse these metabolic sequences appear to be produced by a wide variety of rumen bacteria (Harfoot, 1981). In particular, the anaerobic bacterium, *Butyrivibrio fibrisolvens*, contains a number of enzymes of importance in the biohydrogenation of linoleic acid. With this organism, the required hydrogen atoms are supplied by water (Rosenfeld and Tove, 1971), while it has recently been shown that the electrons could be supplied by NADH, methyl viologen and endogenous electron donors such as α-tocopherolquinol or deoxy-α-tocopherolquinol (Hughes and Tove, 1980a, b). The *cis*-9,*trans*-11-octadecadienoate reductase has now been purified to near homogeneity and characterized (Hughes, Hunter and Tove, 1982). As a reductant, it utilizes α-tocopherolquinol, which is oxidized in the process to the corresponding quinone.

Although this pathway of linoleate biohydrogenation has been well characterized for *B. fibrisolvens*, it is possible that with other microorganisms at sub-optimal concentrations of hydrogen donor, the more stable *trans*- and conjugated isomers might be formed from isomers with a *cis*-,*cis*-methylene-interrupted double bond system by dissociation of enzyme–substrate complexes. In this respect, it is of interest that Harfoot, Noble and Moore (1975) found that although 75% of trilinolein was hydrolysed in the cell-free supernatant fraction of rumen contents and the resulting linoleic acid hydrogenated to stearic acid via *cis*-9,*trans*-11-18:2 and *trans*-11-18:1, about 25% of the acylglycerol was taken up by bacteria and appeared to undergo intracellular hydrolysis and hydrogenation. Under these conditions, when it might be supposed that concentration of hydrogen donor was not limiting, the hydrogenation of linoleic acid to stearic acid seemed to involve *cis*-monoenoic acids as intermediates.

Because linoleic acid is an essential fatty acid, it is important to know the extent to which this polyunsaturated acid is hydrogenated in the rumen. In one of the few quantitative studies that has been reported, Bickerstaffe, Noakes and Annison (1972) found that about 90% of the dietary linoleic acid was hydrogenated in the rumen of the goat. Approximate calculations show that the grazing cow ingests 60–140 g linoleic acid per day; if 90% of this linoleic acid is hydrogenated in the rumen, the amount that is presented for absorption from the small intestine will vary between 6 g and 14 g per day. A lactating cow may secrete 5–10 g linoleic acid per day in its milk (Moore, 1974); this leaves only 1–4 g of linoleic acid to account for the admittedly small turnover of this essential fatty acid in ruminant tissues (Annison *et al.*, 1967; Leat, Lindsay and Valerio, 1975; Lindsay and Leat, 1977) and to allow for the secretion of polyunsaturated fatty acids by the sebaceous glands (Noble, Crouchman and Moore, 1974; Noble *et al.*, 1975). These calculations, in spite of their approximate nature, do stress the need for more quantitative information on the extent of hydrogenation of polyunsaturated fatty acids in the rumen.

The possible relationship between the dietary intake of polyunsaturated fatty acids and cardiovascular disease in man led to the exploration of means whereby the polyunsaturated fatty acid content of ruminant milk and tissue lipids could be increased; it was reasoned that this could be achieved if lipolysis and thus hydrogenation of polyunsaturated fatty acids in the rumen could be inhibited. It was found that when a casein/vegetable-oil emulsion was treated with formaldehyde before being spray-dried, the product provided an encapsulated form of lipid that was protected against lipolysis and hydrogenation in the rumen but was well digested and absorbed from the lower gastrointestinal tract (reviewed by Christie, 1981b). More recently, analogous techniques have been used to produce protected non-polyunsaturated fat supplements for inclusion in the diets of ruminants, when high inputs of dietary energy are required (Storry *et al.*, 1980).

Fatty acid synthesis

There is evidence from a number of laboratories that rumen microorganisms can synthesize long-chain saturated and mono-unsaturated fatty acids *de novo*. For example, Patton, McCarthy and Griel (1968, 1970) showed that mixed rumen bacteria and protozoa could utilize the ^{14}C of [1-^{14}C] acetate or [U-^{14}C] glucose for the synthesis of a variety of fatty acids, but especially those containing 15, 16, 17 and 18 carbon atoms. From *Selenomonas ruminantium*, a strict anaerobe obtained from sheep rumen contents, Kanegasaki and Numa (1970) and Kanegasaki and Takahashi (1970) isolated a fatty acid synthetase that required malonyl CoA, NADH or NADPH for activity, and either butyryl CoA, valeryl CoA, hexanoyl CoA, heptanoyl CoA or octanoyl CoA as a 'primer'. Rumen bacteria were shown to incorporate the ^{14}C derived from [1-^{14}C] isobutyrate or [4-^{14}C]-DL-valine into branched long-chain fatty acids (Tweedie, Rumsby and

Hawke, 1966). Analogous findings have been reported from similar experiments with rumen protozoa (Coleman, 1969; Emmanuel, 1974). The fatty acids synthesized *de novo* by rumen micro-organisms were found to be incorporated into a range of neutral and polar lipid classes (Patton *et al.*, 1970; Tweedie, Rumsby and Hawke, 1966).

The apparent incorporation of ^{14}C from labelled acetate or glucose into linoleic acid by rumen micro-organisms cannot be accepted as evidence for the synthesis *de novo* of polyunsaturated fatty acids until the position of the incorporated ^{14}C along the carbon chain of the labelled linoleic acid has been determined (see Harfoot, 1981).

Uptake of fatty acids by micro-organisms

It has been shown that rumen bacteria (Hawke, 1971) and protozoa (Gutierrez *et al.*, 1962; Williams, Gutierrez and Davis, 1963; Emmanuel, 1974; Broad and Dawson, 1975) are capable of taking up long-chain fatty acids from their environment, and of incorporating these fatty acids into cellular complex lipids. Although this process could account for the appreciable concentrations of linolenic and linoleic acids that are found in the phospholipids of rumen protozoa (Emmanuel, 1974), it seems more likely that these fatty acids originate from intact chloroplasts ingested by the protozoa. Polyunsaturated fatty acids in plant chloroplasts would be in esterified form largely and would thus be protected from hydrogenation in the rumen.

Digestion and absorption of lipids in the small intestine

Amount and composition of lipids entering the duodenum

In contrast to the extensive absorption of short-chain acids that occurs in the rumen, virtually no long-chain fatty acids are absorbed from the digesta before it reaches the small intestine of the ruminant. Thus in an experiment with lactating goats, it was observed that less than 0.05% of an intraruminally administered dose of isotopically labelled linoleic, linolenic and oleic acids was adsorbed from the rumen, omasum and abomasum (Bickerstaffe, Noakes and Annison, 1972). Indeed, because of the synthesis of lipids *de novo* by rumen micro-organisms, and since little or no degradation of fatty acids occurs in the rumen (Garton, Lough and Vioque, 1961; Garton, Moorehouse and Lough, 1961; Wood *et al.*, 1963), the amounts of lipid passing through the omasum and abomasum into the duodenum can be greater than those ingested. For instance, in sheep given a diet containing a high proportion of concentrates, the amount of fatty acids reaching the duodenum exceeded that ingested by as much as 104% (Sutton, Storry and Nicholson, 1970; Czerkawski *et al.*, 1975). On the other hand, the amount of fatty acids reaching the duodenum was virtually the same as that ingested in sheep given diets of dried grass or dried grass plus linseed oil fatty acids (Scott *et al.*, 1969).

Little or no change ocurs in the composition of the lipids in the digesta as it passes through the omasum and abomasum (Bath and Hill, 1967; Lennox, Lough and Garton, 1968; Leat and Harrison, 1969), but the admixture with the gastric secretions of the abomasum causes the disintegration of any intact bacterial and protozoal cells that are transported from the rumen (Smiles and Dobson, 1956; Hoogenraad and Hird, 1970). Thus the lipid entering the duodenum consists mainly of unesterified saturated fatty acids adsorbed on to particulate matter (Scott et al., 1969): a smaller but variable proportion of the lipid is composed of phospholipids and other complex lipids released from intact microbial cells that enter the abomasum from the rumen. Triacylglycerols will be present in the digesta entering the duodenum in ruminant animals given diets supplemented with protected oils or fats.

Bile and pancreatic secretions

In ruminant animals, the pancreatic duct joins the bile duct 5–10 cm from the point of entry of the common duct into the duodenum. The flow rate of the biliary secretion is much greater than that of the pancreatic secretion: in sheep, for example, flow rates of 1.45 and 0.33 ml/h/kg body weight have been reported for bile (Harrison and Hill, 1960; Harrison, 1962) and pancreatic juice (Harrison and Hill, 1962) respectively. When expressed per kg body weight, flow rates for bile and pancreatic juice in cattle (McCormick and Stewart, 1967) were similar to the corresponding values for sheep.

BILIARY LIPIDS

The concentration of total lipid in ruminant bile is of the order of 1400 mg per 100 ml (Yamamoto and Rouser, 1967; Lennox et al., 1968): phosphatidylcholine accounted for about 80% of this lipid whereas lysophosphatidylcholine, phosphatidylethanolamine, cholesterol and cholesterol esters amounted to about 6.3%, 2.7%, 4.7% and 2.0% respectively. The molar percentages of the principal fatty acids present in the phosphatidylcholine of sheep bile are as follows: palmitic, 36.0%; stearic, 9.8%; oleic, 27.9%; linoleic, 6.6%; linolenic, 4.9%; also present were cis-9, trans-11 octadecadienoic acid (4.7%) and cis-9, trans-11, cis-15 octadecatrienoic acid (1.4%) which are intermediates or by-products in the biohydrogenation of dietary polyunsaturated fatty acids in the rumen (Christie, 1973). The saturated fatty acids in sheep biliary phosphatidylcholine were esterified mainly to position sn-1 whereas the unsaturated fatty acids were esterified to position sn-2 (Christie, 1973). The composition and positional distributions of the fatty acids in the biliary phosphatidylcholine of cattle are similar to those reported for sheep (Yamamoto and Rouser, 1967; Lennox et al., 1968; Christie, 1973). The biliary phosphatidylcholine in ruminant animals normally contains much lower proportions of linoleic and arachidonic acid and much higher proportions of oleic than does the same lipid in non-ruminant animals (Balint et al., 1965; Christie, 1973). However,

Christie et al. (1975) showed that the linoleic acid content of biliary phosphatidylcholine obtained from sheep given a diet containing protected linoleic acid was similar to that of non-ruminants.

The concentration of bile acids in ruminant bile varies between 5000 mg and 8000 mg per 100 ml (Lennox et al., 1968; Scott and Lough, 1971). The composition (g per 100 g) of sheep bile salts reported by Peric-Golia and Socic (1968) is as follows: taurocholate, 54; glycocholate, 21; taurodeoxycholate, 11; glycodeoxycholate, 6; taurochenodeoxycholate, 4; glycochenodeoxycholate, 2; cholate, 2. Thus, in sheep bile, the ratio taurine conjugates:glycine conjugates is about 2.4:1 whereas in non-ruminant herbivores the corresponding ratio is about 1:3 (Sjovall, 1960).

PANCREATIC LIPASE

Contrary to earlier reports (Green, Hirs and Palade, 1963; Keller, Cohen and Neurath, 1958), the presence of lipase activity in ruminant pancreatic juice has now been firmly established (Heath and Morris, 1963; Julien et al., 1972; Arienti, Harrison and Leat, 1974). Although it is clear that pancreatic lipase activity in the ruminant is much less than that in the non-ruminant, the properties of the ruminant enzyme are very similar to those of the more widely studied lipase from non-ruminant animals. Thus, the molecular weights and amino-acid compositions of the pancreatic lipases obtained from sheep and cattle were similar to those recorded for that from pig pancreas (Canioni et al., 1975; Julien et al., 1972; Khan, Chandan and Shahari, 1976). Optimum activity of the ruminant enzyme was found between pH 7.5 and 7.8 with little activity below pH 5.0, but in the presence of bile salts, hydrolytic activity occurred under acid conditions below pH 5.0 (Taylor, 1962; Heath and Morris, 1963; Arienti et al., 1974). Treatment of sheep pancreatic juice for 10 minutes at pH 3.0 reduced lipolytic activity by 75%, and there was an almost complete loss of enzyme activity at pH 2.5 (Arienti et al., 1974). For maximum hydrolytic activity, bovine pancreatic lipase required the presence of the polypeptide cofactor, colipase (Julien et al., 1972). From comparative studies on the rates of hydrolysis of a wide variety of triacylglycerols, it would appear that the specificities of sheep pancreatic lipase, with respect to fatty acid chain-length and degree of unsaturation, are similar to those of the pig (Johnson et al., 1974; Frobish et al., 1971).

PANCREATIC PHOSPHOLIPASES

Sheep pancreatic juice contains several phospholipases: one of these, phospholipase A_1, was inactivated by treatment with heat or acid while another, phospholipase A_2, was not affected (Arienti, Leat and Harrison, 1975). Optimum activity for both was observed at pH 5.6. The phospholipase A_1 hydrolysed the acyl ester linkage in position 1 of phosphatidylcholine or phosphatidylethanolamine and thus released mainly saturated fatty acids, whereas phospholipase A_2 hydrolysed the acyl ester linkage in position 2 and released mainly unsaturated fatty acids. There is also evidence that sheep pancreatic juice will hydrolyse the remaining ester linkage in the resulting lysophospholipids, but it is not known whether this is because of the presence of a specific lysophospholipase. A phospholipase A_1, that is specific for phosphatidylinositol, has recently been described (Dawson et al., 1982).

PANCREATIC GLYCOLIPASE

Bajwa and Sastry (1974) showed that sheep pancreas contained an enzyme that catalysed the hydrolysis of the acyl ester bonds in plant galactosyl-diacylglycerols; they did not demonstrate unequivocally that the enzyme was distinct from pancreatic lipase, however. It is difficult to assess the contribution that this enzyme makes to lipid digestion for it seems unlikely that significant amounts of dietary galactosylglyceride ever reach the ruminant small intestine.

pH of intestinal contents

As a consequence of the comparatively low concentration and rate of secretion of HCO_3^- in ruminant pancreatic juice (Kay and Pfeffer, 1970; Caple and Heath, 1972), the degree of neutralization of acid digesta as it passes into the duodenum of the ruminant is much less than it is in the non-ruminant animal. This is clear from the following representative values obtained for the pH of the gastrointestinal contents of the sheep (Noble, 1981) and pig (Moore and Tyler, 1955): abomasum or stomach, sheep 2.0, pig 2.4; proximal duodenum, sheep 2.5, pig 6.1; distal duodenum, sheep 3.5, pig 6.8; proximal jejunum, sheep 3.6–4.2, pig 7.4; distal jejunum, sheep 4.7–7.6, pig 7.4; ileum, sheep 8.0, pig 7.5.

Metabolism of lipids in the lumen of the small intestine

As the 30–40 g of dietary fatty acids that enter the ovine duodenum per day (Harfoot et al., 1974) may be augmented with a daily influx of as much as 10–15 g of biliary lipids (Adams and Heath, 1963), the lipid composition of the digesta changes markedly as it passes the point of entry of the common bile/pancreatic duct. The principal change is an increased proportion of phosphatidylcholine, and this is maintained as the digesta passes through the remainder of the duodenum and the proximal jejunum. The activities of pancreatic phospholipases A_1 and A_2 would certainly be inhibited by the acidic conditions in the duodenum (pH 2–3.5) and the proximal jejunum (pH 3.6–4.2), and it is distinctly possible that the pancreatic phospholipase A_1 is irreversibly inactivated by the extremely low pH (2.5) in the proximal duodenum. Appreciable hydrolysis of phospholipids (mainly phosphatidylcholine) begins only when the digesta reaches the mid-jejunum (pH 4.7–6.0) and continues throughout the distal jejunum (pH 6–7.6): the digesta in these two sections of ovine small intestine thus contain elevated proportions of lysophosphatidylcholine (mainly the 1-acyl isomer) and unsaturated unesterified fatty acids (Leat and Harrison, 1969; Lennox *et al.*, 1968).

In spite of the changes in the composition of the lipid in the digesta as it passes along the small intestine, the major proportion remains associated with particulate matter. For instance, in sheep jejunal contents, 70% of the total phosphatidylcholine, 60% of the total lysophosphatidylcholine and 78% of the total unesterified fatty acids are adsorbed on to the surface of the particulate phase (Lennox *et al.*, 1968; Leat and Harrison, 1969). Before any absorption can occur, this adsorbed lipid of exogenous and endogenous origin must be transferred to the soluble micellar phase, and the function of biliary constituents in this transfer would appear to be particularly important in the ruminant animal. Whereas bile and pancreatic juice are both required for optimal absorption of fatty acids in sheep, and while this absorption is greatly reduced in the presence of bile and in the absence of pancreatic juice, it is virtually eliminated in the absence of bile and in the presence of pancreatic juice (Heath and Morris, 1963; Harrison and Leat, 1970, 1972). Experiments *in vitro* showed that when sheep bile was added to sheep duodenal contents, there was a transfer of unesterified fatty acids from particulate matter into micellar solution (Scott and Lough, 1971), an effect that increased with pH and was later shown to be due to the combined action of the bile salt and phospholipid components (Smith and Lough, 1976). Above pH 4.0, phosphatidylethanolamine and phosphatidylcholine were equally effective in enhancing the solubilization of fatty acids by sheep bile salts, but at pH 3.0 and below, phosphatidylethanolamine was more effective than phosphatidylcholine. It is not without significance that ruminant bile is characterized by an excess of taurine- over glycine-conjugated bile acids: even at pH 2.5, taurine-conjugated bile acids are soluble and partly ionized, whereas glycine-conjugated bile acids are insoluble at pH 4.5 (Hofmann and Small, 1967). Thus, in the acidic digesta in the proximal jejunum of sheep, about 60% of the bile salts were partitioned in the soluble micellar phase and 40% in the particulate phase (Lennox *et al.*, 1968; Harrison and Leat, 1972). The solubilization of fatty acids by bile lipids does not appear to be affected adversely by the hydrolysis of phospholipids that occurs in the mid-jejunum. In the presence of sheep bile acids, lysophosphatidylcholine (1-acyl or 2-acyl isomers) or an equimolar mixture of lysophosphatidylcholine (1-acyl isomer) and oleic acid achieved the same degree of micellar solubilization of palmitic acid as did an equivalent amount of phosphatidylcholine (Lough and Smith, 1976).

Of the total fatty acids adsorbed from the small intestine of sheep given normal diets, about 20% was absorbed from the upper jejunum, where the pH of the digesta varied between 3.6 and 4.2, and about 60% was adsorbed from the middle and lower jejunum where the pH of the digesta was between 4.7 and 7.6; fatty acid absorption was virtually complete by the time that the digesta reached the ileum (Lennox and Garton, 1968; Lennox *et al.*, 1968; Leat and Harrison, 1969; Hogan, 1973; Johnson *et al.*, 1974). In the upper jejunum, the fatty acids would be taken up by the mucosal cells from a mixed micellar solution consisting mainly of unesterified fatty acids (principally saturated acids of exogenous origin), taurine-conjugated bile acids and phosphatidylcholine; the small amounts of phosphatidylethanolamine that are available may make an important contribution to the fatty-acid-carrying capacity of the micelles under these acidic conditions. In the middle and lower jejunum the fatty acids would be taken up by the mucosal cells from mixed micellar solutions consisting of unesterified fatty acids (saturated acids of exogenous origin and unsaturated acids of endogenous origin), taurine-conjugated bile acids and decreasing proportions of phosphatidylcholine and increasing proportions of lysophosphatidylcholine. There is ample evidence to suggest that lysophosphatidylcholine is also absorbed from the middle and lower jejunum (Leat and Harrison, 1974); under normal conditions, digesta reaching the ileum were found to contain little or no phospholipid (Lennox et al., 1968), but in the absence of pancreatic juice, phosphatidylcholine accumulated in the ileum (Leat and Harrison, 1969).

When ruminant animals are given diets containing supplements of 'protected' fats or oils, large amounts of triacylglycerols enter the duodenum; under these unusual circumstances, the mechanisms of digestion of the dietary triacylglycerols and the solubilization of the liberated fatty acids will be closer to those observed in the non-ruminant small intestine where 2-monoacylglycerols have an important role in the micellar solubilization of fatty acids (Thomson and Dietschy, 1981).

Quantitative aspects of fat absorption

Ruminant animals absorb fats with a high degree of efficiency; digestion or absorption coefficients of between 80% and 90% have been reported for a variety of fats, oils and fatty acids (Balch et al., 1952; Lennox and Garton, 1968; Heath and Hill, 1969; Andrews and Lewis, 1970a, b; Outen, Beever and Osbourn, 1974). This high efficiency was maintained even when the dietary intake of fatty acids was greatly increased (Heath and Hill, 1969). In general, the ability of ruminant animals to absorb C16 and C18 fatty acids is greater than that observed for non-ruminant animals (Noble, 1981). It has been suggested that this difference between species might be due to the greater degree of dispersion of long-chain saturated fatty acids in ruminant intestinal contents, and also to the greater solubilization of saturated fatty acids by bile-salt/lysophosphatidylcholine micelles than by bile-salt/2-monoacylglycerol micelles (Lough and Smith, 1976; Freeman, 1969). Another important difference is that in the ruminant small intestine, dietary fatty acids are absorbed principally as unesterified fatty acids, whereas in non-ruminants they are absorbed both as unesterified and esterified fatty acids (i.e. as 2-monoacylglycerols). It is known, for example, that 2-monopalmitin is taken up into micellar solution and is absorbed from the small intestine of the rat far more readily than is unesterified palmitic acid (Renner and Hill, 1961). Therefore, in non-ruminant animals, the digestibility of palmitic acid depends to some extent on its positional distribution within the molecules of dietary triacylglycerols (Tomarelli et al., 1968).

Metabolism of lipids in the mucosal cells of the small intestine

The synthesis of glycerolipids and cholesterol

Quantitatively, the major lipids that enter the rumen enterocyte from the intestinal lumen consist mainly of unesterified fatty acids (much of which are saturated), and lysophospholipids, and only under unusual circumstances are 2-monoacylglycerols present. The mechanism of absorption of these lipids in ruminant animals is unknown, but in non-ruminant animals fatty acids and 2-monoacylglycerols enter the intestinal cells by passive diffusion through the microvillus membrane; metabolic activity in the cells is not obligatory (Johnson and Borgström, 1964). It is thought that the transport of fatty acids from the inner surface of the microvillus membrane to the smooth endoplasmic reticulum, the site of triacylglycerol synthesis, is mediated by a specific fatty acid-binding protein (Ockner et al., 1972; Ockner and Manning, 1974, 1976). However, there have been no reports as yet of the isolation of a fatty acid-binding protein from ruminant small-intestinal tissues.

Before utilization for acylation reactions, the absorbed fatty acids must be converted to their CoA derivatives, the formation of which is catalysed by acyl-CoA synthetase in the presence of CoA and ATP. This enzyme occurs in the microsomal fraction of the small-intestinal mucosa of all species examined (Brindley and Hubscher, 1966; Hubscher, 1970), but there have been no detailed studies of the enzyme in the ruminant. The microsomal fraction obtained from the small-intestinal mucosa of ruminant animals also contains a fatty acid desaturase that converts stearic acid into cis-9,10-octadecenoic acid (Bickerstaffe et al., 1972; Wahle, 1974); there is evidence that up to 10% of the stearic acid that enters the ruminant intestinal mucosa is desaturated to oleic acid before it appears in the lymph.

The synthesis of triacylglycerols in the ruminant enterocyte can occur by the α-glycerophosphate pathway or by the monoacylglycerol pathway, but under normal dietary conditions, the nature of the available substrates dictates that the principal operational pathway is that involving α-glycerophosphate, phosphatidic acid and diacylglycerols as intermediates (Bickerstaffe and Annison, 1969a; Cunningham and Leat, 1969; Skrdlant et al., 1973). It is perhaps not surprising, therefore, that glycerokinase activities in the ruminant intestinal mucosa are greater than those in the non-ruminant intestinal mucosa (Bickerstaffe and Annison, 1969b). However, it is known that the potential capacity of ruminant enterocytes to synthesize triacylglycerols by the monoacylglycerol pathways is considerable (Cunningham and Leat, 1969; Skrdlant et al., 1973) and this capacity is undoubtedly utilized when ruminant animals are given diets containing protected fat supplements. The synthesis of phospholipids in the mucosal cells of the small intestine in ruminant animals probably occurs mainly by the acylation of 1-lysophospholipids absorbed from the intestinal lumen; de novo synthetic pathways (e.g. those involving cholinephosphotransferase) appear to be utilized to only a small extent (Leat and Harrison, 1974). Since ruminant diets rarely contain more than trace amounts of cholesterol, this essential lipoprotein component is actively synthesized in the mucosal cells of the ruminant small intestine (Scott and Cook, 1975). The mechanism of cholesterol esterification in the ruminant enterocyte has not been reported, but microsomes from rat intestinal mucosal cells contain an enzyme (acyl CoA:cholesterol acyl transferase) that catalyses the synthesis of cholesterol esters from cholesterol and fatty acyl CoA derivatives (Haugen and Norum, 1976).

There is an appreciable selectivity in the incorporation of different fatty acids into the various lipid classes synthesized in the mucosal cells of the ruminant small intestine, and this selectivity is reflected in the fatty acid composition of the lymph lipids. For example, in the intestinal lymph of sheep given a normal diet, the triacylglycerols contained only 7% of polyunsaturated fatty acids (linoleic plus linolenic), compared with 27% in the phosphatidylcholine and 24.5% in the cholesterol esters; the greatest proportion of saturated fatty acids was observed in the triacylglycerols, and the greatest proportion of mono-unsaturated fatty acids (mainly oleic) in the cholesterol esters (Christie and Hunter, 1978). In this respect, it is of interest that the acyl CoA:cholesterol acyl transferase in rat enterocytes has a high specificity for oleoyl CoA (Haugen and Norum, 1976). When abnormally large amounts of polyunsaturated fatty acids are absorbed from the ruminant small intestine, these fatty acids may be incorporated into lymph triacylglycerols (Heath, Caple and Redding, 1970; Harrison, Leat and Forster, 1974).

The synthesis of lipoproteins and their secretion into lymph

The lipid components (triacylglycerols, phospholipids, cholesterol and cholesterol esters) and a number of apoproteins synthesized in the bovine enterocyte are there assembled into lipoprotein particles. Newly assembled lipoprotein particles accumulate in the Golgi apparatus in the supranuclear region of the enterocyte. Migration of the Golgi vesicles to the basolateral region of the cell is followed by discharge of the lipoprotein particles into the intercellular space by reverse pinocytosis involving fusion of the Golgi membrane with the basolateral cell membrane. The lipoproteins then pass into the lacteals, and from there into the blood system via the intestinal and thoracic lymph ducts (Sterzing, McGilliard and Allen, 1971).

Because of the intense interest in lipoprotein metabolism in the medical field, much of the knowledge in this area has been accumulated from investigations with man and conventional experimental animals such as the rat; far less is known about lipoprotein metabolism in ruminants. The two major lipoproteins that are synthesized in the mammalian enterocyte are chylomicrons (density less than 0.93 g/ml, diameter 75–1000 nm) and very-low-density lipoproteins (VLDL, density 0.93–1.006 g/ml, diameter 25–75 nm); these two classes of lipoproteins are concerned with the transport of triacylglycerols. In sheep lymph, there is a continuous spectrum in particle size from chylomicrons to VLDL (Gooden et al., 1979) and attempts to distinguish between the two may perhaps be artificial. Chylomicron-like particles from bovine plasma are composed of triacylglycerols, 87%; phospholipids, 4%; cholesterol, 4%; cholesterol esters, 2%; protein, 3%: similarly, bovine VLDL are composed of triacylglycerols,

74%; phospholipids, 7%; cholesterol, 7%; cholesterol esters, 5%; protein, 8% (Ferreri and Elbein, 1982). Chylomicron and VLDL phospholipids consist mainly of phosphatidylcholine with smaller proportions of sphingomyelin and phosphatidylethanolamine (Christie and Hunter, 1978; Ferreri and Elbein, 1982).

In chylomicron and VLDL particles, the central core contains virtually all of the constituent triacylglycerols and cholesterol esters, about 30% of the free cholesterol but no phospholipid; the surface film consists of phospholipids, apoproteins and most of the free cholesterol which together form a monolayer with a constant thickness of about 2.2 nm (Eisenberg and Levy, 1975). As the thickness of the surface film remains constant, it is clear that there must be some variation in composition within the chylomicron and VLDL classes. Thus, as the diameters of the chylomicron and VLDL particles decrease, the proportions of the constituent apoproteins, phospholipids and free cholesterol increase, and the proportion of triacylglycerols decreases (Eisenberg and Levy, 1975). The apoproteins of chylomicrons and VLDL obtained from the intestinal lymph of rats and from the thoracic duct lymph of man have been studied in some detail (Green and Glickman, 1981). Chylomicron apoproteins consist of apo-A_I (38–50%), apo-A_{IV} (7–13%), apo-B (about 10%), apo-C_I, -C_{II} and -C_{III} (about 40%) and apo-E (about 5%). The apoprotein composition of intestinal VLDL (as distinct from plasma VLDL) resembles that of chylomicrons. With the possible exceptions of apo-C_I and apo-E, the results of experiments with rats indicate that all of these apoproteins can be synthesized in the enterocyte, even though the liver is almost certainly the major site for the synthesis of apo-C_{II} and -C_{III}. The more limited studies with ruminant animals have shown that the range of apoproteins associated with ruminant plasma lipoproteins is analogous to that found in chylomicrons and intestinal VLDL in rats (Puppione *et al.*, 1970; Jonas, 1972; Lim and Scanu, 1976; Puppione, 1978; Swaney, 1979; Patterson and Jonas, 1980a, b; Tall *et al.*, 1981; Puppione *et al.*, 1982b). However, at present little is known about the site of synthesis of these of these apoproteins, or their detailed distributions in ruminant lymph and plasma lipoproteins.

The assimilation of dietary fat in non-ruminant animals is characterized normally by the preferential incorporation by the intestinal cells of newly synthesized triacylglycerols into chylomicrons: smaller amounts only of triacylglycerols are incorporated into intestinal VLDL, but this process does assume greater importance during starvation or between meals, and provides a mechanism for the reabsorption and transport of endogenous lipids. In ruminant animals, on the other hand, fatty acids absorbed by the enterocytes are preferentially incorporated into VLDL triacylglycerols and as a result of this the phospholipid:triacylglycerol ratio in lymph lipids of ruminants is usually greater than that in non-ruminant lymph lipids (Hartmann, Harris and Lascelles, 1966; Leat and Hall, 1968; Leat and Harrison, 1974; Christie and Hunter, 1978).

There are probably two reasons for this difference between species. First, it is known that a large flux of triacylglycerols through the intestinal cells results in the formation of chylomicrons, whereas a small flux results in the formation of VLDL (Redgrave and Dunne, 1975). The absorption of lipids in ruminant animals tends to be a continuous process that operates at low rates, whereas in non-ruminant animals, lipid absorption tends to be intermittent and is relatively rapid during the period immediately after a meal. During this period, the rapid influx of large amounts of fatty acids and 2-monoacylglycerols into the non-ruminant enterocyte induces synthetic rates of core material (triacylglycerols) that are in excess of those of surface film components (phospholipids, cholesterol and apoproteins); this results in the assembly of the larger intestinal lipoproteins (i.e. chylomicrons). In ruminant enterocytes, the synthesis of surface film components normally keeps pace with the synthesis of core material and the newly synthesized triacylglycerols are incorporated predominantly into the smaller intestinal lipoproteins (i.e. VLDL). Second, the fatty acids absorbed by ruminant enterocytes tend to be more saturated than those absorbed in non-ruminants. In experiments with rats, Ockner, Hughes and Isselbacher (1969) found that the absorption of saturated fatty acids resulted in the formation of VLDL whereas the absorption of equivalent amounts of unsaturated fatty acids resulted in the formation of chylomicrons. Analogous findings on the synthesis of hepatic VLDL have been reported by Heimberg and Wilcox (1972) and Wilcox, Dunn and Heimberg (1975) who observed that the VLDL secreted by rat livers after perfusion with unsaturated fatty acids consisted of larger particles containing higher proportions of triacylglycerols and smaller proportions of cholesterol and phospholipids than did the VLDL secreted after perfusion with equivalent amounts of saturated fatty acids. The effects of the magnitude of the triacylglycerol flux and the degree of unsaturation of the absorbed fatty acids on the size of the lipoprotein particles synthesized by the ruminant enterocyte are illustrated by the experiments of Harrison et al. (1974). In sheep given a control diet, 72.6% of the lipid in the thoracic duct lymph was transported as VLDL and only 27.4% as chylomicrons, the total lymph lipid contained 18% phospholipids and the major fatty acids in the lymph triacylglycerols were stearic plus palmitic (61.6%) and oleic (14.9%). When maize oil was infused into the duodenum, there was a threefold increase in the rate of flow of lymph fatty acids and 61.5% of the lymph lipid was then transported as chylomicrons and 38.5% as VLDL: under these circumstances, the lymph lipid contained only 11% phospholipid and the major fatty acids in the lymph triacylglycerols were linoleic (42.9%) and oleic (24.7%).

It is not known whether ruminant lymph contains lipoproteins other than chylomicrons and VLDL, but rat mesenteric lymph contains small amounts of low-density lipoproteins (LDL, density 1.006–1.063 g/ml, diameter 17–26 nm) and high-density lipoproteins (HDL, density 1.063–1.21 g/ml, diameter 6–10 nm); it seems likely that these LDL and HDL particles are derived by a process of slow filtration from the rat plasma compartment into the lymph (Green, Tall and Glickman, 1978; Schaefer, Eisenberg and Levy, 1978).

Thoracic duct lymph from cattle and sheep respectively contains 0.54 –1.95 and 0.89–1.07 g lipid per 100 ml (Christie, 1981a).

Metabolism of triacylglycerol-rich lipoproteins in the blood

The metabolism of chylomicrons and intestinal VLDL after entry into the blood stream has been elucidated in some detail in man and experimental rats (Havel, 1975; Eisenberg and Levy, 1975; Eisenberg et al., 1975; Eisenberg, 1976; Green and Glickman, 1981). However, some re-evaluation of the available evidence on the nature and metabolism of the triacylglycerol-rich lipoproteins of ruminants may be necessary following a recent observation that during the separation of lipoproteins some crystallization of triacylglycerols can occur giving rise to fractions with abnormal shapes and density distributions (Puppione et al., 1982a; Small et al., 1981). Such evidence as is available suggests that the metabolism of the triacylglycerol-rich lipoproteins is similar in ruminants and non-ruminants (Palmquist, 1976; Puppione, 1978). On entering the plasma, chylomicrons and VLDL acquire apo-E and apo-C by transfer from plasma HDL synthesized in the liver (Havel, Kane and Kashyap, 1973; Imaizumi, Fainaru and Havel, 1978; Green et al., 1979; Blum, 1980). This acquisition of additional apo-C ensures that the catabolism of chylomicrons and VLDL is diverted away from the liver towards extrahepatic tissues; all of the apo-C components (I, II and III) inhibit the removal of chylomicrons and VLDL by the liver, whereas apo-C_{II} specifically activates the enzyme lipoprotein lipase which is bound to the endothelial surfaces of the capillaries that permeate extrahepatic tissues, such as skeletal muscle, fat depots and the mammary gland. The core triacylglycerols of plasma chylomicrons and VLDL particles fortified with apo-C_{II} are hydrolysed by lipoprotein lipase to fatty acids and partial acylglycerols which are then taken up at the site of hydrolysis and utilized for energy production, the synthesis of adipose tissue triacylglycerols or the synthesis of milk fat. The role of lipoprotein lipase in the provision of substrates for the synthesis of ruminant milk and depot fat has been reviewed by Moore and Christie (1981) and Vernon (1981). Hydrolysis by lipoprotein lipase results in the removal of about 80% of the core triacylglycerols from chylomicrons and VLDL, and in the conversion of these lipoproteins into the much smaller remnant particles or intermediate-density lipoproteins (IDL; mean diameter 30 nm) which contain most of the original core cholesterol esters. This conversion is accompanied by pronounced losses of surface components; apo-C, apo-A_I, apo-A_{IV}, cholesterol and phospholipids are transferred to HDL circulating in the plasma, but all of the apo-B and most of the apo-E are retained by the triacylglycerol-depleted particles. Evidence from experiments with rats indicates that during this conversion some of the surface phospholipid components may be hydrolysed by extrahepatic phospholipases and the products of hydrolysis taken up by the tissues (Eisenberg and Levy, 1975): if this process occurs in ruminant animals it might provide a means whereby small amounts of essential fatty acids (virtually absent from ruminant plasma triacylglycerols and unesterified fatty acids) are distributed to the various tissues of the body.

In man, the remaining core triacylglycerols in remnant or IDL particles are hydrolysed by hepatic lipase, an enzyme probably located at the surface of the capillary endothelium in the liver; the products of hydrolysis together with some surface protein (but not apo-B) are taken up by the hepatocyte, and the particles remaining in the plasma are known as low-density lipoproteins (LDL, density 1.063–1.006 g/ml, diameter 17–26 nm) in which cholesterol esters account for about 80% of the core lipid. In the rat, on the other hand, the major proportion of the plasma remnant or

IDL particles are taken up as complete units by the liver cells where core and surface components are catabolized; the process appears to be initiated by the binding of apo-E on the surface of the remnant particles to a specific hepatic remnant receptor. This difference in the metabolism of remnant or IDL particles probably explains why LDL constitute only a small fraction of rat plasma lipoproteins but a relatively large proportion of human plasma lipoproteins. Since LDL accounts for only a small fraction of ruminant plasma lipoproteins, it seems reasonable to suppose that the fate of remnant particles in ruminant animals is similar to that in the rat. Recent evidence suggests that heparin-releasable hepatic lipase may not be present in the cow (Etienne *et al.*, 1981).

Many aspects of lipoprotein metabolism in the ruminant animal remain to be resolved. It is, however, clear that polyunsaturated fatty acids are directed to lipid classes other than the triacylglycerols in the plasma lipoproteins, with the consequence that the lipid storage tissues in the animal are also relatively deficient in these components.

References

ADAMS, E.P. and HEATH, T.J. (1963). The phospholipids of ruminant bile. *Biochimica et biophysica acta* **70**, 688–690

ANDREWS, R.J. and LEWIS, D. (1970a). The utilization of dietary fats by ruminants. I. The digestibility of some commercially available fats. *Journal of Agricultural Science* **75**, 47–54

ANDREWS, R.J. and LEWIS, D. (1970b). The utilization of dietary fats by ruminants. II. The effect of fatty acid chain length and unsaturation on digestibility. *Journal of Agricultural Science* **75**, 55–60

ANNISON, E.F., BROWN, R.E., LENG, R.A., LINDSAY, D.B. and WEST, C.E. (1967). Rates of entry and oxidation of acetate, glucose, $D(-)-\beta$-hydroxybutyrate, palmitate, oleate and stearate, and rates of production and oxidation of propionate and butyrate in fed and starved sheep. *Biochemical Journal* **104**, 135–147

ARIENTI, G., HARRISON, F.A. and LEAT, W.M.F. (1974). The lipase activity of sheep pancreatic juice. *Quarterly Journal of Experimental Physiology* **59**, 351–359

ARIENTI, G., LEAT, W.M.F. and HARRISON, F.A. (1975). The phospholipase activity of sheep pancreatic juice. *Quarterly Journal of Experimental Physiology* **60**, 15–24

BAJWA, S.S. and SASTRY, P.S. (1974). Degradation of monogalactosyldiglyceride and digalactosyldiglyceride by sheep pancreatic enzymes. *Biochemical Journal* **144**, 177–187

BALCH, C.C., BALCH, D.A., BARTLETT, S., JOHNSON, V.W. and ROWLANDS, S.J. (1952). Factors affecting the utilization of food by dairy cows. 5. The digestibility and rate of passage during L-thyroxine administration. *British Journal of Nutrition* **6**, 356–365

BALINT, J.A., KYRIAKIDES, E.C., SPITZER, H.L. and MORRISON, E.S. (1965). Lecithin fatty acid composition in bile and plasma of man, dogs, rats and oxen. *Journal of Lipid Research* **6**, 96–99

BATH, I.H. and HILL, K.J. (1967). The lipolysis and hydrogenation of lipids in the digestive tract of the sheep. *Journal of Agricultural Science* **68**, 139–148

BICKERSTAFFE, R. and ANNISON, E.F. (1969a). Triglyceride synthesis by the small intestinal epithelium of the pig, sheep and chicken. *Biochemical Journal* **111**, 419–429

BICKERSTAFFE, R. and ANNISON, E.F. (1969b). Glycerokinase and desaturase activity in pig, chicken and sheep intestinal epithelium. *Comparative Biochemistry and Physiology* **31**, 47–54

BICKERSTAFFE, R., NOAKES, D.E. and ANNISON, E.F. (1972). Quantitative aspects of fatty acid biohydrogenation, absorption and transfer into milk fat in the lactating goat, with special reference to the *cis*- and *trans*-isomers of octadecenoate and linoleate. *Biochemical Journal* **130**, 607–617

BLUM, C.B. (1980). Dynamics of apolipoprotein E metabolism in humans. *Circulation* **62**, 181–194

BRINDLEY, D.N. and HUBSCHER, G. (1966). The effect of chain length on the activation and subsequent incorporation of fatty acids into glycerides by the small intestinal mucosa. *Biochimica et biophysica acta* **125**, 92–105

BROAD, T.E. and DAWSON, R.M.C. (1975). Phospholipid biosynthesis in the anaerobic protozoan *Entodinium caudatum*. *Biochemical Journal* **146**, 317–328

CANIONI, P., BENAJIBA, A., JULIEN, R., RATHELOT, J., BENABDELJLIL, A. and SARDA, L. (1975). Ovine pancreatic lipase. Purification and some properties. *Biochimie* **57**, 35–41

CAPLE, I. and HEATH, T. (1972). Regulation of output of electrolytes in bile and pancreatic juice in sheep. *Australian Journal of Biological Sciences* **25**, 155–165

CHRISTIE, W.W. (1973). The structures of bile phosphatidylcholines. *Biochimica et biophysica acta* **316**, 204–211

CHRISTIE, W.W. (1981a). The composition, structure and function of lipids in the tissue of ruminant animals. In *Lipid Metabolism in Ruminant Animals* (W.W. Christie, Ed.), pp. 95–189. Oxford, Pergamon Press

CHRISTIE, W.W. (1981b). The effects of diet and other factors on the lipid composition of ruminant tissues and milk. In *Lipid Metabolism in Ruminant Animals* (W.W. Christie, Ed.), pp. 193–226. Oxford, Pergamon Press

CHRISTIE, W.W. and HUNTER, M.L. (1978). The composition and structure of the lipids of sheep lymph. *Journal of the Science of Food and Agriculture* **29**, 442–446

CHRISTIE, W.W., MOORE, J.H., NOBLE, R.C. and VERNON, R.G. (1975). Changes with diet in the composition of phosphatidylcholine of sheep bile. *Lipids* **10**, 645–648

COLEMAN, G.S. (1969). The metabolism of starch, maltose, glucose and some other sugars by the rumen ciliate *Entodinium caudatum*. *Journal of General Microbiology* **57**, 303–332

CUNNINGHAM, H.M. and LEAT, W.M.F. (1969). Lipid synthesis by the monoglyceride and α-glycerophosphate pathways in sheep intestine. *Canadian Journal of Biochemistry* **47**, 1013–1020

CZERKAWSKI, J.W. (1967). Effect of storage on the fatty acids of dried grass. *British Journal of Nutrition* **21**, 599–608

CZERKAWSKI, J.W., CHRISTIE, W.W., BRECKENRIDGE, G. and HUNTER, M.L. (1975). Changes in the rumen metabolism of sheep given increasing amounts of linseed oil in their diet. *British Journal of Nutrition* **34**, 25–44

DAWSON, R.M.C. (1959). Hydrolysis of lecithin and of lysolecithin by rumen micro-organisms of the sheep. *Nature* **183**, 1822–1823

DAWSON, R.M.C. and HEMINGTON, N. (1974). Digestion of grass lipids and pigments in the sheep rumen. *British Journal of Nutrition* **32**, 327–340

DAWSON, R.M.C., HEMINGTON, N., GRIME, D., LAUDER, D. and KEMP, P. (1974). Lipolysis and hydrogenation of galactolipids and the accumulation of phytanic acid in the rumen. *Biochemical Journal* **144**, 169–171

DAWSON, R.M.C., IRVINE, R.F., HIRASAWA, K. and HEMINGTON, N.L. (1982). Hydrolysis of phosphatidylinositol by pancreas and pancreatic secretions. *Biochimica et biophysica acta* **710**, 212–220

DAWSON, R.M.C. and KEMP, P. (1967). The aminoethylphosphonate-containing lipids of rumen protozoa. *Biochemical Journal* **105**, 837–842

DAWSON, R.M.C. and KEMP, P. (1969). The effect of defaunation on the phospholipids and on the hydrogenation of unsaturated fatty acids in the rumen. *Biochemical Journal* **115**, 351–352

EISENBERG, S. (1976). Mechanisms of formation of low density lipoproteins: metabolic pathways and their regulation. In *Low Density Lipoproteins* (C.E. Day and R.S. Levy, Eds), pp. 73–92. New York, Plenum Press

EISENBERG, S. and LEVY, R.I. (1975). Lipoprotein metabolism. *Advances in Lipid Research* **13**, 1–89

EISENBERG, S., RACHMILEWITZ, D., LEVY, R.I., BILHEIMER, D.W. and LINDGREN, F.T. (1975). Pathways of lioprotein metabolism: integration of structure, function and metabolism. *Advances in Experimental Medicine and Biology* **63**, 61–67

EMMANUEL, B. (1974). On the origin of rumen protozoan fatty acids. *Biochimica et biophysica acta* **337**, 404–413

ETIENNE, J., NOE, L., ROSSIGNOL, M., DOSNE, A.-M. and DEBRAY, J. (1981). Post-heparin lipolytic activity with no hepatic triacylglycerol lipase involved in a mammalian species. *Biochimica et biophysica acta* **663**, 516–523

FERRERI, L.F. and ELBEIN, R.C. (1982). Fractionation of plasma triglyceride-rich lipoproteins of the dairy cow: evidence of chylomicron sized particles. *Journal of Dairy Science* **65**, 1912–1920

FREEMAN, C.P. (1969). Properties of fatty acids in dispersions of emulsified lipid and bile salts and the significance of these properties in fat absorption in the pig and sheep. *British Journal of Nutrition* **23**, 249–263

FROBISH, L.T., HAYS, V.W., SPEER, V.C. and EWAN, R.C. (1971). Effect of fat source on pancreatic lipase activity and specificity and performance of baby pigs. *Journal of Animal Science* **33**, 385–389

GARTON, G.A. (1959). Lipids in relation to rumen function. *Proceedings of the Nutrition Society* **18**, 112–117

GARTON, G.A. (1960). Lipid metabolism in herbivorous animals. *Nutrition Abstracts and Reviews* **30**, 1–16

GARTON, G.A., HOBSON, P.N. and LOUGH, A.K. (1958). Lipolysis in the rumen. *Nature* **182**, 1151–1152

GARTON, G.A., LOUGH, A.K. and VIOQUE, E. (1961). Glyceride hydrolysis and glycerol fermentation by sheep rumen content. *Journal of General Microbiology* **25**, 215–225

GARTON, G.A., MOREHOUSE, M.G. and LOUGH, A.K. (1961). The effect of rumen micro-organisms on fatty acids and fatty acid esters. *Abstract Communications of the Vth International Congress in Biochemistry, Moscow*. p. 279

GOODEN, J.M., FRASER, R., BOSANQUET, A.G. and BICKERSTAFFE, R. (1979). Size of lipoproteins in intestinal lymph of sheep and suckling lambs. *Australian Journal of Biological Science* **32**, 533–542

GREEN, H.O., HIRS, C.H.W. and PALADE, G.E. (1963). On the protein composition of bovine pancreatic zymogen granules. *Journal of Biological Chemistry* **238**, 2054–2070

GREEN, P.H.R. and GLICKMAN, R.M. (1981). Intestinal lipoprotein metabolism. *Journal of Lipid Research* **22**, 1153–1173

GREEN, P.H.R., TALL, A.R. and GLICKMAN, R.M. (1978). Rat intestine secretes discoid high density lipoprotein. *Journal of Clinical Investigation* **61**, 528–534

GREEN, P.H.R., GLICKMAN, R.M., SAUDEK, C.D., BLUM, C.B. and TALL, A.R. (1979). Human intestinal lipoproteins: studies in chyluric subjects. *Journal of Clinical Investigation* **64**, 233–242

GUTTIERREZ, J., WILLIAMS, P.P., DAVIS, R.E. and WARWICK, E.J. (1962). Lipid metabolism of rumen ciliates and bacteria. I. Uptake of fatty acids by *Isotrichia prostoma* and *Entodinium simplex*. *Journal of Applied Microbiology* **10**, 548–551

HARFOOT, C.G. (1981). Lipid metabolism in the rumen. In *Lipid Metabolism in Ruminant Animals* (W.W. Christie, Ed.), pp. 25–55. Oxford, Pergamon Press

HARFOOT, C.G., NOBLE, R.C. and MOORE, J.H. (1973). Food particles as a site for biohydrogenation of saturated fatty acids in the rumen. *Biochemical Journal* **132**, 829

HARFOOT, C.G., NOBLE, R.C. and MOORE, J.H. (1975). The role of plant particles, bacteria and cell-free supernatant fractions of rumen contents in the hydrolysis of trilinolein and the subsequent hydrogenation of linoleic acid. *Antonie van Leeuwenhoek* **41**, 533–542

HARFOOT, C.G., CROUCHMAN, M.L., NOBLE, R.C. and MOORE, J.H. (1974). Competition between food particles and rumen bacteria in the uptake of long-chain fatty acids and triglyceride. *Journal of Applied Bacteriology* **37**, 633–641

HARRISON, F.A. (1962). Bile secretion in the sheep. *Journal of Physiology* **162**, 212–224

HARRISON, F.A. and HILL, K.L. (1960). Bile secretion in the conscious sheep. *Journal of Physiology* **154**, 61P–62P

HARRISON, F.A. and HILL, K.J. (1962). Digestive secretions and the flow of digesta along the duodenum of the sheep. *Journal of Physiology* **162**, 225–243

HARRISON, F.A. and LEAT, W.M.F. (1970). Effect of bile and pancreatic juice on the absorption of long-chain fatty acids in the sheep. *Biochemical Journal* **118**, 3P

HARRISON, F.A. and LEAT, W.M.F. (1972). Absorption of palmitic, stearic and oleic acids in the sheep in the presence or absence of bile and/or pancreatic juice. *Journal of Physiology* **225**, 565–576

HARRISON, F.A., LEAT, W.M.F. and FORSTER, A. (1974). Absorption of maize oil infused into the duodenum of the sheep. *Proceedings of the Nutrition Society* **33**, 101A–102A

HARTMANN, P.E., HARRIS, J.G. and LASCELLES, A.K. (1966). The effect of oil-feeding and starvation on the composition and output of lipid thoracic duct lymph in the lactating cow. *Australian Journal of Biological Science* **19**, 635–644

HAUGEN, R. and NORUM, K.R. (1976). Coenzyme A-dependent esterification of cholesterol in rat intestinal mucosa. *Scandinavian Journal of Gastroenterology* **11**, 615–621

HAVEL, R.J. (1975). Lipoproteins and lipid transport. *Advances in experimental Medicine and Biology* **63**, 37–59

HAVEL, R.J., KANE, J.P. and KASHYAP, M.L. (1973). Interchange of apolipoproteins between chylomicrons and high density lipoproteins during alimentary lipemia in man. *Journal of Clinical Investigation* **52**, 32–38

HAWKE, J.C. (1971). The incorporation of long-chain fatty acids into lipids by rumen bacteria and the effect on biohydrogenation. *Biochimica et biophysica acta* **248**, 167–170

HAWKE, J.C. and SILCOCK, W.R. (1970). The in vitro rates of lipolysis and biohydrogenation in rumen contents. *Biochimica et biophysica acta* **218**, 201–212

HAZLEWOOD, G.P. and DAWSON, R.M.C. (1975). Isolation and properties of a phospholipid-hydrolysing bacterium from ovine rumen fluid. *Journal of General Microbiology* **89**, 163–174

HEATH, T.J. and HILL, L.N. (1969). Dietary and endogenous long-chain fatty acids in the intestine of sheep, with an appendix on their estimation in feeds, bile and faeces. *Australian Journal of Biological Science* **22**, 1015–1029

HEATH, T.J. and MORRIS, B. (1963). The role of bile and pancreatic juice in the absorption of fat in ewes and lambs. *British Journal of Nutrition* **17**, 465–474

HEATH, T.J., CAPLE, I.W. and REDDING, P.M. (1970). Effect of the enterohepatic circulation of bile salts on the flow of bile and its contents of bile salts and lipids in sheep. *Quarterly Journal of Experimental Physiology* **55**, 93–103

HEIMBERG, M. and WILCOX, H.G. (1972). The effect of palmitic and oleic acids on the properties and composition of the very low density lipoproteins secreted by the liver. *Journal of Biological Chemistry* **247**, 875–880

HENDERSON, C. (1968). *A Study of the Lipase of* Anaerovibrio lipolytica, *a Rumen Bacterium*. PhD thesis, University of Aberdeen, Scotland

HENDERSON, C. (1971). A study of the lipase produced by *Anaerovibrio lipolytica*, a rumen bacterium. *Journal of General Microbiology* **65**, 81–89

HENDERSON, C. and HODGKISS, W. (1973). An electron microscopic study of *Anaerovibrio lipolytica* and its lipolytic enzyme. *Journal of General Microbiology* **76**, 389–393

HOBSON, P.N. and MANN, S.O. (1961). The isolation of glycerol-fermenting and lipolytic bacteria from the rumen of sheep. *Journal of General Microbiology* **25**, 227–240

HOBSON, P.N. and SUMMERS, R. (1966). Effect of growth rate on the lipase activity of a rumen bacterium. *Nature* **209**, 736–737

HOFMANN, A.F. and SMALL, D.M. (1967). Detergent properties of bile salts: correlation with physiological function. In *Annual Review of Medicine* (A.C. De Graff and W.B. Onegar, Eds), pp. 333–376. California, Annual Reviews Incorporated

HOGAN, J.P. (1973). Intestinal digestion of subterranean clover by sheep. *Australian Journal of Agricultural Research* **24**, 587–598

HOOGENRAAD, N.J. and HIRD, F.J.R. (1970). Factors concerned in the lysis of bacteria in the alimentary tract of sheep. *Journal of General Microbiology* **62**, 261–269

HOWARD, B.H. (1963). Hydrolysis of naturally occurring galactosidase by some rumen protozoa. *Biochemical Journal* **89**, 90P

HUBSCHER, G. (1970). Glyceride metabolism. In *Lipid Metabolism* (S.J. Wakil, Ed.), pp. 280–370. New York, Academic Press Incorporated

HUGHES, P.E. and TOVE, S.B. (1980a). Identification of an endogenous electron donor for biohydrogenation as α-tocopherolquinol. *Journal of Biological Chemistry* **255**, 4447–4452

HUGHES, P.E. and TOVE, S.B. (1980b). Identification of deoxy-α-tocopherolquinol as another endogenous electron donor for biohydrogenation. *Journal of Biological Chemistry* **255**, 11802–11806

HUGHES, P.E., HUNTER, W.J. and TOVE, S.B. (1982). Biohydrogenation of unsaturated fatty acids. Purification and properties of *cis*-9, *trans*-11-octadecadienoate reductase. *Journal of Biological Chemistry* **257**, 3643–3649

IMAIZUMI, K., FAINARU, M. and HAVEL, R.J. (1978). Composition of proteins of mesenteric lymph chylomicrons in the rat and alterations produced on exposure of chylomicrons to blood serum and serum proteins. *Journal of Lipid Research* **19**, 712–722

JOHNSON, J.M. and BORGSTRÖM, B. (1964). The intestinal absorption and metabolism of micellar solutions of lipids. *Biochimica et biophysica acta* **84**, 412–423

JOHNSON, T.O., MITCHELL, G.E., TUCKER, R.E. and SCHELLING, G.T. (1974). Pancreatic lipase secretion by sheep. *Journal of Animal Science* **39**, 947–951

JONAS, A. (1972). Physicochemical properties of bovine serum high density lipoproteins. *Journal of Biological Chemistry* **247**, 7767–7772

JULIEN, R., CANIONI, P., RATHELOT, J., SARDA, L. and PLUMMER, T.H. (1972). Studies on bovine pancreatic lipase and colipase. *Biochimica et biophysica acta* **280**, 215–224

KANEGASAKI, S. and NUMA, S. (1970). Medium-chain fatty acyl-CoA requirement for long-chain fatty acids synthesis in some anaerobic bacteria. *Biochimica et biophysica acta* **202**, 436–446

KANEGASAKI, S. and TAKAHASHI, H. (1970). Function of growth factors for rumen micro-organisms. II. Metabolic fate of incorporated fatty acids in *Selenomonas ruminantium. Biochimica et biophysica acta* **152**, 40–49

KATZ, I. and KEENEY, M. (1966). Characterization of the octadecenoic acids in rumen digesta and rumen bacteria. *Journal of Dairy Science* **49**, 962–966

KAY, R.N.B. and PFEFFER, E. (1970). Movements of water and electrolytes into and from the intestine of the sheep. In *Physiology of Digestion and Metabolism in the Ruminant* (A.T. Phillipson, Ed.), pp. 390–402. Newcastle upon Tyne, Oriel Press

KEENEY, M. (1970). Fat metabolism in the rumen. In *Physiology of Digestion and Metabolism in the Ruminant* (A.T. Phillipson, Ed.), pp. 489–503. Newcastle upon Tyne, Oriel Press

KELLER, P.J., COHEN, E. and NEURATH, H. (1958). The proteins of bovine pancreatic juice. *Journal of Biological Chemistry* **233**, 344–349

KHAN, I.M., CHANDAN, R.C. and SHAHARI, K.M. (1976). Bovine pancreatic lipase. II. Stability and effect of activators and inhibitors. *Journal of Dairy Science* **59**, 840–846

KUNSMAN, J.E. (1973). Characterization of the lipids of six strains of *Bacteroides ruminicola. Journal of Bacteriology* **113**, 1121–1126

LEAT, W.M.F. and HALL, J.G. (1968). Lipid composition of lymph and blood plasma of the cow. *Journal of Agricultural Science* **71**, 189–194

LEAT, W.M.F. and HARRISON, F.A. (1969). Lipid digestion in the sheep: effect of bile and pancreatic juice on the lipids of intestinal contents. *Quarterly Journal of Experimental Physiology* **54**, 187–201

LEAT, W.M.F. and HARRISON, F.A. (1974). Origin and formation of lymph lipids in the sheep. *Quarterly Journal of Experimental Physiology* **59**, 131–139

LEAT, W.M.F., LINDSAY, D.B. and VALERIO, G. (1975). Oxidation and metabolism of linoleic acid in the sheep. *Proceedings of the Nutrition Society* **34**, 88A–89A

LENNOX, A.M. and GARTON, G.A. (1968). The absorption of long-chain fatty acids from the small intestine of sheep. *British Journal of Nutrition* **22**, 247–254

LENNOX, A.M., LOUGH, A.K. and GARTON, G.A. (1968). Observations on the nature and origin of lipids in the small intestine of the sheep. *British Journal of Nutrition* **22**, 237–246

LIM, C.T. and SCANU, A.M. (1976). Apoproteins of bovine serum high density lipoproteins: isolation and characterization of the small molecular weight components. *Artery* **2**, 483–491

LINDSAY, D.B. and LEAT, W.M.F. (1977). Oxidation and metabolism of linoleic acid in fed and fasted sheep. *Journal of Agricultural Science* **89**, 215–221

LOUGH, A.K. and ANDERSON, L.J. (1973). Effect of ensilage on the lipids of pasture grasses. *Proceedings of the Nutrition Society* **32**, 61A–62A

LOUGH, A.K. and SMITH, A. (1976). Influence of phospholipolysis of phosphatidylcholine on micellar solubilization of fatty acids in the presence of bile salts. *British Journal of Nutrition* **35**, 89–96

McCORMICK, R.J. and STEWART, W.E. (1967). Pancreatic secretion in the bovine calf. *Journal of Dairy Science* **50**, 568–571

MENKE, W. (1966). The structure of the chloroplast. In *Biochemistry of*

Chloroplasts (T.W. Goodwin, Ed.), Vol. 1, pp. 3–18. New York, Academic Press

MOORE, J.H. (1974). Lipid biochemistry—from forage to milk. In *Industrial Aspects of Biochemistry* (B. Spencer, Ed.), pp. 853–863. Amsterdam, North-Holland Publishing Company

MOORE, J.H. and CHRISTIE, W.W. (1981). Lipid metabolism in the mammary gland of ruminant animals. In *Lipid Metabolism in Ruminant Animals* (W.W. Christie, Ed.), pp. 227–277. Oxford, Pergamon Press

MOORE, J.H. and TYLER, C. (1955). Studies on the intestinal absorption and excretion of calcium and phosphorus in the pig. *British Journal of Nutrition* **9**, 63–80

NOBLE, R.C. (1981). Digestion, absorption and transport of lipids in ruminant animals. In *Lipid Metabolism in Ruminant Animals* (W.W. Christie, Ed.), pp. 57–93. Oxford, Pergamon Press

NOBLE, R.C., CROUCHMAN, M.L. and MOORE, J.H. (1974). The presence of linoleic acid in the skin surface lipids of the ox. *Research in Veterinary Science* **17**, 372–376

NOBLE, R.C., CROUCHMAN, M.L., JENKINSON, D.McE. and MOORE, J.H. (1975). Relationship between lipids in plasma and skin secretions of the neonatal calf with particular reference to linoleic acid. *Lipids* **10**, 128–133

OCKNER, R.K. and MANNING, J.M. (1974). Fatty acid binding protein in small intestine: identification, isolation and evidence for its role in cellular fatty acid transport. *Journal of Clinical Investigation* **54**, 326–338

OCKNER, R.K. and MANNING, J.M. (1976). Fatty acid binding protein. Role in esterification of absorbed long chain fatty acids in rat intestine. *Journal of Clinical Investigation* **58**, 632–641

OCKNER, R.K., HUGHES, F.B. and ISSELBACHER, K.J. (1969). Very low density lipoproteins in intestinal lymph: role in triglyceride and cholesterol transport during fat absorption. *Journal of Clinical Investigation* **48**, 2367–2373

OCKNER, R.K., MANNING, J.M., POPPENHAUSEN, R.B. and HO, W.K.L. (1972). A binding protein for fatty acids in cytosol of intestinal mucosa, liver, myocardium and other tissues. *Science* **177**, 56–58

OUTEN, G.E., BEEVER, D.E. and OSBOURN, D.F. (1974). Digestion and absorption of lipids by sheep fed chopped and ground dried grass. *Journal of the Science of Food and Agriculture* **25**, 981–987

PALMQUIST, D.L. (1976). A kinetic concept of lipid transport in ruminants. *Journal of Dairy Science* **59**, 355–363

PATTERSON, B.W. and JONAS, A. (1980a). Bovine apolipoprotein. I. Isolation and spectroscopic investigations of the phospholipid binding properties. *Biochimica et biophysica acta* **619**, 572–586

PATTERSON, B.W. and JONAS, A. (1980b). Bovine apolipoprotein. II. Isolation and partial physicochemical characterization of complexes with L-α-dimyristoyl phosphatidyl choline. *Biochimica et biophysica acta* **619**, 587–603

PATTON, R.A., McCARTHY, R.D. and GRIEL, L.C. (1968). Lipid synthesis by rumen micro-organisms. I. Stimulation by methionine *in vitro*. *Journal of Dairy Science* **51**, 1310–1311

PATTON, R.A., McCARTHY, R.D. and GRIEL, L.C. (1970). Lipid synthesis by rumen micro-organisms. II. Further characterization of the effects of methionine. *Journal of Dairy Science* **53**, 460–465

PERIC-GOLIA, L. and SOCIC, C. (1968). Biliary bile acids and cholesterol in developing sheep. *American Journal of Physiology* **215**, 1284–1287

PUPPIONE, D.L. (1978). Implications of unique features of blood lipid transport in the lactating cow. *Journal of Dairy Science* **61**, 651–659

PUPPIONE, D.L., FORTE, G.M., NICHOLS, A.V. and STRISOWER, E.H. (1970). Partial characterization of serum lipoproteins in the density interval 1.04–1.06 g/ml. *Biochimica et biophysica acta* **202**, 392–395

PUPPIONE, D.L., KUNITAKE, S.T., HAMILTON, R.L., PHILLIPS, M.L., SCHUMAKER, V.N. and DAVIS, L.D. (1982a). Characterization of unusual intermediate density lipoproteins. *Journal of Lipid Research* **23**, 283–290

PUPPIONE, D.L., KUNITAKE, S.T., TOOMEY, M.L., LOH, E. and SCHUMAKER, V.N. (1982b). Physicochemical characterization of ten fractions of bovine alpha lipoproteins. *Journal of Lipid Research* **23**, 371–379

REDGRAVE, T.G. and DUNNE, K.B. (1975). Chylomicron formation and composition in unanesthetized rabbits. *Atherosclerosis* **22**, 389–400

REISER, R. (1951). Hydrogenation of polyunsaturated fatty acids by the ruminant. *Federation Proceedings* **10**, 236

RENNER, R. and HILL, F.W. (1961). Factors affecting the absorbability of saturated fatty acids in the chick. *Journal of Nutrition* **74**, 254–258

ROSENFELD, I.S. and TOVE, S.B. (1971). Biohydrogenation of unsaturated fatty acids. VI. Source of hydrogen and stereospecificity of reduction. *Journal of Biological Chemistry* **246**, 5025–5030

SCHAEFER, E.J., EISENBERG, S. and LEVY, R.I. (1978). Lipoprotein apoprotein metabolism. *Journal of Lipid Research* **19**, 667–686

SCOTT, A.M. and LOUGH, A.K. (1971). The influence of biliary constituents in an acid medium on the micellar solubilization of unesterified fatty acids of the duodenal digesta of sheep. *British Journal of Nutrition* **25**, 307–315

SCOTT, A.M., ULYATT, M.J., KAY, R.N.B. and CZERKAWSKI, J.W. (1969). Measurement of the flow of long-chain fatty acids into the duodenum of sheep. *Proceedings of the Nutrition Society* **28**, 51A

SCOTT, T.W. and COOK, L.J. (1975). Effect of dietary fat on lipid metabolism in ruminants. In *Digestion and Metabolism in the Ruminant* (I.W. McDonald and A.C.I. Warner, Eds), pp. 510–523. Armidale, NSW, University of New England Publishing Unit

SHORLAND, F.B., WEENINK, R.O. and JOHNS, A.T. (1955). Effect of the rumen on dietary fat. *Nature* **175**, 1129

SJOVALL, J. (1960). Bile acids in man under normal and pathological conditions. *Clinica chimica acta* **5**, 33–41

SKRDLANT, H.B., YOUNG, J.W., ALLEN, R.S. and McGILLIARD, D.A. (1973). Pathways of triglyceride synthesis by bovine jejunum during lipid absorption. *Journal of Dairy Science* **56**, 1305–1311

SMALL, D.M., PUPPIONE, D.L., PHILLIPS, M.L., ATKINSON, D., HAMILTON, J.A. and SCHUMAKER, V.N. (1981). Crystallization of a metastable lipoprotein. Massive change of lipoprotein properties during routine preparation. *Arteriosclerosis* **1**, 96

SMILES, J. and DOBSON, M.J. (1956). Direct ultra-violet and ultra-violet negative phase-contrast micrography of bacteria from the stomachs of sheep. *Journal of the Royal Microscopical Society* **75**, 244–256

SMITH, A.M. and LOUGH, A.K. (1976). Micellar solubilization of fatty acids in aqueous media containing bile salts and phospholipids. *British Journal of Nutrition* **35**, 77–87

STERZING, P.R., McGILLIARD, A.D. and ALLEN, R.S. (1971). Ultrastructural aspects of lipid absorption by bovine intestinal mucosa. *Journal of Dairy Science* **54**, 1436–1448

STORRY, J.E., BRUMBY, P.E. and DUNKLEY, W.L. (1980). Influence of nutritional factors on the yield and content of milk fat: protected non-polyunsaturated fat in the diet. In *Factors Affecting the Yields and Contents of Milk Constituents of Commercial Importance* (J.H. Moore and J.A.F. Rook, Eds), pp. 105–125. Brussels, International Dairy Federation

SUTTON, J.D., STORRY, J.E. and NICHOLSON, J.W.G. (1970). The digestion of fatty acids in the stomach and intestines of sheep given widely different rations. *Journal of Dairy Science* **37**, 97–105

SWANEY, J.B. (1979). Characterization of the high density lipoprotein and its major apoprotein from human, canine, bovine and chicken plasma. *Biochimica et biophysica acta* **573**, 489–502

TALL, A.R., PUPPIONE, D.L., KUNITAKE, S.T., ATKINSON, D., SMALL, D.M. and WAUGH, D. (1981). Organisation of the core lipids of high density lipoproteins in the lactating bovine. *Journal of Biological Chemistry* **256**, 165–170

TAYLOR, R.B. (1962). Pancreatic secretion in sheep. *Research in Veterinary Science* **3**, 63–67

THOMSON, A.B.R. and DIETSCHY, J.M. (1981). Intestinal lipid absorption: major extracellular and intracellular events. In *Physiology of the Gastrointestinal Tract* (L.R. Johnson, Ed.), pp. 1147–1220. New York, Raven Press

TOMARELLI, R.M., MEYER, B.J., WEABER, J.R. and BERNHART, F. (1968). Effect of positional distribution on the absorption of the fatty acids in human milk and infant formulas. *Journal of Nutrition* **95**, 583–590

TWEEDIE, J.W., RUMSBY, M.G. and HAWKE, J.C. (1966). Studies on rumen metabolism. V. Formation of branched long-chain fatty acids in cultures of rumen bacteria. *Journal of the Science of Food and Agriculture* **17**, 241–244

VERNON, R.G. (1981). Lipid metabolism in the adipose tissue of ruminant animals. In *Lipid Metabolism in Ruminant Animals* (W.W. Christie, Ed.), pp. 279–362. Oxford, Pergamon Press

VIVIANI, R. (1970). Metabolism of long-chain fatty acids in the rumen. *Advances in Lipid Research* **8**, 267–346

VIVIANI, R., BORGATTI, A.R., CORTESI, P. and CRISETIG, G. (1968). Lipid components of sheep rumen bacteria and protozoa. *Nuova Veterinaria* **44**, 279–283

WAHLE, K.W.J. (1974). Desaturation of long-chain fatty acids by tissue preparations of the sheep, rat and chicken. *Comparative Biochemistry and Physiology* **48B**, 87–105

WILCOX, H.S., DUNN, G.D. and HEIMBERG, M. (1975). Effects of several common long chain fatty acids on the properties and lipid composition of the very low density lipoproteins secreted by the liver. *Biochimica et biophysica acta* **398**, 39–54

WILLIAMS, P.P., GUTIERREZ, J. and DAVIS, R.E. (1963). Lipid metabolism of rumen ciliates and bacteria. II. Uptake of fatty acids and lipid analysis of *Isotrichia intestinales* and rumen bacteria and further information on *Entodinium simplex*. *Applied Microbiology* **11**, 260–264

WOOD, R.D., BELL, M.C., GRAINGER, R.B. and TEEKELL, R.A. (1963). Metabolism of labelled linoleic-1-^{14}C in the sheep rumen. *Journal of Nutrition* **79**, 62–68

YAMAMOTO, A. and ROUSER, G. (1967). Quantitative analysis of bile lipids by dextran gel column chromatography and thin-layer chromatography. *Biochimica e Biologia Sperimentale* **6**, 135–145

III

Role of Essential Fats

7

ESSENTIAL FATTY ACIDS IN POULTRY NUTRITION

C.C. WHITEHEAD*
Agricultural Research Council's Poultry Research Centre, Roslin, Midlothian EH25 9PS, Scotland, UK

Essential fatty acids and their interconversions

Fat has two important roles in nutrition. Firstly, it is a highly concentrated form of metabolizable energy and high-energy diets are an important feature of current practice in poultry nutrition. Secondly, however, fatty acids are important in their own right, playing fundamental parts in basic metabolism, and are synthesized and modified in the body even if provided in the diet. In the course of normal metabolism, fatty acids are interconverted by two processes, chain elongation and desaturation. These reactions take place at the carboxyl end of the molecule, producing a characteristic system of methylene-interrupted *cis* double bonds (*cis, cis,* 1,4-pentadiene), with the configuration at the other end of the molecule remaining unaltered. Unsaturated fatty acids can thus be grouped into families depending upon the number of carbon atoms from the terminal methyl group to the first double bond, the n-number.

Members of the families with greatest nutritional and metabolic significance are shown in *Table 7.1*. Animals can synthesize *de novo* fatty acids of the palmitoleate (n-7) and oleate (n-9) families but the families based on linoleate (n-6) and linolenate (n-3) can be synthesized only if the basic fatty acids are provided in the diet. Because interconversions within families take place at the carboxyl end of the fatty acid, they are independent of the n- number of the acid and members of all families are thus modified by the same enzymes. These reactions are part of a dynamic system where the conversions influence the metabolism of other acids through competition for the enzymes. The formation of the long-chain polyunsaturated fatty acid (PUFA) is regulated by competitive inhibition of the enzymes and the conversions within each family depend upon the concentrations of the substrates and the products. Thus the presence of an excess of linolenic acid can suppress synthesis of the higher members of the linoleic acid family and vice versa. The absence of members of both these families stimulates the conversion of oleic acid, with concentrations of $\Delta^{5,8,11}$

---

*Review prepared while author was Visiting Scientist at CSIRO Food Research Laboratory, North Ryde, NSW, 2113, Australia

Table 7.1 MEMBERS OF POLYENOIC FATTY ACID FAMILIES

Linoleic acid	9,12–18:2	Linolenic acid	9,12,15–18:3
(n-6 series)	6,9,12–18:3	(n-3 series)	6,9,12,15–18:4
	8,11,14–20:3		8,11,14,17–20:4
	5,8,11,14–20:4		5,8,11,14,17–20:5
	7,10,13,16–22:4		7,10,13,16,19–22:5
	4,7,10,13,16–22:5		4,7,10,13,16,19–22:6
Oleic acid	9–18:1	Palmitic acid	9–16:1
(n-9 series)	6,9–18:2	(n-7 series)	11–18:1
	8,11–20:2		8,11–18:2
	5,8,11–20:3		

eicosatrienoic acid (of the n-9 series) becoming especially elevated. These interrelationships can be expressed mathematically: equations have been developed that can estimate dietary intake from body composition in rats (Caster, Hill and Holman, 1963).

Although they are the only ones that animals can synthesize *de novo*, fatty acids with the n-7 and n-9 configurations are not capable of sustaining alone the wide range of metabolic processes requiring fatty acids. Fatty acids with other configurations are also required. These must therefore be produced via the diet and are termed essential fatty acids (EFAs). However, the concept of essentiality as applied to a fatty acid is more complex than the simple interpretation for other nutrients, such as amino acids or vitamins, for several reasons.

Firstly, because of interconversion, the fatty acid provided in the diet need not be the acid ultimately responsible for alleviating the deficiency; it need only have appropriate double-bond configuration to allow synthesis of the metabolically active higher members of the series. A fatty acid may thus have EFA activity without being a necessary dietary constituent.

A further complication arises from the wide range of metabolic processes in which fatty acids are involved. There are thus many manifestations of a dietary EFA deficiency, but these may respond differently to treatment with different fatty acids.

Finally, when assessing responses of various parameters to different dietary fatty acids, it is sometimes difficult to differentiate between responses to a fatty acid acting in the metabolic role of an EFA rather than to other advantageous nutritional properties which, however, may not be unique to that fatty acid or family of acids. For these and other reasons there are thus still important gaps in our knowledge relating to EFAs.

EFA activity of fatty acids

Relative activities of fatty acids

Care must be taken when comparing the EFA activity of different fatty acids since the various metabolic processes involving EFAs can have different structural requirements for these acids. Thus relative EFA activity may vary depending upon which particular effect of a deficiency is used as a criterion of activity. For instance in the rat, the animal for which comparisons are most detailed, linolenic acid is relatively better at sustaining growth than reproduction (Mohrhauer and Holman, 1963; Quackenbush, Kummerow and Steenbock, 1942).

One of the most readily quantifiable criteria of EFA activity in animals is water intake. This changes markedly with changes in skin permeability and water loss induced by EFA deficiency (Thomasson, 1953). The relative EFA potencies of a range of fatty acids are shown in *Table 7.2*. Those with the highest potency belong to the linoleic acid (n-6) family, with greatest activity being shown by arachidonic acid. Although this acid is synthesized from lower members of the family, these precursors can have EFA activity on the basis of their own structural characteristics; they do not have to be converted to arachidonic acid to become active. Thus $\Delta^{8,11,14}$ eicosatrienoic acid is a starting point for the synthesis of some prostaglandins and linoleic acid can heal dermal lesions by direct application to the skin, where it cannot be converted to arachidonic acid (Houtsmuller and van der Beek, 1981). Linoleic acid may thus be thought of as the quintessential fatty acid for poultry, since provision of it in adequate amounts ultimately eliminates all symptoms of EFA deficiency.

Table 7.2 RELATIVE EFA POTENCIES OF FATTY ACIDS

Fatty acid	Position of terminal double bond	Relative potency
5,8,11,14-20:4	n-6	139
6,9,12-18:3	n-6	115
8,11,14-20:3	n-6	102
9,12-18:2	n-6	100
11,14-20:2	n-6	46
3,6,9,12-18:4	n-6	34
4,7,10,13-19:4	n-6	20
10,13-19:2	n-6	9
5,8,11,14-19:4	n-5	49
5,8,11,14-21:4	n-7	62
9,12,15-18:3	n-3	9
7,10,13-19:3	n-6	6
8,11,14-18:3	n-4	0
5,8,11-20:3	n-9	0
8,11,14-22:3	n-8	0

From Holman, 1978

Fatty acids with double-bond configurations other than n-6 also show some EFA activity (*Table 7.2*). Activity of two acids with n-5 and n-7 chains is comparatively high by the criterion of water intake, but these odd-chain fatty acids seldom occur in foods in nutritionally significant amounts. In contrast, n-3 linolenic acid shows only comparatively low EFA activity.

All the double bonds in the structures in *Table 7.2* have the *cis* configuration. Corresponding fatty acids with a *trans* double bond are not active as EFAs themselves and cannot be converted into *cis* derivatives. Thus Lanser, Mounts and Emken (1978) found that, when linoleic acid and its *trans, trans* isomer linoelaidic acid were fed to hens, arachidonic acid was synthesized solely from linoleic acid. *Trans* fatty acids occur in nature and as by-products of industrial processes. In small amounts they are harmless and can be used in metabolic systems. However, when fed in high concentration they can distort phospholipid composition and inhibit prostaglandin synthesis (Kinsella et al., 1979).

An interesting fatty acid containing a *trans* double bond is columbinic acid ($\Delta^{6,9}$,*trans*12-octadecatrienoic acid (of the n-6 series)) which occurs in a concentration of 60% in columbine (*Aquilegia*) seed oil. It has similar EFA activity to linoleic acid in rats as measured by growth and skin permeability although it inhibits prostaglandin synthesis and seriously aggravates EFA deficiency lesions in the kidney (Houtsmuller, 1981).

Biochemical role of EFAs

EFAs are important in two aspects of metabolism. PUFAs are preferentially incorporated into phospholipids: in general a phospholipid molecule contains one saturated and one polyunsaturated fatty acid. Phospholipids are important structural constituents of membrane systems and also of lipid transport systems where they are essential for lioprotein formation. The structure of arachidonic acid, shown in *Figure 7.1*, indicates it has a hooked

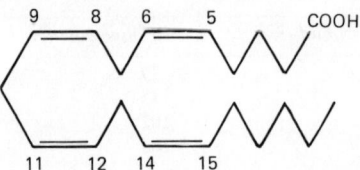

Figure 7.1 Structure of $\Delta^{5,8,11,14}$-eicosatetraenoic acid (of the n-6 series) arachidonic acid

structure. This may be especially favourable for binding in membranes. The low melting points of PUFAs may also contribute to fluidity of membranes. The n-6 chain appears to be important for EFA activity and may be involved in specific lipid–enzyme interactions.

The second vital role of EFAs is in the synthesis of prostaglandins. There are many different derivatives within this class of compound and the related thromboxanes. They are formed only transiently but are nevertheless vital regulators of a wide range of metabolic processes with hormone-like actions. They have special influences on kidney function and reproduction. The dominant prostaglandins so far encountered in mammals are derivatives of n-6, with those of initial chain length C20, and having either 3 or 4 double bonds, being most favourable precursors.

Effects of EFA deficiency

EFA deficiency can be induced by feeding a suitable diet to young chicks. The speed with which the signs develop depends upon the initial EFA status of the chick: they develop sooner in chicks from EFA-depleted hens and in faster-growing chicks.

At the biochemical level, the first changes are in fatty acid composition. The relative proportions of fatty acids of the linoleic acid family fall to low levels and this is particularly noticeable in phospholipids. In response, the proportions of the unsaturated acids in the oleic and palmitoleic acid families increase, with eicosatrienoic acid (n-9) showing the greatest relative increase. The ratio of this acid to arachidonic acid (the triene: tetraene ratio) is thus a good indicator of the degree of EFA deficiency. Holman (1960) has suggested that a ratio of 0.4 represents the threshold of deficiency.

The metabolic function of phospholipids deteriorates with these changes in fatty acid composition. Membranes become less cohesive and less fluid, leading to a wide range of structural and organic abnormalities. Lipoprotein formation and lipid transport are also impaired. The severity of these abnormalities can be intensified by increasing the metabolic demand for phospholipids. Thus the provision of extra dietary cholesterol or fat, especially saturated fatty acids, enhances the severity of a deficiency (Machlin and Gordon, 1961) by increasing the requirement for lipid transport. This necessitates the mobilization of the limited tissue reserves of PUFA for phospholipid synthesis.

Prostaglandins can have such profound effects on metabolic processes that any factors which affect their synthesis must have important consequences. It is thus possible that many of the effects of EFA deficiency are the results of inadequacies in the synthesis of prostaglandins. For instance, prostaglandins are involved in inflammatory reactions and changes in membrane permeability and it has been shown that topical application of some prostaglandins can clear the dermal lesions of EFA deficiency in the rat (Ziboh and Hsia, 1972). However, the field of prostaglandins is so complex that it is not possible to generalize that the only function of EFA is to act as a precursor for prostaglandin synthesis. Indeed, this is unlikely since body growth is more sensitive than prostaglandin synthesis to EFA deficiency. Triene to tetraene ratios of 6 are required before substantial inhibition of synthesis occurs (Parnham *et al.*, 1979).

The first outward sign of EFA deficiency is retarded growth. This can be apparent within one week of feeding a deficient diet (Machlin and Gordon, 1961) although, in slower-growing chicks, not until 6 weeks of age. As in rats, male chicks seem to be more susceptible to EFA deficiency (Roland and Edwards, 1972).

As the deficiency intensifies, the deterioration of membrane structures leads to increased capillary fragility and dermal problems. The skin has a rough, flaky appearance and its increased permeability leads to enhanced water loss and consequently greater water consumption. The birds also have a decreased resistance to disease (Ross and Adamson, 1961), a poorer efficiency of feed utilization and faulty feathering. The impairment of lipid transport leads to the formation of fatty livers and, in males, testis size is depressed and the development of secondary sex characteristics is delayed (Bieri *et al.*, 1956; Edwards, 1967).

Signs of deficiency are much slower to appear if birds are first fed a diet containing EFA. The birds then build up reserves which must be depleted before a deficiency can develop. Thus Rolands and Edwards (1972) found that the time taken for signs to occur in growing broilers fed a deficient diet was proportional to the period during which an adequate diet was fed initially.

Because of their large EFA reserves, adult birds rarely show signs of abnormality if fed a deficient diet. However, if they are reared on a deficient diet or their reserves become depleted, adverse effects occur. Hens suffer from reductions in rate of egg production and egg size while fertility and hatchability of eggs are depressed (Menge, Miller and Denton, 1963). Chicks hatching from deficient eggs are small and weak and show the characteristic abnormalities in fatty acid composition. They also have impaired viability and growth potential, even when fed adequate diets (Menge and Richardson, 1968).

Requirements for EFAs

Growing birds

Although arachidonic acid possesses the highest EFA activity, linoleic acid is the most common EFA from the nutritional point of view. Hence, virtually all practical studies on poultry centre on linoleic acid requirements. As with other nutrients, several criteria can be used to assess linoleic acid requirements of poultry: these include health, freedom from deficiency lesions, performance and biochemical criteria. However, some are unreliable. For instance, lesions can be influenced by environmental factors such as humidity and usually appear only after growth retardation has occurred. The use of growth response is direct, but careful experimental design is required to enable differentiation between responses to linoleic acid or fat in general.

Criticisms can be levelled at several studies where linoleic acid requirements have been assessed on the basis of growth response. For instance, Carew and Foss (1973) added graded levels of safflower or maize oils to a purified fat-free broiler diet. The oils were added at the expense of glucose, but the nutrient densities of the diets apparently changed, as did their fat contents. The authors attributed growth responses solely to linoleic acid and not surprisingly proposed a comparatively high requirement of 1.9% of the diet (5.3% of dietary metabolizable energy, ME). Indeed, dose-response analysis of the data would suggest an even higher requirement (about 4% of the diet). This study ignored the growth-stimulating effects of dietary fat in general, demonstrated in an earlier study by Hopkins and Nesheim (1967). These authors used a basal diet containing 10% hydrogenated coconut oil as a linoleate-free source of fat. However, even with this diet, growth responses of broilers at 3 weeks to the addition of linoleic or oleic acids prevented a meaningful interpretation of growth data. Liver weights and lipid contents were minimal with 1.35% linoleic acid in the diet, but even this value may be unduly high as a practical requirement since the diet contained an unusually high proportion of saturated fatty acids. As has already been shown, these fatty acids can enhance the metabolic requirement for PUFA and can also depress the absorbability of fat in young chickens (Renner and Hill, 1961).

The most uniform interpretation of linoleic acid requirements is provided by measurement of the ratio of eicosatrienoic to eicosatetraenoic acids in tissues. Holman (1978) has shown that the response of this ratio to

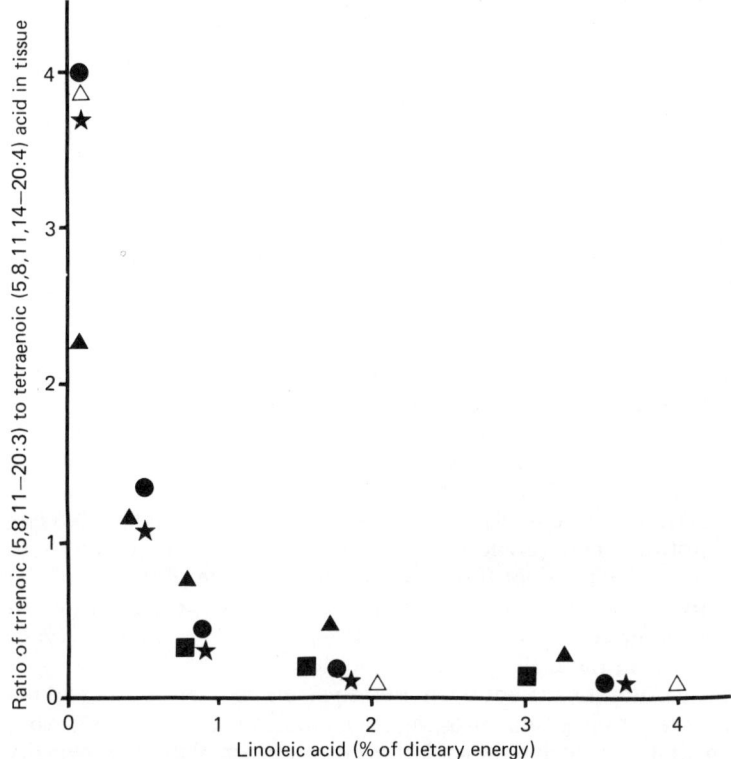

Figure 7.2 Triene:tetraene ratios in tissues from young chickens and turkeys fed diet containing different amounts of linoleic acid. ● (male), ★ (female) chick heart (Menge, 1970); ▲ chick liver (Hopkins and Nesheim, 1967); △ chick heart (Hill, Silbernick and McMeans, 1967); ■ poult liver (Ketola, Young and Nesheim, 1973)

dietary linoleic acid is similar in a wide range of tissues and species, implying that the physiological requirement for EFA is broadly the same for many animals. He suggested that a ratio of 0.4 indicated a minimum requirement and that this corresponded to a dietary linoleic acid content equivalent to 1% of the dietary energy. Data from several authors are plotted in *Figure 7.2* and show that similar relationships exist for young poultry, both chickens and turkeys. The triene: tetraene ratio falls to low values with amounts of linoleic acid providing between 1% and 2% of dietary ME. Using this criterion, the requirement of growing poultry would appear not to exceed 2% of dietary ME, or about 0.8% by weight of a diet containing 12.5 MJ/kg of ME.

This conclusion is applicable to poults and chicks, both males and females. Nevertheless some caveats are in order. Triene to tetraene ratio did not indicate a sex difference in the linoleic acid requirement of broilers (Menge, 1970) but growth data suggested the requirement of females might be lower. This observation could be accounted for by greater responsiveness of males to dietary fat; however it would be consistent with findings that males are more sensitive to EFA deficiency. Secondly, the suggested requirement may be enhanced if the diet contains high levels of saturated fats. Finally, these conclusions are based on experiments carried out many years ago. The author is unaware of any recent satsifactory reports on the linoleic acid requirements of growing birds.

Adult birds

The linoleic acid requirements of adult birds are difficult to quantify because of reserves accumulated during rearing. Experiments involving production responses are therefore best based on birds that have been reared on diets containing only low levels of EFA. However, even with EFA-depleted hens, some production responses are difficult to detect. Thus Hopkins and Nesheim (1962) and Jensen and Shutze (1963) observed no depression in rate of egg production in depleted birds, although other workers have observed responses to feeding small amounts of linoleic acid (Marion and Edwards, 1962; Menge et al., 1963; Menge, Calvert and Denton, 1964). Unfortunately the linoleic acid requirement for egg number cannot be quantified accurately from the data presented by these authors but it is probably not too dissimilar to the value of 0.12% of the diet found by Menge (1968) to be necessary for eggs to be fertile. In contrast, the requirement for hatchability is higher: data of Menge (1968) suggests that it is in the order of 1% of the diet.

Biochemical criteria have not been used to assess requirements in adults to the same extent as in young birds, although Marion and Edwards (1964) have shown that eicosatrienoic acid concentrations in tissues, especially heart, are sensitive to changes in the linoleic acid content of diets. However, the physiological requirement for linoleic acid has been assessed by turnover studies. On this basis, Balnave (1971a) has estimated the requirement of the laying hen to be 0.9% of the diet.

Few studies on requirements of breeding turkeys have been reported. However, Cooper and Barnet (1968), working with non-depleted birds, did not observe any increase in rate of egg production or hatchability when a diet containing 1.2% linoleic acid was further supplemented.

Egg size is one aspect of production where considerable uncertainty has existed over linoleic acid requirements. This followed observations that addition of 3–5% of maize oil to practical diets could markedly increase egg size in hens (Jensen et al., 1958; Shutze and Jensen, 1963) and turkeys (Cooper and Barnet, 1968). In subsequent years a large number of studies were carried out to investigate this response. Balnave (1971b) confirmed that the response was to some property of the fat other than its energy content, by comparing dietary supplements of maize oil and maize starch. Many workers were able to demonstrate responses to vegetable oils (e.g. Edwards and Morris, 1967) but responses to fish oil have been less marked (Marion and Edwards, 1964). The responses have been greatest when EFA-depleted hens have been studied (Menge et al., 1963) and have therefore been attributed to the linoleic acid content of the oils. As a result it was widely held that the laying hen had two requirements for linoleic acid—0.9% for physiological function and a higher requirement, of 2–4%, for maximum egg size. It might be expected that deficient birds would have a high EFA requirement, initially at least, in order to reverse the effects of the deficiency and build up reserves. However the concept of two widely disparate requirements in normal birds is less easy to sustain, especially since many of the studies involved adding differing amounts of oils to diets. The responses could therefore not be attributed unequivocally to linoleic acid alone.

To resolve this problem, two large-scale experiments were carried out at the Poultry Research Centre. The first involved four diets in which the fat supplements contained different ratios of safflower and olive oil. The diets were thus identical in all respects except the relative proportions of linoleic and oleic acid. The linoleic acid contents of the diets ranged from 0.8% to 2.3%, but did not influence egg size or any other aspect of performance when fed from day-old to the end of the laying year (Shannon and Whitehead, 1974).

Table 7.3 PERFORMANCE DURING THE PERIOD 20–73 WEEKS OF AGE OF HENS FED DIETS CONTAINING DIFFERENT AMOUNTS OF FAT AND LINOLEIC ACID

Dietary fat contents					
Total fat (%)	2.5	2.9	5.5	5.5	
Linoleic acid (%)	0.7	0.9	0.9	2.3	
Performance					*Significance of treatment effect*
Egg production (per 100 bird d)	77.9	78.5	78.1	77.3	NS
Mean egg weight (g)	56.7	57.3	58.8	59.2	$P<0.05$
Weight of eggs produced (g/bird d)	44.2	44.9	45.9	45.8	$P<0.001$
Food consumption (g/bird d)	115	116	116	116	NS

From Whitehead, 1981

In the second experiment, isonitrogenous and isoenergetic diets containing differing amounts of maize or olive oils were formulated to give the total dietary fat and linoleic acid contents shown in *Table 7.3*. These diets were fed to hens that had been reared on practical diets containing low or moderate amounts of linoleic acid. The results in *Table 7.3* showed that the responses in egg weight were attributable to the oil content of the diet rather than the linoleic content. Thus egg size with a diet containing 5.5% fat and 0.9% linoleic acid was similar to that with a diet containing the same amount of fat but 2.3% linoleic acid and greater than that with a diet containing 0.9% linoleic acid but only 2.9% fat (Whitehead, 1981). On the basis of these studies, the specific requirement of laying hens for linoleic acid is not greater than 0.9% of the diet.

General conclusions on EFA requirements

On the basis of the conclusions in the previous section, it appears that the EFA requirements of immature and mature poultry are broadly uniform and that under most circumstances a dietary linoleic acid content of 0.9% is adequate. Aspects of performance will respond to higher levels of this or other, non-essential, fatty acids but these responses cannot be ascribed to EFA activity.

This conclusion should not be taken to imply that EFAs cannot be considered in isolation from other aspects of the diet, particularly fat. Efficient digestion and transport of fat is facilitated by fatty acids whose degrees of unsaturation and melting points give them advantageous properties. Thus oleic and linoleic acids are not only highly absorbable in themselves but can enhance the absorbability of other fatty acids (Young and Garrett, 1963). Phospholipids are vital for fat transport but, as has been discussed earlier, require PUFA. It is thus likely that there are ranges of optimal dietary fatty acid compositions and that these may change depending upon the amount or type of fat in the diet.

Meeting EFA requirements

The linoleic acid contents of most poultry feed ingredients are known: the values obtained in one of the more recent surveys are given in *Table 7.4*. Calculation of dietary content is not complicated by questions of availability since the fatty acid is highly digestible, even in young chicks.

The fatty acid compositions of a range of fats used in poultry feeding are given in *Table 7.5*. Vegetable oils such as safflower and maize oils are rich sources of linoleic acid, whereas tallow and fish oils contain only small proportions. However fish oils are rich sources of PUFA of the n-3 linolenic acid family and the extent to which acids in this family can contribute towards EFA requirements is of interest.

As already discussed, linolenic acid can show variable EFA activity in rats, depending upon the criterion used for assessment. Comparable studies have not been carried out in chicks. However Machlin and Gordon (1961) did not observe a growth response when 1.5% linolenic acid was added to a diet deficient in linoleic acid. This comparison extended over one week only, so the evidence is not convincing, but is likely nevertheless that linolenic acid itself has low EFA activity in chicks.

Table 7.4 TOTAL FATTY ACID LINOLEIC ACID CONTENTS OF SOME POULTRY FEED INGREDIENTS

Ingredient	Fatty acids (wt %)	Linoleic acid (wt %)
Barley	1.65	0.94
Maize	3.52	2.14
Maize germ	20.43	12.75
Milo	2.85	1.03
Oats	3.52	1.22
Wheat	1.63	0.83
Wheatfeed	2.96	1.67
Wheatgerm	4.31	2.35
Field bean	1.30	0.70
Groundnut	8.98	1.97
Soybean meal (ext)	1.15	0.59
Fish meal	4.60	0.06
Meat and bone meal	2.23	0.04
Poultry feather, offal and blood meal	19.10	2.30

From Whitehead, 1972

Table 7.5 FATTY ACID PROFILE (%) OF SOME FATS AND OILS

Fatty acid	Position of terminal double bond	Tallow	Lard	Olive oil	Maize oil	Safflower oil	Linseed oil	Herring oil[a]	Menhaden
14:0								3	12
16:0		34	28	6	12	7	5	10	20
18:0		37	20	4	3	3	3	1	4
16:1		3	2		1			7	11
18:1		25	41	83	27	13	21	12	13
20:1								10	
22:1								9	2
9,12-18:2	n-6	1	9	7	56	77	18	2	3
6,9,12-18:3	n-6							0.5	
8,11,14-20:3	n-6							0.5	
5,8,11,14-20:4	n-6								2
7,10,13,16-22:4	n-6							0.5	
9,12,15-18:3	n-3			1			53		3
6,9,12,15-18:4	n-3							3	3
8,11,14,17-20:4	n-3							0.5	
5,8,11,14,17-20:5	n-3							17	12
7,10,13,16,19-22:5	n-3							1	2.5
4,7,10,13,16,19-22:6	n-3							18	7

[a] Residual fat in stabilized herring meal (Gunstone and Wijesundera, 1978)

A more complex situation arises with the PUFA in fish oils or stabilized fish meals (PUFAs are destroyed in unstabilized or cured fish meals). The high concentration of PUFA in these products means that these acids can be provided via the diet in higher quantities than could be synthesized in the bird from moderate amounts of linolenic acid. There have been several studies to explore whether these acids have EFA activity in poultry.

Engster, Carew and Foss (1975) found that diets containing 20% maize oil (12.07% linoleate) or 20% herring oil (0.55% linoleate) supported similar rates of growth in chicks. However, the amounts of unsaturated fatty acids involved were too large to make a meaningful quantitative comparison. In contrast Menge (1971) fed lower levels of oil and found that 2.58% of menhaden oil, providing 0.13% linoleate plus arachidonate and 0.87% other PUFA, gave as good growth as a control diet containing the same amount of vegetable oil. Menge, Calvert and Denton (1965) have also observed that menhaden oil can stimulate reproduction in the hen.

Biochemical criteria are not so useful for assessing the contribution of PUFAs to EFA status. The presence of substantial amounts of n-3 PUFA inhibits the synthesis of $\Delta^{5,8,11}$-eicosatrienoic acid (of the n-9 series); thus ratios involving this fatty acid do not give a meaningful indication of EFA status when fish oils are fed. However, on the basis of production data it seems that n-3 PUFAs can partly replace linoleic acid in meeting some of the metabolic requirements for EFA. They cannot totally replace n-6 fatty acids since there is no evidence that they can act as precursors for prostaglandins derived from n-6 fatty acids; indeed they have been found to inhibit prostaglandin synthesis from arachidonic acid in rats (Lands et al., 1973). $\Delta^{5,8,11,14,17}$-eicosanpentaenoic acid (of the n-3 series) is a precursor of PGE_3 and thromboxanes, but n-3 prostaglandins and derivatives have a minor role in prostaglandin metabolism compared with derivatives of the n-6 EFAs.

Excess of EFAs

EFAs and PUFAs have no direct toxic effects when fed in excess of requirements. However, double-bond systems are susceptible to oxidation and development of rancidity may lead to harmful effects. Oxidation can be inhibited by the presence of antioxidants, which are themselves consumed in the process. Vitamin E is nature's antioxidant and the requirement for it is increased in the presence of high levels of unsaturated fatty acids. In the absence of sufficient vitamin E, these fats can induce the typical signs of a deficiency of this vitamin (Yu, Yu and Young, 1974; Bartov and Bornstein, 1980).

PUFA have also been observed to enhance the severity of zinc deficiency in the chick (Bettger et al., 1980). However this is opposite to the effect observed in the rat and the mechanism is uncertain.

References

BALNAVE, D. (1971a). The contribution of absorbed linoleic acid to the metabolism of the mature laying hen. *Comparative Biochemistry and Physiology* **40A**, 1097–1105

BALNAVE, D. (1971b). Response of laying hens to dietary supplementation with energetically equivalent amounts of maize oil or maize starch. *Journal of the Science of Food and Agriculture* **22**, 125–128

BARTOV, I. and BORNSTEIN, S. (1980). Susceptibility of chicks to nutritional encephalopathy: effect of fat and α-tocopherol content of the breeder diet. *Poultry Science* **59**, 264–267

BETTGER, W.J., REEVES, P.G., MOSCATELLI, E.A., SAVAGE, J.E. and O'DELL, D.L. (1980). Introduction of zinc and polyunsaturated fatty acids in the chick. *Journal of Nutrition* **110**, 50–58

BIERI, J.G., BRIGGS, G.M., FOX, M.R.S., POLLARD, C.J. and ORTIZ, L.O. (1956). Essential fatty acids in the chick. 1. Development of fat deficiency. *Proceedings of the Society for Experimental Biology and Medicine* **93**, 237–240

CAREW, L.B. and FOSS, D.C. (1973). High dietary requirement of male chicks for linoleic acid. *Poultry Science* **52**, 1676–1678

CASTER, W.O., HILL, E.G. and HOLMAN, R.T. (1963). Estimation of essential fatty acid intake in swine. *Journal of Animal Science* **22**, 389–392

COOPER, J.B. and BARNET, B.D. (1968). Response of turkey hens to dietary linoleic acid fed as corn oil. *Poultry Science* **47**, 671–677

EDWARDS, D.G. and MORRIS, T.R. (1967). The effect of maize and maize oil on egg weight. *British Poultry Science* **8**, 163–168

EDWARDS, H.M. (1967). Studies of essential fatty acid deficiency of the growing domestic cock. *Poultry Science* **46**, 1128–1133

ENGSTER, H.M., CAREW, L.B. and FOSS, D.C. (1975). Effect of herring oil on body weight, comb size and gonadal development in the chick. *Poultry Science* **54**, 2118–2121

GUNSTONE, F.D. and WIJESUNDERA, R.C. (1978). The component acids of the lipids in four commercial fish meals. *Journal of the Science of Food and Agriculture* **29**, 28–32

HILL, E.G., SILBERNICK, C.L. and McMEANS, E. (1967). Dietary linoleate and methionine in chicks. *Poultry Science* **46**, 523–526

HOLMAN, R.T. (1960). The ratio of trienoic: tetraenoic acids in tissue lipids as a measure of essential fatty acid requirement. *Journal of Nutrition* **70**, 405–410

HOLMAN, R.T. (1978). Essential fatty acid deficiency. In *CRC Handbook Series in Nutrition and Food. Section E: Nutritional Disorders*, vol 2. (M. Rechcigl, Ed.), pp. 491–515. West Palm Beach, Florida, CRC Press Inc.

HOPKINS, D.T. and NESHEIM, M.C. (1962). The effect of linoleic acid depletion on performance of hens and their progeny. *Poultry Science* **41**, 1651

HOPKINS, D.T. and NESHEIM, M.C. (1967). The linoleic acid requirement of chicks. *Poultry Science* **46**, 872–881

HOUTSMULLER, U.M.T. (1981). Columbinic acid, a new type of essential fatty acid. *Progress in Lipid Research* **20**, 889–896

HOUTSMULLER, U.M.T. and VAN DER BEEK, A. (1981). Effects of topical application of fatty acids. *Progress in Lipid Research* **20**, 219–224

JENSEN, L.S. and SHUTZE, J.V. (1963). Essential fatty acid deficiency in the laying hen. *Poultry Science* **42**, 1014–1019

JENSEN, L.S., ALLRED, J.B., FRY, R.E. and McGINNIS, J. (1958). Evidence for an unidentified factor necessary for maximum egg weight in chickens. *Journal of Nutrition* **65**, 219–223

KETOLA, H.G., YOUNG, R.J. and NESHEIM, M.C. (1973). Linoleic acid requirement of turkey poults. *Poultry Science* **52**, 597–603

KINSELLA, J.E., HWANG, D.H., YU, P., MAI, J. and SHIMP, J. (1979). Prostaglandins and their precursors in tissues from rats fed on *trans, trans*-linoleate. *Biochemical Journal* **184**, 701–704

LANDS, W.E.M., LE TELLIER, P.R., ROME, L.W. and VANDERHOCK, J.Y. (1973). In *Advances in the Biosciences*, vol 9 (S. Bergström and S. Bernhard, Eds), p. 15. New York, Pergamon Press

LANSER, A.C., MOUNTS, T.L. and EMKEN, E.A. (1978). Metabolism of linoleate versus linoelaidate in the laying hen. *Lipids* **13**, 103–109

MACHLIN, L.J. and GORDON, R.S. (1961). Effect of dietary fatty acids and cholesterol on growth and fatty acid composition of the chicken. *Journal of Nutrition* **75**, 157–164

MARION, J.E. and EDWARDS, H.M. (1962). The response of fat deficient hens to corn oil supplementation. *Poultry Science* **41**, 1785–1792

MARION, J.E. and EDWARDS, H.M. (1964). The response of laying hens to dietary oils and purified fatty acids. *Poultry Science* **43**, 911–918

MENGE, H. (1968). Linoleic acid requirement of the hen for reproduction. *Journal of Nutrition* **95**, 578–582

MENGE, H. (1970). Comparative requirements of linoleic acid for male and female chicks. *Poultry Science* **49**, 178–183

MENGE, H. (1971). The influence of dietary oils in chick growth rate. *Poultry Science* **50**, 261–266

MENGE, H. and RICHARDSON, G.V. (1968). The influence of a linoleic-acid-deficient maternal diet on growth of progeny. *Poultry Science* **47**, 542–547

MENGE, H., CALVERT, C.C. and DENTON, C.A. (1964). Further studies on the effect of a low-fat diet on reproduction in the hen. *Poultry Science* **43**, 1341

MENGE, H., CALVERT, C.C. and DENTON, C.A. (1965). Influence of dietary oils on reproduction in the hen. *Journal of Nutrition* **87**, 365–370

MENGE, H., MILLER, E.C. and DENTON, C.A. (1963). Effect of an essential fatty acid deficient diet on the reproductive performance of chickens. *Poultry Science* **42**, 1291

MOHRHAUER, H. and HOLMAN, R.T. (1963). The effect of dietary essential fatty acids upon composition of polyunsaturated fatty acids in depot fat and erythrocytes of the rat. *Journal of Lipid Research* **4**, 346–350

PARNHAM, M.J., VINCENT, J.E., ZILSTRA, F.J. and BONTA, I.L. (1979). The use of essential fatty acid deficient rats to study pathophysiological roles of prostaglandins. Comparison of prostaglandin production with some parameters of deficiency. *Lipids* **14**, 407–412

QUACKENBUSH, F.W., KUMMEROW, F.A. and STEENBOCK, H. (1942). The effectiveness of linoleic, arachidonic and linolenic acids in reproduction and lactation. *Journal of Nutrition* **24**, 213–224

RENNER, R. and HILL, F.W. (1961). Factors affecting the absorbability of saturated fatty acids in the chick. *Journal of Nutrition* **74**, 254–258

ROLAND, D.A. and EDWARDS, H.M. (1972). Effect of linoleic acid reserves on essential fatty acid deficiency of the chick. *Poultry Science* **51**, 382–389

ROSS, E. and ADAMSON, L. (1961). Observations on the requirements of young chicks for dietary fat. *Journal of Nutrition* **74**, 329–334

SHANNON, D.W.F. and WHITEHEAD, C.C. (1974). Lack of a response in egg weight or output to increasing levels of linoleic acid in practical layer's diets. *Journal of the Science of Food and Agriculture* **25**, 553–561

SHUTZE, J.V. and JENSEN, L.S. (1963). Influence of linoleic acid on egg weight. *Poultry Science* **42**, 921–924

THOMASSON, H.J. (1953). Biological standardization of essential fatty acids (a new method). *Internationale Zeitschrift für Vitaminforschung* **25**, 62–82

WHITEHEAD, C.C. (1972). The linoleic acid contents of some British poultry foods. *Journal of the Science of Food and Agriculture* **23**, 1503–1507

WHITEHEAD, C.C. (1981). The response of egg weight to the inclusion of different amounts of vegetable oil and linoleic acid in the diet of laying hens. *British Poultry Science* **22**, 525–532

YOUNG, R.J. and GARRETT, R.L. (1963). Effect of oleic and linoleic acids on the absorption of saturated fatty acids in the chick. *Journal of Nutrition* **81**, 321–329

YU, W.H.A., YU, M.C. and YOUNG, P.A. (1974). Ultrastructural changes in the cerebrovascular endothelium induced by a diet high in linoleic acid and deficient in vitamin E. *Experimental and Molecular Pathology* **21**, 289–299

ZIBOH, V.A. and HSIA, S.L. (1972). Effects of prostaglandin E_2 on rat skin: inhibition of sterol ester biosynthesis and clearing of scaly lesions in essential fatty acid deficiency. *Journal of Lipid Research* **13**, 458–467

8

ESSENTIAL FATTY-ACID/MINERAL INTERACTIONS WITH REFERENCE TO THE PIG

STEPHEN C. CUNNANE
Efamol Research Institute, Kentville, Nova Scotia, Canada B4N 4H8

Introduction

At the outset, I wish to disclaim a specific knowledge of general pig nutrition. Rather, my interest is in the development of an understanding of the importance of zinc, copper, and essential fatty acids (EFA) to nutrition both as discrete nutrients and as they relate to each other. Because the pig has been the animal model from which a good deal of information about the interaction of these nutrients has been learned, it is fitting that this subject should be reviewed with reference to the pig.

Without a practical application, the discussion of the relation between zinc, copper and EFA may seem esoteric to the nutritionist studying farm animals. The application of this interaction of nutrients in the pig is orientated around the difficulties which have been reported with the intensive rearing of pigs. These include parakeratosis, and the potential problems associated with reproduction and early weaning. In young rapidly growing pigs which nowadays are fed a copper-supplemented diet, the anticipated rate of growth is not always achieved, and this may often be associated with parakeratosis. From detailed literature descriptions of these problems, an association with metabolic defects in zinc and EFA metabolism is a strong possibility. This review seeks to address the subject from this point of view.

The EFAs have been recognized as a distinct group of essential nutrients for over 50 years but their true relevance in nutritional studies has not been fully recognized. Two reasons for this are apparent. First, in many instances it is not generally recognized that there are two types of body fat: storage fat and structural fat. Storage fat is made up mainly of fats which can be made by the body and functions primarily as an energy source or depot. Storage fat composition, which usually reflects dietary fat composition, also contains relatively low amounts of EFA. Structural fat comprises the fatty acids which make up cell membranes and its composition changes more slowly than that of storage fat in relation to changes in dietary fat intake. Unlike storage fat, cell membranes are made up to a large extent of EFA. Hence the formation of new cells during growth requires a constant supply of EFA. Cell growth is not only a function of protein availability but equally of EFA availability. As such, EFA availability may be a limiting factor in growth. Cells which divide rapidly or are constantly replaced, e.g. skin, suffer rapidly from EFA deficiency. Therefore, young growing animals or those which are pregnant are most susceptible to EFA deficiency and it is in these rather than in mature animals that the symptoms of EFA deficiency are most often observed.

EFAs also contribute to membrane fluidity and therefore, to a large extent, regulate cell function. This property alone makes EFAs important but the fact that they cannot be synthesized *de novo* by animals, makes them essential. Hence, EFAs are a highly specific sub-group of fatty acids, the function of which does not relate to increasing energy availability as the term 'fat' implies. The term 'crude fat', as is used in dietary analyses or in tables of nutritional requirements, is therefore very misleading and gives no information about EFA content. Thus, a provision for adequate EFA must be made independent of the total fat in the diet. Failure to provide adequate EFA causes symptoms of EFA deficiency: in the pig, for instance, symptoms of parakeratosis such as poor growth, skin lesions and decreased viability of piglets, result.

The second reason why the importance of EFAs in nutrition is not fully recognized is that their metabolism is a function of a number of dietary cofactors. These cofactors, which include zinc and selenium and possibly some of the vitamins, are essential in their own right. Part of their essentiality has been shown to be due to their function in regulating EFA metabolism. Thus, EFAs must be present in the diet at a level of 2–5% of the dietary energy content. This alone does not assure their utilization. The cofactors must also be present to regulate functions such as desaturation and elongation, esterification, and prostaglandin synthesis. Deficiency of these cofactors, particularly zinc, may induce a secondary metabolic deficiency of EFAs, even in the presence of adequate EFA intake. Thus, zinc deficiency, for instance, appears to induce a secondary EFA deficiency, and therapies to correct zinc deficiency have been shown to work more effectively when they include EFAs (Hanson, Sorensen and Kernkamp, 1958; Cash and Berger, 1969; Bettger *et al.*, 1979; Cunnane and Horrobin, 1980).

Inhibitory factors in the diet which affect EFA metabolism also exist. The non-essential fatty acids (non-EFA) are the most important. Saturated fatty acids and mono-unsaturated fatty acids are non-essential, e.g. can be synthesized *de novo* by animals. The metabolism of the long-chain saturated fatty acids to the mono-unsaturated fatty acids involves enzymes common to EFA metabolism. Thus, excess amounts of the non-EFA in the diet or synthesized in the body compete with EFA and may induce an EFA deficiency (Holman, 1971). Similarly, cofactors in non-EFA metabolism, such as copper, may promote EFA deficiency by competitive inhibition.

These conditions affecting true availability of EFAs are relevant to farm animal nutrition. This is especially true in pigs where the dietary zinc:copper ratio is often artificially manipulated without an understanding of the metabolic implication of these alterations. Although most commercially prepared diets are probably adequate in EFAs, there may be some doubt as to whether the ratio of copper to zinc in pig diets is appropriate. By reviewing the effects of deficiency of zinc and excess of copper and the relation of these imbalances to EFA metabolism, it is hoped that a better understanding of the potential hazards of such imbalances may be possible.

Essential fatty acids

The EFAs were established as essential 'vitamins' over 50 years ago. At that time linoleic acid and arachidonic acid were considered to be the only two EFAs. By definition, it is strictly only linoleic which is the EFA since animals have the enzymatic capacity to synthesize arachidonic and the intermediate fatty acids—γ-linolenic acid and dihomo-γ-linolenic acid. α-Linolenic acid and its derivatives, $\Delta^{5,8,11,14,17}$-eicosapentaenoic acid and $\Delta^{4,7,10,13,16,19}$-docosahexaenoic acid are also EFA but are of a different class. They are quantitatively less important than the EFA of the n-6 class

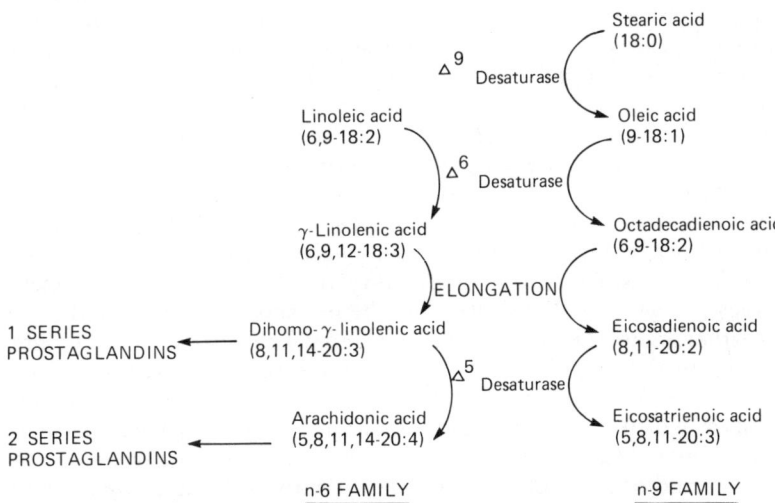

Figure 8.1 Outline of the stepwise parallel metabolism of linoleic acid and stearic acid via common enzymes. The conversion of dihomo-γ-linolenic and arachidonic acids to their respective series of prostaglandins occurs via the intermediate endoperoxides. The other families of fatty acids not shown are the n-3 derived from α-linolenic acid and the n-7 derived from palmitic acid

and are found predominantly in neural tissue and in highly specialized membranes. The parallel metabolic relationships between these fatty acids are shown in *Figure 8.1*. Also included in *Figure 8.1* is the metabolism of the long-chain non-EFAs—palmitic acid, palmitoleic acid, stearic acid and oleic acid. It is evident that their metabolism follows steps parallel to those of the EFA utilizing the same enzymes at the various steps.

The difficulty in supplying adequate EFA metabolites from linoleic acid arises from two factors relating to this unique metabolic relationship: (1) the fact that the enzymes are utilized by common and therefore competing substrates, e.g. the Δ^6 desaturase can accept α-linolenic acid, linoleic acid

Table 8.1 FACTORS AFFECTING ESSENTIAL FATTY ACID METABOLISM

Stimulatory	(A) Minerals	zinc
		selenium
	(B) Vitamins	vitamin B6
		vitamin C
		vitamin E
Inhibitory	(A) Minerals	copper
		calcium
	(B) Other	mono-unsaturated fatty acids
		saturated fatty acids
		age
		corticosteroids

or oleic acid but in decreasing order of preference; (2) the fact that these enzymes are dependent on cofactors which are themselves essential micronutrients, e.g. zinc and iron. These elements can themselves be inhibited by less essential elements such as copper (*Table 8.1*).

The EFAs have two recognized functions: (1) they are membrane components which help to maintain the functional integrity of those membranes; (2) they are precursors of the prostaglandins. The ability of erythrocytes to resist rupture in response to haemolytic stress is an example of membrane integrity. When the relative amount of linoleic to oleic acid is high, the erythrocytes have a greater resistance to haemolysis (Huang *et al.*, 1980). The same is true when the ratio of zinc to copper is increased (Bettger, Fish and O'Dell, 1978). Erythrocyte stability is decreased, however, when the ratio of zinc to copper is decreased, e.g. in zinc deficiency or when supplemental copper is given (Bettger *et al.*, 1978). Thus, EFAs have an important regulatory function in cell membranes, which does not necessarily involve them being metabolically altered.

Their other main function, as prostaglandin precursors, does require enzymatic conversion. A lipase is needed to release free EFA (dihomo-γ-linolenic, arachidonic or $\Delta^{5,8,11,14,17}$eicosapentaenoic acid) from membranes after which the cycloxygenase enzyme converts these acids to prostaglandins. Alternatively, conversion via the lipoxygenase to leukotrienes and other hydroxy fatty acids is possible. The complexity of the actions of prostaglandins and leukotrienes defies a comprehensive summary here; suffice it to say that they regulate cell functions and responsiveness at very low concentrations (10^{-13}–10^{-10} M), that they are ubiquitously synthesized and that many disease processes involve deranged synthesis of prostaglandins and/or leukotrienes.

Typical symptoms of EFA deficiency include the following: growth retardation, dermal lesions, capillary fragility, increased water loss through the skin, immunoincompetence resulting in easy infection, reproductive failure including sterility, disrupted parturition, increased perinatal mortality, and cardiovascular defects. The majority of these defects affect mainly the young rapidly growing or pregnant animal. Mature adult animals generally have sufficient stores of EFA in muscle, liver and adipose tissue to resist EFA deficiency for long periods. Neither young nor pregnant animals have these reserves. In both these cases, cell growth is rapid and constant, hence the demand for additional EFA is also constant and the susceptibility to EFA deficiency is greater.

In the pig, EFA deficiency has been induced in weaners (Witz and Beeson, 1951; Hill et al., 1957; Hanson et al., 1958; Howard et al., 1965). The characteristics include hair loss, scaly dermatitis, skin necrosis, poor growth, inanition and early mortality. Aortic lesions involving calcification but not lipid deposition have also been reported in conjunction with markedly decreased amounts of linoleic and arachidonic acid in heart lipids (Hill et al., 1957). Using older animals, Leat (1961) was less successful at inducing EFA deficiency in the pig, but still demonstrated increased erythrocyte haemolysis (Leat, 1962) and skin lesions in pigs up to 90 kg in weight. Leat (1962) also demonstrated that the requirement of linoleic acid in the growing pig is greatest at an age from 12 to 16 weeks, the age at which parakeratosis in pigs due to EFA deficiency is common (Hanson et al., 1958).

Zinc

Zinc has been known for over 50 years to be essential to animals (for reviews see National Research Council, 1979 and Underwood, 1977). Its essentiality was first established in the enzyme carbonic anhydrase, and more recently for many other enzymes including DNA and RNA polymerase, alcohol dehydrogenase, alkaline phosphatase and the carboxypeptidases. However, it is more the exception than the rule when these enzymes are found to be reduced in activity associated with zinc deficiency. Similarly, tissue zinc levels generally remain within normal limits except under extreme conditions of zinc deficiency. Artificially prepared zinc-deficient diets induce a rapid decline in plasma zinc levels in most species, but there are many situations in which subclinical zinc deficiency exists without a significant decrease in plasma zinc. Hence, two major problems haunt the zinc physiologist: (1) how to assess what zinc really does; (2) how to assess true zinc status. The answer to the latter is not within the scope of this chapter although measurement of zinc in nucleated cells such as peripheral blood leucocytes is showing promise (Jones et al., 1981; Meadows et al., 1981). The answer to the former may involve EFAs (Cunnane, 1982a, b).

Zinc deficiency has been induced experimentally in many species including pigs (Hanson et al., 1958; Miller et al., 1968; Thompson, Gilbreath and Bielk, 1975; Wegger and Palludan, 1978). Zinc deficiency causes reduced growth, skin lesions, immunological and reproductive defects, increased permeability of the skin to water and increased capillary fragility. Tissue zinc levels may be depleted and activities of enzymes such as serum alkaline phosphatase decreased but these effects appear to bear no relation to the severity of the deficiency (Burch et al., 1974). In the pig, zinc deficiency is characterized by parakeratosis, hyperkeratosis, seborrhoea, hyperaemia, skin infection and often poor growth (*Table 8.2*). It is most often observed in young growing animals 4–12 weeks of age. In younger pigs, greasy pig disease is more commonly found than parakeratosis (Hanson et al., 1958). The beneficial effect of zinc in parakeratosis is now well established (Tucker and Salmon, 1955; Hanson et al., 1958). Thus, parakeratosis appears to occur mainly during the period

Table 8.2 EFFECTS OF ZINC DEFICIENCY COMMON TO MOST SPECIES

1. Inhibition of growth
2. Loss of appetite
3. Skin lesions
4. Increased water loss across the skin
5. Hair loss
6. Decreased immune response
7. Reproductive sterility in both sexes
8. Impaired parturition
9. Increased neonatal mortality
10. Increased capillary fragility

of rapid growth and the gross lesions involve particularly the skin, a rapidly and continuously dividing tissue. Zinc, like the EFAs, is crucial for cell division and protein synthesis; hence the role of its deficiency in this disorder is understandable.

The relation of excess dietary calcium to parakeratosis has also been reported (Hanson *et al.*, 1958; Forbes, 1960). This fits with zinc deficiency being a component of parakeratosis because of the inhibitory effect of calcium on zinc absorption and metabolism (Whiting and Bezeau, 1958; Hoekstra, 1964; Norrdin *et al.*, 1973), especially in the presence of marginal zinc intake (Huber and Garshoff, 1970) or during pregnancy (Tao and Hurley, 1975).

Reproductive aspects of zinc deficiency in pigs are also well documented (Wegger and Palludan, 1978). Zinc deficiency which extends through pregnancy to parturition was shown both to delay and to prolong farrowing, effects which are well recognized in other species as well (Hurley, Gowen and Swenerton, 1971; Hurley and Mutch, 1973). The litter weight and size, however, may be relatively unaffected by maternal zinc deficiency. Zinc transfer to the fetus is severely reduced, thus becoming responsible for the higher percentage of stillbirths and lower viability of the piglets (Wegger and Palludan, 1978).

The pig may have a relatively high requirement for zinc, making it easier to induce zinc deficiency in this species. This suggestion follows reports that zinc deficiency symptoms in the pig could be induced on a diet containing as much as 20 mg/kg zinc and that as much as 90 mg/kg zinc was required to maintain adequate zinc status (Thompson *et al.*, 1975).

Zinc requirements of pigs depend not only on the amounts of the competitive elements—copper and calcium—but also on the type of dietary protein. Casein-based diets generally reduce the zinc requirement (Shanklin *et al.*, 1968) whereas soyabean-based diets may increase it to as much as 50 mg/kg (National Research Council, 1979). In view of the fact that pig diets may have an excess of calcium (Hanson *et al.*, 1958) and that feeding is directed at achieving maximal rate of growth, which puts a high demand on available zinc, it is possible that zinc levels in pig rations may be sub-optimal. In situations of artificial rearing, such as in early weaning units, in which milk substitutes are fed from birth, zinc deficiency may be even easier to induce. This is because zinc absorption from formulated diets is well known to be as much as 50% less than from maternal milk (Widdowson, Dauncey and Shaw, 1974; Hambidge *et al.*, 1979). Hence, the zinc in formulated pig diets fed in the early post-natal period cannot be considered to be totally available and should be increased accordingly.

Copper

Copper is also well established as an essential trace element (O'Dell, 1976; Underwood, 1977). Its deficiency is associated with defects in myelination, connective tissue formation and catecholamine synthesis. Its role in connective tissue synthesis was first demonstrated in copper-deficient pigs (Shields *et al.*, 1962). The defect was a dramatic reduction in aortic tensile strength, allowing the formation of aneurysms which burst easily causing death due to internal haemorrhaging (Shields *et al.*, 1962; O'Dell *et al.*, 1966).

The copper requirement of pigs has been suggested to be in the range of 5–10 mg/kg (Underwood, 1977; Okonkwo *et al.*, 1979) for normal growth, an amount easily obtained from most commercial diets. However, following the observation over 50 years ago that copper supplementation increased pig growth, the addition of copper to pig diets to the order of 250 mg/kg has become standard practice, although current legislation in the United Kingdom has reduced this. The beneficial effect of copper at this level on growth and feed conversion efficiency was established (Allen *et al.*, 1961) but the mechanism of its effect still remains to be elucidated.

Copper supplementation at 250 mg/kg has not always proved beneficial to carcass composition or growth (Barber *et al.*, 1961; Suttle and Mills, 1966a; Gipp, Pond and Kallfelz, 1974a; Gipp *et al.*, 1974b). In fact, in some areas it may be detrimental and in areas of restricted availability of other minerals, may even cause death (Underwood, 1977). It is interesting to note that self-selection of dietary copper level in pigs showed that the preference was for about 170 mg/kg, also suggesting that 250 ppm may be excessive (Braude, 1966).

The attempt to achieve maximal feed conversion and growth has overshadowed the possible interaction of high levels of copper with other trace elements. In fact, copper interacts with the utilization of many minerals, especially zinc and iron. It is the amount of zinc and iron in the diet relative to copper which determines whether increasing copper to 250 mg/kg will be detrimental (*Table 8.3*). When iron or zinc levels are low

Table 8.3 COMPARISON OF THE EFFECTS OF COPPER SUPPLEMENTATION ALONE VS COPPER SUPPLEMENTATION WITH ADDITIONAL ZINC AND IRON, ON GROWTH, FOOD CONVERSION EFFICIENCY AND SKIN LESIONS. THE DATA REFER TO SWINE FED OVER A PERIOD OF 60–70 DAYS. ADAPTED WITH PERMISSION FROM SUTTLE AND MILLS (1966b)

	Control	Copper supplementation	Copper, zinc and iron supplementation
Live weight (kg)	42	39	53
Food conversion efficiency (kg food eaten/kg weight gained)	3.43	3.29	2.93
Skin lesions[a]	0.2	1.3	0.2

[a] Scale: 0 = none; 3 = severe lesions

or marginal, high levels of copper are more likely to be detrimental, but when they are above the inflated minimal requirement as a consequence of the presence of added copper, then copper toxicity is unlikely (Gipp et al., 1974b; Suttle and Mills, 1966b). The effect of excess copper on iron is to induce iron-deficient anaemia and to cause increased lipid peroxidation (Jain et al., 1983), effects which are inhibited by increasing iron to the same level as copper. Similarly, increasing zinc is protective (Suttle and Mills, 1966b). The effect of excess copper on zinc is to inhibit zinc absorption and utilization by enzymes. Indirectly, copper also inhibits zinc effects on EFAs by stimulating the synthesis of non-EFAs. Thus, two main problems exist with the supplementation of copper to pigs: (1) it is done without due consideration of the inhibitory effect of copper on iron and zinc-dependent functions; (2) it is started at a very early age when the requirement for zinc, in particular, is high.

Interaction of zinc in essential fatty acid metabolism

To understand the interaction of zinc in EFA metabolism it is important to understand the nature of fatty acid desaturation and elongation (*Figure 8.1*). This process occurs through a series of enzymes (desaturases and elongases) which are used jointly by EFAs and non-EFAs. Hence there is competition, e.g. between α-linolenic acid, linoleic acid and oleic acid for the Δ^6-desaturase, with the more unsaturated fatty acids having preference (α-linolenic acid) when equal amounts of these fatty acids are present. However, in a situation in which excess oleic acid may be present, then its conversion by the Δ^6-desaturase to its non-EFA product—$\Delta^{5,8,11}$-eicosatrienoic acid may proceed in preference to linoleic acid desaturation. This results in metabolic EFA deficiency and appears to occur with zinc deficiency. In the testes lipids of zinc-deficient rats, there is a 50% increase in the activity of the enzyme which synthesizes oleic acid from stearic acid (Δ^9-desaturase) and a 25% decrease in the activity of the Δ^6-desaturase (Clejan et al., 1982). Less dramatic effects of zinc deficiency on enzyme activities controlling fatty acid metabolism were reported for the liver, as has been previously reported elsewhere (Cunnane and Wahle, 1981a). Hence, there appears to be a reduction in the availability of linoleic acid metabolites—γ-linolenic acid, dihomo γ-linolenic acid and arachidonic acid—in zinc deficiency (*Table 8.4*).

When EFA deficiency is of dietary origin, Δ^6-desaturase activity is increased in an attempt to synthesize more linoleic acid metabolites. When EFA deficiency is metabolic, the Δ^6-desaturase appears to be inhibited,

Table 8.4 EFFECTS OF ZINC DEFICIENCY ON ESSENTIAL FATTY ACID METABOLISM

1. Decreased Δ^6-desaturation of linoleic acid except in mammary tissue
2. Increased Δ^9-desaturation of stearic acid
3. Accumulation of linoleic acid in essential fatty acid supplemented animals above what is normally seen
4. Accumulation of arachidonic acid in skin, testes, and mammary tissue
5. Decreased esterification of essential fatty acids into phospholipids

e.g. in zinc deficiency. Three main observations substantiate the effect of zinc deficiency on the metabolism of linoleic acid via the Δ^6-desaturase: (1) EFA supplementation ameliorates the effect of zinc deficiency in humans (Cash and Berger, 1969), in pigs (Hanson et al., 1958) and in rats (Bettger et al., 1979; Cunnane and Horrobin, 1980; Cunnane, 1982c; Huang et al., 1982); (2) EFA tissue levels are abnormal in zinc deficiency in that linoleic acid accumulates when zinc-deficient rats are given supplemental EFA, and arachidonic acid accumulates in various tissues such as testes, liver and skin (Huang et al., 1982); (3) the Δ^6-desaturase is abnormally active in mammary tissue of zinc-deficient lactating rats (Cunnane and Wahle, 1981b). Zinc is therefore well established as a functional cofactor in linoleic acid metabolism. It is not a component of the Δ^6-desaturase enzyme itself; that is a role for iron. Rather, it has an ancillary function, either in relation to the NADP–NADPH cycle which is zinc dependent (Ludwig et al., 1980) or by regulating the availability of free linoleic acid from membrane lipids (Cunnane and Huang, 1982; Cunnane et al., 1982). Enzymes involved in prostaglandin synthesis are also zinc dependent (Cunnane, 1982b) and defects in prostaglandin metabolism are now recognized in zinc deficiency (Cunnane et al., 1983).

In the pig, parakeratosis is well documented as a disease involving the interaction of zinc and EFA (Tucker and Salmon, 1955; Hanson et al., 1958; Holman, 1971). In fact, plasma zinc levels have been shown to be lowest at the age (12–16 weeks) when the pigs are also most susceptible to EFA-dependent parakeratosis (Suttle and Mills, 1966b). What is particularly interesting is that this disease is aggravated by the identical factors which aggravate EFA and zinc deficiency: (1) growth stimulation (which increases EFA demand) in the presence of marginal EFA intake; (2) hydrogenated coconut oil which competitively inhibits EFA metabolism; (3) elevated calcium in the diet which inhibits absorption and utilization of zinc; and (4) excess copper in the diet. When these four exacerbating factors are eliminated or are counteracted, the disease symptoms are relieved. Hence, reduction of calcium intake, supplementation with zinc and feeding with polyunsaturated oils are all beneficial in parakeratosis. Studies corroborating the interactions of zinc, calcium and EFA in pig nutrition have been undertaken using the rat (Holman, 1971). EFA deficiency symptoms developed more rapidly in those rats fed a diet with a high calcium-to-zinc ratio.

Interaction of copper in essential fatty acid metabolism

Zinc deficiency causes an increase in the synthesis of non-EFAs and a decrease in the synthesis of EFA from linoleic acid. Copper supplementation also causes an increase in the synthesis of mono-unsaturated non-EFAs which, by inference, causes a decrease in linoleic acid metabolism. Thus, in the presence of marginal zinc intake, copper may be considered detrimental to EFA metabolism and may induce metabolic EFA deficiency if present in excess. Copper is well established as a physiological antagonist of zinc (Cunnane, 1982b) and this appears to be reflected in its effect on fatty acid metabolism (*Figure 8.2*).

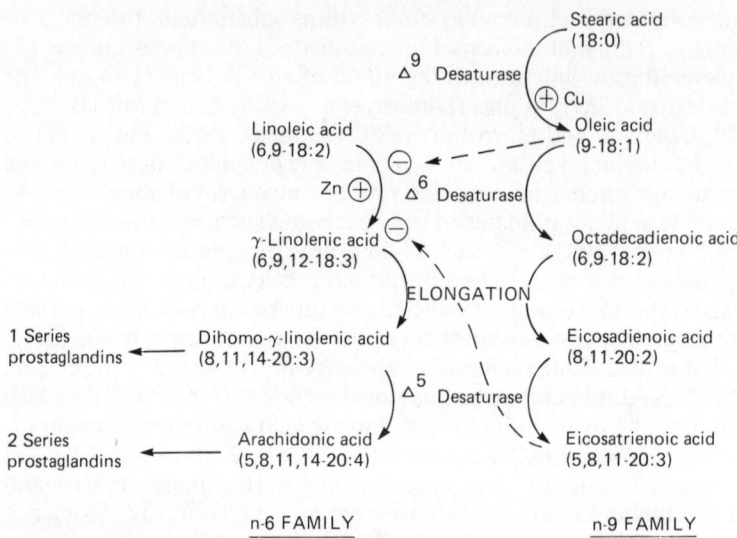

Figure 8.2 Outline of the suggested interaction of zinc and copper in the metabolism of both linoleic acid and stearic acid. An increase in dietary zinc-to-copper ratio favours linoleic acid desaturation whereas the opposite favours stearic acid desaturation. In situations in which both zinc and linoleic acid may be deficient, increased synthesis of oleic acid and its metabolites will further inhibit metabolism of linoleic acid and will inhibit synthesis of prostaglandins

Copper was first shown to be involved in fatty acid metabolism 35 years ago with the observation that, in copper deficiency, phospholipid synthesis was impaired (Gallagher, 1957). In the 1960s, coincident with the reports of its beneficial effect on growth and feed conversion efficiency in pigs, copper at 250 mg/kg was also shown to affect the softness of the back fat of pigs (Taylor and Thomke, 1964). This led to the observation that copper increased mono-unsaturated fatty acids in the back fat (palmitoleic acid and oleic acid) (Elliot and Bowland, 1968). Specific measurement of the enzyme Δ^9-desaturase which is responsible for mono-unsaturation of long-chain fatty acids confirmed the increase in its activity (Ho and Elliot, 1973; Thompson, Allen and Meade, 1973; Wahle and Davies, 1975). Copper deficiency was also shown to decrease the activity of this enzyme (Wahle and Davies, 1975). In conjunction with these effects of copper deficiency on mono-unsaturated fatty acids, copper deficiency also increases the proportion of polyunsaturated fatty acids in tissues such as the liver (Gallagher and Reeve, 1971). Hence, these effects exactly oppose the effects of zinc or zinc deficiency.

As oleic acid is non-essential and, when present in excess, competes with linoleic acid for further metabolism and also destabilizes membranes (Huang *et al.*, 1980), excess oleic acid synthesis in copper-supplemented pigs should be considered detrimental. Not only oleic acid, but also its metabolites ($\Delta^{6,9}$-octadecadienoic acid and $\Delta^{5,8,11}$-eicosatrienoic acid) inhibit both EFA metabolism from linoleic acid to arachidonic acid and the synthesis of prostaglandins (*Figure 8.2*) (Ziboh, 1975).

The interaction of copper, iron and zinc in lipid peroxidation

Lipid membranes rely partly on the long-chain polyunsaturated fatty acids (which include the EFAs and EFA derivatives such as $\Delta^{4,7,10,13,16}$-docosapentaenoic acid, $\Delta^{7,10,13,16,19}$-docosapentaenoic acid and $\Delta^{4,7,10,13,16,19}$-docosahexaenoic acid) for fluidity. These fatty acids contain up to six double bonds which may be attacked by oxygen if insufficiently protected by naturally occurring antioxidants such as vitamin E. The result of such peroxidation is membrane damage and tissue dysfunction. Copper tends to increase such peroxidation: (1) in erythrocytes (Bettger et al., 1978); (2) in hepatocytes previously damaged by carbon tetrachloride (Wills, 1965); (3) in haemolysis involving zinc deficiency (Bettger et al., 1978); and (4) in haemolysis involving iron deficiency (Jain et al., 1983). The widely reported effect of an increased copper–iron ratio causing anaemia (Suttle and Mills, 1966b; Gipp et al., 1974; Underwood, 1977), may therefore be a result of increased peroxidation of long-chain polyunsaturated fatty acids in the erythrocyte membrane causing membrane instability and lysis (Jain et al., 1983). Because the regulation of cellular osmotic balance is a major energy consumer, conditions chronically stressing this function, e.g. copper excess, may, in conjunction with anaemia, cause the ill health frequently reported in copper-supplemented pigs. The incidence of iron-dependent anaemia in copper-supplemented pigs is sufficient to warrant consideration of iron and/or zinc supplementation with the additional copper. Where these three elements interact in a common function such as regulation of membrane lipid peroxidation and EFA metabolism, serious consideration of these factors in animal nutrition is essential.

Role of essential fatty acids in zinc absorption in the neonatal period

In order to increase sow productivity and decrease piglet mortality, early weaning units for pig rearing have been developed. In these units, the piglets are initially housed in incubators from which the nourishment is obtained in the form of commercial milk analogues. Specially formulated pellets also based on skim milk are then fed to the piglets between the ages of 2–6 weeks, after which the pigs are weaned on to a conventional creep diet.

The absorption of trace elements from milk-based diets such as these must be carefully evaluated since clinical studies have established that, when comparing availability from diets containing equal proportions of zinc, a significantly greater amount of zinc will be absorbed by infants fed on maternal milk in contrast to those fed artificial milk-based diets (Widdowson et al., 1974; Hambidge et al., 1979). In the childhood disease, acrodermatitis enteropathica, in which a specific defect in zinc absorption exists, milk substitutes and cow's milk exacerbate the defect but both human breast milk or an infusion of EFA alleviate the symptoms of zinc deficiency (Cash and Berger, 1969). These observations strongly implicate EFAs as important stimulants of zinc absorption. This implication was substantiated in studies of ^{65}Zn absorption in rats (Cunnane, 1982d). Both

Table 8.5 EFFECT OF ESSENTIAL FATTY ACIDS ON ^{65}Zn ABSORPTION IN NEONATAL RAT PUPS. (a) – ESSENTIAL FATTY ACID INJECTION INTO THE MOTHER DURING PREGNANCY. (b) – ESSENTIAL FATTY ACID (1 μg) INJECTION INTO THE RAT PUPS AT THE TIME OF ^{65}Zn INJECTION. ADAPTED FROM CUNNANE (1982d), WITH PERMISSION

(a)	^{65}Zn absorption (%)
Control	18.8
Essential fatty acid supplemented	25.1*
% increase	34

(b)	^{65}Zn absorption (%) compared with control (100%)
Linoleic acid	150**
γ-linolenic acid	180**
Dihomo-γ-linolenic acid	155**
Arachidonic acid	93

*$P<0.05$; **$P<0.01$

EFA supplementation of the mother during pregnancy and injection of the neonatal rat pups with linoleic acid, γ-linolenic acid or dihomo-γ-linolenic acid at the time of injection of the ^{65}Zn, increased ^{65}Zn absorption in the neonates by amounts up to twice that found in controls (Cunnane, 1982d) (*Table 8.5*).

It is, therefore, essential to provide early weaned piglets with artificial milk-based diets which conform as closely as possible to that of sow's milk, including its EFA composition. Some successful preparations are very low in linoleic acid and contain no linoleic acid derivatives such as γ-linolenic acid or dihomo-γ-linolenic acid. Hence, it is very possible that early weaned piglets do not absorb all the zinc available in other preparations and may be marginally zinc deficient. Added to which, the diet on which the piglets will be fed from 2–6 weeks of age contains 180–200 mg/g of copper. Hence, it is important that the role of supplemental copper be established unequivocally and that the optimal amount of copper, zinc and iron in piglet diets be determined.

Conclusions

The question of adequacy and excess of dietary zinc and copper, and their interaction with EFAs, has been discussed in relation to pig nutrition. This relationship is particularly important in pig nutrition for two reasons: (1) the use of copper supplementation in pig diets; and (2) the advent of early weaning of piglets on to diets based on cow's milk. Copper supplementation is a popular, inexpensive and apparently beneficial means of enhancing pig growth. The frequency of the reports of the detrimental effects of copper supplementation is not very great. Furthermore, it appears to be limited to situations involving either mineral imbalances of elements such as zinc and calcium, or possibly to excessive use of growth stimulants which place a high demand on EFAs. Hence, in situations in which copper supplementation is not having the desired effect, its toxic role in zinc and iron metabolism should be considered, as has been frequently suggested (Suttle and Mills, 1966b). In addition, the overall influence of the interaction of these elements on the metabolism of EFAs is also a critical factor. Similarly, in the early weaning of pigs, zinc homeostasis appears to be critical. As it is doubtful that early weaned piglets absorb as much zinc as is intended, because of the relative lack of EFA derived from linoleic acid, immune function in these animals may not be optimal. Rapid transfer to a copper-enriched diet may be further detrimental to zinc availability in the piglet while at the same time it is without confirmed benefit at this age (2–3 weeks). Thus, further research which addresses the relationship of mineral interactions and their relevance to EFA metabolism, should improve the effectiveness of modernized pig rearing.

Acknowledgement

I would like to acknowledge the enthusiasm and advice of Dr M.A. Varley, Rowett Institute, Aberdeen, UK; his concern about this subject was instrumental in my contributing to this conference.

References

ALLEN, M.M., BARBER, R.S., BRAUDE, R. and MITCHELL, K.G. (1961). Further studies on various aspects of the use of high copper supplements for pigs. *British Journal of Nutrition* **15**, 507–515

BARBER, R.S., BOWLAND, J.P., BRAUDE, R., MITCHELL, K.G. and PORTER, J.W.G. (1961). Copper sulphate and copper sulphide (CuS) as supplements for growing pigs. *British Journal of Nutrition* **15**, 189–197

BETTGER, W.G., FISH, T.J. and O'DELL, B.L. (1978). Effects of copper and zinc status of rats on erythrocyte stability and superoxidedismutase activity. *Proceedings of the Society for Experimental Biology and Medicine* **158**, 279–282

BETTGER, W.J., REEVES, P.G., MOSCATELLI, E.A., REYNOLDS, G. and O'DELL, B.L. (1979). Interaction of zinc and essential fatty acids in the rat. *Journal of Nutrition* **109**, 480–488

BRAUDE, R. (1966). The effect of changes in feeding pattern on the performance of pigs. *Proceedings of the Nutrition Society* **26**, 163–181

BURCH, R.E., WILLIAMS, R.V., HAHN, H., NAYAK, R.V. and SULLIVAN, J.F. (1974). Trace element content and enzymatic activities in tissue of zinc deficient pigs. In *Trace Element Metabolism in Animals, Volume 2* (W.G. Hoekstra, J.W. Suttie, G.E. Ganther and W. Mertz, Eds), pp. 513–515. Baltimore, University Park Press

CASH, R. and BERGER, C.K. (1969). Acrodermatitis enteropathica: Defective metabolism of unsaturated fatty acids. *Journal of Pediatrics* **74**, 717–729

CLEJAN, S., CASTRO-MAGANA, M., COLLIPP, P.J., JONAS, E. and MADDAIAH, V.T. (1982). Effects of zinc deficiency and castration on fatty acid composition and desaturation in rats. *Lipids* **17**, 129–135

CUNNANE, S.C. (1982a). The effects of evening primrose oil (Efamol) in zinc-deficient rats. In *Clinical Uses of Essential Fatty Acids* (D.F. Horrobin, Ed.), pp. 97–112. Montreal, Eden Press

CUNNANE, S.C. (1982b). Differential regulation of essential fatty acid metabolism to the prostaglandins: possible basis for the interaction of zinc and copper in biological systems. *Progress in Lipid Research* **21**, 73–90

CUNNANE, S.C. (1982c). Foetal mortality in moderately zinc-deficient rats is strictly related to the process of parturition: effect of maternal essential fatty acid supplementation. *British Journal of Nutrition* **47**, 495–504

CUNNANE, S.C. (1982d). Maternal essential fatty acid supplementation increases zinc absorption in neonatal rats: relevance to the defect in zinc absorption in acrodermatitis enteropathica. *Pediatric Research* **16**, 599–604

CUNNANE, S.C. and HORROBIN, D.F. (1980). Parenteral linoleic and gamma-linoleic acids ameliorate the gross effects of zinc deficiency. *Proceedings of the Society for Experimental Biology and Medicine* **164**, 583–588

CUNNANE, S.C. and HUANG, Y.S. (1982). Incorporation of dihomo-gamma-linolenic acid into tissue lipids in zinc deficiency. Presented at the V International Conference on Prostaglandins, Florence, May, 1982. (Abstracts available)

CUNNANE, S.C. and WAHLE, K.W.J. (1981a). Differential effects of zinc deficiency on delta-6-desaturase activity and fatty acid composition of various tissues in lactating rats. Presented at the XII International Congress on Nutrition, San Diego, August, 1981. (Abstracts available)

CUNNANE, S.C. and WAHLE, K.W.J. (1981b). Zinc deficiency increases delta-6-desaturation of linoleic acid in rat mammary tissue. *Lipids* **16**, 771–774

CUNNANE, S.C., KEELING, P.W.N., THOMPSON, R.P.H. and CRAWFORD, M.A. (1982). Leucocyte essential fatty acid metabolism in zinc-deficient pregnant rats. *Proceedings of the Nutrition Society* **42**, 72A

CUNNANE, S.C., MAJID, E., SENIOR, J. and MILLS, C.F. (1983). Uteroplacental dysfunction and prostaglandin metabolism in zinc deficient pregnant rats. *Life Sciences* **32**, 2471–2478

ELLIOT, J.I. and BOWLAND, J.P. (1968). The effects of dietary copper sulphate on the fatty acid composition of porcine depot fat. *Journal of Animal Science* **27**, 956–960

FORBES, R.M. (1960). Nutritional interactions of zinc and calcium. *Federation Proceedings* **19**, 643–647

GALLAGHER, C.H. (1957). Pathology and biochemistry of copper deficiency. *Australian Veterinary Journal* **33**, 311–317

GALLAGHER, C.H. and REEVE, V.E. (1971). Copper deficiency in the rat: Effect on liver and brain lipids. *Australian Journal of Experimental Biology and Medical Science* **49**, 453–461

GIPP, W.F., POND, W.G. and KALLFELZ, F.A. (1974a). Effect of dietary copper, iron and ascorbic levels on hematology and blood and tissue copper, iron and zinc concentrations. *Journal of Nutrition* **104**, 532–541

GIPP, W.F., POND, W.G., TASKER, J., VAN CAMPEN, D., KROOK, L. and VISEK, W.J. (1974b). Influence on level of dietary copper on weight gain, hematology and liver copper and iron storage of young pigs. *Journal of Nutrition* **103**, 713–719

HAMBIDGE, K.M., WALRAVENS, P.A., CASEY, C.E., BROWN, R.M. and BENDER, C. (1979). Plasma zinc concentrations in breast-fed infants. *Journal of Pediatrics* **94**, 607–611

HANSON, L.J., SORENSEN, D.K. and KERNKAMP, H.C.H. (1958). Essential fatty acid deficiency: Its role in parakeratosis. *American Journal of Veterinary Research* **30**, 921–930

HILL, E.G., WARMANEN, E.L., HAYES, H. and HOLMAN, R.T. (1957). Effects of essential fatty acid deficiency in young swine. *Proceedings of the Society for Experimental Biology and Medicine* **95**, 274–278

HO, S.K. and ELLIOT, J.I. (1973). Supplemental dietary copper and the desaturation of 14-C-stearoyl-CoA by porcine hepatocytes and adipose microsomes. *Canadian Journal of Animal Sciences* **53**, 537–545

HOEKSTRA, W.G. (1964). Observations on mineral interrelationships. *Federation Proceedings* **23**, 1068–1076

HOLMAN, R.T. (1971). Essential fatty acid deficiency. In *Progress in the Chemistry of Fats and other Lipids* (R.T. Holman, Ed.), pp. 275–348. New York, Pergamon

HOWARD, A.N., LEAT, W.M., GRESHAM, G.A., BOWYER, W.D. and DALTON, E.R. (1965). Studies on pigs reared on semi-synthetic diets containing no fat, beef tallow, or maize oil: Husbandry and serum biochemistry. *British Journal of Nutrition* **19**, 383–395

HUANG, Y.S., CUNNANE, S.C., HORROBIN, D.F. and DAVIGNON, J. (1982). Most biological effects of zinc deficiency corrected by gamma-linolenic acid (18:3n-6) but not by linoleic acid (18:2n-6). *Atherosclerosis* **41**, 193–207

HUANG, Y.S., MARCEL, Y.L., VEZINA, C., BARBEAU, A. and DAVIGNON, J. (1980). Lecithin:Cholesterol acyl-transferase activity and fatty acid composition of erythrocyte phospholipids in Friedreich's ataxia. *Canadian Journal of Neurological Sciences* **7**, 429–434

HUBER, A.M. and GERSHOFF, S.N. (1970). Effects of dietary zinc and calcium on the retention and distribution of zinc in rats fed semi-purified diets. *Journal of Nutrition* **100**, 949–954

HURLEY, L.S. and MUTCH, P.B. (1973). Prenatal and postnatal development after transitory gestational zinc deficiency in rats. *Journal of Nutrition* **103**, 649–655

HURLEY, L.S., GOWEN, J. and SWENERTON, H. (1971). Teratogenic effects of short-term and transitory zinc deficiency in rats. *Teratology* **4**, 199–204

JAIN, S.K., YIP, R., HOESCH, R.M., PRAMINAK, R.M., DALLMAN, P.R. and SHOHET, S.B. (1983). Evidence of peroxidative damage to the erythrocyte membrane in iron deficiency. *American Journal of Clinical Nutrition* **37**, 26–30

JONES, R.B., KEELING, P.W.N., HILTON, P.J. and THOMPSON, R.P.H. (1981). The relationship between leucocyte and muscle zinc in health and disease. *Clinical Science* **60**, 237–239

LEAT, W.M.F. (1961). Studies on pigs raised on diets low in tocopherol and essential fatty acids. *British Journal of Nutrition* **15**, 259–270

LEAT, W.M.F. (1962). Studies on pig diets containing different amounts of linoleic acid. *British Journal of Nutrition* **16**, 559–569

LUDWIG, J.C., MISIOROWSKI, R.L., CHVAPIL, M. and SEYMOUR, M.D. (1980).

Interaction of zinc irons with electron carrying NADPH and NADH. *Chemical and Biological Interactions* **30**, 25–34

MEADOWS, N.J., SMITH, M.F., KEELING, P.W.N., RUSE, W., DAY, J., SCOPES, J.W., THOMPSON, R.P.H. and BLOXAM, D.L. (1981). Zinc and small babies. *Lancet* **ii**, 1135–1136

MILLER, E.R., LUECKE, R.W., ULLREY, D.E., BALTZER, B.V., BRADLEY, B.L. and HOEFER, J.A. (1968). Biochemical, skeletal and allometric changes due to zinc deficiency in the baby pig. *Journal of Nutrition* **95**, 278–286

NATIONAL RESEARCH COUNCIL (1979). *Zinc*. Baltimore, University Park Press

NORRDIN, R.L., KROOK, L., POND, W.G. and WALKER, E.F. (1973). Experimental zinc deficiency in weaning pigs on high and low calcium diets. *Cornell Veterinarian* **63**, 264–290

O'DELL, B.L. (1976). Biochemistry and physiology of copper in vertebrates. In *Trace Elements in Human Health and Disease, Volume I: Zinc and Copper* (A.S. Prasad and D. Oberleas, Eds), pp. 391–413. New York, Academic Press

O'DELL, B.L., BIRD, D.W., RUGGLES, D.L. and SAVAGE, J.E. (1966). Composition of aortic tissue from copper-deficient chicks. *Journal of Nutrition* **88**, 9–14

O'DELL, B.L., HARDWICK, B.C., REYNOLDS, G. and SAVAGE, J.E. (1961). Connective tissue defect resulting from copper deficiency. *Proceedings of the Society for Experimental Biology and Medicine* **108**, 402–405

OKONKWO, A.C., KU, P.K., MILLER, E.R., KEAHEY, K.K. and ULLREY, D.E. (1979). Copper requirement of baby pigs fed purified diets. *Journal of Nutrition* **109**, 939–948

SHANKLIN, S.H., MILLER, E.R., ULLREY, D.E., HOEFER, J.A. and LUECKE, R.W. (1968). Zinc requirement of baby pigs on casein diets. *Journal of Nutrition* **96**, 101–108

SHIELDS, G.S., COULSON, W.F., KIMBALL, D.A., CARNES, W.H., CARTWRIGHT, G.E. and WINTROBE, W.W. (1962). Studies on copper metabolism. XXXII. Cardiovascular lesions in copper-deficient swine. *American Journal of Pathology* **41**, 601–621

SUTTLE, N.F. and MILLS, C.F. (1966a). Studies of the toxicity of copper to pigs. I. Effects of oral supplements of zinc and iron salts on the development of copper toxicosis. *British Journal of Nutrition* **20**, 135–148

SUTTLE, N.F. and MILLS, C.F. (1966b). Studies of the toxicity of copper to pigs. II. Effect of protein source and other dietary components on the response to high and moderate intakes of copper. *British Journal of Nutrition* **20**, 149–161

TAO, S.-H. and HURLEY, L.S. (1975). Effect of dietary calcium deficiency during pregnancy on zinc mobilization in intact and parathyroidectomised rats. *Journal of Nutrition* **105**, 220–225

TAYLOR, M. and THOMKE, S. (1964). Effect of high-level copper on the depot fat of bacon pigs. *Nature* **201**, 1246

THOMPSON, E.H., ALLEN, C.E. and MEADE, R.J. (1973). Influence of copper on stearic acid desaturation and fatty acid composition in the pig. *Journal of Animal Science* **36**, 868–873

THOMPSON, R.W., GILBREATH, R.L. and BIELK, F. (1975). Alterations of porcine skin mucopolysaccharides in zinc deficiency. *Journal of Nutrition* **105**, 154–160

TUCKER, H.F. and SALMON, W.D. (1955). Parakeratosis or zinc deficiency disease in the pig. *Proceedings of the Society for Experimental Biology and Medicine* **88**, 613–616

UNDERWOOD, E.J. (1977). *Trace Elements in Human Nutrition*, 4th Edn. New York, Academic Press

WAHLE, K.W.J. and DAVIES, N.T. (1975). Effect of dietary copper deficiency in the rat on fatty acid composition of adipose tissue and desaturase activity of liver microsomes. *British Journal of Nutrition* **34**, 105–112

WEGGER, I. and PALLUDAN, B. (1978). Zinc metabolism in swine with special emphasis on reproduction. In *Trace Element Metabolism in Man and Animals, Volume 3* (M. Kirchgessner, Ed.), pp. 428–435. Institut für Ernahrungsphysiologie, Technische Universität München, Freising-Weihenstaphan

WHITING, F. and BEZEAU, L.M. (1958). Calcium, phosphorus and zinc balance in pigs as influenced by the weight of the pig and the level of calcium, zinc and vitamin D in the ration. *Canadian Journal of Animal Sciences* **38**, 109–117

WIDDOWSON, E.M., DAUNCEY, J. and SHAW, J.C.L. (1974). Trace elements in foetal and early post-natal development. *Proceedings of the Nutrition Society* **33**, 275–283

WITZ, W.M. and BEESON, W.M. (1951). A study of fatty acid deficiency in swine. *Journal of Animal Science* **10**, 112–117

WILLS, E.D. (1965). Mechanisms of lipid peroxide formation in tissues: Role of metals and hematin proteins in the catalysis of the oxidation of unsaturated fatty acids. *Biochimica et biophysica acta* **98**, 238–251

ZIBOH, V.A. (1975). Prostaglandins and their biological significance in the skin. *International Journal of Dermatology* **14**, 485–493

9

ESSENTIAL FATTY ACIDS IN THE RUMINANT

R.C. NOBLE
Hannah Research Institute, Ayr KA6 5HL, Scotland, UK

Introduction

Little metabolism of lipids occurs in simple-stomached animals before the small intestine is reached, but this is not so in ruminant animals. Events which occur within the complex polygastric arrangement of the alimentary tract, principally those associated with the rumen, have a profound effect on the chemical and physical nature of the lipids subsequently presented to the small intestine for digestion. Indeed, many of the unusual features which characterize ruminant lipids and their metabolism can be related directly to the exposure of the dietary lipids to the ruminal processes. Among the main ruminal effects on the dietary lipids are those upon the unsaturated fatty acid constituents. Through the process of lipolysis and hydrogenation, the high proportion of naturally occurring C18 polyunsaturated fatty acids in the diet are converted to a range of fatty acids consisting predominantly of stearic acid but accompanied also by a spectrum of geometrical and positional di- and monoenoic isomers. Thus, in the simple-stomached animal, inclusion of linoleic and other polyunsaturated fatty acids into the diet will readily increase the content of such acids in the tissue lipids, but this is not so in the ruminant where the difference between the unsaturated fatty acid composition of the diet and absorbed lipid is very marked. The process of rumen biohydrogenation has, therefore, been much investigated. However, the present review concerns itself primarily with that part of the dietary linoleic acid which escapes the process of biohydrogenation. It has become clear that the metabolism of the linoleic acid displays many features that are unique to the ruminant animal. The process of biohydrogenation is frequently held responsible, both directly and indirectly, for many of the observed features. This influence, undoubtedly important as it is, may be an oversimplification of ruminant essential fatty acid metabolism.

The diet

Ruminants are herbivores and therefore the lipids of the diet are derived from forages, either fresh or conserved, supplemented under circumstances of intensive production with a wide variety of oil-containing seeds and cereals of high caloric value. Although leaf tissue does not, in general, contain large amounts of lipid, the presence within the lipid of high proportions of glycolipids and phospholipids with characteristically high contents of linoleic and linolenic acids, and the very large bulk of dietary material consumed, ensures a considerable intake of linoleic acid (see Kates, 1970). In those seed oils normally included in concentrate rations, the fatty acids present generally contain a high proportion of linoleic acid as well. Although the processes of herbage conservation so widely adopted under the more intensive systems of animal production may have a considerable effect upon lipid composition, there is an absence of any great deterioration in the polyunsaturated fatty acid content (Czerkawski, 1967).

Unsaturated fatty acid changes in the rumen

The presence of lipolytic activity in rumen contents has been known for many years. Since the original observations using linseed and olive oil (Garton, Hobson and Lough, 1958; Garton, Lough and Vioque, 1959), it has been shown that lipolytic action extends to a wide range of esterified substrates which may or may not be considered as constituents of a natural diet. Although it is considered that lipolysis is a microbial process, it is not certain whether both bacteria and protozoa participate (see Viviani, 1970, and Harfoot, 1978). Attention has also been directed towards extracellular lipolysis and the contribution to the overall process made by the lipases present in the herbage (Henderson, 1971; Dawson and Hemington, 1974). Through a combination of incubations *in vitro* of the rumen contents (Reiser, 1951; Shorland, Weenink and Johns, 1955; Reiser and Reddy, 1956) and comparisons of the fatty acid compositions of the contents of various parts of the ruminant digestive tract (Bath and Hill, 1967), it was established that the rumen displayed a great ability to hydrogenate linoleic acid. It is now accepted that the conversion of linoleic acid and other polyunsaturated fatty acids of the diet to stearic acid, together with smaller quantities of a wide range of unsaturated positional and geometrical isomers of C18 components, is one of the most significant features of rumen metabolism (see Viviani, 1970, and Harfoot, 1978). It has also been demonstrated that hydrolysis of the ester linkages is a prerequisite of the biohydrogenation (Hawke and Silcock, 1969).

The mechanism of biohydrogenation is complex and, in spite of extensive investigations, the sequence of events still remains a subject of considerable argument (see Viviani, 1970, and Harfoot, 1978). Although it is acccepted that the cell-free ruminal fluid is unable to bring about any biohydrogenation, the relative contributions of the bacterial, protozoal and fine food particles, all of which have over the years been credited with biohydrogenation abilities, remain largely unresolved. However, no doubt exists as to the final outcome of the dietary lipid changes in the rumen. Unesterified fatty acids, adsorbed largely on to the surface of the particulate matter in the rumen, become the predominant lipid fraction of the digesta (Ward, Scott and Dawson, 1964). Linoleic acid is reduced to a minor component only of the fatty acids and the concentration of stearic acid is increased to become the major component (Bath and Hill, 1967; Lennox, Lough and Garton, 1968). Although the long-chain unesterified fatty acids constitute by far the major lipid class within the digesta, some 20% of the total lipid can be found in association with the protozoal (16%) and bacterial (4%) populations in the form of both neutral and phospholipids, thereby making a significant contribution to the lipid supply of the host animal (see Viviani, 1970, and Harfoot, 1978). Whereas in the protozoal lipid, linoleic acid can account for up to 20% of the total fatty acids present (Emmanuel, 1974), in the bacteria the levels of linoleic acid, although significant, are relatively low (Keeney, Katz and Allison, 1962; Tweedie, Rumsby and Hawke, 1966). The contribution of the microbial population to the linoleic acid content of the digesta is open to wide variation depending upon the state of a host of chemical and physical conditions prevailing within the rumen (Ifkovits and Ragheb, 1968). The capacity of the rumen bacteria to contribute to the linoleic acid supply of the host animal through synthesis *de novo* (Sklan, Budowski and Volcani, 1972) appears to have no significance under the predominantly anaerobic conditions that prevail within the rumen.

Linoleic acid changes in the small intestine

Although there is evidence that some absorption of long-chain fatty acids, including linoleic acid, occurs before they reach the small intestine, the quantitative significance of the process in the combined areas of the rumen, reticulum, omasum and abomasum is of minimal importance (Bickerstaffe, Noakes and Annison, 1972). The lipids are mainly in association with the particulate material of the digesta, and their fatty acid compositions remain virtually unaltered during the rapid passage through the omasum and abomasum (Bath and Hill, 1967). In the acid environment of the abomasum, the bacteria and protozoa are disintegrated and their lipid contents released, thereby facilitating subsequent digestion further down the gastrointestinal tract. In the digesta leaving the abomasum and entering the duodenum, it has been estimated (Bickerstaffe *et al.*, 1972; Leat and Harrison, 1973) that linoleic acid accounts for only 0.3–0.5% of the energy available, that is much less than the level of 1–2% of the dietary energy considered to be the minimum requirement in many animal species to prevent essential fatty acid deficiency symptoms (see Holman, 1971). Upon entering the small intestine the digesta is augmented by the addition of the various digestive secretions associated with the duodenum. Levels of linoleic acid in both the unesterified and especially the esterified components of the digesta are therefore increased significantly through the addition of the biliary phosphatidylcholine (Lennox, Lough and Garton, 1968; Leat and Harrison, 1969); however, stearic acid remains the principal fatty acid in the digesta with linoleic acid accounting for no more than 7% of the total.

In the absence of any substantial quantities of diet-derived partial glycerides, the ruminant animal possesses a distinctive lipid-solubilizing system which, nevertheless, results in an efficient absorption of all the

C14–C18 fatty acids (see Noble, 1978). Absorption of the fatty acids, including linoleic acid, occurs maximally in the middle and lower jejunum and is virtually complete by the time the digesta reaches the ileum (Lennox and Garton, 1968). Although it has been suggested that some discrimination in favour of the absorption of linoleic acid relative to stearic acid may operate, any quantitative effect is marginal; compared with stearic acid, the amount of linoleic acid absorbed remains small. Fatty acid uptake, therefore, is dictated by the events which occurred in the rumen and which marked the beginning of dietary lipid digestion.

Transport and tissue metabolism of linoleic acid

Degree of unsaturation is undoubtedly extremely important in determining the distribution and rate of fatty acid incorporation into the various lipid fractions of the tissues. The fatty acid composition of the lipids in the lymph draining the gastrointestinal tract is clearly dominated by the alimentary lipids with triglycerides, esterified predominantly to stearic acid, constituting the major lipid class (Leat and Hall, 1968; Wadsworth, 1968; Christie and Hunter, 1978). A distinctive preference for available linoleic acid to be incorporated into the phospholipids was evident. A specific role for the phospholipids of the lymph in the transport and turnover of the available linoleic acid was therefore suggested. Thus, whereas the fatty acids of the triglycerides can be considered as predominantly of exogenous or dietary origin, the fatty acids of the phospholipids are derived partly from the diet but also to a large extent from endogenous sources, in particular through the re-acylation of the hydrolytic products of the biliary lipids and recycling via capillary filtrate (Leat and Hall, 1968; Wadsworth, 1968). Only under deliberately contrived conditions, where large amounts of linoleic acid are absorbed, does it appear that synthesis of phospholipids is insufficient to accommodate the polyunsaturated fatty acid supply and extensive incorporation into the triglycerides becomes apparent (Heath, Caple and Redding, 1970). The small contribution that the cholesteryl esters make to the lipids of the intestinal lymph, also appears to involve a preferential association with linoleic acid.

For many years the distribution of the major plasma lipids and their content of polyunsaturated fatty acids proved to be one of the principal anomalies with respect to lipid metabolism in ruminant animals. In spite of the extensive absorption of saturated fatty acids and the small uptake of linoleic and other polyunsaturated fatty acids, the plasma was found to have an exceedingly high content of linoleic acid (Duncan and Garton, 1963; Garton and Duncan, 1964), e.g. in cow plasma, linoleic acid accounts for up to 55% of the total fatty acids present. The principal components of the plasma are the cholesteryl esters and phospholipids, which together comprise up to 80% of the plasma lipid. Almost all the linoleic acid is esterified to the cholesteryl ester and phospholipid fractions. It has become clear subsequently that this extreme selectivity in the distribution of linoleic acid is disturbed only under conditions such that absorption of linoleic acid is far in excess of that normally encountered, when the metabolic processes responsible for the segregation of the fatty acid species are unable to cope (Cook *et al.*, 1972). The ability to increase substantially the linoleic acid content of the cholesteryl ester and phospholipid fractions, without having any effect upon the triglycerides, has been explained in terms of the limited supply of linoleic acid from the small intestine and the differential rates of turnover and hydrolysis of the various plasma lipid fractions. For example in the ruminant animal, as in most other species, the triglycerides and the major triglyceride-carrying lipoprotein fractions, have a rapid turnover compared with that of other plasma lipid fractions (Lascelles *et al.*, 1964). In the sheep and probably the cow, the high content of linoleic acid in the plasma cholesteryl esters is determined by their biosynthetic origin. Virtually all the plasma cholesteryl esters originate through the lecithin:cholesterol acyltransferase system which operates within the plasma, and this has been shown to possess a high specificity for the transfer of linoleic acid (Noble, O'Kelly and Moore, 1972; Noble, Crouchman and Moore, 1975a).

The distribution of linoleic acid between the major plasma lipids has important consequences for the lipid composition and metabolism of ruminant tissues. The triglycerides and unesterified fatty acids are the most active metabolically in supplying fatty acids to the peripheral tissues. Their content of predominantly saturated and monoenoic fatty acids, in conjunction with their rapid metabolic turnover, accounts for the extremely low concentration of linoleic acid in the triglycerides of storage lipids (see Christie, 1978). The structural lipids of ruminant tissues, on the other hand, display the high levels of linoleic and other polyunsaturated fatty acids similar to those of non-ruminant species. Because of their commercial importance, the most widely studied ruminant triglycerides are those of the adipose tissue and milk. Their extremely low content of linoleic acid when compared with non-ruminant species is well known and has been highlighted by the current concern with respect to the role of ruminant fats in the human diet.

A notable exception to the presence of low levels of linoleic acid in ruminant tissue triglycerides is that of the triglycerides found on the skin surface of cattle. These triglycerides, which comprise about 20% of the skin surface lipids, are remarkable for their appreciable content (greater than 20% of the total fatty acids present) of linoleic acid (Noble, Crouchman and Moore, 1974). From comparative analyses with the lipids of dermal and epidermal regions, it is apparent that the origin of these uniquely unsaturated triglycerides is the sebaceous glands (Jenkinson, McMaster and Noble, 1982). Although no definitive reason can be given to account for the presence of these triglycerides on the skin surface, a bacteriostatic function through the lipolytic release of free linoleic acid seems most plausible. It seems strange, however, that a species which receives such small amounts of linoleic acid because of biohydrogenation should secrete a significant amount on to its skin but not into its milk lipids.

Dietary manipulation of the linoleic acid content of tissues

Dietary fat supplements are normally subjected to the same processes within the ruminant animal as is the fat in the basal ration, i.e. hydrolysis followed rapidly by hydrogenation of the unsaturated components, before passing to the intestine and into the tissues. When basal diets are supplemented with linoleic acid there is, therefore, little effect on the fatty acid compositions of the tissues. Some exceptions are occasionally reported. For instance, vegetable oils fed with diets containing high amounts of barley have been found to give depot fats with increased levels of linoleic acid (Gibney and L'Estrange, 1975). It seems possible that under such circumstances different microbial populations arise in the rumen with a diminished capacity for biohydrogenation. Intellectual curiosity apart, in view of the current concern over the association between saturated fatty acid intake and human health, there are now sound commercial reasons for attempting to increase the linoleic acid content of ruminant fats. The effect of rumen micro-organisms on dietary linoleic acid can be eliminated or avoided by a number of experimental procedures. New-born ruminants to a large degree do not possess a rumen. The development of rumen function can be delayed by continual feeding with liquids which, by stimulating the closure of the oesophageal groove, allows the liquid diets to pass directly into the abomasum so that any linoleic acid escapes biohydrogenation (Ørskov and Benzie, 1969). Provided that suitable precautions are taken, the techniques can be employed for long periods, producing dramatic changes in the linoleic acid content of the tissues. With adult ruminants, the rumen and its micro-organisms have been by-passed by infusing rat suspensions directly into the abomasum or duodenum following appropriate surgery or, indeed, directly into the bloodstream (Tove and Mochrie, 1963). These procedures, useful as they are for experimental purposes, have little practical value. Diets that contain fats that are protected against hydrolysis and hydrogenation have been developed. This has usually consisted of encapsulating the fat (Scott et al., 1970) or the vegetable seed itself (Scott et al., 1972) in a coating which is resistant to microbial action during passage through the rumen but enabling subsequent release of the fat in the small intestine. Some dramatic increases in the linoleic acid contents of ruminant tissues and milk fats have been achieved routinely by this technique (see Scott and Cook, 1975). The procedure does not appear to have been successful commercially.

The ability of the plasma lipids to respond to the large changes in the uptake of linoleic acid determines the consequential effects upon the linoleic acid content of the tissues. Although the most abundant single fatty acid circulating in the plasma of the ruminant is linoleic, only 1% of it is associated with the triglycerides, the fraction which is available for adipose and milk-fat synthesis. However, when there is dietary supplementation with protected linoleic acid, considerable increases occur in the proportions of the very-low-density lipoproteins, the primary transporter of the triglycerides, and their content of linoleic acid-containing triglyceride species (Noble et al., 1977).

Little information exists about the metabolic side effects of feeding protected polyunsaturated fats to an animal which is normally adapted to an existence of low linoleic acid availability. Reduced tissue lipogenesis is certainly evident under such conditions (Vernon, 1975). Possible effects on tissue peroxidation have also to be considered. Over short-term feeding periods the advantage of feeding such diets would appear to outweigh the disadvantages but whether this applies to long-term feeding has yet to be assessed.

Miscellaneous factors affecting the linoleic acid content of tissues

The linoleic acid content of the tissues can be affected in varying degrees by a host of factors other than through a straightforward dietary supplementation, e.g. the overall general composition of the diet (roughage content in particular), breed, pregnancy, lactation, season and climate. The nature and biological activity of the rumen microflora can also vary considerably, depending upon outside influences that are not readily quantifiable. There is a great deal of evidence to show that exposure of the ruminant animal to high environmental temperatures results in substantial alterations to the plasma and tissue levels of linoleic acid. Thus, in heat-stressed cattle, the plasma concentrations of the cholesteryl esters and phospholipids, the main carriers of linoleic acid in the plasma, were very much reduced compared with those of animals maintained at a more normal environmental temperature (Noble, O'Kelly and Moore, 1973). Exposure to the high environmental temperatures also diminished the proportions of linoleic acid in both the cholesteryl ester and phospholipid fractions of the plasma, a finding which was related to a diminished activity of lecithin cholesterol acyltransferase activity. Seasonal variations in the plasma concentrations have also been observed (O'Kelly, 1972). The magnitude of this effect is dependent upon the breed as it is more apparent in breeds of cattle from temperate zones (*Bos taurus*) than in Zebu cattle (*Bos indicus*) (O'Kelly, 1968). At similar environmental temperatures, Zebu cattle display higher plasma levels of linoleic acid than the temperate breeds. Sebum output in young steers has been observed to increase under elevated environmental temperatures (Smith, Noble and Jenkinson, 1975). The sebum fatty acid composition was altered too, and displayed a greater than average increase in the secretion of linoleic acid in triglyceride form. It has been suggested that this may be involved both in controlling water loss and in protecting the animal against bacterial action in warm environments.

The essential fatty acid composition of fetal and new-born tissues

In so far as the fetus can be considered as part of the ruminant tissue pool, the new-born ruminant also is exposed during development to the limitation in linoleic acid availability. There is now much evidence that the plasma and tissues of fetal and new-born ruminants contain small concentrations only of essential fatty acids (see Noble, 1980; Noble and Shand, 1982; Christie, 1978). The concentrations found are very much lower than those of the plasma and tissues of fetuses and new-born of non-ruminant species and are much lower than those found in adult ruminants. For instance, it has been shown that the total carcass of a new-born lamb weighing 5.2 kg contained only 0.3 g of linoleic acid and

1.3 g of arachidonic acid (Noble, Steele and Moore, 1972). Much higher proportions of essential fatty acids have been found, for example, in the tissue of the new-born rat. In contrast to the adult, in the plasma of the ruminant fetus and new-born the linoleic acid content did not exceed 2–3% of the total fatty acids present in any of the major plasma lipid fractions (Leat, 1966; Noble, Steele and Moore, 1971a). Furthermore, in fetal and new-born tissues, high levels of palmitoleic acid were observed, the accumulation of which in tissues of non-ruminant species has also been reported to occur when there is a dietary inadequacy of linoleic acid. In spite of a lack of C18 polyunsaturated fatty acids, the plasma and tissues of fetal and new-born ruminants contain appreciable concentrations of C20 and C22 polyunsaturated fatty acids, but in particular $\Delta^{5,8,11}$-eicosatrienoic acid which is known to accumulate under conditions of inadequate linoleic acid supply (see Holman, 1971). High ratios of the $\Delta^{5,8,11}$-eicosatrienoic acid to arachidonic acid (known as the triene:tetraene ratio) are displayed. In the human and other non-ruminant species it is now accepted that a measure of the essential fatty acid status of an animal can be obtained from this ratio and when it exceeds a value of 0.4 it is usually accompanied by external symptoms of essential fatty acid deficiency (see Holman, 1971). The triene:tetraene ratios in fetal and new-born ruminants greatly exceeds this value. In the plasma lipids of the new-born lamb and kid, respective triene:tetraene ratios of 1.7 and 2.8 were found (Leat, 1966). Similarly high values have been observed in the plasma and tissues of new-born and fetal lambs and new-born calves (Noble, Steele and Moore, 1971a, b; Noble *et al.*, 1975; Noble, Shand and Calvert, 1982). Thus, on this biochemical criterion, the fetal and new-born ruminant could be classified as being deficient in essential fatty acids.

Maternal and fetal relationships in essential fatty acid supply

The features displayed by the tissues of the fetal and new-born ruminant with regard to their essential fatty acid content is explained by the relationship between the maternal lipid metabolism and that of the developing fetus. The only possible source of linoleic acid and its derivatives available to the fetus is the maternal circulation. From the direct and indirect evidence currently available, it would appear that of the major lipid fractions present in the maternal plasma, only the unesterified fatty acids contribute directly to the fetus (see Noble, 1980), and this to a much more limited extent than in non-ruminant species. Not only are the concentrations of the unesterified fatty acids in the maternal plasma of the ruminant very low, but their content of polyunsaturated fatty acids is small. Thus during the gestation period, the ruminant fetus is presented with a far more difficult problem than its non-ruminant counterpart in accumulating essential fatty acids.

In its function in enabling the passage of all the nutrients required to nourish the fetus, the placenta has some opportunity to play a role in modifying the lipid supply to the fetus. Evidence has now been obtained for such a role in the supply of essential fatty acids to the fetus. The presence of much higher proportions of arachidonic acid in the fatty acids of the fetal plasma compared with that of the unesterified fatty acids of the maternal plasma in the ruminant animal has been shown (Shand and Noble, 1979; Noble, Shand and Calvert, 1982); the proportion of arachidonic acid in the placenta was also shown to be very much higher than that in the unesterified fatty acids of the maternal plasma. In the ruminant, there is also a marked difference between the relative proportions of linoleic and arachidonic acids in the plasma lipids of the mother and fetus (Noble et al., 1978a; Shand and Noble, 1979; Noble, Shand and Calvert, 1982). In the sheep, for instance, the arachidonic acid : linoleic acid ratio in the fetal plasma was 2.72 compared with 0.21 for the maternal plasma. The placenta also showed a high arachidonic acid : linoleic acid ratio and negligible levels of the $\Delta^{5,8,11}$-eicosatrienoic acid. The higher proportions of arachidonic acid in the placenta and fetal plasma are not due to either a preferential uptake and transfer of the acid from the maternal plasma or synthesis and subsequent release by the fetal tissues. Evidence that the placenta is responsible for the increased fetal levels of arachidonic acid has been obtained through the identification in sheep of higher levels of Δ^6-desaturase activity in the placenta (Shand and Noble, 1979, 1981) than had previously been found in other tissues from either adult or new-born animals (Wahle, 1974; Shand, Noble and Moore, 1978). In view of the essential role of arachidonic acid in the synthesis of prostaglandins, an association between the high levels of arachidonic acid in the placental tissue and the increased importance of prostaglandins in the reproductive processes, especially during the latter part of pregnancy, should perhaps be considered.

Changes in the essential fatty acid status after birth

Rectification of the unsatisfactory essential fatty acid status of the new-born ruminant is rapid (see Noble, 1980, and Noble and Shand, 1982). Within 3–4 days of birth, the concentrations of linoleic acid in the plasma and tissues are increased markedly, the proportions of the $\Delta^{5,8,11}$-eicosatrienoic acid are decreased and there are large reductions in the triene:tetraene ratios. In the plasma, the changes in the linoleic acid content during the period immediately after birth have been correlated with significant changes in the activity of specific enzymes associated with its metabolism (Noble, Crouchman and Moore, 1975b). All this occurs in spite of the fact that the diet during this period consists of colostrum and milk in which the concentration of linoleic acid rarely exceeds 1% of the total fatty acids present and provides only about 0.4–0.6% of the total calories available (Noble, Steele and Moore, 1970). In the human and other non-ruminant species it has been shown that the minimum requirement for essential fatty acids is met only when 1% of the dietary calories is provided by linoleic acid (see Holman, 1971). Indeed, where cows' milk has been fed to non-ruminant species for extended periods, progressive increases in the triene : tetraene ratios into the range of deficiency have been observed. For instance, when new-born calves were fed a diet in which the linoleic acid provided 0.10, 0.32 and 1.00% of the total calories, the linoleic and arachidonic acid concentrations in the erythrocytes and plasma increased rapidly. On the other hand, when rats were given the diet in which linoleic acid provided 0.32% of the total calories, the concentrations of the essential fatty acids decreased and the triene:tetraene ratios increased (Sklan, Volcani and Budowski, 1972).

In the absence of any specific stores of essential fatty acids for mobilization during the immediate period after birth, all the indications are that the young ruminant animal is extremely efficient in utilizing and conserving the limited amount of linoleic acid obtained from the diet during the initial period after birth. That the level of linoleic acid in the diet needed to satisfy the minimum requirement for essential fatty acids in the young ruminant is somewhat less than that established for non-ruminant species, is also indicated. The results from comparative balance studies on neonatal lambs would substantiate an immediate and high requirement for essential fatty acids during the period just after birth (Noble, Steele and Moore, 1972). Attempts to induce classic symptoms of essential fatty acid deficiency in ruminants have met with mixed success as there is some doubt that the lesions observed were directly related to a deficient intake of linoleic acid (Lambert *et al.*, 1954). The new-born ruminant can respond to diets of extremely low essential fatty acid content: when lambs were fed from birth a diet in which linoleic acid provided only 0.001% of the energy, the triene:tetraene ratios in the plasma and tissues increased to values well above those observed at birth (Noble *et al.*, 1971a, b). More recently it has been claimed that both gnotobiotically and conventionally reared lambs deprived of linoleic acid showed only slow weight gains and exhibited classic essential fatty acid deficiency symptoms (Bruckner *et al.*, 1983).

Improvement of the essential fatty acid status of the new-born

Although little is known about a specific metabolic role for the essential fatty acids in the tissues of the ruminant, their importance in the control and regulation of a variety of metabolic and physiological parameters, such as maintenance of membrane structure and function, control of water balance, involvement in disease resistance, in other species is well established (see Holman, 1971). It is possible that conditions could arise under which the lowly essential fatty acid status might place the new-born ruminant at risk. There are many circumstances, such as conditions of high infection, environmental stresses, the occurrence of multiple births, and lack of vigour at birth, under which beneficial effects could arise from an improved essential fatty acid status. It is known, for instance, that exposure to elevated environmental temperatures dramatically reduces the polyunsaturated fatty acid levels in tissues (O'Kelly, 1972; Noble *et al.*, 1973) and resistance of the body to invasion by bacteria and other exogenous agents (Pan, 1970). Evidence is now available to suggest that, in the ruminant, the stresses which could accompany the poor essential fatty acid status at birth might be exacerbated under such environmental conditions. Thus exposure to high environmental temperatures severely limits the increase normally observed in the linoleic acid content of the plasma immediately after birth (Noble and Moore, 1974). Under normal circumstances it is some 4 weeks after birth before the sebaceous secretions on the skin attain their natural content of linoleic acid (Noble et al., 1975). Delay in attaining the correct levels of linoleic acid through exposure to an elevated environmental temperature may reduce even further the ability of the new-born to regulate both microbiological infestation of the skin and moisture loss through the skin.

It could therefore be argued that some practical benefits might be derived from an improvement of the essential fatty acid status of the young ruminant during the period immediately after birth. Whether or not any response of practical importance could be observed would depend upon the degree of adversity encountered by the new-born animal. Increased growth rates during the first 28 days after birth have been observed in simple feeding trials of new-born lambs whose diets had been supplemented with linoleic acid, the response being significantly greater in those lambs which presented the highest risk of non-survival at birth (K. Jaguch, personal communication). In an investigation of the essential fatty acid requirement of neonatal colostrum-deprived lambs based on growth and performance factors under gnotobiotic and normal conditions, exposure to a microbial stress increased the linoleic acid requirement (Bruckner et al., 1983). A relationship between the essential fatty acid status of the new-born lamb and its ability to withstand a high environmental temperature has also been shown (Noble, McLean and Downie, 1981); in new-born lambs supplemented with linoleic acid there was a great increase in cutaneous moisture loss and an absence of signs of thermal stress. Clear evidence has also been obtained in new-born lambs whose essential fatty acid status had been improved that erythrocyte fragility, which is related to membrane structure and function, was much reduced (Noble et al., 1978b).

Improvements in the essential fatty acid status of the young ruminant can most simply be obtained by incorporation of linoleic acid into the postparturient diet. An improvement to the essential fatty acid supply during fetal development would seem to present the greatest advantage. However, the capacity to achieve this is diminished not only by the limited nature of the fatty acid transfer between mother and fetus but also by the fact that any linoleic acid that escapes rumen biohydrogenation in the mother would be preferentially incorporated into the cholesteryl ester and phospholipid fractions of the plasma, thereby rendering it virtually unavailable to the fetus. The capacity to increase the linoleic acid content of the milk during early lactation through simple manipulation of the diet is also limited by biohydrogenation and lack of incorporation of available linoleic acid into the correct plasma lipid precursor for milk fat synthesis (Barry et al., 1963). Both of these problems may be overcome through the inclusion into the maternal diet, over an extended period of time, of linoleic acid protected from biohydrogenation. Thus in sheep, not only were the concentrations of linoleic and arachidonic acids increased and the triene:tetraene ratios reduced to less than 0.4 in the plasma and the tissues of the new-born lambs, but the concentrations of linoleic acid in the colostrum and milk were increased to well above 1% of the total caloric content (Noble et al., 1978b; Shand et al., 1978).

In summary

Obvious differences exist in the apparent efficiency of utilization of linoleic acid between ruminant and non-ruminant species. In the adult ruminant on a normal diet the highly selective incorporation of any linoleic acid that escapes rumen biohydrogenation into the cholesteryl esters and phospholipid fractions of plasma ensures that it is in a form that enables maximum benefit to be derived from its specialized physiological role. It is possible that the highly efficient utilization of essential fatty acids demonstrated by ruminant animals may have been evolved as a process to counteract the extensive biohydrogenation of linoleic acid which occurs in the rumen. There is also a conflict between the ability of the adult ruminant to overcome its own problem of essential fatty acid availability and its capacity to supply sufficient essential fatty acids to the young neonatal ruminant both during development and immediately after birth. At best there exists a fine balance between maternal supply and new-born requirement.

References

BARRY, J.M., BARTLEY, W., LINZELL, J.L. and ROBINSON, D.S. (1963). The uptake from the blood of triglyceride fatty acids of chylomicra and low density lipoproteins by the mammary gland of the goat. *Biochemical Journal* **89**, 6–11

BATH, I.H. and HILL, K.J. (1967). The lipolysis and hydrogenation of lipids in the digestive tract of the sheep. *Journal of Agricultural Science* **68**, 139–148

BICKERSTAFFE, R., NOAKES, D.E. and ANNISON, E.F. (1972). Quantitative aspects of fatty acid biohydrogenation, absorption and transfer into milk fat in the lactating goat, with reference to the *cis*- and *trans*-isomers of octadecanoate and linoleate. *Biochemical Journal* **130**, 607–617

BRUCKNER, G.G., GRUNEWALD, K.K., TUCKER, R.E. and MITCHELL, G.E. (1983). Essential fatty acid status and characteristics associated with colostrum-deprived gnotobiotic and conventional lambs. Part 1. Growth, organ development, cell membrane integrity and parameters associated with lower bowel function. *Journal of Animal Science* (in press)

CHRISTIE, W.W. (1978). The composition, structure and function of lipids in the tissues of ruminant animals. *Progress in Lipid Research* **17**, 111–205

CHRISTIE, W.W. and HUNTER, M.L. (1978). The composition and structure of the lipids of sheep lymph. *Journal of the Science of Food and Agriculture* **29**, 442–446

COOK, L.J., SCOTT, T.W., FAICHNEY, G.T. and LLOYD-DAVIES, H. (1972). Fatty acid interrelationships in plasma, liver, muscle and adipose tissues of cattle fed safflower oil protected from ruminal hydrogenation. *Lipids* **7**, 83–89

CZERKAWSKI, J.W. (1967). Effect of storage on the fatty acids of dried ryegrass. *British Journal of Nutrition* **21**, 599–608

DAWSON, R.M.C. and HEMINGTON, N. (1974). Digestion of grass lipids and pigments in the sheep rumen. *British Journal of Nutrition* **32**, 327–340

DUNCAN, W.R.H. and GARTON, G.A. (1963). Plasma lipids of the cow during pregnancy and lactation. *Biochemical Journal* **89**, 414–419

EMMANUEL, B. (1974). On the origin of rumen protozoan fatty acids. *Biochemica et biophysica acta* **337**, 404–413

GARTON, G.A. and DUNCAN, W.R.H. (1964). The lipids of sheep plasma. *Biochemical Journal* **92**, 472–475

GARTON, G.A., HOBSON, P.N. and LOUGH, A.K. (1958). Lipolysis in the rumen. *Nature* **182**, 1511–1512

GARTON, G.A., LOUGH, A.K. and VIOQUE, E. (1959). The effect of sheep rumen contents on triglycerides *in vitro*. *Biochemical Journal* **73**, 46P

GIBNEY, M.J. and L'ESTRANGE, J.L. (1975). Effects of dietary unsaturated fat and of protein source on melting point and fatty acid composition of lamb fat. *Journal of Agricultural Science* **84**, 291–296

HARFOOT, C.G. (1978). Lipid metabolism in the rumen. *Progress in Lipid Research* **17**, 21–54

HAWKE, J.C. and SILCOCK, W.R. (1969). Lipolysis and hydrogenation in the rumen. *Biochemical Journal* **112**, 131–132

HEATH, T., CAPLE, I.W. and REDDING, P.M. (1970). Effect of the enterohepatic circulation of bile salts on the flow of bile and its contents of bile salts and lipids in sheep. *Quarterly Journal of Experimental Physiology* **55**, 93–103

HENDERSON, C. (1971). A study of the lipase produced by Anaerovibrio lipolytica, a rumen bacterium. *Journal of General Microbiology* **65**, 81–89

HOLMAN, R.T. (1971). Essential fatty acid deficiency. *Progress in the Chemistry of Fats and Other Lipids* **9**, 275–348

IFKOVITS, R.W. and RAGHEB, H.S. (1968). Cellular fatty acid composition and identification of rumen bacteria. *Applied Microbiology* **16**, 1406–1413

JENKINSON D. McEWAN, McMASTER, J.D. and NOBLE, R.C. (1982). Bovine sebum and dermal lipid composition. *Journal of Physiology* **329**, 38P–39P

KATES, M. (1970). Plant phospholipids and glycolipids. *Advances in Lipid Research* **8**, 225–265

KEENEY, M., KATZ, I. and ALLISON, M.J. (1962). On the probable origin of some milk fat acids in rumen microbial lipids. *Journal of the American Oil Chemists Society* **39**, 198–201

LAMBERT, M.R., JACOBSON, N.L., ALLEN, R.S. and ZALETEL, J.H. (1954). Lipid deficiency in the calf. *Journal of Nutrition* **52**, 259–272

LASCELLES, A.K., HARDWICK, D.C., LINZELL, J.L. and MEPHAM, T.B. (1964). The transfer of [H^3] stearic acid from chylomicron to milk fat in the goat. *Biochemical Journal* **92**, 36–42

LEAT, W.M.F. (1966). Fatty acid composition of the plasma lipids of new-born and maternal ruminants. *Biochemical Journal* **98**, 598–603

LEAT, W.M.F. and HALL, J.G. (1968). Lipid composition of lymph and blood plasma of the cow. *Journal of Agricultural Science* **71**, 189–194

LEAT, W.M.F. and HARRISON, F.A. (1969). Lipid digestion in the sheep: effect

of bile and pancreatic juice on the lipids of intestinal content. *Quarterly Journal of Experimental Physiology* **54**, 187–201
LEAT, W.M.F. and HARRISON, F.A. (1973). Intake and absorption of essential fatty acids by the sheep. *Proceedings of the Nutrition Society* **31**, 70A–71A
LENNOX, A.M. and GARTON, G.A. (1968). The absorption of long-chain fatty acids from the small intestine of the sheep. *British Journal of Nutrition* **22**, 247–254
LENNOX, A.M., LOUGH, A.K. and GARTON, G.A. (1968). Observations on the nature and origin of lipids in the small intestine of the sheep. *British Journal of Nutrition* **22**, 237–246
NOBLE, R.C. (1978). Digestion, absorption and transport of lipids in ruminant animals. *Progress in Lipid Research* **17**, 55–91
NOBLE, R.C. (1980). Lipid metabolism in the neonatal ruminant. *Progress in Lipid Research* **18**, 179–216
NOBLE, R.C. and MOORE, J.H. (1974). Heat exposure and the fatty acid composition of the plasma of the young lamb. *Research in Veterinary Science* **17**, 204–209
NOBLE, R.C. and SHAND, J.H. (1982). Fatty acid metabolism in the neonatal ruminant. *Advances in Nutritional Research* **4**, 287–337
NOBLE, R.C., CROUCHMAN, M.L. and MOORE, J.H. (1974). The presence of linoleic acid in the skin surface lipids of the ox. *Research in Veterinary Science* **17**, 372–376
NOBLE, R.C., CROUCHMAN, M.L. and MOORE, J.H. (1975a). Synthesis of cholesterol esters in the plasma and liver of sheep. *Lipids* **10**, 790–799
NOBLE, R.C., CROUCHMAN, M.L. and MOORE, J.H. (1975b). Plasma cholesterol ester formation in the neonatal lamb. *Biology of the Neonate* **26**, 117–121
NOBLE, R.C., McCLEAN, J.A. and DOWNIE, A.J. (1981). The linoleic acid status of the new-born lamb and thermoregulation. *Research in Veterinary Science* **30**, 129–130
NOBLE, R.C., O'KELLY, J.C. and MOORE, J.H. (1972). Observations on the lecithin:cholesterol acyltransferase system in bovine plasma. *Biochemica et biophysica acta* **270**, 519–528
NOBLE, R.C., O'KELLY, J.C. and MOORE, J.H. (1973). Observations on changes in lipid composition and lecithin-cholesterol acyltransferase reaction of bovine plasma induced by heat exposure. *Lipids* **8**, 216–223
NOBLE, R.C., SHAND, J.H. and CALVERT, D.T. (1982). The role of the placenta in the supply of essential fatty acids to the fetal sheep: studies of lipid compositions at term. *Placenta* **3**, 287–296
NOBLE, R.C., STEELE, W. and MOORE, J.H. (1970). The composition of ewe's milk fat during early and late lactation. *Journal of Dairy Research* **37**, 297–301
NOBLE, R.C., STEELE, W. and MOORE, J.H. (1971a). Diet and the fatty acids in the plasma of lambs during the first eight days after birth. *Lipids* **6**, 26–34
NOBLE, R.C., STEELE, W. and MOORE, J.H. (1971b). Fatty acid composition of liver lipids of young lambs. *British Journal of Nutrition* **26**, 97–105
NOBLE, R.C., STEELE, W. and MOORE, J.H. (1972). The metabolism of linoleic acid by the young lamb. *British Journal of Nutrition* **27**, 503–508
NOBLE, R.C., CROUCHMAN, M.L., JENKINSON, D.McEWAN and MOORE, J.H.

(1975). Relationships between lipids in plasma and skin secretions of neonatal calf with particular reference to linoleic acid. *Lipids* **10**, 128–133

NOBLE, R.C., SHAND, J.H., BELL, A.W., THOMPSON, G.E. and MOORE, J.H. (1978a). The transfer of free palmitic and linoleic acids across the ovine placenta. *Lipids* **13**, 610–615

NOBLE, R.C., SHAND, J.H., DRUMMOND, J.T. and MOORE, J.H. (1978b). 'Protected' polyunsaturated fatty acid in the diet of the ewe and the essential fatty acid status of the neonatal lamb. *Journal of Nutrition* **108**, 1868–1876

NOBLE, R.C., VERNON, R.G. CHRISTIE, W.W., MOORE, J.H. and EVANS, A.J. (1977). The effect of dietary fats on the plasma lipid composition of sheep. *Lipids* **12**, 423–433

O'KELLY, J.C. (1968). Comparative studies of lipid metabolism in Zebu and British cattle in a tropical environment. *Australian Journal of Biological Science* **21**, 1013–1024

O'KELLY, J.C. (1972). Seasonal variations in the plasma lipids of genetically different types of cattle: grazing steers. *Comparative Biochemistry and Physiology* **43B**, 283–294

ØRSKOV, E.R. and BENZIE, D. (1969). Studies on the oesophageal groove reflex in sheep and on the potential use of the groove to prevent the fermentation of food in the rumen. *British Journal of Nutrition* **23**, 415–420

PAN, Y.S. (1970). Breed and seasonal differences in quantities of lipids on skin surface and hair in cattle. *Journal of Agricultural Science* **75**, 41–46

REISER, R. (1951). Hydrogenation of polyunsaturated fatty acids by the ruminant. *Federation Proceedings* **10**, 236

REISER, R. and REDDY, H.G.R. (1956). The hydrogenation of dietary unsaturated fatty acids by the ruminant. *Journal of the American Oil Chemists Society* **33**, 155–156

SCOTT, T.W. and COOK, L.J. (1975). Effect of dietary fat on lipid metabolism in ruminants. In *Digestion and Metabolism in the Ruminant* (I.W. McDonald and A.C.I. Warner, Eds), pp. 510–523. Armidale, University of New England Publishing Unit

SCOTT, T.W., BREADY, P.J., ROYAL, A.J. and COOK, L.J. (1972). Oil seed supplements for the production of polyunsaturated ruminant milk fat. *Search* **3**, 170–171

SCOTT, T.W., COOK, L.J., FERGUSON, K.A., McDONALD, I.W., BUCHANAN, R.A. and HILLS, G.L. (1970). Production of polyunsaturated milk fat in domestic ruminants. *Australian Journal of Science* **32**, 291–293

SHAND, J.H. and NOBLE, R.C. (1979). $\Delta 9$- and $\Delta 6$-desaturase activities of the ovine placenta and their role in the supply of fatty acids to the fetus. *Biology of the Neonate* **36**, 298–304

SHAND, J.H. and NOBLE, R.C. (1981). The metabolism of 18:0 and 18:2(n-6) by the ovine placenta at 120 and 150 days of gestation. *Lipids* **16**, 68–71

SHAND, J.H., NOBLE, R.C. and MOORE, J.H. (1978). Dietary influences on fatty acid metabolism in the liver of the neonatal lamb. *Biology of the Neonate* **34**, 217–224

SHORLAND, F.B., WEENINK, R.O. and JOHNS, A.T. (1955). Effect of the rumen on dietary fat. *Nature* **175**, 1129–1130

SKLAN, D., BUDOWSKI, P. and VOLCANI, R. (1972). Synthesis *in vitro* of linoleic acid by rumen liquor in calves. *British Journal of Nutrition* **28**, 239–248

SKLAN, D., VOLCANI, R. and BUDOWSKI, P. (1972). Effects of diets low in fat or essential fatty acids on the fatty acid composition of blood lipids of calves. *British Journal of Nutrition* **27**, 365–374

SMITH, M.E., NOBLE, R.C. and JENKINSON D. McEWAN (1975). The effect of environment on sebum output and composition in cattle. *Research in Veterinary Science* **19**, 253–258

TOVE, S.B. and MOCHRIE, R.D. (1963). Effect of dietary and injected fat on the fatty acid composition of bovine depot fat and milk fat. *Journal of Dairy Science* **46**, 686–689

TWEEDIE, J.W., RUMSBY, M.G. and HAWKE, J.C. (1966). Formation of branched long-chain fatty acids in cultures of rumen bacteria. *Journal of the Science of Food and Agriculture* **17**, 241–244

VERNON, R.G. (1975). Effect of safflower oil upon lipogenesis in neonatal lamb. *Lipids* **10**, 284–289

VIVIANI, R. (1970). Metabolism of long-chain fatty acids in the rumen. *Advances in Lipid Research* **8**, 267–346

WADSWORTH, J.C. (1968). Fatty acid composition of lipid in the thoracic duct lymph of grazing cows. *Journal of Dairy Science* **51**, 876–881

WAHLE, K.J. (1974). Desaturation of long-chain fatty acids by tissue preparations of the sheep, rat and chicken. *Comparative Biochemistry and Physiology* **48B**, 87–105

WARD, P.F.V., SCOTT, T.W. and DAWSON, R.M.C. (1964). The hydrogenation of unsaturated fatty acids in the ovine digestive tract. *Biochemical Journal* **92**, 60–68

10

PROTECTIVE FUNCTIONS OF FAT-SOLUBLE VITAMINS

ALFRED W. KORMANN and HARALD WEISER
F. Hoffmann-La Roche & Co. Ltd., Central Research Units, 4002 Basle, Switzerland

The group of fat-soluble vitamins is composed of vitamins A, D, E and K. This designation is derived from their common lipophilic nature, and accordingly their biological availabilities depend on an intact intestinal absorption of dietary oils and fats. Impaired uptake of lipids may lead to an insufficient supply of these vitamins and corresponding deficiency symptoms.

The following chapter describes some of the main aspects concerning biological functions of the fat-soluble vitamins. Because of space limitations this review cannot attempt to cover the whole field extensively and concentrates on a few selected highlights. In consideration of the topic of this volume the main emphasis has been put on vitamin E and its interactions with lipids. For more detailed reports the reader will be referred to several reviews.

Vitamin E

Naturally occurring vitamin E is composed of several tocopherols and tocotrienols, some plant seeds and vegetable oils being particularly rich sources. D-α-Tocopherol (R,R,R-α-tocopherol; IUPAC IUB, 1982) has the highest biopotency *in vivo*. The main vitamin E component for enrichment of feeds is DL-α-tocopheryl acetate (RS,RS,RS or all-rac), stabilized as suitable formulations; 1.49 mg have the same biological activity as 1.0 mg of D-α-tocopherol (USP XX, 1980). Two detailed discussions concerning these activity factors have been published recently (Weiser and Vecchi, 1981, 1982).

Vitamin E deficiency symptoms in animals

Functions of vitamin E in living organisms are based on its antioxidant properties and its structural role in cellular membranes. The former are especially important with regard to lipid peroxidation and subsequent processes which will cause tissue damage. Scott (1980) has summarized the known vitamin E deficiency diseases of all species.

In animals, a dietary lack of vitamin E may cause several well-known pathological conditions. They include reproductive failure in both sexes, recognizable as testicular degeneration or death and subsequent resorption of fetuses *in utero*. The latter events can be utilized for a standardized biological assay of vitamin E activity (Weiser and Vecchi, 1981, 1982). Vitamin E deficiency may also affect reproduction in birds, e.g. quails (Kling and Soares, 1980). Dietary lack of this vitamin can also cause nutritional muscular dystrophy in domestic animals (McMurray, 1980; Vos, Hulstaert and Molenaar, 1981; Hutchinson, Scholz and Drake, 1982). Erythrocyte haemolysis in several species (including man) or encephalomalacia, a neurological impairment in chicks, are other familiar deficiency syndromes. With regard to the latter, Muller, Lloyd and Wolff (1983) postulated that Vitamin E is important for normal neurological function in man also.

Protective functions of vitamin E and selenium

Biological activities of vitamin E are often linked with those of selenium because the latter is an essential part of enzymes such as glutathione peroxidase which participate in degradation pathways for damaging radicals:

$$O_2^- \rightarrow H_2O_2 \rightarrow HO\cdot + OH^- + O_2 \xrightarrow{\text{vitamin E}} \text{free radicals}$$
$$\downarrow \text{GHS-Px (Se)}$$
$$H_2O$$

This simplified scheme indicates that formation of superoxide anions (O_2^-) leads to H_2O_2 which may be converted by Se-containing glutathione peroxidase (GSH-Px) to a harmless product, water. If it is not thus converted, H_2O_2 may give rise to free radicals which will attack unsaturated lipids and lead to tissue damage. Vitamin E is an inhibitor of free radical formation and thus an efficient protectant of lipids, especially those containing polyunsaturated fatty acids (PUFAs). Extensive reviews concerning these reactions have been published, for example by Simic and Karel (1980), Machlin (1980), or Demopoulos *et al.* (1980).

These concerted actions of vitamin E and selenium affect several species and a wide range of conditions. In addition to correction of simple nutritional deficiencies, administration of vitamin E and/or selenium may prevent or cure disease states caused by different toxicants. A few examples concerning poultry are given in *Table 10.1*. It should be emphasized that mammals, too, may be similarly affected and treated (see McMurray, 1980) but poultry have been used more frequently for this type of investigation. Hill (1981) has reviewed the role of vitamin E and selenium in poultry nutrition.

Vitamin E–selenium deficiencies are always connected with disturbances of lipid metabolism in general and with imbalances of PUFAs in particular. Nutritionists have therefore proposed that human and animal

Table 10.1 EXAMPLES OF PROTECTIVE ACTIONS OF VITAMIN E AND/OR SELENIUM IN POULTRY AFFECTED BY TOXICANTS

Species	Disorder	Toxicant	Reference
Chicks	Encephalomalacia	PUFAs (excess or oxidized)	Bartov and Bornstein, 1980
Chicks	Reduced liver capacity for detoxification	Aflatoxin B_1	Chen et al., 1982
Chickens	Infection	E. coli	Likoff et al., 1981
Chickens, laying	Disturbances of fertility, hatchability of eggs etc.	Vicine (bean constituents)	Muduuli, Marquardt and Guenter, 1982
Jap. quails	Increased mortality	Mercury compounds	Kling and Soares, 1981
Ducklings	Necrosis of several organs	Metals and other elements	Van Vleet, 1982

diets ought to contain minimal doses of vitamin E and selenium. These doses have been expressed as amounts ingested per day (Scott, 1980) or as optimal ratios of tocopherol per total lipid (Farrell et al., 1978). However, it must be borne in mind that different diets contain widely differing percentages of unsaturated lipids; the required intakes of protectants should be calculated accordingly. For example, it has been determined in a rat model that the animals needed about 0.5 mg DL-α-tocopheryl acetate per gram of linoleic acid consumed (Weber, Weiser and Wiss, 1964).

Vitamin E and arachidonic acid metabolism

The intimate involvement of vitamin E and fatty acids is evident in other physiological areas as well. Arachidonic acid is a PUFA of special relevance as it is the starting point for pathways leading to prostaglandins, thromboxanes and leukotrienes. These compounds are extremely potent regulators of many biological processes including blood pressure, inflammation and uterine contractions. Some of the biosynthetic steps involve peroxidations which make antioxidants such as vitamin E likely candidates for intervention: vitamin E does indeed influence these reactions, and a few examples are given in *Table 10.2*. It is evident that vitamin E can influence a diversity of physiological reactions and pathological conditions. The last example in *Table 10.2* shows that this

Table 10.2 EXAMPLES OF INVOLVEMENT OF VITAMIN E WITH PATHWAYS LEADING TO PROSTAGLANDINS (PG)

Species, tissue	Effect of vitamin E deficiency	Reference
Rabbit platelets	Increased PGE_2 production	Tangney and Driskell, 1981
Rabbit aorta	Decreased prostacyclin synthesis	Chan and Leith, 1981
Rat	Aorta: decreased prostacyclin synthesis Lung: higher amounts of PGE_2 and $PGF_{2\alpha}$ in perfusate	Valentovic, Gairola and Lubawy, 1982
Rat testis	Decreased $PGF_{2\alpha}$ synthesis (prior to any overt morphological changes)	Carpenter, 1981
Rat foot	More severe inflammation after treatment with irritant	Dillard, Kunert and Tappel, 1982a
Chickens (with E. coli infection)	Bursa with higher contents of $PGF_{2\alpha}$, PGE_2 and PGE_1	Likoff et al., 1981

includes immunological manifestations (see Beisel, 1982, for a review). Although the mechanism is not known, the involvement of vitamin E in the thyroid also has been documented (Weiser and Salkeld, 1977).

Vitamin E and other vitamins

The discussion above documented that the functions of vitamin E are not isolated occurrences; they affect the biological activities of many other substances including other vitamins, particularly if the latter are administered in high doses (*Table 10.3*). The results cited in this Table are

Table 10.3 INTERACTIONS OF VITAMIN E AND HIGH DOSES OF OTHER VITAMINS (–E = VITAMIN E DEFICIENCY)

Other vitamin(s)	Species	Biological effects	Reference
A	Chick	Enhanced oxidation and increased turnover of dietary α-tocopherol in gastrointestinal tract	Sklan and Donoghue, 1982
A	Chick and rat	Reduced α-tocopherol levels in liver and plasma	Weiser, 1982 (unpublished)
A	Rat	Retina: extensive lipofuscin accumulation, dependent on status of vitamins A and E	Robison, Kuwabara and Bieri, 1980
A, C	Rat	Liver and kidney: vitamin A inhibited lipid peroxidation in –E animals Liver and brain: vitamin C inhibited lipid peroxidation in –E animals Erythrocytes: vitamins A and C increased lysis in –E animals	Bai *et al.*, 1982
C	Rat	Erythrocytes: increased haemolysis, reduced glutathione	Chen, 1981
C	Rat	Alloxan-treated animals Liver, pancreas, plasma: increased lipid peroxidation due to vitamin C in –E animals Plasma glucose: elevated in –E animals	Dillard, Kunert and Tappel, 1982b
C	Guinea pig	Liver: enhancement of ability to detoxicate after oral administration of vitamins E and C	Ginter *et al.*, 1982
C	Man	Reduced nitrosation of proline (*in vivo*) after oral administration of vitamins E or C	Ohshima and Bartsch, 1981
C	Man	Reduced mutagen levels in faeces after oral administration of vitamins E and C	Dion *et al.*, 1982

clear-cut evidence of the varied consequences of a vitamin E deficiency or imbalance. A review of the progress in cancer chemoprevention by vitamins and carotenoids has been published recently (Bianchi *et al.*, 1982).

Agents other than vitamins also may interfere with vitamin E (see *Table 10.1*), as well as dietary factors such as types and amount of protein, fats and micronutrients (trace elements and other factors). These 'hidden' differences may possibly explain some contradictory results obtained with seemingly identical animal models.

Vitamin E as an in vitro *antioxidant*

Nutritionists and consumers have been aware for a long time that unprotected lipids oxidize to unpalatable and even dangerous products. It is also known that suitable storage measures and addition of vitamin E or other antioxidants will prevent these reactions. These problems and their solutions have been discussed in detail in several publications (e.g. Machlin, 1980; Simic and Karel, 1980). Combinations of antioxidants often may be more effective than a single compound (Battná, Parízková and Kučerová, 1982; Pongracz, 1982 and unpublished results).

Values of relative antioxidant activity of different compounds depend strongly on the test system utilized. For example, Boguth, Patzelt-Wenczler and Repges (1971) applied a simple assay with a radical source and β-carotene as substrate and indicator to study relative activities of some natural and synthetic antioxidants; γ-tocopherol was the most active of all vitamin E compounds tested. Burton and Ingold (1981) used the same radical initiator in a different experimental programme to measure autoxidation of biological molecules: they concluded from their findings that α-tocopherol was the most reactive chain-breaking phenolic antioxidant known. In addition, they reported a similar conclusion with regard to vitamin E as an antioxidant in blood plasma (Burton, Joyce and Ingold, 1982). Autoxidation in model membrane systems is achieved under *in vitro* conditions but findings may be of relevance to *in vivo* situations (see Weenen and Porter, 1982).

Dietary supplements of vitamin E may also have a favourable influence on the properties of food products derived from animals. Bartov and Bornstein (1981) reported that increasing vitamin E levels in broiler feeds resulted in higher α-tocopherol contents of abdominal fat and corresponding improvements in meat properties. Similar observations of better stability due to vitamin E have been made for other animal products (Uebersax, Dawson and Uebersax, 1978; Tagwerker, 1981) whereas in trout the response of muscle α-tocopherol was too small to be economically feasible (Hung and Slinger, 1982). In addition to better storage properties these applications have considerable nutritional implications for the human consumer (see Tagwerker, 1981; Sauberlich *et al.*, 1982).

Vitamin A

Vitamin A occurs in nature mainly as retinyl fatty acid esters; the highest levels are found in the livers of some fish and of mammals. The all-*trans* form of retinol is the most important natural and synthetic isomer, whereas 13-*cis* vitamin A has a lower biological activity. Retinyl esters are absorbed in the intestine and transferred to the liver, the main storage organ for this vitamin. Mobilization from the liver, transport by the blood circulation and uptake by organs are rigidly controlled processes, involving distinct plasma and cellular retinol-binding proteins (RBP) and cell surface receptors for RBP (see Ganguly *et al.*, 1980; Goodman, 1980; Chytil, 1982).

Vitamin A deficiency symptoms

Vitamin A participates in a number of physiological functions, a classic example being the visual process. Deficiency of vitamin A may lead to reversible or permanent eye damage and even to total blindness. This is still a major health problem in some Third World countries, particularly if combined with general malnutrition (Anonymous, 1982a). The complexity of the reactions in the bovine retina and the chemical processes of vision in general have been reviewed by Saari, Bredberg and Garwin (1982) and by O'Brien (1982).

Another pronounced effect of insufficient intake of vitamin A is growth retardation in humans (Hamdy *et al.*, 1982) and in animals (Ganguly *et al.*, 1980). As vitamin A deficiency will also cause degenerative changes in a variety of organs, e.g. respiratory and gastrointestinal systems, affected organisms will be more susceptible to infections. Similar disturbances of epithelial cells and membranes occur in the reproductive tracts of male and female animals, and subsequently these organs will atrophy and lose their reproductive capabilities. For example, Catignani and Bieri (1980) and Steinberg and Sgoutas (1981) demonstrated damaging effects on rat testis, while Sietsema and DeLuca (1982) induced vaginal cornification with vitamin A deficiency. The latter symptom was utilized for a quantitative bioassay of vitamin A-like compounds (see also Weiser, 1978).

A few less familiar syndromes associated with vitamin A deficiency have also been reported, and include malfunctions of the inner ear (Chole, 1980; Biesalsky and co-workers, unpublished data), bladder calculi and tumours (Gershoff and McGandy, 1981; Bichler *et al.*, 1983), or disturbances of thyroid hormone metabolism (Higueret and Garcin, 1982).

Interactions of vitamin A with other vitamins

The influence of high doses of vitamin A on the availability of vitamin E has been mentioned above. Relationships between these two vitamins and ascorbic acid during the development of encephalomalacia in chicks have also been studied by Dror, Bartov and Bubis (1982). Saito *et al.* (1982) investigated the effects of dietary antioxidants (including DL-α-tocopheryl acetate) on polychlorinated biphenyl-induced hepatic lipid peroxidation and vitamin A reduction. Although vitamin E reduced lipid peroxides, it failed to elevate vitamin A levels in liver (see also *Table 10.3*). With regard to another vitamin combination, Ferrando, Fourlon and Clech (1977) concluded from their studies that the ratio of dietary vitamins D_3 and A may influence hepatic stores of the latter.

In contrast to the multitude of deficiency symptoms, it should be remembered that an excess of this vitamin may be harmful. In addition to the effects on vitamin E it has been reported that pharmacological doses of vitamin A increase parathyroid hormone levels (Chertow *et al.*, 1977). Hypervitaminosis A symptoms after administration of multi-vitamin preparations have been described, e.g. by Macapinlac and Olson (1981) or Kozicki, Silva and Barnabe (1981).

Provitamins A and retinoids

A number of provitamins may be converted to vitamin A in the body, β-carotene being the most prominent example. In addition to its precursor function it appears to have different pharmacological functions. It seems to influence reproduction in cattle and possibly also in other species (Jackson, 1981; Lotthammer, 1981; see also Kiatoko *et al.*, 1982), or to be effective in enhancing wound strength (Gerber and Erdman, 1982). In man, increased intake of food rich in β-carotene may offer some protection against stomach and lung cancer (Wolf, 1982; Kolonel *et al.*, 1983). The latter findings supplement earlier results indicating that low levels of serum retinol were associated with an increased risk of cancer (Wald *et al.*, 1980).

Retinoids are a class of vitamin A-like compounds with promising antineoplastic and dermatological applications in man and animals. Their potential as chemopreventive and anticancer agents in animals has been reviewed recently (Hill and Grubbs, 1982; Bollag and Hartmann, 1983).

Vitamin D

Several forms of vitamin D have been found in nature, but D_3 (cholecalciferol) and D_2 (ergocalciferol) are the only ones of nutritional significance. Some fish-liver oils contain extremely high amounts, whereas the majority of other animal products and plants contain little or no vitamin D. It can be acquired by dietary means or through UV-dependent synthesis in the skin. With regard to domestic animals it should be noted that vitamin D_2 is several times less active in the fowl than corresponding amounts of D_3 (Horst, Napoli and Littledike, 1982).

The following section will concentrate on cholecalciferol and its metabolites because the major part of vitamin D research has been carried out with D_3 forms. An exhaustive comparison of D_3 and D_2 has been made by Horst *et al.* (1981) who determined plasma levels in the mature turkey, chicken, cow, sheep and pig.

Metabolism of vitamin D_3 to hormonal forms

The biological pathways leading to vitamin D activities are unique among the vitamins (*Figure 10.1*). An initial hydroxylation of D_3 in the liver yields 25-hydroxy-D_3 [25(OH)D_3]. This metabolite is further hydroxylated by kidney enzymes to either 1,25-dihydroxy-D_3 or 24,25-dihydroxy-D_3. Synthesis and reactions of these extremely potent compounds are strictly controlled by several regulating mechanisms which act according to established criteria for hormones; consequently these dihydroxylated D_3 metabolites are viewed as hormones. This, of course, deviates strongly from the classic image of vitamins as simple cofactors.

In addition to the major events indicated in *Figure 10.1*, a number of other organs and regulators are involved, including the placenta, pancreas, egg-shell gland, oestrogens, growth hormone and prolactin. Catabolic inactivation of the dihydroxylated hormonal D metabolites occurs via

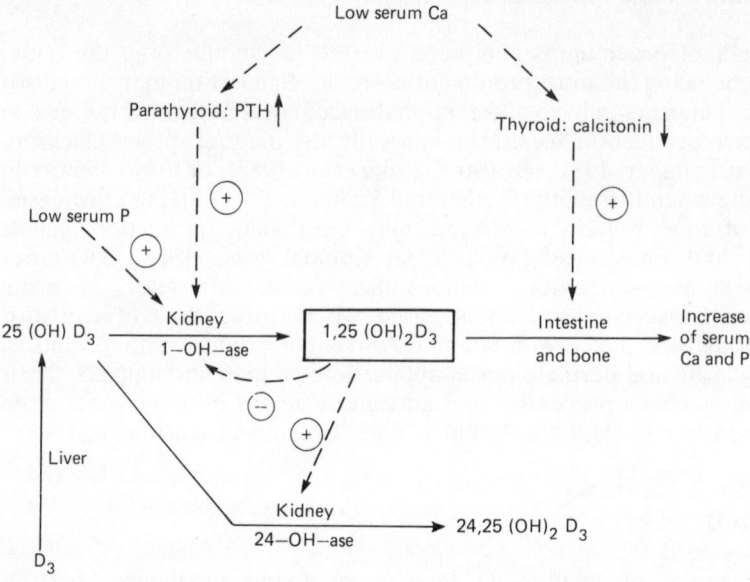

Figure 10.1 Regulation of calcium and phosphorus homeostasis by vitamin D metabolites (simplified model). Low blood levels of calcium (via PTH) and of phosphate activate renal synthesis of $1,25(OH)_2D_3$. This results in increased mineral transport from intestine and bone, aided by a simultaneous drop of calcitonin secretion, and a normalization of blood calcium and phosphate. These processes are regulated by several feedback mechanisms, e.g. an increased $1,25(OH)_2D_3$ level will inhibit its own synthesis and stimulate renal 24-hydroxylase to produce less potent $24,25(OH)_2D_3$. The parathyroid and thyroid glands will be influenced correspondingly. In addition to those mentioned above a number of other organs and regulators are also involved in these events (see text)

further hydroxylations and side-chain modifications; these products are then excreted through the bile. Among the hundreds of recent papers dealing with all aspects of vitamin D metabolism there are several comprehensive reviews, e.g. Lawson (1981), Bronner and Coburn (1982), DeLuca (1982), Norman, Roth and Orci (1982).

Vitamin D, calcium homeostasis and calcified tissues

Maintenance of a balanced calcium and phosphorus homeostasis is the main function of vitamin D and its hormonal metabolites. Their principal target organs are intestine and bone, one being the major site for uptake of dietary minerals, the other representing the bulk of calcified tissue.

Rickets and osteomalacia are classic examples of vitamin D deficiency: most cases can be prevented or cured by administration of vitamin D and/or suitable calcium–phosphorus feedstuffs. Any defect of liver and renal functions which involves the vitamin D hydroxylases may also affect calcium metabolism. Pregnancy and lactation are situations of increased calcium mobilization for dams. Goff, Horst and Littledike (1982) investigated the effect of maternal vitamin D status at parturition on the neonatal calf. Vitamin D status of the cow had a strong influence on the plasma levels of D metabolites, calcium and phosphorus of the neonatal calf. Initiation of lactation may result in parturient paresis (milk fever) if calcium and phosphorus fall below critical levels: this can be prevented by adequate diets or supplementation of 1-hydroxylated vitamin D forms (see Kichura et al., 1982). Various aspects of vitamin D metabolism have been studied in several domestic animals (Horst et al., 1981; Thomas, Ely and Boling, 1981a; Eagle et al., 1982; Winkler, Grabe and Harmeyer, 1982).

One of the most dramatic events of calcium homeostasis is the process of egg-laying: a hen loses 1–2 g of calcium within less than a day without any problem. For this purpose mature birds develop a special calcium store, medullary bone, to meet the requirements for shell calcification. After ovulation, production of $1,25(OH)_2D_3$ is stimulated and results in increased intestinal calcium uptake. At the same time medullary bone is mobilized, and these concerted processes allow a rapid transfer of sufficient calcium to the shell gland. Functions of vitamins D and K in poultry have been reviewed already (Weiser and Kormann, 1981; see also Abe et al., 1982).

Despite the beneficial effects of vitamin D compounds their potential toxicity should not be disregarded. Although amounts greatly in excess of the recommended dose of vitamins D_2, D_3 or their 25-hydroxy forms (see Klein, 1980; Thomas, Boling and Muir, 1981b; Harrington, 1982) are necessary to induce hypervitaminosis D, the therapeutic range of 1-hydroxylated D metabolites is small. Chicks on a vitamin D-deficient diet with $2.17 \mu g$ $1,25(OH)_2D_3$ per kg feed grew optimally, whereas the body weights of a group with $4.34 \mu g/kg$ were 36% lower (Weiser and Kormann, unpublished work). In grazing animals an excessive intake of certain calcinogenic plants containing vitamin D-like compounds may cause extensive calcification (Rambeck and Zucker, 1982).

Actions of vitamin D on other organs

As indicated above, the major targets of vitamin D are intestine and bone. In addition, receptors and/or binding proteins for $1,25(OH)_2D_3$ and vitamin D-dependent calcium-binding proteins (CaBP) have been demonstrated in a wide variety of tissues from several species (Norman et al., 1982). They include a number of glands such as the parathyroid, pituitary, parotid and thymus, but also reproductive organs like uterus, placenta, mammary gland and testes. This distribution is indicative of some unexpected biological actions of vitamin D and its metabolites. For example, Clark, Stumpf and Sar (1981) and Wilson, Horst and Schedl (1982) reported evidence of a relationship between $1,25(OH)_2D_3$ and insulin in the rat; similar processes may operate in chick pancreas (Pike, 1982; see also Bikle et al., 1982). $1,25(OH)_2D_3$ receptors are also present in several human and animal cancer cells, but the relevance of these findings is not yet clear.

Vitamin K

The main forms of vitamin K are phylloquinone (K_1), menaquinones (K_2 series), and menadione (K_3). Two of these are important in animal nutrition: K_1 is the predominant species in green plants and K_3 formulations are major products for feed. Menadione can be converted to phylloquinone by intestinal bacteria.

Johnson (1981) summarized the current status of the physiological functions of vitamin K; an overview by Gallop, Lian and Hauschka (1980) emphasized clinical aspects, while Uotila and Suttie (1982) discussed mainly biochemical reactions involving vitamin K. Results of activity tests and their implications for feed supplementation have been presented by Weiser and Tagwerker (1981).

Vitamin K and blood coagulation

It has been known for a long time that vitamin K is essential for the functional integrity of the blood coagulation system. The mechanisms of these processes have been elucidated during the past few years. Inactive precursors of the coagulation factors II (prothrombin), VII, IX and X are converted to their active forms by the 'vitamin K cycle' in the liver. This occurs by enzymatic γ-carboxylation of specific glutamate residues of the protein backbone. The resulting γ-carboxyglutamate (Gla) residues are strong chelators for calcium ions, a prerequisite for interactions of these factors with the other components of the coagulation cascade.

Vitamin K deficiency or anticoagulant administration will lead to factors with a reduced number of Gla residues, a corresponding loss of activity, and prolonged bleeding times. Malhotra (1982) isolated several prothrombin varieties from cows treated with dicoumarol. Whereas normal bovine prothrombin contains 10 Gla residues per molecule, these atypical factors had 0-7 Gla per molecule. By a one-stage coagulation assay it could be shown that minimal clotting time increased proportionally with the decrease of Gla residues. Activation time for normal prothrombin was 7 minutes, for factors with five or less Gla residues it was 3 hours or longer.

A test for determination of biopotency of vitamin K compounds and formulations in chicks has been developed in our laboratory (Weiser and Kormann, 1983). It is based on truly vitamin K-deficient animals without any administration of anticoagulant or bactericidal compounds.

Vitamin K-dependent proteins in other organs

In addition to the four vitamin K-dependent coagulation factors a number of other proteins and peptides with Gla residues have been found. Vitamin K-dependent proteins apparently have an important role in all major aspects of calcium metabolism, under normal as well as pathological conditions (*Table 10.4*).

Table 10.4 EXAMPLES OF PROTEINS AND PEPTIDES CONTAINING VITAMIN K-DEPENDENT γ-CARBOXYGLUTAMATE (Gla)

Tissue, protein or peptide	Function
Blood: coagulation factors II, VII, IX, X	Maintenance of blood coagulation
Bone: osteocalcin	Bone maturation
	Informational role, hormonal properties?
Egg-shell gland, chorioallantoic membrane	Calcium transfer: from shell gland to egg shell
	from egg shell to chick embryo
Placenta, kidney, pancreas, spleen, lung; corals, oysters, bacteria	Unknown
Calcified atherosclerotic plaques, renal calculi, calcified tendons, tumour cells	Unknown

The examples in *Table 10.4* demonstrate the broad distribution of γ-carboxylation in different types of organisms ranging from mammals to corals and bacteria (Johnson, 1981; Hamilton *et al.*, 1982). The involvement of vitamin K in egg laying and embryonic development of poultry has been reviewed separately (Weiser and Kormann, 1981).

Osteocalcin is the chief non-collagenous protein in bone of mammals and birds. It has 47–51 amino-acid residues including three Gla residues, and the structures of several species are highly homologous. In the long bones of chick embryos it appears coincident with the earliest signs of mineralization on day 7–8. In these events of bone maturation an optimal interplay of vitamin D metabolites and vitamin K-dependent proteins is required. This has been demonstrated in vitamin D-deficient chickens: the ratio of Gla to calcium ions and osteocalcin contents were significantly different in rachitic bone (Carr, Hauschka and Biemann, 1981; Lian *et al.*, 1982; Hauschka *et al.*, 1983).

Price, Parthemore and Deftos (1980) showed a relationship between these two vitamins in man. They measured plasma osteocalcin in patients with bone diseases which included disturbances of calcium homeostasis and vitamin D metabolism. Plasma osteocalcin levels of all patients deviated significantly from normal values, and the authors postulated that osteocalcin may also have informational or hormonal functions (Price and Williamson, 1981; Price, Williamson and Lothringer, 1981).

Pharmacological properties of vitamin K compounds

Vitamin K and related compounds may also act as antibiotics, antiphlogistics or analgesics. The mechanism(s) of these phenomena have not yet been elucidated, but it appears that they are different from those involving γ-carboxyglutamic acid. Bearing in mind the quinone structure of the K vitamins and their structural similarities to tocopherols, these mechanisms are likely to involve pathways of arachidonic acid metabolism (the vitamin E–arachidonic acid relationship has been referred to above).

One of the major urinary vitamin K_1 metabolites in man and mammals is a phylloquinone analogue with a shortened side-chain ending with a carboxy group (McBurney, Shearer and Barkhan, 1980). It has no antihaemorrhagic activity (Weiser and Kormann, 1983), but it has been claimed to have anti-inflammatory properties (Morimoto and Watanabe, 1973). We investigated this compound in a rat model with carrageenin-induced paw oedema and confirmed its strong antiphlogistic potency (Weiser, unpublished work). Anti-inflammatory, analgesic and other pharmacological effects of vitamin K_1 have been discussed previously (Hanck and Weiser, 1983). A number of additional protective functions of vitamin K analogues have been reported, including antiallergic, antiulcer, hypotensive and diuretic actions (Shinji, Shiraishi and Maki, 1981).

Vitamin K metabolism and subsequent events may be impaired during certain stress situations. For example, administration of some antibiotics, or mycotoxin T-2 intoxication, can interfere with vitamin K-dependent functions (Deyl, Vančíková and Macek, 1981; Doerr, Hamilton and Burmeister, 1981). T-2 will also impair vitamin E status (Coffin and Combs, 1981). On the other hand, anticoagulant treatment may have additional effects such as changes of thyroid hormone metabolism (Goswami, Leonard and Rosenberg, 1982). The interplay between anticoagulants and vitamin K may also be influenced by high doses of vitamin E (Corrigan and Ulfers, 1981; Anonymous, 1982b).

In view of the vitamin K-dependent protein found in calcified atherosclerotic plaques (*Table 10.4*) it may also be mentioned that certain calcium antagonists suppress diet-induced atherogenesis in experimental animals (Henry and Bentley, 1981; Kramsch, Aspen and Rozler, 1981).

Concluding remarks

Fat-soluble vitamins have well-known, recognized functions such as the involvement of vitamin A in vision, prevention of rickets by vitamin D, antioxidant actions of vitamin E, or maintenance of blood coagulation by vitamin K. This short review has tried to show that these vitamins have additional important activities which influence other vital processes during normal and stress conditions such as infections, toxicoses and food imbalances. Furthermore, it has been emphasized that the functions of these vitamins are not isolated events: they can affect each other, as seen by the interplay of vitamins A and E or by the concerted actions of vitamins D and K in calcified tissues, or they participate in reactions involving a variety of agents such as hormones, prostaglandins, dietary ingredients, toxicants and others. To sum up, it must be recognized that fat-soluble vitamins are far more than simple feed or food additives.

References

ABE, E., HORIKAWA, H., MASUMURA, T., SUGAHARA, M., KUBOTA, M. and SUDA, T. (1982). Disorders of cholecalciferol metabolism in old egg-laying hens. *Journal of Nutrition* **112**, 436–446
ANONYMOUS (1982a). Xerophthalmia control (Editorial). *Lancet* **ii**, 28–29
ANONYMOUS (1982b). Vitamin K, vitamin E and the coumarin drugs. *Nutrition Reviews* **40**, 180–182
BAI, N.J., KUMAR, P.S., GEORGES, T. and KRISHNAMURTHY, S. (1982). Effect of dietary protein and hypervitaminosis A or C on tissue peroxidation and erythrocyte lysis of vitamin E deficiency. *International Journal for Vitamin and Nutrition Research* **52**, 386–392

BARTOV, I. and BORNSTEIN, S. (1980). Susceptibility of chicks to nutritional encephalopathy: effect of fat and α-tocopherol content of the breeder diet. *Poultry Science* **59**, 264-267

BARTOV, I. and BORNSTEIN, S. (1981). Stability of abdominal fat and meat of broilers: Combined effect of dietary vitamin E and synthetic antioxidants. *Poultry Science* **60**, 1840-1845

BATTNÁ, J., PARÍZKOVÁ, H. and KUČEROVÁ, Z. (1982). Fat and vitamin A stability in the presence of Ronoxan A^R and other antioxidants. *International Journal for Vitamin and Nutrition Research* **52**, 241-247

BEISEL, W.R. (1982). Single nutrients and immunity. *American Journal of Clinical Nutrition* **35**, 417-468

BIANCHI, A., SANTAGATI, G., ANDREONI, L., BERMOND, P. and SANTAMARIA, L. (1982). Perspectives in cancer chemoprevention by vitamins and carotenoids. Updating of progress. *Médicine, Biologie et Environnement* **10**, 357-364

BICHLER, K.-H., KIRCHNER, C., WEISER, H., KORN, S., STROHMAIER, W., SCHMITZ-MOORMANN, P., HANCK, A. and NELDE, H.J. (1983). Influence of vitamin A deficiency on the excretion of uromucoid and other substances of the urine in rats. *Clinical Nephrology* **20**, 32-39

BIKLE, D.D., PECK, C.C., HOLFORD, N.H.G., ZOLOCK, D.T. and MORISSEY, R.L. (1982). Pharmacokinetics and pharmacodynamics of 1,25-dihydroxyvitamin D_3 in the chick. *Endocrinology* **111**, 939-946

BOGUTH, W., PATZELT-WENCZLER, R. and REPGES, R. (1971). Prüfung der Antioxidansaktivität mit Azo-bis-isobutyronitril und β-Carotin. *International Journal for Vitamin and Nutrition Research* **41**, 21-32

BOLLAG, W. and HARTMANN, H.R. (1983). Prevention and therapy of cancer with retinoids in animals and man. *Cancer Surveys* **2**, 295-314

BRONNER, F. and COBURN, J.W. (Eds) (1982). *Disorders of Mineral Metabolism. Volume II: Calcium Physiology*. New York and London, Academic Press

BURTON, G.W. and INGOLD, K.U. (1981). Autoxidation of biological molecules. 1. The antioxidant activity of vitamin E and related chain-breaking phenolic antioxidants in vitro. *Journal of the American Chemical Society* **103**, 6472-6477

BURTON, G.W., JOYCE, A. and INGOLD, K.U. (1982). First proof that vitamin E is major lipid-soluble, chain-breaking antioxidant in human blood plasma. *Lancet* **ii**, 327

CARPENTER, M.P. (1981). Antioxidant effects on the prostaglandin endoperoxide synthetase product profile. *Federation Proceedings* **40**, 189-194

CARR, S.A., HAUSCHKA, P.V. and BIEMANN, K. (1981). Gas chromatographic mass spectrometric sequence determination of osteocalcin, a γ-carboxyglutamic acid-containing protein from chicken bone. *Journal of Biological Chemistry* **256**, 9944-9950

CATIGNANI, G.L. and BIERI, J.G. (1980). Ineffectiveness of testosterone and FSH in maintaining the vitamin A deficient rat testis. *Nutrition and Metabolism* **24**, 255-260

CHAN, A.C. and LEITH, M.K. (1981). Decreased prostacyclin synthesis in vitamin E-deficient rabbit aorta. *American Journal of Clinical Nutrition* **34**, 2341-2347

CHEN, J., GOETCHIUS, M.P., COMBS, G.F. Jr and CAMPBELL, T.C. (1982). Effects of dietary selenium and vitamin E on covalent binding of aflatoxin to chick liver cell macromolecules. *Journal of Nutrition* **112**, 350–355

CHEN, L.H. (1981). An increase in vitamin E requirement induced by high supplementation of vitamin C in rats. *American Journal of Clinical Nutrition* **34**, 1036–1041

CHERTOW, B.S., WILLIAMS, G.A., NORRIS, R.M., BAKER, G.R. and HARGIS, G.K. (1977). Vitamin A stimulation of parathyroid hormone: interactions with calcium, hydrocortisone, and vitamin E in bovine parathyroid tissues and effects of vitamin A in man. *European Journal of Clinical Investigation* **7**, 307–314

CHOLE, A.R. (1980). Autoradiographic localization of vitamin A in the stria vascularis of the rat cochlea. *Archives of Otolaryngology* **106**, 741–743

CHYTIL, F. (1982). Liver and cellular vitamin A binding proteins. *Hepatology* **2**, 282–287

CLARK, S.A., STUMPF, W.E. and SAR, M. (1981). Effect of 1,25-dihydroxyvitamin D_3 on insulin secretion. *Diabetes* **30**, 382–386

COFFIN, J.L. and COMBS, G.F. Jr (1981). Impaired vitamin E status of chicks fed T-2 toxin. *Poultry Science* **60**, 385–392

CORRIGAN, J.J. Jr and ULFERS, L.L. (1981). Effect of vitamin E on prothrombin levels in warfarin-induced vitamin K deficiency. *American Journal of Clinical Nutrition* **34**, 1701–1705

DELUCA, H.F. (1982). Metabolism and molecular mechanism of action of vitamin D: 1981. *Biochemical Society Transactions* **10**, 147–158

DEMOPOULOS, H.B., PIETRONIGRO, D.D., FLAMM, E.S. and SELIGMAN, M.L. (1980). The possible role of free radical reactions in carcinogenesis. *Journal of Environmental Pathology and Toxicology* **3**, 273–303

DEYL, Z., VANČÍKOVÁ, O. and MACEK, K. (1981). The effect of oxytetracycline and some related antibiotics upon the γ-carboxyglutamic acid level in bone and kidney cortex. *Biochemical and Biophysical Research Communications* **100**, 79–85

DILLARD, C.J., KUNERT, K.-J. and TAPPEL, A.L. (1982a). Lipid peroxidation during chronic inflammation induced in rats by Freund's adjuvant: effect of vitamin E as measured by expired pentane. *Research Communications in Chemical Pathology and Pharmacology* **37**, 143–146

DILLARD, C.J., KUNERT, K.-J. and TAPPEL, A.L. (1982b). Effects of vitamin E, ascorbic acid and mannitol on alloxan-induced lipid peroxidation in rats. *Archives of Biochemistry and Biophysics* **216**, 204–212

DION, P.W., BRIGHT-SEE, E.B., SMITH, C.C. and BRUCE, W.R. (1982). The effect of dietary ascorbic acid and α-tocopherol on fetal mutagenicity. *Mutation Research* **102**, 27–37

DOERR, J.A., HAMILTON, P.B. and BURMEISTER, H.R. (1981). T-2 toxicosis and blood coagulation in young chickens. *Toxicology and Applied Pharmacology* **60**, 157–162

DROR, Y., BARTOV, I. and BUBIS, J.J. (1980). Exacerbative effect of vitamin A on the development of nutritional encephalomalacia in chicks. *Nutrition Reports International* **21**, 769–778

EAGLE, M.E., KOCH, D.B., WHELAN, J.P., HINTZ, H.F. and KROOK, L. (1982). Mineral metabolism and immobilization osteopenia in ponies treated with 25-hydroxycholecalciferol. *Cornell Veterinarian* **72**, 372–393

FARRELL, P.M., LEVINE, S.L., MURPHY, M.D. and ADAMS, A.J. (1978). Plasma tocopherol levels and tocopherol-lipid relationships in a normal population of children as compared to healthy adults. *American Journal of Clinical Nutrition* **31**, 1720–1726

FERRANDO, R., FOURLON, C. and CLECH, I. (1977). Rapport vitamine A–vitamine D_3 chez le rat en croissance. *International Journal for Vitamin and Nutrition Research* **47**, 157–161

GALLOP, P.M., LIAN, J.B. and HAUSCHKA, P.V. (1980). Carboxylated calcium-binding proteins and vitamin K. *New England Journal of Medicine* **302**, 1460–1466

GANGULY, J., RAO, M.R.S., MURTHY, S.K. and SARADA, K. (1980). Systemic mode of action of vitamin A. *Vitamins and Hormones* **38**, 1–54

GERBER, L.E. and ERDMAN, J.W. Jr (1982). Effect of dietary retinyl acetate, β-carotene and retinoic acid on wound healing in rats. *Journal of Nutrition* **112**, 1555–1564

GERSHOFF, S.N. and McGANDY, R.B. (1981). The effects of vitamin A-deficient diets containing lactose in producing bladder calculi and tumors in rats. *American Journal of Clinical Nutrition* **34**, 483–489

GINTER, E., KOŠINOVÁ, A., HUDECOVÁ, A. and MADARIČ, A. (1982). Synergism between vitamin C and E: effect on microsomal hydroxylation in guinea pig liver. *International Journal for Vitamin and Nutrition Research* **52**, 55–59

GOFF, J.P., HORST, R.L. and LITTLEDIKE, E.T. (1982). Effect of the maternal vitamin D status at parturition on the vitamin D status of the neonatal calf. *Journal of Nutrition* **112**, 1387–1393

GOODMAN, DeW.S. (1980). Vitamin A metabolism. *Federation Proceedings* **39**, 2716–2722

GOSWAMI, A., LEONARD, J.L. and ROSENBERG, I.N. (1982). Inhibition by coumadin anticoagulants of enzymatic outer ring monodeiodination of iodothyronines. *Biochemical and Biophysical Research Communications* **104**, 1231–1238

HAMDY, B.H., ELNOKALY, F., GAAFAR, S., ELNAGGAR, B. and HUSSEIN, L. (1982). Effectiveness of periodic oral vitamin A dosage on hypovitaminosis in Egypt. *International Journal for Vitamin and Nutrition Research* **52**, 235–240

HAMILTON, S.E., KING, G., TESCH, D., RIDDLES, P.W., KEOUGH, D.T., JELL, J. and ZERNER, B. (1982). γ-Carboxyglutamic acid in invertebrates: Its identification in hermatypic corals. *Biochemical and Biophysical Research Communications* **108**, 610–613

HANCK, A. and WEISER, H. (1983). Physiological and pharmacological effects of vitamin K. *International Journal for Vitamin and Nutrition Research*, supplement no. 24, 155–170

HARRINGTON, D.D. (1982). Acute vitamin D_2 (ergocalciferol) toxicosis in horses: Case report and experimental studies. *Journal of the American Veterinary Medical Association* **180**, 867–873

HAUSCHKA, P.V., FRENKEL, J., DEMUTH, R. and GUNDBERG, C.M. (1983). Presence of osteocalcin and related higher molecular weight 4-carboxyglutamic acid-containing proteins in developing bone. *Journal of Biological Chemistry* **258**, 176–182

HENRY, P.D. and BENTLEY, K.I. (1981). Suppression of atherogenesis in cholesterol-fed rabbits treated with nifedipine. *Journal of Clinical Investigation* **68**, 1366–1369

HIGUERET, P. and GARCIN, H. (1982). Peripheral metabolism of thyroid hormones in vitamin A-deficient rats. *Annals of Nutrition and Metabolism* **26**, 191–200

HILL, D.L. and GRUBBS, C.L. (1982). Retinoids as chemopreventive and anticancer agents in intact animals (review). *Anticancer Research* **2**, 111–124

HILL, R. (1981). Vitamin E and selenium in poultry nutrition. In *3rd European Symposium on Poultry Nutrition* (D.W.F. Shannon and I.E. Wallace, Eds), pp. 16–22. Published by the Organising Committee (ISBN 0-9508013-0-5)

HORST, R.L., LITTLEDIKE, E.T., RILEY, J.L. and NAPOLI, J.L. (1981). Quantitation of vitamin D and its metabolites and their plasma concentrations in five species of animals. *Analytical Biochemistry* **116**, 189–203

HORST, R.L., NAPOLI, J.L. and LITTLEDIKE, E.T. (1982). Discrimination in the metabolism of orally dosed ergocalciferol and cholecalciferol by the pig, rat and chick. *Biochemical Journal* **204**, 185–189

HUNG, S.S.O. and SLINGER, S.J. (1982). Effect of dietary vitamin E on rainbow trout (*Salmo gairdneri*). Muscle α-tocopherol and storage stability. *International Journal for Vitamin and Nutrition Research* **52**, 120–125

HUTCHINSON, L.J., SCHOLZ, R.W. and DRAKE, R.T. (1982). Nutritional myodegeneration in a group of Chianina heifers. *Journal of the American Veterinary Medical Association* **181**, 581–584

IUPAC . IUB (1982). Joint Commission on Biochemical Nomenclature, Recommendations 1981. *European Journal of Biochemistry* **123**, 473–475

JACKSON, P.S. (1981). A note on a possible association between plasma β-carotene levels and conception rate in a group of winter-housed dairy cattle. *Animal Production* **32**, 109–111

JOHNSON, B.C. (1981). Post-translational carboxylation of preprothrombin. *Molecular and Cellular Biochemistry* **38**, 77–121

KIATOKO, M., McDOWELL, L.R., BERTRAND, J.E., CHAPMAN, H.L., PATE, F.M., MARTIN, F.G. and CONRAD, J.H. (1982). Evaluating the nutritional status of beef cattle herds from four soil order regions of Florida. I. Macroelements, protein, carotene, vitamins A and E, haemoglobin and hematocrit. *Journal of Animal Science* **55**, 28–37

KICHURA, T.S., HORST, R.L., BEITZ, D.C. and LITTLEDYKE, E.T. (1982). Relationships between prepartal dietary calcium and phosphorus, vitamin D metabolism and parturient paresis in dairy cows. *Journal of Nutrition* **112**, 480–487

KLEIN, L. (1980). Direct measurement of bone resorption and calcium conservation during vitamin D deficiency or hypervitaminosis D. *Proceedings of the National Academy of Sciences of the United States of America* **77**, 1818–1822

KLING, L.J. and SOARES, J.H. Jr (1980). Vitamin E deficiency in the Japanese quail. *Poultry Science* **59**, 2352–2354

KLING, L.J. and SOARES, J.H. Jr (1981). The effect of vitamin E and dietary linoleic acid on mercury toxicity. *Nutrition Reports International* **24**, 39–46

KOLONEL, L.N., NOMURA, A.M.Y., HINDS, M.W., HIROHATA, T., HANKIN, J.H. and LEE, J. (1983). Role of diet in cancer incidence in Hawaii. *Cancer Research* **43** (suppl), 2397s–2402s

KOZICKI, L., SILVA, R.G. and BARNABE, R.C. (1981). Effects of vitamins A, D_3, E and C on the characteristics of bull semen. *Zentralblatt für Veterinärmedizin* **A28**, 538–546

KRAMSCH, D.M., ASPEN, A.J. and ROZLER, L.J. (1981). Atherosclerosis: prevention by agents not affecting abnormal levels of blood lipids. *Science* **213**, 1511–1512

LAWSON, D.E.M. (1981). Vitamin D—still an unsolved problem. *Trends in Biochemical Sciences* **6**, 285–287

LIAN, J.B., GLIMCHER, M.L., ROUFOSSE, A.H., HAUSCHKA, P.V., GALLOP, P.M., COHEN-SOLAL, L. and REIT, B. (1982). Alterations of the γ-carboxyglutamic and osteocalcin concentrations in vitamin D-deficient chick bone. *Journal of Biological Chemistry* **257**, 4999–5003

LIKOFF, R.O., GUPTILL, D.R., LAWRENCE, L.M., McKAY, C.C., MATHIAS, M.M., NOCKELS, C.F. and TENGERDY, R.P. (1981). Vitamin E and aspirin depress prostaglandins in protection of chickens against *Escherichia coli* infection. *American Journal of Clinical Nutrition* **34**, 245–251

LOTTHAMMER, K.-H. (1981). *Importance of β-Carotene for the Reproduction of Female Cattle*. Basle, Switzerland, Roche Information Service

MACAPINLAC, M.P. and OLSON, J.A. (1981). A lethal hypervitaminosis A syndrome in young monkeys (*Macacus fascicularis*) following a single intramuscular dose of a water-miscible preparation containing vitamins A, D_2 and E. *International Journal for Vitamin and Nutrition Research* **51**, 331–341

McBURNEY, A., SHEARER, M.J. and BARKHAN, P. (1980). Preparative isolation and characterization of urinary aglycones of vitamin K_1 (phylloquinone) in man. *Biochemical Medicine* **24**, 250–267

MACHLIN, L.J. (Ed.) (1980). *Vitamin E—A Comprehensive Treatise*. New York and Basle, Marcel Dekker

McMURRAY, C.H. (1980). Nutritional supplies, requirements and effects of deficiencies of vitamin E and selenium. In *Proceedings of the Roche Symposium, London, October 23, 1980*. 4002 Basle, Switzerland, Roche Information Service, Animal Nutrition Department

MALHOTRA, O.P. (1982). Partially carboxylated prothrombins. I. Comparison of activation properties and purification of 1- and 0-carboxyglutamyl variants. *Biochemica et biophysica acta* **702**, 178–184

MORIMOTO, H. and WATANABE, M. (1973). Naphthoquinones (patent application). *Chemical Abstracts* **78**, 369 (abstract 15892b)

MUDUULI, D.S., MARQUARDT, R.R. and GUENTER, W. (1982). Effect of dietary vicine and vitamin E supplementation on the productive performance of growing and laying chickens. *British Journal of Nutrition* **47**, 53–60

MULLER, D.P.R., LLOYD, J.K. and WOLFF, O.H. (1983). Vitamin E and neurological function. *Lancet* **i**, 225–228

NORMAN, A.W., ROTH, J. and ORCI, L. (1982). The vitamin D endocrine system: Steroid metabolism, hormone receptors, and biological response (calcium binding proteins). *Endocrine Reviews* **3**, 331–366

O'BRIEN, D.F. (1982). The chemistry of vision. *Science* **218**, 961–966

OHSHIMA, H. and BARTSCH, H. (1981). Quantitative estimation of endogenous nitrosation in humans by monitoring N-nitrosoproline excreted in the urine. *Cancer Research* **41**, 3658–3662

PIKE, J.W. (1982). Receptors for 1,25-dihydroxyvitamin D_3 in chick pancreas: A partial physical and functional characterization. *Journal of Steroid Biochemistry* **16**, 385–395

PONGRACZ, G. (1982). Stabilisierung von Kakaobutterersatzfetten (Stabilisation of cocoa butter substitute fats). *Fette-Seifen-Anstrichmittel* **84**, 269–272

PRICE, P.A. and WILLIAMSON, M.K. (1981). Effects of warfarin on bone. Studies on the vitamin K-dependent protein of rat bone. *Journal of Biological Chemistry* **256**, 12754–12759

PRICE, P.A., PARTHEMORE, J.G. and DEFTOS, L.J. (1980). New biochemical marker for bone metabolism. Measurement by radioimmunoassay of bone Gla protein in the plasma of normal subjects and patients with bone disease. *Journal of Clinical Investigation* **66**, 878–883

PRICE, P.A., WILLIAMSON, M.K. and LOTHRINGER, J.W. (1981). Origin of the vitamin K-dependent bone protein found in plasma and its clearance by kidney and bone. *Journal of Biological Chemistry* **256**, 12760–12766

RAMBECK, W.A. and ZUCKER, H. (1982). Vitamin D-artige Aktivitäten in calcinogenen Pflanzen. *Zentralblatt für Veterinärmedizin* **A29**, 289–296

ROBISON, W.G. Jr, KUWABARA, T. and BIERI, J.G. (1980). Deficiencies of vitamins E and A in the rat: retinal damage and lipofuscin accumulation. *Investigative Ophthalmology and Visual Science* **19**, 1030–1037

SAARI, J.C., BREDBERG, L. and GARWIN, G.G. (1982). Identification of the endogenous retinoids associated with three cellular retinoid-binding proteins from bovine retina and retinal pigment epithelium. *Journal of Biological Chemistry* **257**, 13329–13333

SAITO, M., IKEGAMI, S., ITO, Y. and INNAMI, S. (1982). Influence of dietary antioxidants on polychlorinated biphenyls (PCB)-induced hepatic lipid peroxide formation and vitamin A reduction in rats. *Journal of Nutritional Science and Vitaminology* **28**, 455–466

SAUBERLICH, H.E., KRETSCH, M.J., JOHNSON, H.L. and NELSON, R.A. (1982). Animal products as a source of vitamins. In *Animal Products in Human Nutrition* (D.C. Beitz and R.G. Hansen, Eds), pp. 339–372. New York, Academic Press

SCOTT, M.L. (1980). Advances in our understanding of vitamin E. *Federation Proceedings* **39**, 2736–2739

SHINJI, T., SHIRAISHI, M. and MAKI, Y. (1981). *Para-benzoquinone Derivatives with Unsaturated Side-chain—Useful as Antiasthmatic, Antiallergic and Antiulcer Agents, etc.* European patent application 38160, filed April 7, 1981

SIETSEMA, W.K. and DELUCA, H.F. (1982). A new vaginal smear assay for vitamin A in rats. *Journal of Nutrition* **112**, 1481–1489

SIMIC, M.G. and KAREL, M. (Eds) (1980). *Autoxidation in Food and Biological Systems*. New York and London, Plenum Press

SKLAN, D. and DONOGHUE, S. (1982). Vitamin E response to high dietary vitamin A in the chick. *Journal of Nutrition* **112**, 759–765

STEINBERG, K.K. and SGOUTAS, D.S. (1981). Effect of vitamin A deficiency on luteinizing hormone receptors and adenosine 3', 5'-monophosphate-mediated steroidogenesis in rat testicular tissue. *Proceedings of the Society for Experimental Biology and Medicine* **167**, 110–116

TAGWERKER, F.J. (1981). Vitamins and quality of food products of animal origin. In *Proceedings of Roche Seminars, Osaka and Tokyo, Japan, October 17 and 18, 1981*. 4002 Basle, Switzerland, Roche Information Service, Animal Nutrition Department

TANGNEY, C.C. and DRISKELL, J.A. (1981). Effects of vitamin E deficiency on the relative incorporation of ^{14}C-arachidonate into platelet lipids of rabbits. *Journal of Nutrition* **11**, 1839–1847

THOMAS, R.J., BOLING, J.A. and MUIR, W.M. (1981a). Serum calcium, phosphorus and magnesium responses to massive dosing of cholecalciferol (CC) and 25-OH-CC in young and aged ewes. *International Journal for Vitamin and Nutrition Research* **51**, 365–372

THOMAS, R.J., ELY, D.G. and BOLING, J.A. (1981b). Magnesium, calcium and phosphorus response in the blood and tissues of the ovine injected with cholecalciferol. *International Journal for Vitamin and Nutrition Research* **51**, 16–25

UEBERSAX, M.A., DAWSON, L.E. and UEBERSAX, K.L. (1978). Storage stability (TBA) of meat obtained from turkeys receiving tocopherol supplementation. *Poultry Science* **57**, 937–946

UOTILA, L. and SUTTIE, J.W. (1982). Recent findings in understanding the biological function of vitamin K. *Medical Biology* **60**, 16–24

USP XX (1980). *The United States Pharmacopeia*, 20th edn, p. 846. Rockville, Maryland, USA, United States Pharmacopeial Convention Inc.

VALENTOVIC, M.A., GAIROLA, C. and LUBAWY, W.C. (1982). Lung, aorta and platelet metabolism of ^{14}C-arachidonic acid in vitamin E deficient rats. *Prostaglandins* **24**, 215–224

VAN VLEET, J.F. (1982). Amounts of eight combined elements required to induce selenium–vitamin E deficiency in ducklings and protection by supplements of selenium and vitamin E. *American Journal of Veterinary Research* **43**, 1049–1055

VOS, J., HULSTAERT, C.E. and MOLENAAR, I. (1981). Nutritional myopathy in ducklings: a growth rate-dependent symptom of 'tissue peroxidosis' due to a net nutritional shortage of vitamin E plus selenium in skeletal muscle. *Annals of Nutrition and Metabolism* **25**, 299–306

WALD, N., IDLE, M., BOREHAM, J. and BAILEY, A. (1980). Low serum-vitamin-A and subsequent risk of cancer. *Lancet* **ii**, 813–815

WEBER, F., WEISER, H. and WISS, O. (1964). Bedarf an Vitamin E in Abhängigkeit von der Zufuhr an Linolsäure. *Zeitschrift für Ernährungswissenschaft* **4**, 245–252

WEENEN, H. and PORTER, N.A. (1982). Autoxidation of model membrane systems: cooxidation of polyunsaturated lecithins with steroids, fatty acids, and α-tocopherol. *Journal of the American Chemical Society* **104**, 5216–5221

WEISER, H. (1978). Biologische Qualitätskontrolle mit optimierten Modellen. In *Das Tier im Experiment* (W.H. Wiehe, Ed.), pp. 147–174. Bern-Stuttgart-Wien, Huber

WEISER, H. and KORMANN, A.W. (1981). Functions of vitamins D and K in poultry—a short review. In *3rd European Symposium on Poultry Nutrition* (D.W.F. Shannon and I.E. Wallace, Eds), pp. 8–15. Published by the Organising Committee (ISBN 0-9508013-0-5)

WEISER, H. and KORMANN, A.W. (1983). Biopotency of vitamin K. I. Antihemorrhagic properties of structural analogs of phylloquinone as determined by curative prothrombin time tests. *International Journal for Vitamin and Nutrition Research* **53**, 143–155

WEISER, H. and SALKELD, R.M. (1977). Vitamin E—its interference with the effects of polyunsaturated fatty acids and its influence on thyroid function. *Acta Vitaminologica et Enzymologica* **31**, 143–155

WEISER, H. and TAGWERKER, F. (1981). Optimization of the vitamin K bio-assay of vitamin K forms for feed supplementation. In *Papers*

Dedicated to Prof. J. Moustgaard (K. Brummerstedt, Ed.), pp. 138–146. Copenhagen, Royal Danish Agricultural Society

WEISER, H. and VECCHI, M. (1981). Stereoisomers of α-tocopheryl acetate. Characterization of the samples by physico-chemical methods and determination of biological activities in the rat resorption–gestation test. *International Journal for Vitamin and Nutrition Research* **51**, 100–113

WEISER, H. and VECCHI, M. (1982). Stereoisomers of α-tocopheryl acetate. II. Biopotencies of all eight stereoisomers, individually or in mixtures, as determined by rat resorption-gestation tests. *International Journal for Vitamin and Nutrition Research* **52**, 351–370

WILSON, H.D., HORST, R.L. and SCHEDL, H.P. (1982). Calcium intake regulates 1,25-dihydroxy-vitamin D formation in the diabetic rat. *Diabetes* **31**, 401–405

WINKLER, I., GRABE, C. and HARMEYER, J. (1982). Pseudo vitamin D deficiency rickets in pigs: In vitro measurements of renal 25-hydroxycholecalciferol-1-hydroxylase activity. *Zentralblatt für Veterinärmedizin* **A29**, 81–88

WOLF, G. (1982). Is dietary β-carotene an anti-cancer agent? *Nutrition Reviews* **40**, 257–261

Note added in proof

After submission of our manuscript a large number of papers concerning protective functions of fat-soluble vitamins have been published. The following list contains a few of the most interesting reports.

Several vitamins

AMES, B.N. (1983). Dietary carcinogens and anticarcinogens—oxygen radicals and degenerative diseases. *Science* **221**, 1256–1264

KUNERT, K.-J. and TAPPEL, A.L. (1983). The effect of vitamin C on *in vivo* lipid peroxidation in guinea pigs as measured by pentane and ethane production. *Lipids* **18**, 271–274

McBRIEN, D.C.H. and SLATER, T.F., Eds (1983). *Protective agents in cancer*. London, Academic Press

MILLER, E.R. and KORNEGAY, E.T. (1983). Mineral and vitamin nutrition of swine. *Journal of Animal Science* **57** (suppl. 2), 315–329

SKLAN, D. (1983). Vitamin A absorption and metabolism in the chick—response to high dietary intake and to tocopherol. Effect of high vitamin A or tocopherol intake on hepatic lipid metabolism and intestinal absorption and secretion of lipids and bile acids in the chick. *British Journal of Nutrition* **50**, 401–408, 409–416

Vitamin E

BIERI, J.G., CORASH, L. and HUBBARD, V.S. (1983). Medical uses of vitamin E. *New England Journal of Medicine* **308**, 1063–1071

BRUCKNER, C., INFANTE, J., COMBS, G.F. Jr. and KINSELLA, J.E. (1983). Effects of vitamin E and aspirin on the incidence of encephalomalacia, fatty acids status and serum thromboxane levels in chicks. *Journal of Nutrition* **113**, 1884–1889

BURTON, G.W., JOYCE, A. and INGOLD, K.U. (1983). Is vitamin E the only lipid-soluble, chain-breaking antioxidant in human blood plasma and erythrocyte membranes? *Archives of Biochemistry and Biophysics* **221**, 281–290

CHAN, A.C., PRITCHARD, E.T. and CHOY, P.C. (1983). Differential effects of dietary vitamin E and antioxidants on eicosanoid synthesis in young rabbits. *Journal of Nutrition* **113**, 813–819

DILLARD, C.L., GAVINO, V.C. and TAPPEL, A.L. (1983). Relative anti-oxidant effectiveness of α-tocopherol and γ-tocopherol in iron-loaded rats. *Journal of Nutrition* **113**, 2266–2273

KOTT, R.W., RUTTLE, J.L. and SOUTHWARD, G.M. (1983). Effects of vitamin E and selenium injections on reproduction and preweaning lamb survival in ewes consuming diets marginally deficient in selenium. *Journal of Animal Science* **57**, 553–558

LUBIN, B. and MACHLIN, L.J., Eds (1982). Vitamin E—biochemical, hematological and clinical aspects. *Annals of the New York Academy of Sciences* **393**

LYNCH, G.P. (1983). Changes of tocopherols in blood serum of cows fed hay or silage. *Journal of Dairy Science* **66**, 1461–1465

SLONIM, A.E., SURBER, M.L., PAGE, D.L., SHARP, R.A. and BURR, I.M. (1983). Modification of chemically induced diabetes in rats by vitamin E—supplementation minimizes and depletion enhances development of diabetes. *Journal of Clinical Investigation* **71**, 1282–1288

TANNENBAUM, S.R. (1983). *N*-Nitroso compounds: a perspective on human exposure. *Lancet* **i**, 629–632

TENGERDY, R.P., MEYER, D.L., LAUERMAN, L.H., LUEKER, D.C. and NOCKELS, C.F. (1983). Vitamin E-enhanced humoral antibody response to *Clostridium perfringens* type D in sheep. *British Veterinary Journal* **139**, 147–152

WILLETT, W.C., POLK, B.F., MORRIS, J.S., STAMPFER, M.J., PRESSEL, S., ROSNER, B., TAYLOR, J.O., SCHNEIDER, K. and HAMES, C.G. (1983). Prediagnostic serum selenium and risk of cancer. *Lancet* **ii**, 130–134

Vitamin A

ARBEITER, K., KNAUS, E. and THURNHER, M. (1983). Repetitionstest über die Genitalfunktion von Rindern in Abhängigkeit vom β-Carotingehalt im Blut (Repeat testing of genital function of heifers in relation to blood level of β-carotene). *Zentralblatt für Veterinärmedizin A* **30**, 206–213

DONOGHUE, S., DONAWICK, W.J. and KRONFELD, D.S. (1983). Transfer of vitamin A from intestine to plasma in lambs fed low and high intakes of vitamin A. *Journal of Nutrition* **113**, 2197–2204

ORFANOS, C.E. and BAUER, R. (1983). Evidence for anti-inflammatory activities of oral retinoids: experimental findings and clinical experience. *British Journal of Dermatology* **25**, 55–60

PETO, R. (1983). The marked differences between carotenoids and retinoids: methodological implications for biochemical epidemiology. *Cancer Surveys* **2**, 327–340

SKLAN, D. (1983). Carotene cleavage activity in the corpus luteum of cattle. *International Journal for Vitamin and Nutrition Research* **53**, 23–26

SOMMER, A., TARWOTJO, I., HUSSAINI, G. and SUSANTO, D. (1983). Increased mortality in children with mild vitamin A deficiency. *Lancet* **ii**, 585–588

SYMPOSIUM (1983). 'Vitamin A in nutrition and disease' (several papers). *Proceedings of the Nutrition Society* **42**, 1–101

ZILE, M.H. and CULLUM, M.E. (1983). The function of vitamin A: current concepts. *Proceedings of the Society of Experimental Biology and Medicine* **172**, 139–152

Vitamins D and K

DEKEL, S., SALAMA, R. and EDELSTEIN, S. (1983). The effect of vitamin D and its metabolites on fracture repair in chicks. *Clinical Science* **65**, 429–436

DELMAS, P.D., WAHNER, H.W., MANN, K.G. and RIGGS, B.L. (1983). Assessment of bone turnover in postmenopausal osteoporosis by measurement of serum bone Gla-protein. *Journal of Laboratory and Clinical Medicine* **102**, 470–476

DEYL, Z. and ADAM, M. (1983). Evidence for vitamin D-dependent γ-carboxylation in osteocalcin-related proteins. *Biochemical and Biophysical Research Communications* **113**, 294–300

DURAND, D., BRAITHWAITE, G.D. and BARLET, J.-P. (1983). The effect of 1α hydroxycholecalciferol on the placental transfer of calcium and phosphate in sheep. *British Journal of Nutrition* **49**, 475–480

GUNDBERG, C.M., COLE, D.E.C., LIAN, J.B., READE, T.M. and GALLOP, P.M. (1983). Serum osteocalcin in the treatment of inherited rickets with 1,25-dihydroxyvitamin D_3. *Journal of Clinical Endocrinology and Metabolism* **56**, 1063–1067

HARRINGTON, D.D. and PAGE, E.H. (1983). Acute vitamin D_3 toxicosis in horses. Case reports and experimental studies of the comparative toxicity of vitamins D_2 and D_3. *Journal of the American Veterinary Medical Association* **182**, 1358–1369

HIDIROGLOU, M. and PROULX, J.G. (1983). Effect on calf growth and blood composition of prepartum injection of vitamin D_3 or 25-hydroxyvitamin D_3 to beef heifers. *Canadian Journal of Animal Science* **63**, 251–254

HOLMES, R.P. and KUMMEROV, F.A. (1983). The relationship of adequate and excessive intake of vitamin D to health and disease. *Journal of the American College of Nutrition* **2**, 173–199

HORST, R.L. and LITTLEDIKE, E.T. (1982). Comparison of plasma concentrations of vitamin D and its metabolites in young and aged domestic animals. *Comparative Biochemistry and Physiology B* **73**, 485–489

NOFF, D., SIMKIN, A. and EDELSTEIN, S. (1982). Effect of cholecalciferol derivatives on the mechanical properties of chick bones. *Calcified Tissue International* **34**, 501–505

RIDDELL, C. (1983). Rickets in turkey poults. *Avian Diseases* **27**, 430–441

RONCAGLIONI, M.C., SOUTE, B.A.M., DE BOER, M.A.G. and VERMEER, C. (1983). Warfarin-induced accumulation of vitamin K-dependent proteins—comparison between hepatic and non-hepatic tissues. *Biochemical and Biophysical Research Communications* **114**, 991–997

SOMMERVILLE, B.A., HARVEY, S. and CHADWICK, A. (1983). 83). Early changes in the adaption to a low calcium diet in the chick. *Calcified Tissue International* **35**, 339–343

IV

Fats as Energy-yielding Compounds

11

FATS AS ENERGY SOURCES IN ANIMAL TISSUES

D.W. PETHICK
School of Veterinary Studies, Murdoch University, Western Australia
A.W. BELL
School of Agriculture, La Trobe University, Bundoora, Victoria
and
E.F. ANNISON
Department of Animal Husbandry, University of Sydney, Camden, NSW, Australia

Summary

Long-chain free fatty acids (FFA), triacylglycerols, 3-hydroxybutyrate, acetoacetate and acetate are major energy sources for animal tissues. These energy-yielding nutrients are derived either from the diet or directly or indirectly as a result of the mobilization of adipose tissue when nutrient supply from the alimentary tract fails to meet overall energy requirements. As sources of energy they complement glucose, but the relative importance of glucose and circulating lipids varies from tissue to tissue, and is influenced by both nutritional and physiological status and is species dependent. Quantitative data on the uptake, release and oxidation of lipids by organs and tissues have been obtained by combining arterio-venous difference procedures with isotope dilution based on the use of [^{14}C]-labelled substrates. The isotope dilution procedure also provides information on flux rates in the whole animal, and on the contribution of the metabolites under study to overall oxidative metabolism. The application of these procedures to the measurement of the uptake and oxidation of circulating lipids by the liver, skeletal muscle, and mammary gland is discussed in relation to their overall metabolism.

Introduction

Triacylglycerols, plasma free fatty acids (FFA) and the ketone bodies 3-hydroxybutyrate and acetoacetate, which are major sources of energy in the blood of all animals, are joined by acetate in herbivores. These nutrients are derived either from the diet, or directly or indirectly from adipose tissue mobilized when nutrients absorbed from the alimentary tract fail to meet the energy requirements of the tissues. As energy sources, circulating lipids complement glucose, but the relative importance of these metabolites is influenced by both nutritional and physiological status and is species dependent. In starvation, for example, plasma FFA and ketone bodies derived from FFA assume increasing importance as energy sources relative to glucose. An intriguing example of the response of lipid metabolism to physiological status is the reduction in lipoprotein lipase activity in the adipose tissue of the rat 5 days before parturition, which results in a substantial rise in blood triacylglycerol concentrations. Shortly before parturition, increased lipoprotein lipase activity in mammary tissue gives rise to extensive utilization of triacylglycerols for milk fat synthesis, and blood levels return to normal by the onset of parturition (Bauman and Currie, 1980).

The special features of digestion in ruminants impinge heavily on the supply of energy-yielding nutrients. Dietary starch and other sugars which, in non-ruminants, are hydrolysed to glucose in the small intestine, are fermented to volatile fatty acids (VFA) in the rumen. This severely limits the supply of alimentary glucose, and the absorption of large amounts of VFA, of which the most important is acetate. This is the most striking feature of the pattern of nutrients which supply the energy needs of ruminant tissues. Although this has not led to the development of unique metabolic pathways, the quantitative significance of certain key processes in ruminants differs from that of non-ruminants. Gluconeogenesis, for example, is a major metabolic activity in the fed as well as the fasted ruminant, since in spite of the paucity of alimentary glucose, glucose flux in ruminants is at least 50% of that of non-ruminants (Ballard, Hanson and Kronfeld, 1969). A second major process, lipogenesis, shows marked differences in ruminant tissues. Whereas in many mammalian and avian systems glucose is the major precursor of long-chain fatty acids, in ruminants this role is largely taken over by acetate (Ballard *et al.*, 1969).

Lipid metabolism is dominated by the supply of absorbed nutrients: when this exceeds the immediate energy needs of tissues, triacylglycerols are synthesized and stored in adipose tissue. In some species, including man, substantial fatty acid synthesis occurs in the liver, but in ruminants rates of hepatic fatty acid synthesis are extremely low (Bauman and Davis, 1975; Bell, 1980). In contrast, lipogenesis in birds is almost entirely confined to the liver (Leveille *et al.*, 1975). Hepatic triacylglycerols formed either from FFA taken up from the blood, or synthesized *de novo* in the liver, are incorporated into lipoproteins for transport to extrahepatic tissues (see Bell, 1980). When nutrient supply fails to meet the energy needs of tissues, adipose tissue triacylglycerols are mobilized exclusively as plasma FFA, which is used as an energy source by most tissues.

Glucose is the major source of energy for tissues in non-ruminants, and excess glucose is used for fatty acid synthesis. The energy required for fatty acid synthesis is generated by the pentose cycle activity, and by the decarboxylation of malate via the citrate cleavage pathway. In ruminants, acetate is the major source of energy for tissues, and acetate in excess of immediate energy needs is used for lipogenesis. Glucose is not a significant precursor of long-chain fatty acids in ruminant tissues. This has been thought to stem from the virtual absence of an active citrate cleavage pathway essential for the translocation of acetyl CoA derived from pyruvate within the mitochondrion to sites of fatty acid synthesis in the cytosol (see Bell, 1980). Recent studies have suggested that pyruvate kinase and pyruvate dehydrogenase (Robertson, Faulkner and Vernon,

1982) and the key glycolytic enzymes, especially hexokinase (Smith and Prior, 1981) may have important roles in restricting the utilization of glucose carbon for fatty acid synthesis in ruminant adipose tissue. Glucose oxidation by the pentose cycle pathway is a source of NADPH for fatty acid synthesis. Acetate oxidation by the isocitrate dehydrogenase pathway probably supplies the remainder of the NADPH required for fatty acid synthesis, but the relative importance of the two sources of NADPH is not known (Vernon, 1981).

Extensive quantitative date on the uptake, release and oxidation of blood lipids, including acetate and ketone bodies, have been obtained by combining arterio-venous (AV) difference procedures with isotope dilution measurements. These procedures provide information on the contribution of individual blood lipids to energy expenditure, both in specific tissues and in the whole animal (Linzell and Annison, 1975): their application to the measurement of the uptake and oxidation of circulating lipids by the liver, gut, skeletal muscle, and mammary gland is discussed in this chapter.

Procedures for the *in vivo* study of tissue metabolism

The classic AV difference procedure was refined by Linzell (1960) in his studies on the goat mammary gland, an easily accessible organ with a vasculature that allows an important requirement of the technique to be satisfied, namely, the ability to sample venous blood representative of total venous drainage. In order to obtain quantitative data on substrate uptake, blood flow must be measured by a technique validated for the organ or tissue being studied. A further crucial requirement is the avoidance of excitement or stress during any aspect of the experimental procedure. Even slight excitement will raise plasma FFA levels, and indeed, relatively unchanged levels of this lipid fraction is a useful criterion of absence of stress. Significant stress accompanied by catecholamine release may result in vasoconstriction of the organ or tissue, resulting in reduced blood flow, while even small body movements can be associated with substantial increases in muscle blood flow.

The utility of the AV difference procedure is increased when combined with isotope dilution procedures based on the continuous infusion of [^{14}C]-labelled substrates to achieve constant specific radioactivity of substrate in arterial blood (Linzell and Annison, 1975). A fall in the specific radioactivity of substrate across the tissue provides unequivocal evidence of substrate release. Measurements of O_2 uptake and CO_2 production, and of $^{14}CO_2$ production by the tissue, make it possible to calculate the proportion of substrate taken up that is directly oxidized. The total energy expenditure of the tissue may be calculated from the O_2 uptake and CO_2 output. In a further development, Corbett *et al.* (1971) demonstrated that CO_2 entry rate in sheep measured by the continuous infusion of [^{14}C]-labelled sodium bicarbonate into the whole animal was closely correlated with both total CO_2 production and energy expenditure. The latter procedure, devised as a relatively simple method for the measurement of the energy expenditure of grazing animals, complements the AV difference combined with isotope dilution procedure, since the same animal preparations and assay procedures may be used. In summary, the contribution of blood lipids and other metabolites to the energy expenditure of organs and tissues may be measured by procedures which also provide data on the role of the metabolite in whole-body energy metabolism.

These procedures, which have proved so fruitful in the study of mammary gland (Linzell and Annison, 1975), liver and gut (Bergman, 1975) and hind limb skeletal muscle (Bell, 1980) metabolism, are subject to limitations and intrinsic difficulties which must be borne in mind. When measurements are made in relation to plane of nutrition, it is imperative that the catheterization of blood vessels and blood sampling procedures do not interfere with feed intake, which must be stabilized for at least 2–3 weeks before experimentation. Avoidance of stress is essential, as discussed earlier, but even in unstressed animals there can be considerable minute-to-minute variation in blood flow and in AV difference (Zierler, 1961). This problem may be minimized by the continuous withdrawal of venous blood during the sampling period, which provides an integrated sample. Since the precision of measurement is directly related to that of the measurement of blood flow, every effort must be made to validate the blood flow procedure for each tissue studied.

The measurement of CO_2 entry rate is subject to systematic errors of about 20% before the fixation of $^{14}CO_2$ in reactions which do not return $^{14}CO_2$ to the body CO_2 pool during the course of the experiment, but this difficulty was recognized by Corbett et al. (1971) when the relationship between CO_2 entry rate and energy expenditure was established. Short-term variations in CO_2 entry rate which accompany feeding, postural changes and handling, substantially reduce the precision of the method relative to conventional measurement of heat production by respiration calorimetry, when measurements are made over days, and not hours.

The precision of the measurement of the rates of oxidation of [^{14}C]-labelled substrates in tissues from $^{14}CO_2$ release may also suffer from the involvement of $^{14}CO_2$ in slow-moving metabolic pools, but the problem is much less severe in tissues and organs than in the whole animal. Nevertheless, when carbon balances are established for individual tissues, it has not yet been possible to account for total CO_2 production from the contribution of individual substrates: this is probably more a reflection of underestimates of the contributions of individual substrates to CO_2 production than the oxidation of hitherto unrecognized substrates.

Splanchnic tissues

General aspects

In ruminant and non-ruminant animals, the liver has a central role in fat metabolism, although the quantitative details of intrahepatic lipid metabolism vary considerably between species. For example, in all species studied, the capacity for hepatic uptake and oxidation of plasma FFA, and for ketogenesis, appears to be high, whereas the ability to synthesize fatty acids *de novo* and to secrete large amounts of triacylglycerol for export in VLDL is high in some animals, such as man, and low in others, including the domestic ruminants (see Bell, 1980). The gut is also an important user of lipid substrates, some of which may be taken up from arterial blood (e.g. plasma FFA) while others may be obtained from the gut lumen (e.g. ruminal VFA).

The techniques for *in vivo* study of tissue metabolism, outlined earlier, have been successfully applied to the measurement of fatty acid and ketone metabolism in the liver and extrahepatic splanchnic tissues of conscious animals (see Bergman, 1975). Such studies have provided invaluable data, but caution is required in their interpretation. In particular, the problem of making accurate, dual measurements of large portal venous and small hepatic arterial blood flows has not been completely resolved. In addition, the usual practice of sampling mixed portal venous blood does not define the exact site of metabolism in the portal-drained viscera, which include the stomach(s) and intestines plus associated adipose tissue, pancreas and spleen.

Short-chain fatty acid metabolism

The importance of VFA as energy sources in ruminants (Annison and Armstrong, 1970) has led to increasingly detailed *in vivo* studies of the production and utilization of acetate, propionate and butyrate by the portal-drained viscera and liver. By combining isotope dilution AV difference techniques, Bergman and Wolff (1971) and Pethick *et al.* (1981) showed that acetate is simultaneously produced and utilized by the gut and liver of sheep.

In each of these studies, about 30% of the acetate absorbed by fed sheep was used by the gut, which is remarkably similar to the finding that up to 45% of the acetate absorbed by rumen epithelium *in vitro* may be metabolized (Stevens, 1970). Most of this acetate appears to be directly oxidized, possibly accounting for up to 50% of the gut's energy requirements (*Table 11.1*) (Pethick *et al.*, 1981). In this regard, the gut tissues are typical of the whole body in fed ruminants (see Annison and Armstrong, 1970) and it is encouraging to note the close agreement between independently derived estimates of the contributions of the gut to total acetate utilization (22–25%, Bergman and Wolff, 1971; Pethick *et al.*, 1981) and to total oxygen consumption (22%, Thompson *et al.*, 1978b) in sheep fed a maintenance ration.

Table 11.1 ACETATE PRODUCTION AND UTILIZATION IN SPLANCHNIC TISSUES OF SHEEP[a] (MEAN OF 5 SHEEP)

Tissue	Production (mmol/h)	Utilization (mmol/h)	Contribution to CO_2 production (%)	Fraction oxidized (%)
Gut	119	34	36	77
Liver	31	23	3	6

[a] Maintenance-fed, alloxan-diabetic, insulin-stabilized (Pethick *et al.*, 1981)

The picture for acetate metabolism in the ruminant liver is less clear. In sheep and cattle, hepatic uptake and utilization of acetate from portal blood is low, ensuring a substantial supply of acetate for extrahepatic tissues (see Bell, 1980). Nevertheless, net hepatic utilization of acetate can occur in fed sheep (e.g. Thompson, Gardner and Bell, 1975) and Pethick *et al.* (1981) estimated that the 'true' or 'gross' hepatic uptake of acetate in fed, insulin-stabilized, alloxan-diabetic sheep accounted for about 17% of whole-body utilization. However, very little of this acetate was oxidized to CO_2 and H_2O (*Table 11.1*). This is consistent with the very low mitochondrial activity of acetyl CoA synthetase in sheep liver (Knowles *et al.*, 1974) but difficult to explain in terms of alternative, non-oxidative pathways, particularly since long-chain fatty acid synthesis is minimal in the ruminant liver (see Bell, 1980). Such is not the case in most non-ruminants and Snoswell *et al.* (1982) have recently demonstrated significant lipogenesis from acetate in the isolated rat liver.

The inadequacy of simply measuring blood acetate concentration differences across the liver was highlighted by Pethick *et al.* (1981) who showed that a substantial uptake of acetate was masked by a greater and coincident rate of production in the sheep liver (*Table 11.1*). These authors also examined skeletal muscle, another tissue with the capacity to produce significant amounts of acetate (Knowles *et al.*, 1974) and concluded that the liver accounts for most of the endogenous acetate production in the sheep. The liver also produces acetate in rat (Buckley and Williamson, 1977) and man (Skutches *et al.*, 1979) but it has been suggested that this becomes significant only when the availability of acetate in portal blood is low. It is notable that in fed rats, the hepatic mitochondrial activity of acetyl CoA hydrolase is considerably higher than that in fed sheep, and that in both species, starvation causes only a moderate increase in activity (Knowles *et al.*, 1974).

Propionate and butyrate, the other VFA produced in significant amounts by rumen and/or hind-gut fermentation, are extensively metabolized by the splanchnic tissues in ruminants, to such an extent that only trivial amounts reach the peripheral circulation (Bergman and Wolff, 1971). However, their contributions to oxidative metabolism in these tissues *in vivo* have not been measured. Apart from being potentially oxidizable fuels, especially in the gut tissues which absorb them, both these VFA are precursors of important substrates for extrahepatic metabolism in the fed ruminant. Propionate, via hepatic gluconeogenesis, contributes 30–60% of all glucose produced in the fed ruminant (see Elliot, 1980), while most absorbed butyrate is converted to 3-hydroxybutyrate, mainly in rumen epithelium (Weigand, Young and McGilliard, 1972).

Long-chain fatty acid metabolism

The liver is a major site for the uptake and metabolism of plasma FFA. Bergman *et al.* (1971) simultaneously measured hepatic uptake and total body flux of FFA in conscious fed sheep and post-absorptive dogs: in each case the liver took up about 25% of all FFA entering the bloodstream. In this study the gut tissues in sheep and dogs took up a further 8% of the total supply of FFA, so that the total splanchnic uptake accounted for about one-third of FFA supply. A similar value of about 40% has been reported for post-absorptive humans (Havel et al., 1970).

Hepatic uptake of FFA does not appear to be regulated by intrahepatic metabolism, but rather by the plasma concentration of FFA in ruminant and non-ruminant animals (see Bell, 1980). Thus, in conscious sheep, the fractional extraction of FFA by the liver was about 10% over a wide range of physiological states and arterial FFA concentrations (Katz and Bergman, 1969; Thompson and Darling, 1975; Thompson et al., 1975). There is some evidence that stearic acid (C18:0), the predominant individual FFA in the plasma of fed ruminants, is not extracted as efficiently by the sheep liver as are other individual plasma FFA, especially at higher arterial concentrations (Thompson and Darling, 1975; Thompson et al., 1975). These observations have not been explained, although it has been reported that in the isolated, perfused rat liver, the uptake of individual FFA was inversely related to the chain length of saturated fatty acids and directly related to the degree of saturation (see Bell, 1980).

The rate of FFA supply not only determines hepatic uptake, but also influences FFA metabolism in the liver, although the latter is modulated by metabolic state. Ketogenesis is favoured in fasting and in the diabetic state, whereas esterification predominates in the fed condition. Studies on the perfused rat liver (see Mayes, 1976; McGarry and Foster, 1980) have shown that when plasma FFA concentration is low, most FFA taken up by the liver are esterified and exported in VLDL but as the plasma concentration and hepatic uptake increase, the esterification pathway becomes saturated and an increasing fraction of fatty acids enters the β-oxidation pathway. Some of these are oxidized to completion, depending on the energy requirement and TCA cycle activity of the liver, while the remainder are converted to ketones. This general picture has been confirmed in experiments on conscious humans (Havel et al., 1970) and anaesthetized dogs (Basso and Havel, 1970) in which the splanchnic metabolism of infused ^{14}C-labelled FFA was examined by AV difference and radioactive tracer techniques.

It is unlikely that the above features of intrahepatic lipid metabolism differ qualitatively between species, but there may be important quantitative differences between ruminants and non-ruminants. Net hepatic uptake of FFA and output of ketones have been measured in fed and fasted sheep (Katz and Bergman, 1969) and cows (Baird et al., 1979), while Reid et al. (1979) examined net hepatic uptake of FFA and output of triacylglycerol in fed and fasted cows (*Table 11.2*). Their data were consistent with the notion that during fasting, hepatic ketogenesis is almost entirely due to increased FFA catabolism in ruminants as in other species. In cows fasted for 1–6 days, this was associated with a switch from net hepatic output to net uptake of triacylglycerol (Reid et al., 1979), which contrasts with the direct relation between splanchnic FFA uptake and VLDL release in the post-absorptive human (Havel et al., 1970). Evidence for the importance of the ruminant liver as a source of plasma VLDL is scanty and conflicting (see Bell, 1980). However, the above results suggest that an impaired capacity to secrete VLDL during periods of negative energy balance may be partly responsible for the predisposition to fatty infiltration of the liver

Table 11.2 HEPATIC METABOLISM OF FFA, KETONES AND TRIACYLGLYCEROL IN SHEEP AND COWS

Species	Nutritional/physiological state	Net uptake or production (mmol/h)			
		FFA uptake	3-hydroxybutyrate production	Acetoacetate production	Triacylglycerol production
Sheep	Fed, non-pregnant	11	3	−3	−
	Fed, twin-pregnant	9	15	−4	−
	Fasted (3d), non-pregnant	19	19	8	−
	Fasted (3d), twin-pregnant	13	34	9	−
Cow	Fed, non-lactating	19	176	−47	17
	Fasted (4–6d), non-lactating	74	235	25	−3

and hyperketonaemia shown by undernourished pregnant ewes and spontaneously ketotic or underfed lactating cows. This, of course, does not diminish the likely importance of hepatic carbohydrate insufficiency as a major determinant of the rate of hepatic ketogenesis (Baird, 1981).

Plasma FFA taken up by the liver are probably also the major precursors for hepatic acetate synthesis in fasted ruminants (Palmquist, 1972) but contrary to an earlier suggestion (Costa, McIntosh and Snoswell, 1976), the level of acetate production appears to be unrelated either to FFA supply or to the ratio acetyl CoA:CoA ('acetyl pressure') in the liver of cows (Snoswell *et al.*, 1978) or sheep (Pethick *et al.*, 1981).

Altered metabolic states

In maintenance-fed sheep, it has been calculated that the measured rate of oxidative metabolism in the gut *in vivo* could be more than satisfied by observed uptakes of VFA and glucose, assuming complete combustion of these substrates (Webster, 1980). Similar data for fed non-ruminants are not available. The liver accounts for more than 20% of respiratory oxygen uptake in the sheep at maintenance (Thompson *et al.*, 1978b), but the pattern of hepatic substrate oxidation in the fed state remains to be determined *in vivo*, in any animal species.

In contrast, in fasting or underfed animals of all species, the importance of FFA oxidation as a source of energy in gut and liver is likely to increase progressively as the supply of oxidizable alimentary nutrients (VFA in ruminants, glucose in non-ruminants) declines. Thus Pethick *et al.* (1981) found that acetate utilization in splanchnic tissues was linearly related to arterial supply (or weighted arterial/portal supply for the liver) over the range of arterial concentrations seen in well-fed and severely undernourished ruminants. Oxidation of FFA in splanchnic tissues *in vivo* has not been measured in fed or fasted ruminants, but in post-absorptive man, FFA oxidation accounts for all splanchnic oxygen consumption (Havel *et al.*, 1970).

Other factors which can acutely or chronically alter energy balance, such as exercise and cold exposure, may affect the pattern of substrate utilization in splanchnic tissues. In fed sheep exposed to cold for several hours, there was a significant, sustained increase in the rates of appearance of VFA in portal blood (Thompson *et al.*, 1975; Thompson, Bassett and Bell, 1978a). A possible explanation is that the cold-induced increase in FFA flux (Bell and Thompson, 1979) was associated with a partial substitution of FFA for VFA as oxidizable substrate in rumen epithelium, assuming unchanged rates of ruminal production and absorption of VFA. Similarly, an increase in FFA uptake and oxidation by splanchnic tissues might be expected during exercise. It is notable that in sheep, exercise not only causes a pronounced increase in FFA flux (Pethick, 1982), but also a significant decline in blood flow to the forestomachs (Bell *et al.*, 1983), which could temporarily limit VFA absorption. In post-absorptive man, prolonged exercise causes progressive and proportionately similar increases in the whole-body turnover and splanchnic uptake of oleic acid (Ahlborg *et al.*, 1974).

Fatty acid metabolism in skeletal muscle

General aspects

Skeletal muscle is of major importance in whole-body metabolism. It is the largest organ in the body, accounting for about 25% of the body mass in ruminants (Palsson and Verges, 1952) and as much as 40% in non-ruminants (Vernon, 1970). Despite a relatively low metabolic rate at rest, muscle may account for as much as 90% of overall oxygen consumption during intensive exercise (Vernon, 1970). The AV difference and isotope dilution techniques discussed earlier have proved particularly suitable for the study of hind limb skeletal muscle metabolism since only minor surgical interference to the conscious intact animal, or human subject, is involved and measurements are made on a relatively large muscle mass. Although the tissue studied is predominantly skeletal muscle, it also contains small but metabolically significant amounts of adipose tissue. This implies that values for the net exchange of substrate may simply reflect the net difference between the simultaneous uptake and release of substrate by both tissues.

Substrate utilization by muscle

Until about 20 years ago it was generally held that glucose was the most important energy source of skeletal muscle. Even with the recognition of plasma FFA as a major vehicle for the transport of fat in blood, Baltzan *et al.* (1962) found no significant AV difference across the human forelimb. In a new approach, Friedberg *et al.* (1960) using intravenously infused [^{14}C] FFA showed that there is usually a simultaneous uptake and output of FFA by skeletal muscle. This conclusion was confirmed in subsequent studies reviewed by Zierler (1976), who concluded that FFA are the major energy source for skeletal muscle. A consistent feature in these studies in humans and dogs was that all experiments were carried out during the post-absorptive state, which is characterized by elevated concentrations of FFA. Indeed Zierler and Andres (1956) had shown earlier that glucose made a significant contribution to the energy demand of resting human muscle after a meal and that this declined as the post-absorptive state developed.

Data on the uptake and potential contribution to oxidative metabolism of circulating lipids by the hind limb of several species and their role as fuels are summarized in *Table 11.3*. Estimates of the contribution of particular metabolites show marked differences, which may reflect in part species variation and differences in technique and feeding regimens. Nevertheless, acetate emerges clearly as an important energy source for ruminant skeletal muscle accounting for 30–40% of oxygen consumption. Furthermore Pethick *et al.* (1981) have shown that nearly 80% of this uptake is promptly oxidized. Indeed, acetate's contribution to the energy metabolism of skeletal muscle relates directly to its arterial concentration in sheep (Pethick and Lindsay, 1982a), man (Lundquist *et al.*, 1973) and rat (Karlsson, Fellenius and Kiessling, 1975). This suggests that acetate is only

Table 11.3 METABOLITE CONCENTRATION IN BLOOD (mM) AND MAXIMUM POSSIBLE CONTRIBUTION TO OXIDATION (%) IN HIND LIMB SKELETAL MUSCLE IN THE HUMAN, FED SHEEP AND FED PIG

Metabolite	Human[a]				Sheep[c]		Pig[c]	
	High-carbohydrate diet		High-fat diet					
	Arterial concentration	Potential contribution to oxidation	Arterial concentration	Potential contribution to oxidation	Arterial concentration	Potential contribution to oxidation	Arterial concentration	Potential contribution to oxidation
FFA[f]	0.59	111	1.02	221	0.06	6	0.32[d]	–
Ketone bodies	0.17	2	1.94	8	0.35	15	–	–
Acetate	0.17[b]	–	–	–	1.20	30–40	0.33[d]	–
Glucose[g]	4.80	30	4.20	10	2.95–3.81	30–57	4.32[d]	95[e]

[a]From Jansson (1980).
[b]From Ballard (1972).
[c]From Pethick and Lindsay (1982a); Pethick (1982); Bird, Chandler and Bell (1981).
[d]From Freeman et al. (1970).
[e]From Lindsay (1981).
[f]Potential contribution to oxidation from gross uptake of FFA.
[g]Corrected for lactate release or uptake.

a minor energy source for skeletal muscle of humans in view of the low arterial concentration. The relatively high concentration of acetate in the blood of pigs, however, and the observation that 12% of the respiratory CO_2 was derived from acetate (Freeman, Noakes and Annison, 1970) suggests that this substrate may be more extensively oxidized in muscle in the pig.

Ketone bodies are usually considered to be significant fuels only during fasting or altered metabolic states such as diabetes (Owen and Reichard, 1975), but their relatively high blood concentration in humans fed a high fat diet, and in ruminants indicate that they may account for 8–15% of the oxygen consumed by skeletal muscle (*Table 11.3*). In non-ruminant species fed a diet high in digestible carbohydrate, ketones are a trivial fuel for skeletal muscle, but about half the energy needs of ruminant muscle are derived from ketones and acetate. In contrast, these metabolites probably account for no more than 10% of oxygen consumption in human skeletal muscle (*Table 11.3*).

The other sources of energy for skeletal muscle must be FFA or carbohydrate. In the fed human FFA would appear to be the predominant fuel in view of the relatively elevated plasma FFA concentration, since FFA are utilized by skeletal muscle in direct proportion to arterial concentration in both non-ruminants (Zierler, 1976) and ruminants (Pethick *et al.*, 1983). Relatively elevated plasma FFA in the fed human might be an inherent property or it might simply reflect the fact that experiments were conducted in human subjects (*Table 11.3*) fed only a light breakfast (10% of the daily calorie energy allowance) before measurements of nutrient uptake were made (Jansson, 1980). The relatively small amount of food consumed and the stress associated with surgery might have resulted in elevated circulating FFA. In contrast, farm animals were surgically prepared in advance and measurements made either on continuously fed animals or on animals soon after feeding.

Estimates of the part played by FFA as an energy source for muscle in fed animals are further confounded because a considerable proportion of the FFA utilized by skeletal muscle is not promptly oxidized. Only 20% of FFA uptake is oxidized in human muscle (Dagenais, Tancredi and Zierler, 1976), and about 40% in fed sheep (Pethick, 1982). It would appear that direct oxidation of circulating FFA accounts for less than 5% of the energy requirements of ruminant muscle. The corresponding value in humans is about 20% and 45% on high carbohydrate and fat diets respectively. These estimates must be taken as minimum values due to isotope hold-up in metabolic pools, particularly since CO_2 of muscle origin is in slow equilibrium with the circulating pool (Pethick *et al.*, 1981). Finally, Zierler (1976) has suggested that most of the FFA utilized by skeletal muscle pass through an obligatory triacylglycerol pool of intramuscular origin, and are destined for oxidation. Clearly not all FFA are oxidized, however, because in human subjects on a high fat diet (*Table 11.3*), the potential contribution of FFA to the oxidation of skeletal muscle is an impossible 221%.

A difficulty in the interpretation of FFA utilization by muscle is the possible contribution of FFA derived from hydrolysis of circulating triacylglycerol within muscle capillaries. Skeletal muscle contains modest lipoprotein lipase activity in a wide range of mammalian species (Cryer and

Jones, 1979) and measurable AV differences of exogenous triacylglycerol were observed across the human forearm (Kaijser and Rossner, 1975). This possible problem highlights the utility of studies based on [^{14}C] FFA, which provide specific data on uptake, release and oxidation.

In all species, glucose uptake, corrected for lactate release or uptake, is large enough to make a significant contribution to the oxygen utilized by skeletal muscle (*Table 11.3*). As with FFA, it is possible that not all utilized glucose is directly oxidized. In addition to lactate formation, Felig (1975) has suggested that glucose carbon can be transferred to alanine and subsequently released to the circulation, i.e. the glucose–alanine cycle. The overall process is thought to allow the utilization of muscle-borne amino acids (particularly the branched-chain amino acids) as energy sources while simultaneously recycling glucose carbon for gluconeogenesis in the liver (see Lindsay, 1980). In fed ruminants the net release of alanine is small (Lindsay, Steel and Buttery, 1976) or zero (Bell *et al.*, 1975) and using the figures of Lindsay *et al.* (1976), less than 1% of the glucose carbon could be recycled via this process in sheep. Recent observations (V.H. Oddy, unpublished) on groups of ewes fed a range of diets have shown that about 26% of glucose uptake is promptly oxidized by hind limb muscle, and that glucose accounts for 7% of CO_2 production by the tissue.

Pregnancy appears to promote a shift to FFA metabolism (*Table 11.4*) but this may reflect raised circulating levels of plasma FFA, since the twin pregnant ewes did not consistently consume their entire ration in the last 2 weeks of pregnancy.

On some occasions skeletal muscle may be responsible for a breakdown in homeostasis. Thus twin pregnant ewes are susceptible to ketoacidosis late in pregnancy (Reid, 1968) which is typified by a marked elevation in the concentration of D-3-hydroxybutyrate (Leng, 1965). The capacity of skeletal muscle to utilize circulating D-3-hydroxybutyrate is severely limited above concentrations of 3–4 mM (Pethick and Lindsay, 1982b). At this concentration, small changes in the production of ketones by the liver will promote massive increases in the level of D-3-hydroxybutyrate and result in a decreased blood pH.

A related phenomenon, which on this occasion facilitates homeostasis, is seen during a long-term fast in human subjects. In the initial days of starvation, ketone body uptake increased in parallel with circulating levels to account for 50–85% of the oxygen consumed by skeletal muscle (Owen and Reichard, 1971). As fasting progressed, circulating ketone body concentration increased but the net uptake by skeletal muscle declined (Owen and Reichard, 1971), suggesting that there is a transition to the oxidation of FFA and that this represents a specific adaptation to spare ketone bodies for the energy demands of the central nervous system which can utilize ketone bodies (Owen *et al.*, 1976). Clearly this would decrease the individual's dependence on carbohydrate oxidation and so prolong the reserves of glucose and its precursors. The important feature of this observation is that factors (as yet undefined) other than circulating concentration can modulate the utilization of fatty acids in some metabolic states.

Lactation imposes a severe metabolite drain on the body. Glucose and acetate, two potentially significant fuels for skeletal muscle in non-lactating

Table 11.4 METABOLITE CONCENTRATION IN BLOOD (mM) AND MAXIMUM POSSIBLE CONTRIBUTION TO OXIDATION (%) IN HIND LIMB SKELETAL MUSCLE IN FASTED AND PREGNANT SHEEP

Metabolite	Non-pregnant[a] 6-day fasted		Pregnant[b] Fed		Pregnant[c] 4-day fasted	
	Arterial concentration	Contribution to oxidation	Arterial concentration	Contribution to oxidation	Arterial concentration	Contribution to oxidation
FFA[a]	1.40	80	0.84	52	1.59	91
Ketone bodies	1.17	36	0.95	19	3.23	48
Acetate	0.12	<1	1.2	34	0.16	<1
Glucose[e]	2.32	0	2.50	35	1.41	10

[a] From Jarrett et al. (1976).
[b] From Pethick and Lindsay (1982b) and Faichney et al. (1981).
[c] From Pethick et al. (1983).
[d] From Pethick (1981).
[e] Corrected for lactate release or uptake.

fed ruminants (see *Table 11.4*), are required in large quantities by the mammary gland. In accordance with this, Pethick and Lindsay (1982a) found that the glucose utilized by skeletal muscle declined by 20% and in addition a small lactate uptake in non-lactating animals changed to a significant lactate release. Lactate release by skeletal muscle in lactation was also reported by Mercer *et al.* (1980). A further adaptation was found in lactating sheep which made more acetate available to the mammary gland. At a given arterial concentration the uptake of acetate by the skeletal muscle of lactating ewes was reduced, with the spared acetate being quantitatively removed by the mammary gland (Pethick and Lindsay, 1982a). The significance of this finding needs careful interpretation because the lactating ewes in this study were eating 70% more ME than their non-lactating counterparts. The added food intake doubled the mean daily acetate concentration in the blood of lactating ewes and so allowed acetate to make the same contribution to the oxidation of skeletal muscle as in non-lactating animals while simultaneously sparing acetate for the mammary gland.

Table 11.5 THE MAXIMUM POSSIBLE CONTRIBUTION (%) OF CIRCULATING AND ENDOGENOUS METABOLITES TO OXIDATION IN HIND LIMB SKELETAL MUSCLE DURING EXERCISE IN THE SHEEP AND HUMAN

Metabolite	Sheep[a]		Human[b]	
	Duration of exercise		*Duration of exercise*	
	60 min	120 min	90 min	180 min
FFA	20	38	37	62
Glucose[c]	25	32	41	30
Acetate	8	8	–	–
Endogenous[d] substrates in muscle	47	22	22	8

[a] From Pethick (1982) and Bird *et al.* (1981).
[b] From Ahlborg *et al.* (1974).
[c] Human values uncorrected for uptake or release of lactate.
[d] Computed by difference.

Studies of the major nutrients contributing to the oxidation of skeletal muscle during exercise in both humans and sheep are summarized in *Table 11.5*. Firstly, acetate is a relatively minor energy source for working muscle in the sheep and, as discussed previously, it is unlikely to make a significant contribution in the human. Secondly, the contribution of FFA progressively increases at the expense of endogenous substrates as exercise progresses. This is related to a steady increase in the concentration of circulating FFA in both studies. Furthermore, there is evidence that the FFA utilized by muscle is completely oxidized to CO_2 and water during exercise (Pethick, 1982). The endogenous substrate utilized in these studies is probably glycogen although the role of intramuscular triacylglycerol cannot be excluded (Essen, Hagenfeldt and Kaijser, 1977). The overall trend is for FFA oxidation to predominate as the carbohydrate reserves of muscle decline. It is interesting to note in both humans and sheep that the utilization of circulating carbohydrate is maintained despite the increased uptake of fatty acids. This is in direct contrast to the other physiological states considered.

In conclusion, fatty acids are generally utilized by skeletal muscle when made available. Availability can be determined by dietary supply or by the mobilization of endogenous fatty acids. The role of long-chain fatty acids in the energy metabolism of skeletal muscle is related to a decreased dependence on circulating glucose. Thus the classic hypothesis of fatty acids having a central role in the conservation of glucose for those tissues with an obligatory demand remains unchallenged. However, it is important to remember that in some metabolic states the uptake of fatty acids is controlled by factors other than availability, in such a way that alternative tissues, with a high demand for fatty acids, have an assured supply.

Mammary gland metabolism

General aspects

The concentration of effort on mammary gland metabolism in ruminants reflects the economic importance of these species, and their ideal size for *in vivo* studies. A further important factor is that dairy animals have been selected for milk output, and in the high-yielding gland the proportion of secretory tissue to skin and other tissues is high enough to equate AV differences with metabolic requirements for milk synthesis.

The major biochemical pathways for the synthesis of the large amounts of milk lactose, fat and protein require energy supplied as ATP and reducing equivalents. As pointed out by Smith and Taylor (1977), the subsidiary biochemical pathways generating ATP and reducing equivalents are related and give rise to alternative products of oxidizable substrates. Most of the ATP used in milk synthesis is derived from tricarboxylic acid cycle activity.

Generation of reducing equivalents

Fatty acid synthesis, which occurs in the cytosol, has a specific requirement for reducing equivalents in the form of reduced nicotinamide adenine dinucleotide phosphate (NADPH). A major source of NADPH in both ruminant and non-ruminant mammary tissue is the oxidation of glucose by the pentose cycle pathway (see Bauman and Davis, 1974). A second source of NADPH in non-ruminants, which are able to synthesize fatty acids from glucose, arises from the processes involved in the translocation of the acetyl CoA formed within the mitochondrion from pyruvate. The mitochondrial membrane is impermeable to acetyl CoA, which moves into the cytosol as citrate, where it is split into oxaloacetate and acetyl CoA by the enzyme ATP-citrate lyase. The low activity of this enzyme in ruminant tissue contributes to the failure of glucose carbon to contribute to fatty acid synthesis in these species (see Introduction). In non-ruminant tissue, however, the action of ATP-citrate lyase generates NADH in the cytosol, which is used for the production of NADPH by the coupled operation of NAD-malate dehydrogenase and NADP-malate dehydrogenase.

Bauman, Brown and Davis (1970) suggested that in ruminant tissue the operation of the isocitrate dehydrogenase pathway in the cytosol is an important alternative pathway for NADPH generation. The evidence for this pathway is that fatty acids are synthesized from acetate in ruminant mammary tissue *in vitro* in the absence of glucose, and that isocitrate dehydrogenase activity is high in this tissue relative to that shown by non-ruminant mammary tissue. This scheme was later extended by Gumaa, Greenbaum and McLean (1973) to include the reduction of 2-oxoglutarate to isocitrate, with the generation of a further mole of NADPH. The production of two mole of NADPH per mole acetate oxidized is more in accord with *in vitro* observations, as discussed by Smith and Taylor (1977). Bauman and Davis (1975) estimated that in the bovine mammary gland the isocitrate dehydrogenase pathway generated about 50% of the cytoplasmic NADPH required for fatty acid synthesis, the remainder being derived from glucose metabolism by the pentose cycle pathway. The observation of Chaiyabutr, Faulkner and Peaker (1980) that in lactating goats, pentose cycle activity accounted for only 34% of the required NADPH, suggested an even greater role for the isocitrate dehydrogenase pathway in this tissue.

Acetate and plasma oxidation in mammary tissue

Measurements of the contribution of acetate and plasma FFA to the oxidative metabolism of the lactating mammary gland by the AV difference and isotope dilution procedures in goats (Annison and Linzell, 1964;

Table 11.6 OXIDATION OF ACETATE, PLASMA FFA AND GLUCOSE IN THE MAMMARY GLAND OF THE FED GOAT, COW AND EWE

	Contribution of substrate to CO_2 production (%)			Substrate oxidized (% uptake)		
	Goat	Cow	Ewe	Goat	Cow	Ewe
Acetate	21	30	25	29	29	25
Plasma FFA	<1	–	–	<1	–	–
Glucose	49	24	20	34	11	18

Annison, Linzell and West, 1968), cows (Bickerstaffe, Annison and Linzell, 1974) and ewes (K.R. King, J.M. Gooden and E.F. Annison, unpublished work) have revealed that in fed animals, acetate, but not FFA, is extensively oxidized (*Table 11.6*). Plasma FFA oxidation was not measured in the cow and ewe, but the negligible net mammary uptakes of FFA precluded a significant contribution to CO_2 production. Data for glucose, included for comparison, reveal that although not a precursor of milk fatty acids, glucose is a major source of energy for the lactating mammary gland.

Table 11.7 METABOLISM OF BLOOD ACETATE, FFA, TRIACYLGLYCEROL, 3-HYDROXYBUTYRATE AND GLUCOSE BY THE MAMMARY GLAND OF THE GOAT STARVED FOR 24 h (MEAN OF 3 EXPERIMENTS)

	Mammary uptake (mg C/min)	Contribution to CO_2 production (%)	Substrate oxidized (%)
Acetate	1.8	11	63
FFA	17.5	22	48
3-Hydroxybutyrate	3.4	–	–
Triacylglycerol	5.5	–	–
Glucose	11.2	9	7

Studies on lipid metabolism in the mammary gland of the starved (24 h), lactating goat (Annison *et al.*, 1968) revealed a dramatic shift in the relative importance of acetate and plasma FFA in mammary metabolism. Data on the uptake and oxidation of acetate, FFA, 3-hydroxybutyrate (uptake only), triacylglycerol (uptake only) and glucose are shown in *Table 11.7*.

The failure to account for more than 42% of CO_2 production in the starved goat mammary gland reflects both the substantial utilization of udder tissue in starvation (Annison *et al.*, 1968), and the absence of information on the contribution of 3-hydroxybutyrate to oxidative metabolism in mammary tissue.

Overall contribution of lipids to oxidative metabolism

Comprehensive data on the role of circulating lipids in oxidative metabolism in the whole animal are available only for ruminants. Collated data for fed and fasted sheep (*Table 11.8*) show clearly the dominant effect of plane of nutrition on the contribution of individual lipids to total CO_2 production. In particular, the reversal of the importance of acetate and other VFA relative to plasma FFA is in marked contrast to the slight change in the contribution of glucose to oxidative metabolism.

Table 11.8 FRACTIONAL CONTRIBUTION OF MAJOR METABOLITES TO RESPIRATORY CO_2 OF FED NON-PREGNANT AND FASTED (3–4 d) PREGNANT EWES (FROM PETHICK AND LINDSAY, 1982b)

	Fed	Fasted
Acetate	32	5
Propionate, direct	19	–
Propionate, via glucose	4	–
Glucose, direct	5	9
Butyrate, direct	8	–
Butyrate, via ketones	8	–
FFA, direct	2	30
FFA, via ketones	2	30
Total	81	74

References

AHLBORG, G., FELIG, P., HAGENFELDT, L., HENDLER, R. and WAHREN, J. (1974). Substrate turnover during prolonged exercise in man. Splanchnic and leg metabolism of glucose, free fatty acids and amino acids. *Journal of Clinical Investigation* **53**, 1080–1090

ANNISON, E.F. and ARMSTRONG, D.G. (1970). Volatile fatty acid metabolism and energy supply. In *Physiology of Digestion and Metabolism in the Ruminant* (A.T. Phillipson, Ed.), pp. 422–437. Newcastle upon Tyne, Oriel Press

ANNISON, E.F. and LINZELL, J.L. (1964). The oxidation and utilization of glucose and acetate by the mammary gland of the goat in relation to their overall metabolism and to milk formation. *Journal of Physiology* **175**, 372–385

ANNISON, E.F., LINZELL, J.L. and WEST, C.E. (1968). Mammary and whole animal metabolism of glucose and fatty acids in fasting lactating goats. *Journal of Physiology* **197**, 445–459

BAIRD, G.D. (1981). Metabolic modes indicative of carbohydrate status in the dairy cow. *Federation Proceedings* **40**, 2530–2535

BAIRD, G.D., HEITZMAN, R.J., REID, I.M., SYMONDS, H.W. and LOMAX, M.A. (1979). Effects of food deprivation on ketonaemia, ketogenesis and hepatic intermediary metabolism in the non-lactating dairy cow. *Biochemistry Journal* **178**, 35–44

BALLARD, F.J. (1972). Supply and utilisation of acetate in mammals. *American Journal of Clinical Nutrition* **25**, 773–779

BALLARD, F.J., HANSON, R.W. and KRONFELD, D.S. (1969). Gluconeogenesis and lipogenesis in tissue from ruminant and non-ruminant animals. *Federation Proceedings* **28**, 218–231

BALTZAN, M.A., ANDRES, R., CADER, G. and ZIERLER, K.L. (1962). Heterogeneity of forearm metabolism with special reference to the free fatty acids. *Journal of Clinical Investigation* **41**, 116–125

BASSO, L.V. and HAVEL, R.J. (1970). Hepatic metabolism of free fatty acids in normal and diabetic dogs. *Journal of Clinical Investigation* **49**, 537–547

BAUMAN, D.E. and CURRIE, B. (1980). Partitioning of nutrients during pregnancy and lactation: a review of mechanisms involving homeostasis and homeorhesis. *Journal of Dairy Science* **63**, 1514–1529

BAUMAN, D.E. and DAVIS, C.L. (1974). Biosynthesis of milk fat. In *Lactation: A Comprehensive Treatise*, Vol. 2 (B.L. Larson and V.R. Smith, Eds), pp. 31–75. New York, Academic Press

BAUMAN, D.E. and DAVIS, C.L. (1975). Regulation of lipid metabolism. In *Digestion and Metabolism in the Ruminant* (I.W. McDonald and A.C.I. Warner, Eds), pp. 496–509. Armidale, University of New England Publishing Unit

BAUMAN, D.E., BROWN, R.E. and DAVIS, R.L. (1970). Pathways of fatty acid synthesis and reducing equivalent generation in mammary gland of rat, sow and cow. *Archives of Biochemistry and Biophysics* **140**, 237–244

BELL, A.W. (1980). Lipid metabolism in liver and selected tissues and in the whole body of ruminant animals. *Progress in Lipid Research* **18**, 117–164

BELL, A.W. and THOMPSON, G.E. (1979). Free fatty acid oxidation in bovine muscle *in vivo*: effects of cold exposure and feeding. *American Journal of Physiology* **237**, E309–E315

BELL, A.W., HALES, J.R.S., KING, R.B. and FAWCETT, A.A. (1983). Influence of heat stress on exercise-induced changes in regional blood flow in sheep. *Journal of Applied Physiology* (in press)

BELL, A.W., GARDNER, J.W., MANSION, W. and THOMPSON, G.E. (1975). Acute cold exposure and the metabolism of blood glucose, lactate and pyruvate, and plasma amino acids in the hind leg of the fed and fasted young ox. *British Journal of Nutrition* **33**, 207–127

BERGMAN, E.N. (1975). Production and utilization of metabolites by the alimentary tract as measured in portal and hepatic blood. In *Digestion and Metabolism in the Ruminant* (I.W. McDonald and A.C.I. Warner, Eds), pp. 292–305. Armidale, University of New England Publishing Unit

BERGMAN, E.N. and WOLFF, J.E. (1971). Metabolism of volatile fatty acids by liver and portal-drained viscera in sheep. *American Journal of Physiology* **221**, 586–592

BERGMAN, E.N., HAVEL, R.J., WOLFE, B.M. and BØHMER, T. (1971). Quantitative studies of the metabolism of chylomicron triglycerides and cholesterol by liver and extrahepatic tissues of sheep and dogs. *Journal of Clinical Investigation* **50**, 1831–1839

BICKERSTAFFE, R., ANNISON, E.F. and LINZELL, J.L. (1974). The metabolism of glucose, acetate, lipids and amino acids in lactating dairy cows. *Journal of Agricultural Science* **82**, 71–85

BIRD, A.R., CHANDLER, K.D. and BELL, A.W. (1981). Effects of exercise and plane of nutrition on nutrient utilisation by the hind limb of the sheep. *Australian Journal of Biological Science* **34**, 541–550

BUCKLEY, B.M. and WILLIAMSON, D.H. (1977). Origins of blood acetate in the rat. *Biochemical Journal* **166**, 539–545

CHAIYABUTR, N., FAULKNER, A. and PEAKER, M. (1980). The utilisation of glucose for the synthesis of milk components in the fed and starved lactating goat *in vivo*. *Biochemical Journal* **186**, 301–308

CORBETT, J.L., FARRELL, D.J., LENG, R.A., McCLYMONT, G.L. and YOUNG, B.A. (1971). Determination of energy expenditure of penned and grazing sheep from estimates of carbon dioxide entry rate. *British Journal of Nutrition* **26**, 277–291

COSTA, N.D., McINTOSH, G.H. and SNOSWELL, A.M. (1976). Production of endogenous acetate by the liver in lactating ewes. *Australian Journal of Biological Science* **29**, 33–42

CRYER, A. and JONES, H.M. (1979). The distribution of lipoprotein lipase (clearing factor lipase) activity in the adiposal, muscular and lung tissues of ten animal species. *Comparative Biochemistry and Physiology* **63B**, 501–505

DAGENAIS, G.R., TANCREDI, R.G. and ZIERLER, K.L. (1976). Free fatty acid oxidation by forearm muscle at rest, and evidence for an intramuscular lipid pool in the human forearm. *Journal of Clinical Investigation* **58**, 421–431

ELLIOT, J.M. (1980). Propionate metabolism and vitamin B_{12}. In *Digestive*

Physiology and Metabolism in Ruminants (P. Thivend, Ed.), pp. 485–503. Lancaster, MTP

ESSEN, B., HAGENFELDT, L. and KAIJSER, L. (1977). Utilization of blood-borne and intramuscular substrates during continuous and intermittent exercise in man. *Journal of Physiology* **265**, 489–506

FAICHNEY, G.J., BARKER, P.J., SETCHELL, B.P. and LINDSAY, D.B. (1981). Lactate utilization by pregnant ewes. *Quarterly Journal of Experimental Physiology* **66**, 195–201

FELIG, P. (1975). Amino acid metabolism in man. *Annual Review of Biochemistry* **44**, 933–955

FREEMAN, C.P., NOAKES, D.E. and ANNISON, E.F. (1970). The metabolism of glucose, palmitate, stearate and oleate in pigs. *British Journal of Nutrition* **24**, 705–716

FREIDBERG, F.J., KLEIN, R.F., TROUT, D.L., BOGDONOFF, M.D. and ESTES, E.H. Jr (1960). The characterization of the peripheral transport of ^{14}C-labelled palmitic acid. *Journal of Clinical Investigation* **39**, 1511–1515

GUMAA, K.A., GREENBAUM, A.L. and McLEAN, P. (1973). Adaptive changes in satellite systems related to lipogenesis in rat and sheep mammary gland and in adipose tissue. *European Journal of Biochemistry* **34**, 188–198

HAVEL, R.J., KANE, J.P., BALASSE, E.O., SEGEL, N. and BASSO, L.V. (1970). Splanchnic metabolism of free fatty acids and production of triglycerides of very low density lipoproteins in normotriglyceridemic and hypertriglyceridemic humans. *Journal of Clinical Investigation* **49**, 2017–2035

JANSSON, E. (1980). Diet and muscle metabolism in man with reference to fat and carbohydrate utilization and its regulation. *Acta Physiologica Scandinavica* Supplement 487

KAIJSER, L. and ROSSNER, S. (1975). Removal of exogenous triglycerides in human forearm muscle and subcutaneous tissue. *Acta Medica Scandinavica* **197**, 289–294

KARLSSON, N., FELLENIUS, E. and KIESSLING, K.H. (1975). The metabolism of acetate in the perfused hindquarter of the rat. *Acta Physiologica Scandinavica* **93**, 391–400

KATZ, M.L. and BERGMAN, E.N. (1969). Hepatic and portal metabolism of glucose, free fatty acids and ketone bodies in the sheep. *American Journal of Physiology* **216**, 953–960

KNOWLES, S.E., JARRETT, I.G., FILSELL, O.H. and BALLARD, F.J. (1974). Production and utilization of acetate in mammals. *Biochemical Journal* **142**, 401–411

LENG, R.A. (1965). Ketone body metabolism in normal and underfed pregnant sheep and in pregnancy toxemia. *Research in Veterinary Science* **6**, 433–441

LEVEILLE, G.A., ROMSOS, D.R., YEH, Y-Y and O'TLEA, E.K. (1975). Lipid biosynthesis in the chick. A consideration of site of synthesis, influence of diet and possible regulatory mechanisms. *Poultry Science* **54**, 1075–1093

LINDSAY, D.B. (1980). Amino acids as energy sources. *Proceedings of the Nutrition Society* **39**, 53–59

LINDSAY, D.B. (1981). Characteristics of the metabolism of carbohydrate in ruminants compared with other animals. *Current Topics in Veterinary Medicine and Animal Science* **10**, 101–121

LINDSAY, D.B., STEEL, J.W. and BUTTERY, P.J. (1976). The net exchange of amino acids from muscle of fed and starved sheep. *Proceedings of the Nutrition Society* **36**, 33A

LINZELL, J.L. (1960). Mammary gland blood flow and oxygen, glucose and volatile fatty acid uptake in the conscious goat. *Journal of Physiology* **153**, 498–509

LINZELL, J.L. and ANNISON, E.F. (1975). Methods of measuring the utilization of metabolites absorbed from the alimentary tract. In *Digestion and Metabolism in the Ruminant* (I.W. McDonald and A.C.I. Warner, Eds), pp. 306–319. Armidale, University of New England Publishing Unit

LUNDQUIST, F., SESTOFT, L., DAMGAARD, S.E., CLAUSEN, J.P. and TRAP-JENSEN, J. (1973). Utilisation of acetate in the human forearm during exercise after ethanol ingestion. *Journal of Clinical Investigation* **52**, 3231–3235

MCGARRY, J.D. and FOSTER, D.W. (1980). Regulation of hepatic fatty acid oxidation and ketone body production. *Annual Review of Biochemistry* **49**, 395–420

MAYES, P.A. (1976). Control of hepatic triacylglycerol metabolism. *Biochemical Society Transactions* **4**, 575–580

MERCER, J.R., GOODEN, J.M., TELENI, E., HOUGH, G.M., McDOWELL, G.H. and ANNISON, E.F. (1980). Uptake and oxidation of glucose by the hind limb muscle and mammary gland of the lactating ewe. *Proceedings of the Nutrition Society of Australia* **5**, 170

OWEN, O.E. and REICHARD, G.A. (1971). Human forearm metabolism during prolonged starvation. *Journal of Clinical Investigation* **50**, 1536–1545

OWEN, O.E. and REICHARD, G.A. (1975). Ketone body metabolism in normal, obese and diabetic subjects. *Israel Journal of Medical Science* **11**, 560–570

OWEN, O.E., MORGAN, A.O., KEMP, H.B., SULLIVAN, J.M., HERRERA, M.G. and CAHILL, G.F. (1967). Brain metabolism during fasting. *Journal of Clinical Investigation* **46**, 1589–1595

PALMQUIST, D.L. (1972). Palmitic acid as a source of endogenous acetate and β-hydroxybutyrate in fed and fasted ruminants. *Journal of Nutrition* **102**, 1401–1406

PALSSON, H. and VERGES, J.B. (1952). The effects of the plane of nutrition on growth and the development of carcass quality in lambs. Part 1. The effects of high and low planes of nutrition at different ages. *Journal of Agricultural Science* **42**, 1–92

PETHICK, D.W. (1981). *Metabolism of Fatty Acids in Normal and Ketotic Sheep*. PhD thesis, University of Cambridge

PETHICK, D.W. (1982). Fatty acid metabolism of ovine skeletal muscle at rest and during exercise. *Proceedings of the Nutrition Society of Australia* **7**, 210

PETHICK, D.W. and LINDSAY, D.B. (1982a). Acetate metabolism in lactating sheep. *British Journal of Nutrition* **48**, 319–328

PETHICK, D.W. and LINDSAY, D.B. (1982b). Metabolism of ketone bodies in pregnant sheep. *British Journal of Nutrition* **48**, 549–533

PETHICK, D.W., LINDSAY, D.B., BARKER, P.J. and NORTHROP, A.J. (1981). Acetate supply and utilization by the tissues of sheep *in vivo*. *British Journal of Nutrition* **46**, 97–110

PETHICK, D.W., LINDSAY, D.B., BARKER, P.J. and NORTHROP, A.J. (1983). The metabolism of circulating non-esterified fatty acids by the whole animal, hind limb muscle and uterus of pregnant ewes. *British Journal of Nutrition* **49**, 129–143

REID, R.L. (1968). The physiopathology of undernourishment in pregnant sheep, with particular reference to pregnancy toxaemia. *Advances in Veterinary Science* **12**, 163–238

REID, I.M., COLLINS, R.A., BAIRD, G.D., ROBERTS, C.J. and SYMONDS, H.W. (1979). Lipid production rates and the pathogenesis of fatty liver in fasted cows. *Journal of Agricultural Science* **93**, 253–256

ROBERTSON, J.P., FAULKNER, A. and VERNON, R.G. (1982). Regulation of glycolysis and fatty acid synthesis from glucose in sheep adipose tissue. *Biochemical Journal* **206**, 577–586

SKUTCHES, C.L., HOLROYDE, C.P., MYERS, R.N., PAUL, P. and REICHARD, G.A. (1979). Plasma acetate turnover and oxidation. *Journal of Clinical Investigation* **64**, 708–713

SMITH, S.B. and PRIOR, R.L. (1981). Evidence for a functional ATP-citrate lyase : NADP-malate dehydrogenase pathway in bovine adipose tissue : enzyme and metabolite levels. *Archives of Biochemistry and Biophysics* **211**, 192–201

SMITH, G.H. and TAYLOR, D.J. (1977). Mammary energy metabolism. In *Comparative Aspects of Lactation* (M. Peaker, Ed.), pp. 95–109. London, Academic Press

SNOSWELL, A.M., COSTA, N.D., McLEAN, J.G., BAIRD, G.D., LOMAX, M.A. and SYMONDS, H.W. (1978). Interrelationships between acetylation and the disposal of acetyl groups in the livers of dairy cows. *Journal of Dairy Research* **45**, 331–338

SNOSWELL, A.M., TRIMBLE, R.P., FISHLOCK, R.C., STORER, G.B. and TOPPING, D.L. (1982). Metabolic effects of acetate in perfused rat liver. Studies on ketogenesis, glucose output, lactate uptake and lipogenesis. *Biochemica et biophysica acta* **716**, 290–297

STEVENS, C.E. (1970). Fatty acid transport through the rumen epithelium. In *Physiology of Digestion and Metabolism in the Ruminant* (A.T. Phillipson, Ed.), pp. 101–112. Newcastle upon Tyne, Oriel Press

THOMPSON, G.E. and DARLING, K.F. (1975). The hepatic uptake of individual free fatty acids in sheep during noradrenaline infusion. *Research in Veterinary Science* **18**, 325–327

THOMPSON, G.E., BASSETT, J.M. and BELL, A.W. (1978a). The effects of feeding and acute cold exposure on the visual release of volatile fatty acids, estimated hepatic uptake of propionate and release of glucose, and plasma insulin concentration in sheep. *British Journal of Nutrition* **39**, 219–226

THOMPSON, G.E., GARDNER, J.W. and BELL, A.W. (1975). The oxygen consumption, fatty acid and glycerol uptake of the liver in fed and fasted sheep during cold exposure. *Quarterly Journal of Experimental Physiology* **60**, 107–121

THOMPSON, G.E., MANSON, W., CLARKE, P.L. and BELL, A.W. (1978b). Acute cold exposure and the metabolism of glucose and some of its precursors in the liver of the fed and fasted sheep. *Quarterly Journal of Experimental Physiology* **63**, 189–199

VERNON, R.G. (1981). Lipid metabolism in the adipose tissue of ruminant animals. In *Lipid in Ruminant Animals* (W.W. Christie, Ed.), pp. 280–362). London, Pergamon Press

VERNON, V.R. (1970). The role of skeletal muscle and cardiac muscle in the regulation of protein metabolism. In *Mammalian Protein Metabolism* (H.M. Munro, Ed.), pp. 585–674. New York and London, Academic Press

WEBSTER, A.J.F. (1980). Energy costs of digestion and metabolism in the gut. In *Digestive Physiology and Metabolism in Ruminants* (Y. Ruckebusch and P. Thivend, Eds), pp. 469–484. Lancaster, MTP

WEIGAND, E., YOUNG, J.W. and McGILLIARD, A.D. (1972). Extent of butyrate metabolism by bovine rumino-reticulum epithelium and the relationship to absorption rate. *Journal of Dairy Science* **55**, 589–597

ZIERLER, K.L. (1961). Theory of the use of arterio-venous concentration differences for measuring metabolism in steady and non-steady states. *Journal of Clinical Investigation* **40**, 2111–2125

ZIERLER, K.L. (1976). Fatty acids as substrates for heart and skeletal muscle. *Circulation Research* **38**, 459–463

ZIERLER, K.L. and ANDRES, A. (1956). Carbohydrate metabolism in intact skeletal muscle in man during the night. *Journal of Clinical Investigation* **35**, 991–997

12

FATS AS ENERGY-YIELDING COMPOUNDS IN THE RUMINANT DIET

J.W. CZERKAWSKI and J.L. CLAPPERTON
Hannah Research Institute, Ayr, KA6 5HL, Scotland, UK

The fat content of the ruminant diet is relatively low, ranging from 2–4% in concentrate foods without added fat to 5–7% in roughages. Fats are, however, widely used as ingredients in the preparation of concentrate foods especially for milking cows and it has been estimated (Clapperton and Steele, 1983a) that fats comprise about 1.3% of all the ingredients used in the UK compounding industry.

A large proportion of these dietary fats, both those occurring naturally within the food and those added, are triglycerides but there are also varying proportions of glycosyl lipids. Plant material also contains cuticular lipids (mainly waxes and complex lipids) but there is no evidence that these contribute significantly to the nutrition of the animal.

Because the energy content of the fats is more than twice that of the other ingredients and they are usually highly digestible, their contribution to the energy metabolism of the animal is much higher than the proportion of fat in the diet would suggest. However, it is difficult to allot a single value for energy to the fats because there is often an interaction between the added fat and the other components of the diet (Clapperton and Steele, 1983a). For example, many workers have found that adding fat to the diet of sheep reduces the apparent digestibility of the fibre (Brooks *et al.*, 1954; Ward *et al.*, 1957; Nottle and Rook, 1963; Steele and Moore, 1968; Kowalczyk *et al.*, 1977). Conversely, Palmquist and Conrad (1978) found an apparent increase in fibre digestibility when fat was added to the diet of cows. Further, the effect also depends on the type of fat added—short-chain fatty acids cause a greater depression than long-chain ones (Steele and Moore, 1968) and unsaturated oils have more effect than saturated fats (Macleod and Buchanan-Smith, 1972). The latter also found that free fatty acids cause a larger depression than the corresponding triglycerides.

Different fats are known to have different effects on food intake and this also depends on the other components of the food mixture. For example, when mixtures of tallow and barley were fed to dairy cows, total food intake was hardly affected until more than 750 g/d of tallow were given (Clapperton and Steele, 1983b). When a mixture of soyabean oil and soyabean meal was given to similar cows, food intake was markedly reduced (Steele, unpublished results).

Many of these interactions of fat and the remainder of the diet probably occur within the rumen and the rest of this review will deal exclusively with these effects.

Fate of dietary fat in the rumen

Lipids are largely non-polar compounds, insoluble or only sparingly soluble in water. This creates problems with the incorporation of lipids into feeds and with the availability of these lipids during digestive processes in the rumen. Although lipids cannot readily be dissolved in aqueous media, they can be emulsified before inclusion in the diet. It would appear that the conditions in the rumen (relatively high temperature of 39°C, low surface tension, presence of salts and considerable amount of movement) may facilitate further emulsification and wider distribution within the heterogeneous contents. However, unless present as special compounds, added lipids tend to associate with the solid phase of digesta (Harfoot, 1978). For instance, the fat in tallow particles coated with whey at high temperature, could not readily be extracted with hexane, whereas the same extraction procedure could remove more than 90% of fat from a simple mixture of whey and tallow (Czerkawski and Breckenridge, 1978a). Moreover, the whey 'coat' is utilized by rumen micro-organisms, with the tallow having little or no detrimental effect. When a mixture of whey and tallow was used, about 70% of the recovered lipid was in the washed, undigested food residue and the rest in the effluent in the Rumen Simulation Technique (Rusitec). The effluent is a fairly uniform microbial suspension and it is likely that a large proportion of lipid in this part of the system adheres to microbial cells (Harfoot, 1978). When whey-coated tallow was used, the proportions of fat associated with the washed solid were considerably smaller.

It is difficult to study the fate of dietary fat in the rumen without systematic information about the distribution of the fat in the heterogeneous system and about the factors that may bring about changes in such distribution. The mean residence time of the solid digesta in the rumen is considerably longer than the mean residence time of the small particle pool (Poppi *et al.*, 1980). It follows that the uneven distribution of lipid in the rumen will cause lipids at different locations to leave the rumen at different rates. This is important, since the mean residence time of any given lipid in the rumen will determine the availability of this lipid to microbial attack and the effects on the microbial metabolism.

The concepts of compartmentation of microbial population in the rumen (Czerkawski, 1979) may form the basis of a systematic approach to the very complex problem of distribution and fate of lipids in this organ. The definition of the basic three compartments stems directly from the Rumen Simulation Technique (Czerkawski and Breckenridge, 1979), but many of the features observed under controlled conditions *in vitro* have also been identified in the rumen. The three compartments have been defined as follows:

Compartment 1.   Free suspension of rumen micro-organisms (used extensively as strained rumen contents);

Compartment 2. Space occupied by micro-organisms that are loosely associated with the solid digesta (they can be washed out with buffer solution);

Compartment 3. Space occupied by micro-organisms that cannot be washed out.

In Rusitec, compartments 2 and 3 account for 15–20% of the total volume, whereas in the rumen these compartments make a much larger contribution (more than 50% of total volume). There are large differences in the concentrations and types of micro-organisms in the three compartments.

The lipids in compartment 1 consist mainly of microbial lipids, with small amounts of extraneous lipids adsorbed on to the microbial surface and possibly some emulsified lipids. The population of compartment 2 is at least 10 times more concentrated than the population of compartment 1 and this microbial population is very active metabolically. There is little or no information about the distribution of lipids in this compartment. From what has been said already, a great deal of dietary lipid will reside close to compartment 3; it may influence the relative distribution of micro-organisms in this and in the other two compartments and consequently their metabolic activity. In other words, the preferential association of lipids with the solid phase of digesta may be responsible for some of the detrimental effects of lipids that have been observed, not only by physical masking of the solid substrates, but by redistribution of the microbial population. The longer residence time or sequestration of some micro-organisms may be necessary for their survival. An impairment of this process by detergent action or by preventing attachment may have detrimental effects. Studies in this field would throw a great deal of light on many of the conflicting findings in fat-feeding of ruminant animals.

The first stage in the degradation of dietary lipids is the hydrolysis of the glyceride esters. There is a great deal of lipolytic activity in the rumen, the main products being unesterified fatty acids and glycerol, the latter being rapidly fermented by a variety of micro-organisms (Harfoot, 1978). The released long-chain fatty acids could be incorporated into microbial lipid, mainly as complex lipids; they could form calcium soaps (Jenkins and Palmquist, 1982); or they could simply become absorbed on to the solid surface of the digesta and possibly on to the microbial cell wall which often has an outer lipid coat and would be expected to show affinity to the released lipid material. Both the calcium soaps and the surface layers of lipid will tend to reduce the outflow of the added lipid and possibly decrease its availability to the host animal (White et al., 1958).

The rumen has been considered to be a strictly anaerobic system and it has been assumed that there is no oxidation of fatty acids in this organ. It has been suggested (Czerkawski, 1969) and actually demonstrated more recently, that significant amounts of oxygen can enter the rumen by diffusion from the capillary network in the rumen epithelium, and that the microbial population of the rumen can tolerate and actually use oxygen. Perfusion of 1 ℓ of rumen contents in Rusitec with 100 ml O_2/d resulted in its complete uptake and only a small effect on fermentation (Czerkawski and Breckenridge, 1979). Scott et al. (1983) showed that there are considerable concentrations of oxygen in the rumen, that these concentrations decrease significantly during 15–30 min after feeding, and that certain rumen protozoa can utilize oxygen. Clearly, ideas about anaerobiosis in the rumen and, consequently, about the possibility of microbial oxidation of dietary fat, need revision.

Unlike the input and output of many other dietary components, it is difficult to measure those of lipid in the rumen. Very often the results reported indicate that there is a reasonable recovery of added lipid, sometimes approaching 90–100% (Scott, Ulyatt and Kay, 1969). It must be remembered that, if dietary lipids pass through the rumen undegraded, the recovery should be greater than 100%, because there is considerable microbial synthesis of lipids. If a sheep consumes 1 kg food containing about 40 g lipid and the digestibility of DM is 60% in the rumen, this may result in an output of 143 g microbial dry matter or about 16 g microbial lipid. If all the microbial lipid is synthesized *de novo*, the lipid recovery in the rumen should be about 140%. The sad fact is that it is much easier to analyse accurately the lipid in food than in the material that emerges from the rumen or even from a good artificial rumen. This aspect of the subject will show more progress if there is considerable improvement in the techniques of analysis and by the careful use of labelled compounds.

The unesterified long-chain fatty acids are biohydrogenated in the rumen, so that the supply of fat for the host animal is largely saturated, and this is reflected in the predominantly saturated fat in the ruminant tissues. It is known that there is no biohydrogenation without lipolysis, but there is relatively little information on the actual mechanism of biohydrogenation and on the type of micro-organisms responsible (Garton, 1967). The micro-organism studied most extensively is *Vibrio fibrisolvens*; it was shown that it can carry out the last step in the hydrogenation of polyunsaturated fatty acids (monoenoic to stearic acid) and that it can isomerise the *cis*-form to the more stable *trans*-form. It would appear that the saturated long-chain fatty acids are less toxic to micro-organisms than the polyunsaturated ones (see below), but it is not known whether the biohydrogenation of dietary fatty acids is some kind of detoxification mechanism or whether it is a consequence of the prevailing reducing conditions in the rumen. It can be shown that the fermentation of 1 kg OM in the rumen would result in production of 12.8 mol H_2 as gas or as reduced cofactors. A large part of this is used in VFA production (4.6 mol) and large proportions are used to reduce CO_2 to CH_4 (6.4 mol). The rest (1.8 mol) is used in the synthesis of microbial matter, particularly microbial lipid, and in the biohydrogenation of dietary fatty acids. There is no spontaneous biohydrogenation when the micro-organisms are inactivated by heating (Harfoot, 1978) and the amount of metabolic hydrogen used for biohydrogenation forms only a very small proportion of total hydrogen produced (1–2%). It is unlikely, therefore, that biohydrogenation is simply a consequence of the prevailing conditions in the rumen.

The above arguments concerning the distribution and metabolism of lipids by rumen micro-organisms can be illuminated by some of the results of experiments with the Rumen Simulation Technique. *Table 12.1* shows the distribution of radioactivity in the three compartments 24 h after addition of small amounts of labelled stearic acid with food. It can be seen

Table 12.1 DISTRIBUTION OF VOLUME AND RADIOACTIVITY IN RUSITEC VESSELS AND IN THE EFFLUENT 24 h AFTER ADDITION OF 9-10-^{3}H-STEARIC ACID WITH THE LINSEED OIL (LO) SUPPLEMENT

	Volumes (% total including effluent)		Radioactivity (% total recovered)		Relative concentration of radioactivity	
LO g/d	0.5	1.0	0.5	1.0	0.5	1.0
Effluent (1 d collection)	51.0	51.0	5.6	8.4	0.11	0.16
Liquid reaction mixture (compartment 1)	42.0	40.8	6.8	4.4	0.16	0.11
Washings (compartment 2)						
New bag (24 h)	2.8	3.2	40.5	56.8	14.5	17.8
Old bag (48 h)	3.0	3.0	4.6	4.1	1.5	1.4
Washed solid (compartment 3)						
New bag (24 h)	0.8	1.0	37.9	20.8	47.4	20.8
Old bag (48 h)	0.7	0.9	1.7	1.3	2.4	1.4
Removed from surfaces	–	–	2.7	3.4	–	–
In methane	–	–	0.2	0.1	–	–

Table 12.2 DISTRIBUTION OF LIPID BETWEEN THE EFFLUENT AND THE WASHED UNDIGESTED RESIDUE LEAVING THE SYSTEM IN RUSITEC

			Effluent	Residue
			(% total leaving)	
Expt. 1	Control		29	71
	+ Tallow		37	63
	+ Whey		28	72
	+ Tallow + Whey		29	71
	+ Volac 50:50		52	48
Expt. 2	Control		22	78
	+ Tallow-delactosed whey	L_1	31	69
	,,	L_2	50	50
	+ Megalac	L_1	57	43
	,,	L_2	72	28
	Volac 50:50	L_1	51	49
	,,	L_2	70	30
Expt. 3	Hay		28	72
	+ Blood-Tallow		45	55
	Hay/concentrate		48	52
	+ Blood-Tallow		46	54
Expt. 4	Control		33[a]	67
	+ Linseed oil		28[b]	72

[a] 8% in solution, rest in particulate matter.
[b] 6% in solution.

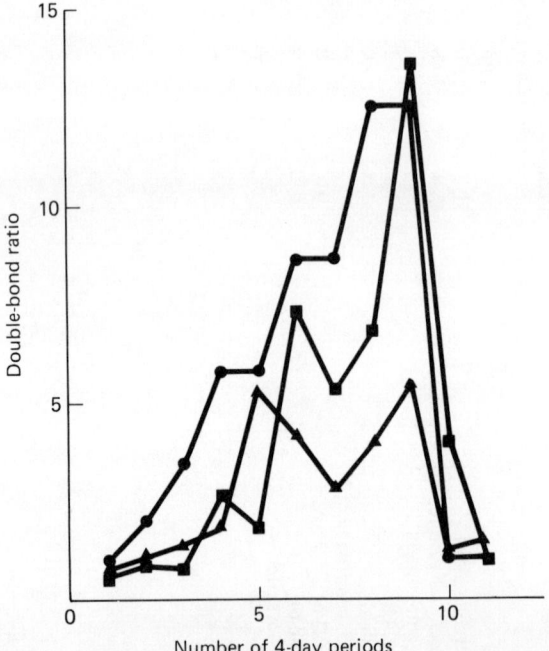

Figure 12.1 The ratio of the number of double bonds in the food (●●), effluent (▲▲) and solid residue (■■) in the test vessel to those in the control vessel when different amounts of linseed oil were used in Rusitec

that the relative concentration of label in the new bag, i.e. the bag to which the label was added, was considerably higher both in compartments 2 and 3, than in the old bag (about 15 times) whereas the concentration in the old bags was more than 12 times higher than in the free liquid. This would suggest, but does not prove, that the transfer of added lipid from the added food to the solid digesta occurs without passage through the liquid phase. *Table 12.2* summarizes the relative amounts of lipid leaving the reaction vessels in the effluent and in the washed residues (the washings being returned to the vessels), during various experiments with added fats. It can be seen that about 2/3 of the lipid in both the controls and after direct addition of fat leaves the reaction vessels in association with the washed solid. However, some tallow preparations, particularly at high concentrations, tended to reverse the distribution in favour of the liquid. *Figure 12.1* shows the ratio of double bonds of long-chain fatty acids in the vessel where increasing amounts of linseed oil were added and in the control vessel. This ratio increased progressively in food and it was usually lower in both the effluent and in the washed undigested residue. With a small addition of linseed oil the difference between the input and output curves was relatively large, indicating an efficient biohydrogenation in both the liquid and solid phases of digesta. With greater amounts of linseed oil the efficiency of biohydrogenation tended to decrease and then to increase again in both phases. The results also suggest that the biohydrogenation of polyunsaturated fatty acids in the free suspension of micro-organisms may be more efficient than in the solid matrix of the digesta. Since the recovery of lipids from incubated samples is rather poor, attempts to calculate the actual extent of biohydrogenation cannot give very meaningful results.

Effect of dietary fat on microbial metabolism

The ratio of microbial surface area to microbial volume is very large and generally increases with decreasing microbial size (see *Table 12.3*). Some protozoa are capable of ingesting food particles, but most of the metabolic exchanges between bacterial cells and the environment must involve the microbial surface, unless, of course, the cell is lysed and the contents are released into the medium. The calculations in *Table 12.3* show that the bacterial surface area is significantly greater than the protozoal surface area. The calculations refer to the usual microbial concentration in the

Table 12.3 TYPICAL MICROBIAL DIMENSIONS IN THE RUMEN

	Protozoa (10^6/ml)	*Bacteria* (10^{10}/ml)
Diameter (μm)	20(10–50)	1.5(0.3–5.0)
Volume		
one cell (μl)	4×10^{-9}	1.7×10^{-12}
in 1 ml (μl)	4	17
Surface area		
one cell (cm^2)	1.26×10^{-5}	7.1×10^{-8}
in 1 ml (cm^2)	13	710
in the rumen of a cow (m^2)	90	5000

strained rumen contents (compartment 1). It is known that the microbial concentrations are much greater in compartments 2 and 3 (possibly by a factor of 10 and 20 respectively, Czerkawski and Breckenridge, 1978b) and that in the rumen, compartments 2 and 3 may account for a significant proportion of the contents. Therefore, the total microbial surface area may be considerably greater than that given in *Table 12.3*.

Any biological system in which metabolic exchanges with the environment involve such a large surface area and in which the cells have a well-developed surface membrane, inevitably involving complex structural lipids, must be influenced by surface-active compounds. Therefore, it is not surprising that lipids, and particularly the polar lipids, such as the unesterified fatty acids, should have an effect on microbial metabolism.

Many years ago, work started on the study of inhibition of methane production in the rumen with a simple premise—polyunsaturated fatty acids are hydrogenated; hydrogen is also used in methanogenesis; therefore, the addition of the polyunsaturated fat should lower the methane production by simple competition for hydrogen. Polyunsaturated fatty acids did indeed lower the methane production (see Blaxter and Czerkawski, 1966) and the effect increased with the number of double bonds in the acid, but the increase per bond was relatively small and the saturated acids (stearic and palmitic) also had a considerable effect. Many of the naturally occurring fatty acids can modify microbial fermentation by depressing some reactions and potentiating others, the effect usually increasing with the chain length of the acid. For reasons that are not understood, two acids in the series of aliphatic acids ranging from C8 to C18, namely lauric and myristic, not only depressed methane production, but had a marked effect on microbial activity in general and resulted in a serious increase in the faecal output of undigested food. Work with sheep (Czerkawski, Blaxter and Wainman, 1966b) showed that long-chain alcohols and paraffins have little or no influence on rumen fermentation and that detergents (e.g. sodium lauryl sulphate) were probably toxic, leading to general depression in microbial activity. Other experiments *in vivo* and *in vitro* (Czerkawski, Blaxter and Wainman, 1966a) showed that linseed oil is as effective as linseed oil fatty acids (because of the lipase activity in the rumen) and that, with care, considerable amounts of linseed oil could be incorporated into the diet of sheep without any ill effects.

The effects of linseed oil on fermentation are best illustrated by some results of experiments with the Rumen Simulation Technique, where it was possible to exert a great deal of control of inputs, outputs and conditions within the reaction vessels. It can be seen in *Table 12.4* that linseed oil depressed the digestion of lipid-free DM. The extent of this depression, like other changes in the Table, increased with the amount of linseed oil added and was about 10% for an incorporation of 1 g lipid/d, i.e. 7% of the ration (w/w) or nearly 16% of the ration on an energy basis. Supplementation with smaller amounts *in vivo* would probably give changes in digestion that are impossible to detect and, of course, these would be confounded by compensatory effects of secondary fermentation in the lower gut.

In agreement with many other reported results, linseed oil depressed the production of acetate and butyrate and increased the production of propionic acid. There was also a disproportionately large decrease in

Table 12.4 EFFECT OF LINSEED OIL ON FERMENTATION OF BASAL DIET OF DRIED GRASS (8 g/d) AND BARLEY (4 g/d) IN RUSITEC

	Control diet (mean for 6 periods of 8 d)	Change per g linseed oil (Actual)	(% of control)
Lipid-free DM digested	7.51 ± 0.01	−0.78	(−10)
Methane (mmol/d)	11.9 ± 0.4	−6.4	(−52)
VFA (mmol/d)	51.2 ± 0.5	−2.8	(−5)
Acetic	30.9	−4.1	(−13)
Propionic	10.8	+1.0	(+9)
Butyric	5.7	−2.4	(−41)
C_5-acids	3.8	+0.1	(+2)
ATP yield	128.9	−16.1	(−12)

Table 12.5 EFFECT OF LINSEED OIL ON PROTEIN AND MICROBIAL OUTPUT IN RUSITEC (mg/d)

	Control diet		Change per g linseed oil	
	Effluent	Residue	Effluent	Residue
Protein (α-NH_2 group)	188	176	+59	+31
Microbial DM				
Bacteria	521	266	+250	+13
Protozoa	130	206	−115	−25
Total		1123		+123

Table 12.6 EFFECT OF LINSEED OIL ON THE HYDROGEN BALANCE IN RUSITEC (mmol/d)

	Control diet	Change per g linseed oil
Hydrogen produced	106.8	−16.5
Hydrogen used		
VFA	48.2	−2.4
CH_4	47.6	−25.6
cells	9.1	+1.0
Total	104.9	−27.0

methane production (52%) and as with other experiments *in vitro* and *in vivo* with fat additives there was no accumulation of gaseous hydrogen. The large decrease in the output of butyric acid was consistent with a decrease in protozoal numbers, particularly in the effluent (*Table 12.5*). This was more than compensated for by the increase in the output of bacterial matter, giving an overall increase in microbial output of some 11% with 1 g linseed oil. It can be shown from the data in *Tables 12.4* and *12.5* that linseed oil increased the efficiency of microbial synthesis (ATP) from 8.7 to 11.0 g/mol.

The recovery of hydrogen in the control experiment was good (98%, *Table 12.6*). Linseed oil decreased the production of hydrogen (mainly because of the depression of acetic and butyric acids), but it also decreased the utilization of hydrogen, giving a discrepancy of some 10–12 mmol/d;

this hydrogen would be used for other purposes not listed in the Table. Linseed oil contains about 7 mmol double bonds/g and as can be seen in *Figure 12.1* the biohydrogenation of the polyunsaturated acids, particularly with larger amounts, was far from complete. It would appear that at most 3–4 mmol of H_2 would be used in biohydrogenation, leaving some 7 mmol still to be accounted for. If this hydrogen is used in the synthesis of palmitic acid (there was a significant net increase in the output of this acid), one would expect an increased synthesis of about 130 mg/d in addition to the increased output of microbial matter (*Table 12.5*), which was based on the measurement of microbial markers (diaminopimelic acid and aminoethylphosphonic acid) and not microbial dry matter.

In earlier experiments *in vivo* (Czerkawski *et al.*, 1975) it was shown that feeding of large amounts of linseed oil to sheep increased the capacity of the bacterial fractions in the rumen to incorporate ^{14}C acetate into lipid 10–15-fold.

The effect of fatty acids on microbial activity has been investigated by a number of authors (e.g. Kodicek and Worden, 1945; Nieman, 1954; Galbraith and Miller, 1973; Henderson, 1973; Maczulak, Dehority and Palmquist, 1981). Many of these studies were made with pure cultures of micro-organisms, adding trace amounts of the acids. Long-chain fatty acids appear to inhibit the growth of certain rumen bacteria, while the growth of others may be stimulated by some acids. For instance, oleic acid stimulated the growth of some strains of *B. fibrisolvens* and *S. ruminantium*, but inhibited the growth of *R. flavefaciens* at similar concentrations (Maczulak *et al.*, 1981). Stimulation of growth of *B. fibrisolvens* was also demonstrated by Henderson (1973). There is some agreement, and also a great deal of disagreement, between various reports on the effect of lipids on microbial activity. However, it would appear that the unsaturated long-chain fatty acids are more active than the saturated ones and that the *cis*-configuration is more potent than the *trans*-configuration produced in the rumen. It would also appear that the Gram-positive bacteria (i.e. many cellulolytic species) may be affected more than the Gram-negative ones. The latter usually have a tough lipid envelope which may offer protection. The effect of long-chain fatty acids is reduced in the presence of fibrous particles and by formation of insoluble calcium soaps (Jenkins and Palmquist, 1982).

The effect of fatty acids on microbial growth depends on the concentration of the acids, often resulting in stimulation of growth at low concentrations and then in depression at higher concentrations. The solubility of lipids is low and the actual proportions of added lipid will be influenced by the concentrations of salts, presence of fibrous food and possibly other surface-active substances and potential emulsifiers. Even if the results of experiments with pure cultures and individual long-chain fatty acids were completely consistent and reproducible, it would be difficult to predict what would happen in various parts of the rumen. We have seen in an earlier section that the distribution of added lipid is uneven and tends to favour the solid. The concentration of the fatty acids in the three compartments identified earlier is not known; nor is it known whether these concentrations are related to the amount of lipid adsorbed to the neighbouring surfaces.

The dietary lipids may have a more passive effect on microbial metabolism. Plant leaves and stems are protected by cuticular lipids and this may be detrimental to microbial colonization of plant tissue. With the entry being gained at the damaged or abraded parts of the plant (Akin, 1979), it is possible that a layer of added dietary fat may seriously impede microbial colonization. Such a layer of lipid may also interfere with the efficient removal of end-products of fermentation—very important in compartments 2 and 3, where the microbial concentrations and activity are much greater than in the free suspension (compartment 1).

Microbial attachment is very important in the rumen, because it brings the micro-organisms close to their substrate and it increases the mean residence time in the rumen, which may actually enable some slow-growing species to survive without being washed out. The concepts of compartmentation are depicted in *Figure 12.2*. Compartment 1 is not very active

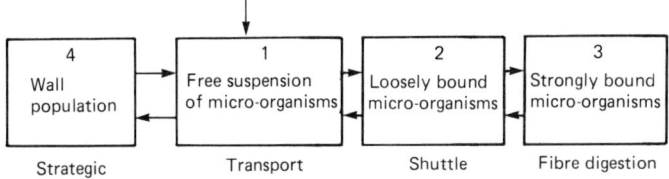

Figure 12.2 Compartmentation in the rumen

metabolically but it is important in communication 'with the outside world'. Food enters through this compartment; it links the two digestive compartments 2 and 3 with the strategic population close to the rumen wall (compartment 4), and it probably represents much of the small-particle pool leaving the rumen. The initial fibre digestion is clearly carried out by the population of compartment 3, but this process would be inefficient without the shuttle compartment 2. This contains a very active microbial population, the function of which is both to process and to remove many of the products of fermentation in compartment 3 and also to supply that compartment with nutrients (Czerkawski and Breckenridge, 1982). It would seem that, in the absence of compartment 2, compartment 3 would soon become stagnant and the digestion of fibre would be impaired. It is conceivable that surface-active agents, such as fatty acids, may diminish the size of compartment 2 by interfering with microbial attachment and it is also conceivable that the same agents may help in the removal of substances from compartment 3 (detergent action), thus compensating for impairment of the shuttle compartment. By increasing the input of linseed oil gradually, it was possible to feed more than 100 g/d in sheep (Czerkawski, 1966). Although there were marked changes in the microbial population in the rumen of these sheep, the animals never refused their food and in fact increased in weight on a seemingly 'maintenance' level of feeding.

Microbial contribution to fat in ruminant diet

Between 5% and 15% of microbial dry mass consists of lipid and, as the microbial mass is about 230 g/kg OM digested, microbial lipid may make a significant contribution to the nutrition of the host animal (Czerkawski,

1976). These microbial lipids are mainly structural and therefore complex and it is not known with certainty what proportion can be digested, absorbed and utilized by the animal.

There is not much information on the mechanisms of synthesis of microbial lipids. It is possible that some preformed units are used by micro-organisms directly (Harfoot, 1978) but this would require an available supply of correct fatty acids in solution and ability of the fatty acids to enter the microbial cell. It has been demonstrated by a number of workers that rumen micro-organisms incorporate ^{14}C acetate into their lipids (Harfoot, 1978); that is, they can synthesize lipids *de novo*. It is usually accepted that lipid synthesis is an energy-expensive process. For instance, in animal tissue the synthesis of 1 mol palmitic acid from 8 mol acetate requires 58 mol ATP; however, 42 of these are used to reduce the oxidized cofactors. The rumen is largely anaerobic, with the prevailing conditions largely reducing. Most of the fermentative processes result in net production of hydrogen either as H_2 gas or in the form of reduced cofactors; thus lipid synthesis in the rumen is relatively cheap. Moreover, the concentrations of acetate in the rumen are considerably higher than in blood and one would not expect any shortage of C2 units. If the acetate used by micro-organisms comes directly from glycolysis via pyruvate, it will be in the form of acetyl-CoA or acetyl-phosphate, again saving energy. Thus the rumen micro-organisms can synthesize their own lipid cheaply and this ensures that the correct lipids are produced. The most plentiful long-chain fatty acid in microbial lipids is palmitic acid, but the microbial lipids also contain considerable proportions of odd-number carbon acids and branched-chain acids (the *iso*- and *ante-iso*-series, with the methyl group at the n-2 or n-3 position). The starting unit in the synthesis of the odd-number carbon acid would be propionic or valeric acid, both ubiquitous end-products of fermentation. Isobutyric and isovaleric acids would give rise to the *iso*-series of long-chain acids and 2-methylbutyric acid, another end-product found in the rumen, would give rise to the *ante-iso*-series.

Methanogenesis is a reductive process which results in generation of energy that can be used by methane-producing organisms for growth. The conversion of carbon dioxide and hydrogen to methane and water is associated with a negative heat of reaction (see *Table 12.7*) as one would expect from an energy-generating process. The conversion of acetic to butyric acid also gives a negative ΔF (as much as 30% acetic acid produced in the rumen is converted to butyric acid, Bergman *et al.*, 1965). It is not known whether the micro-organisms responsible for this reaction can obtain useful energy. *Table 12.7* gives other reductive reactions including

Table 12.7 EXAMPLES OF KNOWN AND POSSIBLE ENERGY-GENERATING REACTIONS (kJ)

Reaction	ΔF	(per mole H_2)
$O_2 + 2H_2 = 2H_2O$	−474	(−237)
$CO_2 + 4H_2 = CH_4 + 2H_2O$	−134	(−33)
$2CO_2 + 4H_2 = CH_3COOH + 2H_2O$	−101	(−25)
$2CH_3COOH + 2H_2 = CH_3CH_2CH_2COOH + 2H_2O$	−62	(−31)
$8CH_3COOH + 14H_2 =$ Palmitic acid $+ 14H_2O$	−302	(−22)

the synthesis of palmitic acid—this too has a negative heat of reaction. It may be postulated that some rumen micro-organisms can synthesize lipids, not for their structural requirements, but to dispose of some of the end-products of fermentation. It is known that some hydrogen-producing micro-organisms can grow more efficiently when the hydrogen tension is kept low by an associated hydrogen utilizer (Bryant *et al.*, 1967; Iannotti *et al.*, 1973). If an organism produces hydrogen and acetic acid in the course of fermentation of glucose and if, as most micro-organisms are, it is equipped with all the enzymes necessary for lipid synthesis, then lipogenesis would be advantageous.

Concluding remarks

It is impossible in this short review to cover comprehensively the subject of fats as energy-yielding compounds in ruminant animals. Ruminant animals consume fats in their food and it is possible to increase their fat intake by supplementing those already in the diet by further addition of fat. However, dietary fats may affect the microbial metabolism in the rumen and the extent of supplementation has to be carefully controlled. It is difficult to recommend any particular level of incorporation of fat in the diet and the reasons for this have been discussed at some length. Fat as a constituent of the ruminant diet is an expensive commodity. The ruminant animal synthesizes a lot of fat from acetate which is primarily produced by micro-organisms in the rumen; this synthesis is an expensive process in animal tissues, but may be less costly in the rumen, and the production of acetate results in the generation of large amounts of hydrogen gas which is used in methanogenesis—a wasteful process as far as the animal is concerned. If some of this hydrogen is rechannelled to lipid synthesis in the rumen, and if the extent of this process can be increased, then the efficiency of feeding could be improved because the animal would be supplied with increasing amounts of fat without adding extra fat in the diet and the loss of energy as methane would be reduced.

References

AKIN, D.E. (1979). Microscopic evaluation of forage digestion by rumen micro-organisms—a review. *Journal of Animal Science* **48**, 701–710

BERGMAN, E.N., REID, R.S., MURRAY, M.G., BROCKWAY, J.M. and WHITELAW, F.G. (1965). Interconversions and production of volatile fatty acids in the sheep rumen. *Biochemical Journal* **97**, 53–58

BLAXTER, K.L. and CZERKAWSKI, J.W. (1966). Modifications of the methane production of the sheep by supplementation of its diet. *Journal of the Science of Food and Agriculture* **17**, 417–421

BROOKS, C.C., GARNER, G.B., GEHRKE, C.W., NUHRER, M.E. and PFANDER, W.H. (1954). The effect of added fat on the digestion of cellulose and protein by ovine rumen micro-organisms. *Journal of Animal Science* **13**, 758–764

BRYANT, M.P., WOLIN, E.A., WOLIN, M.J. and WOLFE, R.S. (1967). *Methanobacillus omelianskii*, a symbiotic association of two species of bacteria. *Archives of Microbiology* **59**, 20–23

CLAPPERTON, J.L. and STEELE, W. (1983a). Fat supplementation in animal production—ruminants. *Proceedings of the Nutrition Society* **42**, 343–350

CLAPPERTON, J.L. and STEELE, W. (1983b). Effects of concentrates with beef tallow on food intake and milk production of cows fed grass silage. *Journal of Dairy Science* **66**, 1032–1038

CZERKAWSKI, J.W. (1966). The effect on digestion in the rumen of a gradual increase in the content of fatty acids in the diet of sheep. *British Journal of Nutrition* **20**, 833–842

CZERKAWSKI, J.W. (1969). The effect of oxygen on fermentation of sucrose by rumen micro-organisms *in vitro*. *British Journal of Nutrition* **23**, 67–80

CZERKAWSKI, J.W. (1976). Chemical composition of microbial matter in the rumen. *Journal of the Science of Food and Agriculture* **27**, 621–632

CZERKAWSKI, J.W. (1979). Compartmentation in the rumen. *Annual Report of the Hannah Research Institute*, pp. 69–85

CZERKAWSKI, J.W. and BRECKENRIDGE, G. (1978a). Effect of tallow and whey powder on fermentation of hay and barley, using the Rumen Simulation Technique (Rusitec). *Proceedings of the Nutrition Society* **37**, 53A

CZERKAWSKI, J.W. and BRECKENRIDGE, G. (1978b). Use of the Rumen Simulation Technique (Rusitec) to study the distribution of microbial matter in the solid and liquid phases of the reaction mixture; sequestration of micro-organisms. *Proceedings of the Nutrition Society* **37**, 70A

CZERKAWSKI, J.W. and BRECKENRIDGE, G. (1979). Experiments with the long-term Rumen Simulation Technique (Rusitec); use of soluble food and an inert solid matrix. *British Journal of Nutrition* **42**, 229–245

CZERKAWSKI, J.W. and BRECKENRIDGE, G. (1982). Distribution and changes in urease (*EC* 3.5.1.5) activity in Rumen Simulation Technique (Rusitec). *British Journal of Nutrition* **47**, 331–348

CZERKAWSKI, J.W., BLAXTER, K.L. and WAINMAN, F.W. (1966a). The effect of linseed oil and of linseed oil fatty acids incorporated in the diet on the metabolism of sheep. *British Journal of Nutrition* **20**, 485–494

CZERKAWSKI, J.W., BLAXTER, K.L. and WAINMAN, F.W. (1966b). The effect of functional groups other than carboxyl on the metabolism of C18 and C12 alkyl compounds by sheep. *British Journal of Nutrition* **29**, 495–508

CZERKAWSKI, J.W., CHRISTIE, W.W., BRECKENRIDGE, G. and HUNTER, M.L. (1975). Changes in the rumen metabolism of sheep given increasing amounts of linseed oil in their diet. *British Journal of Nutrition* **34**, 25–44

GALBRAITH, H. and MILLER, T.B. (1973). Effect of long-chain fatty acids on bacterial respiration and amino acid uptake. *Journal of Applied Bacteriology* **36**, 659–675

GARTON, G.A. (1967). The digestion and absorption of lipids in ruminant animals. *World Review of Nutrition and Dietetics* **7**, 225–250

HARFOOT, C.G. (1978). Lipid metabolism in the rumen. *Progress in Lipid Research* **17**, 21–54

HENDERSON, C. (1973). The effects of fatty acids on pure cultures of rumen bacteria. *Journal of Agricultural Science* **81**, 107–112

IANNOTTI, E.L., KAFKEWITZ, D., WOLIN, M.J. and BRYANT, M.P. (1973). Glucose fermentation products of *Ruminococcus albus* grown in continuous culture with *Vibrio succonogenes*; changes caused by interspecies transfer of hydrogen. *Journal of Bacteriology* **114**, 1231–1240

JENKINS, T.C. and PALMQUIST, D.L. (1982). Effect of added fat and calcium on *in vitro* formation of insoluble soaps and cell wall digestibility. *Journal of Animal Science* **55**, 957–963

KODICEK, E. and WORDEN, A.N. (1945). The effect of unsaturated fatty acids on *Lactobacillus helveticus* and other Gram-positive micro-organisms. *Biochemical Journal* **39**, 78–85

KOWALCYZK, J., ØRSKOV, E.R., ROBINSON, J.J. and STEWART, C.S. (1977). Effect of fat supplementation on voluntary food intake and rumen metabolism in sheep. *British Journal of Nutrition* **37**, 251–257

MacLEOD, G.K. and BUCHANAN-SMITH, J.G. (1972). Digestibility of hydrogenated tallow, saturated fatty acids and soyabean oil-supplemented diets by sheep. *Journal of Animal Science* **35**, 890–895

MACZULAK, A.E., DEHORITY, B.A. and PALMQUIST, D.L. (1981). Effects of long-chain fatty acids on growth of rumen bacteria. *Applied and Environmental Microbiology* **42**, 856–862

NIEMAN, C. (1954). Influence of trace amounts of fatty acids on the growth of micro-organisms. *Bacteriological Reviews* **18**, 147–163

NOTTLE, M.C. and ROOK, J.A.F. (1963). The effect of dietary fat on the production of volatile fatty acids in the rumen of the cow. *Proceedings of the Nutrition Society* **22**, VII

PALMQUIST, D.L. and CONRAD, H.R. (1978). High fat rations for dairy cows. Effects on feed intake, milk and fat production, and plasma metabolites. *Journal of Dairy Science* **61**, 890–901

POPPI, D.P., NORTON, B.W., MINSON, D.J. and HENDRICKSEN, R.E. (1980). The validity of the critical size theory for particles leaving the rumen. *Journal of Agricultural Science* **94**, 275–280

SCOTT, A.M., ULYATT, M.J. and KAY, R.N.B. (1969). Measurement of the flow of long-chain fatty acids into the duodenum of sheep. *Proceedings of the Nutrition Society* **28**, 51A

SCOTT, R.I., YARLETT, N., HILLMAN, K., WILLIAMS, T.N., WILLIAMS, A.G. and LLOYD, D. (1983). The presence of oxygen in rumen liquor and its effects on methanogenesis. *Journal of Applied Bacteriology* **53**, 143–149

STEELE, W. and MOORE, J.H. (1968). The digestibility coefficients of myristic, palmitic and stearic acids in the diet of sheep. *Journal of Dairy Research* **35**, 371–376

WARD, J.K., TEFFT, C.W., SIRNY, R.J., EDWARDS, H.N. and TILLMAN, A.D. (1957). Further studies concerning the effect of alfalfa ash upon the utilization of low-quality roughages by ruminant animals. *Journal of Animal Science* **16**, 633–641

WHITE, T.W., GRAINGER, R.B., BAKER, F.H. and STROUD, J.W. (1958). Effect of supplemental fat on digestion and the ruminal calcium requirement of sheep. *Journal of Animal Science* **17**, 797–803

13

THE EXTRA CALORIC VALUE OF FATS IN POULTRY DIETS

JOHN D. SUMMERS
Department of Animal and Poultry Science, University of Guelph, Guelph, Ontario, Canada N1G 2W1

In the late 1940s fats began appearing on the list of ingredients available to the feed industry. Immediately research took place to investigate their value in animal feeds. Early work showed that the addition of fat to diets was well tolerated by a number of animals (*Table 13.1*) and in many cases, an improvement in the general performance of the animals was noted (Buckner *et al.*, 1947; Pearson and Panzer, 1949; Reiser and Pearson, 1949; Barki *et al.*, 1950; Siedler and Schweigert, 1952). Although the reason for the apparent improved performance of diets containing added fat was not known, Barki *et al.* (1950), working with rats, showed that the response was not due to essential fatty acids while Pearson and Panzer (1949) suggested that the response of the fat diets 'may be due to the

Table 13.1 GROWTH OF RATS ON DIETS WITH AND WITHOUT FAT SUPPLEMENTATION

Diets	4-week gain (g)	Grain per gram of protein consumed (g)
No fat	76.0	1.4
Maize oil	107.8	1.6
Lard	105.2	1.7

Pearson and Panzer (1949) (selected data).

Table 13.2 PERFORMANCE OF BROILERS FED DIETS WITH 3% ADDED ANIMAL GREASE

Animal grease	Age (wk)					
	4		7		10	
	Wt (g)	F/G	Wt (g)	F/G	Wt (g)	F/G
+	454	1.91	1070	2.25	1783	2.63
−	458	1.99	1061	2.44	1769	2.92

Runnels, 1955 (selected data)

Table 13.3 EFFECT OF FATS AND FATTY ACIDS ON BODY WEIGHT, FEED CONVERSION AND FAECES FAT IN 4-WEEK-OLD CHICKS

Diets	Weight (g)	Feed:gain	Fat in dry faeces (%)
Basal	283	2.14	1.18
+ 5% white grease	311	1.92	3.17
+ 10% white grease	289	1.91	5.80
+ 5% oleic acid	292	1.92	3.68
+ 5% hydrogenated fat	309	2.21	9.27

Sunde (1956) selected data.

Table 13.4 CONSUMPTION OF FEED (g) WITH AND WITHOUT 5% SUPPLEMENTAL FAT

Age (days)	Feed intake (g)		
	Fat supplementation		Ratio of non-fat to supplement fat diet
	−	+	
0–7	380	545	1:1.4
0–14	1100	2367	1:2.15
0–21	2315	5380	1:2.3
0–28	4950	9610	1:1.9

Sunde 1956 (selected data).

Table 13.5 WEIGHT OF 27-DAY-OLD BROILERS (g) FED DIETS VARYING IN LEVEL OF DIETARY PROTEIN WITH AND WITHOUT 10% SUPPLEMENTAL TALLOW

Protein level (%)	Tallow	
	+	−
13	190	195
16	266	268
20	307	334
25	314	360
31	292	343

Waibel (1955).

Table 13.6 RATE OF PASSAGE (HOURS:MINUTES) OF DIETS CONTAINING 12% OF VARIOUS SUGARS, FAT AND CELLULOSE IN 5-WEEK-OLD CHICKS

		Supplement			
Nil	Fat	Sucrose	Lactose	Dextrose	Cellulose
2:18	2:33	2:34	2:04	2:32	2:12

Tuckey, March and Biely (1958) (selected data).

Table 13.7 WEIGHT GAIN AND PROTEIN RETENTION OF 3-WEEK-OLD CHICKS WITH EQUALIZED FEED INTAKES OF DIETS VARYING IN LEVEL OF FAT AND PROTEIN

Dietary protein (%)	20	20	30	30
Dietary maize oil (%)	1	8	1	8
Energy intake (MJ of ME)	5356	5356	5484	5484
Weight gain (g)	184	201	202	217
Protein retention (%)	51.6	56.2	38.9	41.7

Rand *et al.* (1958) (selected data).

retention of the feed in the gut for longer periods of time with higher levels of fat which resulted in more complete utilization of the diet'.

As work progressed it was becoming clear that the most consistent response noted with fat additions was an improvement in feed utilization or feed:gain ratios (Siedler and Schweigert, 1953; Yacowitz, 1953; Sunde, 1954a; Runnels, 1955) (*Table 13.2*). It was also becoming apparent that not all types of fat gave similar results. Mixtures of fatty acids and unsaturated fatty acids were well utilized while hydrogenated fat or stearic acid would not improve feed utilization (Carver *et al.*, 1954; Sunde, 1954b; Siedler, Scheid and Schweigert, 1955; Sunde, 1956) (*Table 13.3*). It was suggested that saturated fats were poorly absorbed by the digestive tract and therefore were of little nutritional value. Palatability of diets containing fat may also have been a contributing factor in the better performance of birds receiving supplemental fat because, as demonstrated by Sunde (1956), birds consumed almost twice as much of a fat diet as compared to a non-fat diet if given a choice (*Table 13.4*).

Biely and March (1954) and Waibel (1955) reported that the response to added fat was influenced by the level of dietary protein in the diet and showed that weight gains were more likely to be observed with the addition of fat to diets relatively high in protein (*Table 13.5*).

It was now becoming clear that diet composition was a significant factor in the response to be expected from the addition of fats to poultry diets. It had already been demonstrated that diets containing different carbohydrates varied in the amount of time they took to pass through the digestive tract of birds (*Table 13.6*) and that rate of passage appeared to have an influence on the nutritive value of a diet (Monson, Dietrich and Elvehjem,

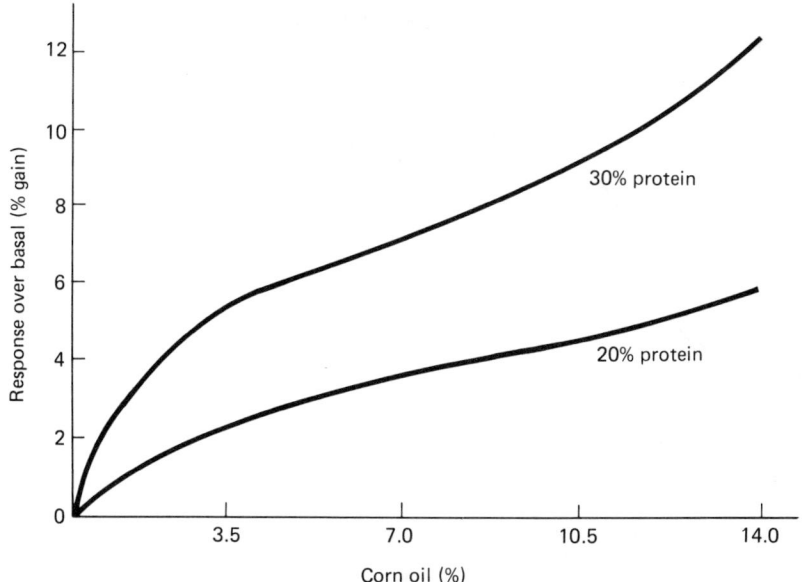

Figure 13.1 Response (% gain) of chicks to maize oil at two levels of dietary protein. Rand *et al.* (1958)

1950; Stokstad, Jukes and Williams, 1953; Tuckey, March and Biely, 1958). Hence, the enhanced nutritive value of diets containing supplemental fat could well be due not only to diet composition but also to physiological factors that influenced either degree or rate of digestion.

In much of the early work with dietary fat supplementation the response to fat had been influenced by differences in the intake of energy, protein and other nutrients. In order to overcome this problem, Rand, Scott and Kummerow (1958) utilized equalized feed intake techniques as well as attempting to compare responses in diets of equal energy and protein level. By such means the above workers were able to demonstrate clearly that the inclusion of maize oil in a diet resulted in improvements in growth (*Figure 13.1*), efficiency of protein and energy utilization and protein retention (*Table 13.7*). Their data indicated that the growth response to dietary fat was more pronounced at higher protein levels and that the best performance occurred when fat contributed between 20% and 38% of the total metabolizable energy content of the diet or between 34% and 48% of the non-protein energy of the diet (*Figure 13.2*). Protein retention improved as the percentage of metabolizable energy originating from corn oil was increased with a maximum response being noted at essentially the same level of 'fat calories' as quoted above for maximizing weight gain. These authors postulated that the major oil effect was due either to: (1) increased efficiency of metabolizable energy utilization; or (2) an improvement in the utilization of dietary protein; or (3) the presence of an unidentified growth factor.

Touchburn and Naber (1966) clearly demonstrated that dietary fat improved the utilization of the metabolizable energy content of turkey

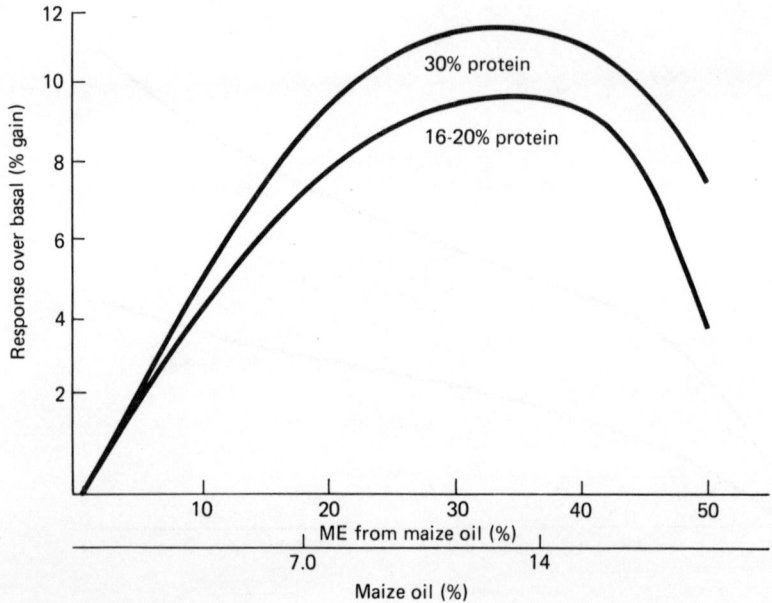

Figure 13.2 Response (% gain) of chicks to maize oil with diets differing in level of protein (combined data of several experiments). Rand *et al.* (1958)

Table 13.8 CALCULATED ENERGY VALUE (MJ OF ME/kg) OF FAT BASED ON FEED:GAIN RATIOS WHEN FED TO MALE TURKEYS[a]

	Age (weeks)		
8–12	12–16	16–20	20–24
ME[b]	ME	ME	ME
41.5	37.5	42.3	48.9

[a] Values shown are averages of three fat and two protein levels.
[b] Overall average ME = 42.5; value used in formulation = 32.3
Jensen, Schumaier and Latshaw (1970) (selected data).

Table 13.9 ABSORBABILITY AND ME OF FATS AND FATTY ACIDS BY POULTS OF DIFFERENT AGES

Fat	Age (wk)	Absorbability (%)			ME (MJ/kg) (determined)
		Fat	Fatty acids		
			16:00	18:1	
Maize oil	2	95.7	90.0	95.4	40.4
	8	97.8	96.2	101.9	45.3
Tallow	2	57.3	27.3	92.7	30.5
	8	73.5	62.5	94.7	33.6
Lard	2	91.4	84.9	97.9	41.3
	8	90.2	86.2	96.2	38.2

Whitehead and Fisher 1975 (selected data).

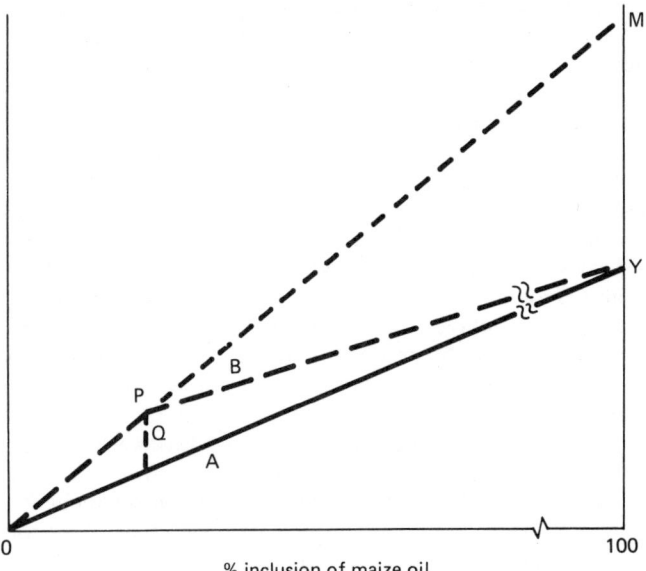

Figure 13.3 Diagrammatic representation of the effect on test diet ME values of the substitution of maize oil into basal diets varying in their degree of fatty acid saturation. A = regression line developed from basal diet containing unsaturated fatty acids; B = regression line developed from basal diet containing saturated fatty acids; Y = ME maize oil

diets and referred to this increased efficiency as the 'extra caloric' effect of supplemental fat. Jensen, Schumaier and Latshaw (1970) confirmed the above work and demonstrated that the calculated metabolizable energy value of fat could markedly exceed its gross energy (*Table 13.8*). Cullen, Rasmussen and Wilder (1962) had previously shown that higher than theoretical values for fat could be obtained in metabolizable energy assays—an effect which has since been demonstrated by a number of workers (e.g. Whitehead and Fisher, 1975; Sell, Tenesaca and Bales, 1979; Mateos and Sell, 1980a).

The metabolizable energy value of a fat is closely related to its absorbability. That fat absorbability in the chick is influenced by the fatty acid composition of the intact fat is well known (Renner and Hill, 1960; Young, 1961). In general, fats high in unsaturated fatty acids show higher absorbability values for the young chick than do fats low in unsaturated fatty acids. Whitehead and Fisher (1975) confirmed that the same was true for turkey poults (*Table 13.9*) and suggested that fat had a synergistic effect on the metabolizable energy value of a diet. Leeson and Summers (1976) attempted to explain how this synergism could work in diets varying in level of saturated and unsaturated fat (*Figure 13.3*).

In an idealized diagram (*Figure 13.3*) line A (OY) represents the effect of substituting corn oil into a basal diet containing only unsaturated fat, i.e. where there is no potential for increased absorption of saturated fatty acids. As the level of corn oil is increased in the test diet there is a linear increase in dietary ME. Extrapolation to 100% corn oil yields an ME for corn oil of Y. If corn oil is substituted into a basal diet containing appreciable quantities of saturated fat, thus providing a potential for synergism, it is proposed that the line B (OPY) would be obtained. With low inclusion levels, increasing the proportion of corn oil produces a disproportionately large increase in ME of the diet due to synergism. This increase would only be evident up to point P, the point at which all saturated fatty acids in the basal diet have been synergistically acted upon. Further increases in corn oil would not increase the synergistic effects but rather decrease it as the increasing proportion of corn oil diluted the basal diet saturated fatty acids. Thus at 100% corn oil inclusion, there are no basal diet saturated fatty acids and thus the ME is derived only from corn oil and the lines A and B meet at Y. The value represented by Q therefore represents the maximum advantage in terms of test diet ME values, to be derived from synergistic effects. In practical situations where basal diets are composed of commonly used feed ingredients, it is evident that some degree of synergism will occur. The problem is that considerably different ME values will be projected depending on the inclusion levels used. If all inclusion levels of oil are less than P, an ME value of M will be extrapolated to 100%. At the other end of the scale, when all inclusion levels are greater than P, an ME value of Y will be projected. Inclusion levels around P will produce ME values in the range of M to Y. These workers suggested that a major portion of the 'extra caloric' values of fat could be explained by the synergistic effect.

Many reports have appeared in the literature, of results demonstrating a synergistic effect with various types of oil mixtures (Young, 1961; Lewis and Payne, 1966; Lall and Slinger, 1973; Sibbald and Kramer, 1978).

Recently Mateos and Sell (1980a) have shown that the 'extra caloric' values of fat show a linear decrease going from 3% to 15% supplemental yellow grease. These findings support the hypothesis of Leeson and Summers (1976).

Kalmback and Potter (1959) reported differences in metabolizable energy for different fat sources depending on the glucose level of the diet. Dal Borgo, Pubols and McGinnis (1967) observed that young chicks had superior growth when fed autoclaved raw soyabean meal diets when starch was used as a carbohydrate source, as compared with glucose or sucrose. Gomez and Polin (1974) noted metabolizable energy values of fat in excess of their gross energy value and suggested that this may be due to enhanced utilization of the non-lipid components of the diet. Horani and Sell (1977) observed differences in the effect of supplemental animal tallow depending on the cereal component of laying hen diets (*Table 13.10*). The above findings suggest that there may be an interrelationship between dietary carbohydrate and fat with regard to energy utilization.

In an attempt to show that the carbohydrate composition of a diet could influence energy utilization, Mateos and Sell (1980b) fed laying hens diets containing either starch or sucrose, in various combinations, as well as a combination of animal fat and soybean oil (*Table 13.11*). Their work

Table 13.10 CALCULATED VERSUS MEASURED CHANGE OF ME/kg (kJ) DUE TO THE ADDITION OF 2 OR 4% ANIMAL FAT

Ration	2% fat			4% fat		
	A (Expected change)	B (Observed change)	B–A (Extra calories)	A (Expected change)	B (Observed change)	B–A (Extra calories)
Maize	314	377	+63	523	774	+251
Oats	439	230	−209	837	1130	+293
Barley	397	167	−230	774	941	+167
Maize + oats	356	460	−104	523	858	+335
Maize + barley	209	251	+42	418	1067	+649

Horani and Sell (1977) (selected data)

Table 13.11 EFFECT OF FAT AND CARBOHYDRATE SOURCE ON ME_n OF DIETS

Fat source	Sucrose:starch ratio	Difference between determined and calculated ME_n (kJ/kg)
6% yellow grease	0:49	+397
	10:39	+310
	30:19	+623
	49:00	+770
		$\bar{x}$ 523
4% yellow grease + 2% soya oil	0:49	+431
	10:39	+414
	30:19	+619
	49:00	+883
		$\bar{x}$ 587

Mateos and Sell (1980b) (selected data).

provided conclusive evidence that the 'extra caloric' effect of fat was attributable to (1) synergism between saturated and unsaturated fatty acids and (2) enhanced energy utilization of the non-lipid portion of the diet. They noted, as did Rao and Clandinin (1970), that the degree of enhanced energy utilization depended on an interaction between fat and the particular carbohydrate in the diet. Mateos and Sell (1980b) concluded as did Rao and Clandinin (1970) that rate of passage through the intestinal tract could be a major contributory factor, in that a slowing-down of passage time would increase the available energy in carbohydrates with a fast passage rate. It is clear, however, that the magnitude of the 'extra caloric' effect would vary considerably depending on the procedure used for evaluating a response (Mateos and Sell, 1981a).

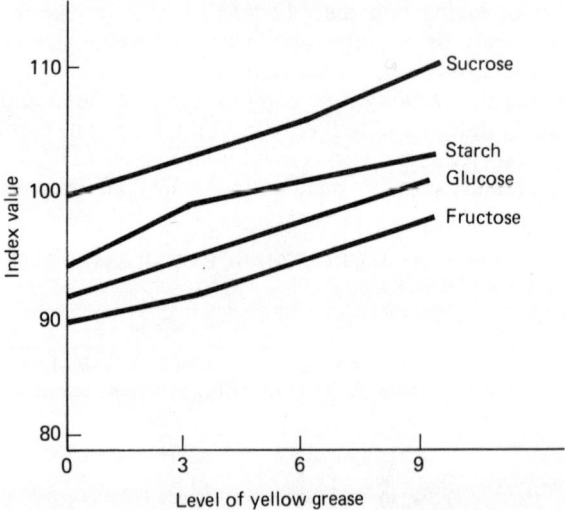

Figure 13.4 Relative ME values of pure carbohydrates at different levels of supplemental grease (assuming an index value of 100 for sucrose in the no-added-fat diet). Selected data from Mateos and Sell (1980c)

Utilizing diets containing 30% of various pure carbohydrates, Mateos and Sell (1980c) studied the effect of supplemental yellow grease (0, 3, 6 and 9%) on the metabolizable energy content of various diets. While there were differences in energy values for the various carbohydrates, all carbohydrates showed an increase in metabolizable energy values (assuming the dietary increase was due to the carbohydrate portion) as dietary fat level increased from 0% to 9% (*Figure 13.4*). While it is true that the above work does not separate out the effects of fat on individual dietary components, it does indicate that various carbohydrates do have different metabolizable energy values and that these may be influenced by level of dietary fat. Vohra (1967) had reported similar results for carbohydrates based on changes in diet composition.

Mateos and Sell (1981b) provided further conclusive evidence that the 'extra caloric' effect of supplemental fat resulted in large part from a reduction in rate of food passage, whereby the overall digestibility of the

Table 13.12 INFLUENCE OF CARBOHYDRATE SOURCE AND FAT SUPPLEMENTATION ON TIME OF FIRST APPEARANCE OF MARKER IN EXCRETA

	Fat (%)	Time after feeding (min)[a]
Starch	0	155
	7	158
Sucrose	0	111
	7	143

[a] Averages of four measurements per hen for five hens.
Selected data Mateos and Sell (1981c).

diet as a whole is increased, as well as a synergistic effect between saturated and unsaturated fats.

Recently, Mateos and Sell (1981c) measured the rate of passage of diets containing 47% of either starch or sucrose in the presence and absence of 7% supplemental yellow grease. They showed (*Table 13.12*) that sucrose had a faster rate of passage (as measured by first appearance of a marker in the faeces) than starch and that yellow grease decreased rate of passage. This confirms and expands on some of the earlier work (Tuckey *et al.*, 1958) in showing that fat not only reduced the rate of passage of the diet as a whole, but also had different effects on different dietary components.

It is obvious that the 'extra caloric' response to dietary fat is a rather complex and variable effect, and there are still many questions to be answered with regard to the effect of diet composition and nutrient level on the response. However, the question that is being asked by the nutritionist formulating poultry diets is: 'How can I programme in this "extra caloric" effect with some degree of confidence?' Sell and Owings (1981), using regression analysis, have shown that for the male turkey, 140-day body weights were increased from 0.6% to 1.5% and feed:gain ratios from 0.8% to 1.9% for each 1% of supplemental dietary fat. Although similar data have been reported by other workers, there is also good evidence to indicate that such values are influenced by such factors as age of bird, environmental temperature and physical form of the ration (Potter, 1976; Waibel, Devegowda and Palarski, 1977; Waibel, 1978; de Albuquerque *et al.*, 1978; Owen and Waldroup, 1979).

From the above it is evident that much work remains to be done before reliable estimates can be given of the enhancement in energy utilization that can be expected when supplemental fat is added to a diet.

In conclusion, one can state, with some degree of confidence, that the 'extra caloric' effect of fat can be explained by two main factors: (1) synergism between saturated and unsaturated fats; and (2) a slowing-down in the rate of passage of food, thus allowing enhanced nutrient absorption. Some of the points yet to be elucidated are:

1. Level of fat required for maximum response;
2. Optimum saturated to unsaturated ratio for optimum response;
3. Effect of age of bird on the response to be expected;
4. Effect on amino acid absorption—are some amino acids affected more than others?

5. More information on the interaction between fats and various carbohydrates in the diet;
6. Can feed utilization be improved even further by the use of other factors, either with or without fat supplementation, to further retard rate of passage?

References

BARKI, V.H., COLLINS, R.A., ELVEHJEM, C.A. and HART, E.B. (1950). The importance of the dietary level of fats on their nutritional evaluation. *Journal of Nutrition* **40**, 383–392

BIELY, J. and MARCH, B. (1954). Fat studies in poultry. 2. Fat supplements in chick and poult rations. *Poultry Science* **33**, 1220–1227

BUCKNER, G.D., INSKO, W.M. Jr, HENRY, A.H. and WACHS, E.F. (1947). Effects of four all-mash rations on laying hens. *Kentucky Agricultural Experimental Station Bulletin* No 495

CARVER, D.S., RICE, E.E., GRAY, R.E. and MONE, P.E. (1954). The utilization of fats of different melting points added to broiler feeds. *Poultry Science* **33**, 1048

CULLEN, M.P., RASSMUSSEN, O.G. and WILDER, O.M.M. (1962). Metabolizable energy value and utilization of different types and grades of fat by the chick. *Poultry Science* **41**, 360–367

DAL BORGO, G.M., PUBOLS, M.H. and McGINNIS, J. (1967). Effect of using sugar or starch in the diet on biological response in the chick to autoclaved hexane-extracted soybean meal. *Poultry Science* **46**, 885–889

DE ALBUQUERQUE, K., LEIGHTON, A.T. Jr, MANSON, J.P. Jr and POTTER, L.M. (1978). The effects of environmental temperature, sex and dietary energy levels on growth performance of large white turkeys. *Poultry Science* **57**, 353–362

GOMEZ, M.X. and POLIN, D. (1974). Influence of cholic acid on the utilization of fats in the growing chicken. *Poultry Science* **53**, 773–781

HORANI, F. and SELL, J.L. (1977). Effect of feed grade animal fat on laying hen performance and on metabolizable energy of rations. *Poultry Science* **56**, 1972–1980

JENSEN, L.S., SCHUMAIER, G.W. and LATSHAW, J.D. (1970). Extracaloric effect of dietary fat for developing turkeys as influenced by calorie:protein ratio. *Poultry Science* **49**, 1697–1704

KALMBACK, M.P. and POTTER, L.M. (1959). Studies in evaluating energy content of feeds for the chick. 3. The comparative values of corn oil and tallow. *Poultry Science* **38**, 1217

LALL, S.P. and SLINGER, S.J. (1973). The metabolizable energy content of rapeseed oil roots and the effect of blending with other fats. *Poultry Science* **52**, 143–151

LEESON, S. and SUMMERS, J.D. (1976). Fat ME values: The effect of fatty acid saturation. *Feedstuffs* **48** (46), 26–28

LEWIS, D. and PAYNE, C.G. (1966). Fats and amino acids in broiler rations. 6. Synergistic relationship in fatty acid utilization. *British Poultry Science* **9**, 13–30

MATEOS, G.G. and SELL, J.L. (1980a). True and apparent metabolizable energy value of fat for laying hens: Influence of level of use. *Poultry Science* **59**, 369–373

MATEOS, G.G. and SELL, J.L. (1980b). Influence of carbohydrate and supplemental fat source on the metabolizable energy of the diet. *Poultry Science* **59**, 2129–2135

MATEOS, G.G. and SELL, J.L. (1980c). Influence of graded levels of fat on utilization of pure carbohydrate by the laying hen. *Journal of Nutrition* **110**, 1894–1903

MATEOS, G.G. and SELL, J.L. (1981a). Metabolizable energy of supplemental fat as related to dietary fat level and methods of estimation. *Poultry Science* **60**, 1509–1515

MATEOS, G.G. and SELL, J.L. (1981b). Nature of the extrametabolic effect of supplemental fat used in semi-purified diets for laying hens. *Poultry Science* **60**, 1925–1930

MATEOS, G.G. and SELL, J.L. (1981c). Influence of fat and carbohydrate source on rate of food passage of semi-purified diets for laying hens. *Poultry Science* **60**, 2114–2119

MONSON, W.J., DIETRICH, L.S. and ELVEHJEM, C.A. (1950). Studies on the effect of different carbohydrates on chick growth. *Proceedings of the Society for Experimental Biology and Medicine* **75**, 256–259

OWEN, J.A. and WALDROUP, P.W. (1979). Response of growing turkeys to diets varying in energy content. *Poultry Science* **58**, 1090–1091

PEARSON, P.B. and PANZER, F. (1949). The effect of fat in the diet of rats on their growth and their excretion of amino acids. *Journal of Nutrition* **38**, 257–265

POTTER, L.M. (1976). Fat in turkey diets: how much is it worth? *Turkey World* **51**, 6–8

RAND, N.T., SCOTT, H.M. and KUMMEROW, F.A. (1958). Dietary fat in the nutrition of the growing chick. *Poultry Science* **37**, 1075–1085

RAO, P.V. and CLANDININ, D.R. (1970). Effect of method of determination on the metabolizable energy value of rapeseed meal. *Poultry Science* **49**, 1069–1074

REISER, R. and PEARSON, P.B. (1949). The influence of high levels of fat with suboptimal levels of riboflavin on the growth of chicks. *Journal of Nutrition* **38**, 247–256

RENNER, R. and HILL, F.W. (1960). The utilization of corn oil, lard and tallow by chickens of various ages. *Poultry Science* **39**, 849–854

RUNNELS, T.D. (1955). Animal fat in combination with various other ingredients in broiler rations. *Poultry Science* **34**, 140–144

SELL, J.L. and OWINGS, W.J. (1981). Supplemental fat and metabolizable energy-to-nutrient ratios for growing turkeys. *Poultry Science* **60**, 2293–2305

SELL, J.L., TENESACA, L.G. and BALES, G.L. (1979). Influence of dietary fat on energy utilization by laying hens. *Poultry Science* **58**, 900–905

SIBBALD, I.R. and KRAMER, J.K.G. (1978). The effect of the basal diet on the true metabolizable energy of fat. *Poultry Science* **57**, 685–691

SIEDLER, A.J. and SCHWEIGERT, B.S. (1952). The effect of the level of fat in the diet on the performance of dogs. *Journal of Nutrition* **48**, 81–90

SIEDLER, A.J. and SCHWEIGERT, B.S. (1953). Effect of feeding graded levels of fat with and without choline and antibiotic + B_{12} supplements to chicks. *Poultry Science* 32, 449–454

SIEDLER, A.J., SCHEID, H.E. and SCHWEIGERT, B.S. (1955). Effects of different grades of animals fats on the performance of chicks. *Poultry Science* 34, 441–414

STOCKSTAD, E.L.R., JUKES, T.H. and WILLIAMS, W.L. (1953). Growth-promoting effect of aureomycin on various types of diets. *Poultry Science* 32, 1054–1058

SUNDE, M.L. (1954a). Use of animal fats in poultry feeds. *Journal of the American Oil Chemists Society* 31, 49–52

SUNDE, M.L. (1954b). The effects of fats and fatty acids on feed conversion in chicks. *Poultry Science* 33, 1084

SUNDE, M.L. (1956). The effect of fats and fatty acids in chick rations. *Poultry Science* 35, 362–368

TOUCHBURN, S.P. and NABER, E.C. (1966). The energy value of fats for growing turkeys. In *Proceedings, 13th World Poultry Congress, Kiev, Russia*, pp. 190–195

TUCKEY, R., MARCH, B.E. and BIELY, J. (1958). Diet and the rate of food passage in the growing chick. *Poultry Science* 37, 786–792

VOHRA, P. (1967). Requirements of poultry for carbohydrates. *World's Poultry Science Journal* 23, 20–31

WAIBEL, P.E. (1955). Effect of dietary protein level and added tallow on growth and carcass composition of chicks. *Poultry Science* 34, 1226

WAIBEL, P.E. (1978). Studies on protein and energy requirements of turkeys during the growing period. In *Proceedings, 39th Minnesota Nutrition Conference, University of Minnesota, St. Paul, MN.*, pp. 143–154

WAIBEL, P.E., DEVEGOWDA, G. and PALARSKI, J. (1977). Estimation of the value of animal fats in diets for turkeys. In *Proceedings, 38th Minnesota Nutrition Conference, University of Minnesota, St. Paul, MN.*, pp. 33–46

WHITEHEAD, G.C. and FISHER, C. (1975). The utilization of fats and fatty acids for chicks. 1. Metabolizable energy. *British Poultry Science* 16, 481–485

YACOWITZ, H. (1953). Supplementation of corn–soybean oil meal rations with penicillin and various fats. *Poultry Science* 32, 930

YOUNG, R.J. (1961). The energy value of fats and fatty acids for chicks. *Poultry Science* 40, 1225–1233

14

ASSESSMENT OF THE DIGESTIBLE AND METABOLIZABLE ENERGY OF FATS FOR NON-RUMINANTS

JULIAN WISEMAN
University of Nottingham School of Agriculture, Sutton Bonington, Loughborough, Leics. LE12 5RD, UK

The basis for feed evaluation in this country is digestible energy (DE) in pigs and metabolizable energy (ME) in poultry. DE is measured as the gross energy minus the corresponding faecal loss, whereas metabolic losses are also accounted for in the determination of ME. Recently, true metabolizable energy (TME) in which allowances are made for digestible and metabolic losses not directly of dietary origin (in contrast to apparent metabolizable energy—AME—which does not account for such losses) has been suggested for poultry. It is essential that reliable data for the dietary energy value of feedstuffs are available for accurate diet formulation. The use of fats in animal nutrition has increased rapidly in recent years, particularly because they are an extremely concentrated source of dietary energy. A considerable amount of attention has been focused on the generation of dietary energy values for fats, not least because they may be influenced by numerous factors ranging from chemical structure of the fat through to the methodology employed in their assessment. This chapter considers the variables associated with the evaluation of fats.

Methodology

Three main methods are available for the determination of the dietary energy value of fats, all of which involve experiments incorporating basal diets containing no added fat, which are then referred to those diets to which a specific quantity of fat has been added. Such an approach is necessary because fats, by virtue of their physical nature, cannot be evaluated independently of any basal component.

The first method is direct energy balance, where the gross energy of a diet is related to the gross energy of the corresponding faecal output. Indirectly, dietary energy values may be determined by measurement of fat digestibility or absorbability, or from knowledge of animal or bird performance arising from feeding diets on the assumption that each unit of liveweight gain requires a specific dietary energy input. An alternative version of this latter technique relates performance obtained with diets containing test fats to that achieved with a diet containing soyabean oil.

Fat digestibility as a means of determining the dietary energy value of a fat (i.e. fat gross energy multiplied by its digestibility) has been widely used. One major advantage of this method when compared with direct calorimetry is that no allowance need be made in respect of any possible interaction between added fat and the basal diet in the utilization of dietary energy. This may explain why comparative studies between the two techniques tend to produce poultry AME data that are lower if determined by lipid digestibility (Whitehead and Fisher, 1975; Lessire and Leclercq, 1982—*Table 14.1*). In addition, Mateos and Sell (1981a) illustrated a consistent superiority in AME values derived from direct calorimetry, although the difference was smaller at higher inclusion levels of fat. Conversely, Renner and Hill (1960) observed the opposite trend, although the differences were small and they concluded that the two methods were, in fact, closely related.

Table 14.1 COMPARISONS OF AME VALUES OBTAINED BY DIRECT CALORIMETRY[1] OR BY LIPID DIGESTIBILITY[2]

Reference	Species	Fat	AME (MJ/kg)	
			(1)	(2)
Whitehead and Fisher (1975)	Turkey poults ($\bar{x}$ of 3 ages)	Maize oil	42.8	38.3
		Tallow	32.1	25.8
		Lard	40.2	36.3
Lessire and Leclercq (1982)	Chickens ($\bar{x}$ of 3 ages)	Beef tallow	31.2	29.3[a] 30.0[b]
		Animal fat	35.6	31.5[a] 32.3[b]
		Poultry fat	38.3	33.5[a] 34.3[b]

[a] Determined with reference to starved birds
[b] Determined with reference to birds on lipid-free diet

A closer investigation of the method based upon lipid digestibility, however, reveals a number of possible disadvantages associated with its use. Carlson and Bayley (1972) considered that it confounded three processes, namely the emulsification and absorption of fatty acids in the small intestine, secretion of endogenous fat and modifications to fatty acids in the large intestine. Examples of the latter point include hydrogenation of fatty acids (e.g. Bayley and Lewis, 1965; Carlson and Bayley, 1972; Just, Anderson and Jorgensen, 1980) and explains why faecal analysis is of only limited application in the study of the digestion and absorption of individual fatty acids.

Quantifying secretion of fat by reference to the faecal fat production after feeding a fat-free diet should allow a correction to true fat digestibility. However, there is evidence (Freeman, Holme and Annison, 1968) that endogenous fat secretion is diet dependent, thus casting doubt on the validity and accuracy of any correction factor. Two means of correction, where estimates of endogenous losses may be achieved either by reference to birds fed a lipid-free diet, or to starved birds, in fact produced different poultry dietary energy values depending upon which method was used (Lessire and Leclercq, 1982—*Table 14.1*).

Young and Artman (1961) considered the performance measured, in terms of weight gain, of birds fed diets containing fats. It was accepted that results obtained could be confounded by the inclusion of fats of varying nutritive value altering the dietary energy:protein ratios. If growth performance were not to be limited by dietary protein content, then a level of 28% ought to be used. Additionally, it was appreciated that responses may be modified by variations in food intake, as a consequence of compensatory mechanisms. Thus it was suggested that food restriction would be appropriate, and a level of 90% of the mean *ad libitum* intake was suggested. Young and Artman (1961) generated 'soybean oil equivalents' (SBE) for a number of fats which were subsequently used to derive 'utilizable energy' (UE). The SBE of a fat was defined as the growth response obtained with a diet containing the test fat related to that with one with added soybean oil. The UE was determined from the SBE, using an assumed figure for the AME of soybean oil (9.07 cal/g or 37.9 MJ/kg).

There was good agreement between UE thus measured, and determinations of dietary energy from the product of gross energy and absorbability (*Table 14.2*). Artman (1964) reported similar results, although he noted that data based upon performance were higher than those based upon fat digestibility (*Table 14.3*). Performance as a criterion for assessing the dietary energy values of fats for poultry has also been suggested by Jensen, Shumaier and Latshaw (1970) who calculated AME on the assumption that each unit of liveweight gain requires the same amount of dietary energy input. In a comparative study of the determination of the AME of fats, Halloran and Sibbald (1979) concluded that data derived using the method of Jensen *et al.* (1970) tended to be higher, but not significantly so, than figures obtained by direct calorimetry (*Table 14.4*) although AME data were corrected on the basis of nitrogen retention.

Table 14.2 COMPARATIVE DATA FOR UTILIZABLE ENERGY (UE, DETERMINED WITH REFERENCE TO SOYABEAN OIL) AND AME (DETERMINED AS GROSS ENERGY × % ABSORBABILITY AT 16% ADDED FAT)

Fat	UE		AME	
	Cal/g	MJ/kg	Cal/g	MJ/kg
Soya bean oil	9.07	37.9	9.07	37.9
Animal fat	7.98	33.4	8.08	33.8
Lard	9.25	38.7	8.65	36.2
Beef tallow	7.62	31.9	7.33	30.7

From Young and Artman (1961)

Table 14.3 COMPARISON OF NUTRITIVE VALUE OF FATS AS DETERMINED BY DIRECT CALORIMETRY[1], FAT DIGESTIBILITY[2] AND FOOD CONVERSION RATIO[3]

	AME					
	(1)		(2)		(3)	
	Cal/g	MJ/kg	Cal/g	MJ/kg	Cal/g	MJ/kg
Soybean oil[a]	9.07	37.9	9.07	37.9	9.07	37.9
Menhaden oil	9.25	38.7	9.12	38.2	7.74	32.4
Tallow	7.26	30.4	6.53	27.3	8.44	35.3

[a] Assumed value
From Artman, 1964

Table 14.4 COMPARISON BETWEEN APPARENT METABOLIZABLE ENERGY (AME) DETERMINED DIRECTLY[1] BY CALORIMETRY AND INDIRECTLY[2] BY THE METHOD OF JENSEN ET AL., 1970

Fat	Species	AME			
		(1)		(2)	
		Cal/g	MJ/kg	Cal/g	MJ/kg
A	Poults	6.44	26.9	6.56	27.4
B	Poults	7.36	30.8	7.30	30.5
C	Chicks	7.82	32.7	7.92	33.1
D	Chicks	8.11	33.9	8.31	34.8
E	Chicks	8.45	35.4	9.09	38.0
F	Chicks	8.26	34.6	8.60	36.0

From Halloran and Sibbald (1979)

However, it could be argued that although data derived from performance are relevant in the assessment of the nutritive value of fats, one major criticism of their use is that they in fact measure net energy (NE) or productive energy (PE) and thus it is conceptually unsound to relate them to ME. Additionally, inherent in the assumption integral to the method of Jensen *et al.* (1970) is that the efficiency of utilization of dietary energy for liveweight gain is constant irrespective of diet composition. De Groote (1974) in a review of the subject indicated that the ratio NE/ME was influenced by type of diet, with those of higher fat content being utilized more efficiently. This explains the observations of Artman (1964) referred to above and also why a system based upon NE has been considered more appropriate in the evaluation of feedstuffs, particularly with the current trends to increased levels of dietary fat.

Animal/bird factors

Two principal animal/bird factors have been identified in influencing determined dietary energy values or absorbability—age, and species or breed.

A number of studies have indicated an age effect on the absorbability of fats and fatty acids in poultry (*Table 14.5*). Renner and Hill (1961b) suggested that the ability of the hen to absorb fatty acids was between 4% and 11% higher than in the 4-week-old chick. Young (1961), however, found no large differences between the 4-week and 8-week-old chick in the absorption of a number of fats or fatty acid mixtures, a trend confirmed by Lessire and Leclercq (1982) in their assessment of three fats with 2-week and 6-week-old chicks. Halloran and Sibbald (1979) had indicated that the effect of age (from 3½ weeks to 6½ weeks) on subsequent dietary energy values was dependent upon fat type and it should also be noted from *Table 14.5* that level of inclusion may also be important.

Carew *et al.* (1972) concluded that young chicks do not possess the full physiological capacity for fat absorption, but that this develops as the birds age. In a comparison of two sources of fat, they also indicated that this development was not at the same rate for all fats. Thus the apparent absorbability of a corn-oil-based diet was 80.2% between 2 and 7 days of age, and 94.3% between 8 and 15 days of age. Corresponding figures for a tallow-based diet were 35% and 77.8% respectively. These observations

Table 14.5 INFLUENCE OF AGE ON ABSORBABILITY AND AME VALUES OF VARIOUS DIETS CONTAINING FAT, OR OF PURE FATS

Reference	Species	Diet, fat or fatty acid	Age	Absorbability (%)	Age	Absorbability (%)	Age	ME (MJ/kg)	Age	ME (MJ/kg)
Fedde et al. (1960)	Chickens	Basal + 10% corn oil Basal + 10% beef tallow Basal + 20% corn oil Basal + 20% beef tallow	7–14 d	85.0 59.1 90.2 46.6	49–56d	95.7 74.3 92.4 78.6				
Young (1961)	Chickens	Corn oil (20% of diet) Beef tallow (10% of diet)	21–28d	91.4 71.7	49–56d	94.7 73.8	21–28d	35.9 28.1	49–56d	36.1 29.0
Carew et al. (1972)	Chickens	Basal + 20% corn oil Basal + 20% beef tallow	2–7d	80.2 35.0	8–15d	94.3 77.8				
Halloran and Sibbald (1979)	Chickens	Animal fat A (12.9% of diet) Animal fat B (12.9% of diet) Animal fat C (12.9% of diet)					22–24d	27.3 31.3 31.5	43–45d	29.3 31.0 33.8
Lessire and Leclercq (1982)	Chickens	Beef tallow (7% of diet) Animal fat (7% of diet) Poultry fat (7% of diet)					13–17d	29.4 35.4 38.0	40–44d	30.6 35.7 37.4
Janssen (1983)	Chickens	Renders fat (12.5% of diet) Vierhouten fat (12.5% of diet)					14–21d	22.5 17.1	35–42d	32.6 28.5
Whitehead and Fisher (1975)	Turkeys	Maize oil (10% of diet) Tallow (10% of diet) Lard (10% of diet)	11–14d	95.7 57.3 91.4	25–28d	97.0 70.0 92.1	53–56d	97.8 73.5 90.2	11–14d	40.4 30.5 41.3
									25–28d	42.6 32.1 40.0
									53–56d	45.3 33.6 38.2
Eusebio et al. (1965)	Pigs	Basal + 10% soyabean oil Basal + 10% coconut oil Basal + 5% soyabean oil Basal + 5% coconut oil	22d	65.8 77.6 55.9 62.3	43d	81.4 88.6 79.2 80.5	64d	88.9 89.9 83.3 83.6	Adult	37.7 35.1

had confirmed those of Fedde, Waibel and Burger (1960) who illustrated an improvement in absorbability with age, the rate of which was markedly influenced by fat type. In general, fats high in polyunsaturated fatty acids (e.g. corn oil, safflower oil) were well absorbed even as early as 2 weeks of age, whereas a tallow-based diet was poorly absorbed at this age, but well absorbed at 8 weeks. Restricted feeding of a 20% beef tallow diet did not improve fat absorbability at either 2 or 4 weeks of age, although levels of fat intake even in these diets would still have been high. Certainly, the absorbability of the beef-tallow-based diets was dramatically improved when the inclusion of fat was reduced from 20% to 10% suggesting that quantity of fat *per se* may be a limiting factor in its utilization in the young chicken.

In general, therefore, it would seem that the young chicken is capable of digesting and absorbing certain fat types (those high in polyunsaturated fatty acids of plant origin) but is unable to utilize adequately those with high levels of saturated fatty acids. This capacity increases gradually with age, reaching an optimum after approximately 6 weeks of age, although Janssen (1982) indicated a considerable improvement in determined AME valus beyond this age. Such a situation applies equally to the turkey poult. Thus Whitehead and Fisher (1975) determined the absorbability of maize oil, tallow and lard (together with their constituent fatty acids). Data for maize oil and lard indicated no age effect, whereas those for tallow increased progressively from 57%, 70% to 74% at 2, 4 and 8 weeks of age respectively.

The influence of age on the utilization of fats has also been investigated in pigs. Eusebio *et al.* (1965) concluded that the young pig is unable to utilize diets containing tallow and, in contrast to the chick, soyabean oil. Diets based upon coconut oil (where average chain length of constituent fatty acids is somewhat shorter than with the other two fats, as it contains a preponderance of lauric acid) were, however, better digested. Availability of all three diets improved significantly as age increased from 3 to 9½ weeks.

Estimates of absorbability and dietary energy values as influenced by breed or species are infrequent. Although Whitehead and Fisher (1975) had indicated a fairly similar degree of utilization, Lall and Slinger (1973) in their measurements of the AME of fats in the young chick and turkey poult illustrated a definite superiority in values obtained with the latter (*Table 14.6*).

Table 14.6 DIFFERENCES BETWEEN THE BROILER CHICK AND TURKEY POULT IN UTILIZATION OF FATS AS MEASURED BY AME

	AME (MJ/kg)	
	Chick 3–4 wk of age	*Poult* 2–3 wk of age
Rapeseed oil	33.4	40.0
Animal/vegetable blend	32.9	36.1
Tallow	29.8	35.3

From Lall and Slinger (1973)

It would appear, therefore, that an important problem in the generation of dietary energy values for fats is the choice of age of the experimental animal or bird. The use of values derived from stock of one particular age should not be applied to those of another. In addition, because of differential rates of improvement of fat utilization with age, the trend obtained with one fat may not apply to another. Finally, values obtained with one breed or species cannot reliably be used for others.

Fat structure

It is generally accepted that chemical structure of a fat has a considerable influence upon its subsequent biological availability. In particular, chain length, degree of saturation, position on the glyceride molecule and the proportions of free fatty acids, are all important determinants of subsequent digestibility and absorption and therefore of dietary energy values.

Thus, Renner and Hill (1961b), confirming the observations of Lloyd and Crampton (1957), illustrated a progressive reduction in both absorbability and subsequent poultry AME values of fats with increasing chain length (*Table 14.7*). The influence of degree of saturation was also

Table 14.7 INFLUENCE OF CHAIN LENGTH ON ABSORBABILITY AND AME OF FATTY ACIDS IN YOUNG CHICKS

Fatty acid	Absorbability (%)	AME (MJ/kg)
Lauric	65	24.4
Myristic	25	9.0
Palmitic	2	2.6
Stearic	−2	−1.5

From Renner and Hill (1961b)

investigated and the absorbability and AME value obtained for stearic acid in the young chick was −0.2% and −1.5 MJ/kg respectively compared with values with oleic acid of 88% and 34.8 MJ/kg respectively—a trend that has also been demonstrated with pigs (Bayley and Lewis, 1965). Similar results have been obtained for entire fats (March and Biely, 1957) when hydrogenation of an animal fat resulted in a lowered subsequent absorbability. The DE of a fully hydrogenated tallow fed to growing pigs was only 2.5 MJ/kg (Tullis and Whittemore, 1980) attributable, it was suggested, to high levels of stearic acid.

Position on the glyceride molecule has also been considered by Renner and Hill (1961a) and Sibbald and Kramer (1977) who suggested that the reason for the relatively high degree of utilization of palmitic acid in lard was that this normally poorly utilized fatty acid was present at the 2 position in this particular fat. Renner and Hill (1961a) also illustrated a reduction in absorbability of tallow, lard and soybean oil when these three fats were hydrolysed and the constituent free fatty acids fed. These observations agreed with those of Young (1961) who had worked with soybean oil, maize oil, lard and tallow (*Table 14.8*). However, it was also concluded in the latter study that the decrease in utilization of fats

Table 14.8 INFLUENCE OF HYDROLYSIS OF FATS ON SUBSEQUENT AME VALUES OBTAINED WITH CHICKS

Fat	AME (MJ/kg)
Soyabean oil	38.7
Soyabean fatty acids	34.6
Corn oil	35.3
Corn oil fatty acids	36.6
Lard	38.5
Lard fatty acids	35.1
Tallow	27.4
Tallow fatty acids	23.9

From Young (1961)

following hydrolysis tended to be greater in animal fats when compared with those fats of vegetable origin—presumably a reflection of the higher proportion of the less-well-utilized saturated fatty acids in the former. Shannon (1971) observed a gradual reduction in the AME of tallow determined with hens, from 34.3 MJ/kg to 30.0 MJ/kg, when the free fatty acid content was raised from 20% to 97%. Conversely Bayley and Lewis (1965) working with semi-purified fat sources. indicated that there was no conclusive evidence to support such a trend in pigs.

Interactions involving fats

It should be remembered that as feed grade fats generally are a mixture of fatty acids of varying chain length and degree of saturation, and tend to be randomly distributed on the glyceride molecule, then large differences in absorbability and subsequent dietary energy values of specific fatty acids outlined previously may not be of much practical significance. This is particularly relevant, additionally, in view of numerous observations which have illustrated a considerable degree of interaction between fats of differing chemical structure when they are fed together. The situation is complicated further by possible interactions between fats and the non-fat component of the diet.

Thus, these two possible sources of interaction involving fats may be of importance during determinations of their dietary energy content. It has been established that the addition of a fat containing a high proportion of unsaturated fatty acids to one with a preponderance of saturated ones will enhance the absorption and subsequent dietary energy value of the latter. Such a response has been demonstrated with individual fatty acids (Renner and Hill, 1961a; Young, 1961; Young and Garrett, 1963) where, generally, the addition of oleic and linoleic acids improved the utilization of palmitic and stearic acids although Young and Garrett (1963) indicated (*Table 14.9*) that the improvement following addition of linoleic acid was less than that obtained with oleic acid.

Similar observations have been observed with complete fats (*Table 14.10*) where the determined dietary energy value of a mixture of a predominantly saturated fat (e.g. tallow) and a relatively unsaturated one (e.g. soybean oil) has been higher than that which would have been

Table 14.9 ABSORBABILITY OF PALMITIC ACID AS INFLUENCED BY PRESENCE OF OLEIC AND LINOLEIC ACIDS

Ratio	Absorbability of palmitic (%)
Palmitic:oleic	
1/0.26	15.0
1/0.43	25.6
1/0.69	35.8
Palmitic:linoleic	
1/0.26	9.1
1/0.41	20.2
1/0.95	20.5

From Young and Garrett (1963)

Table 14.10 ME OF INDIVIDUAL FATS AND MIXTURES THEREOF

| Fat | ME (MJ/kg) | |
	Determined	Calculated
Beef tallow[a]	29.5	–
Soybean oil[a]	35.0	–
Beef tallow/soybean oil (50:50)[a]	34.1	32.3
Tallow[b]	34.0	–
Soybean oil[b]	39.8	–
Corn oil[b]	41.3	–
Tallow/soybean oil (50:50)[b]	40.6	36.9
Tallow/corn oil (50:50)[b]	40.0	37.7

[a] Sibbald et al. (1961b) (values are for AME)
[b] Sibbald and Kramer (1977) (values are for TME)

Table 14.11 AME OF A MIXTURE OF TALLOW AND SOYBEAN OIL

| Components of fat mixture (%) | | AME (MJ/kg) |
Beef tallow	Soybean oil	
10.0	–	29.5
9.5	0.5	34.3
9.0	1.0	36.3
8.0	2.0	37.5

From Lewis and Payne (1966)

Table 14.12 TME OF A MIXTURE OF TALLOW AND SOYBEAN OIL (ADDED AT A LEVEL OF 15% INTO A BASAL)

| Proportions | | TME (MJ/kg) | |
Tallow :	Soya bean oil	Observed	Theoretical
100 :	0	33.14	–
99 :	1	33.14	33.18
98 :	2	33.64	33.26
96 :	4	34.10	33.39
92 :	8	34.81	33.64
84 :	16	35.65	34.10
68 :	32	36.53	35.06
36 :	64	38.53	36.99
0 :	100	38.74	–

From Sibbald (1978)

Table 14.13 AME OF A MIXTURE OF RAPESEED OIL FOOTS AND TALLOW (ADDED AT A LEVEL OF 20% INTO BASAL)

Proportions Tallow : Rapeseed oil foots	AME (MJ/kg) Observed	Theoretical
100 : 0	30.1	–
80 : 20	31.6	30.3
60 : 40	33.4	30.4
50 : 50	33.6	30.5
40 : 60	31.8	30.5
20 : 80	32.2	30.7
0 : 100	30.8	–

From Lall and Slinger (1973)

predicted from knowledge of the two individual fat values (Sibbald, Slinger and Ashton, 1961a,b; Artman, 1964; Sibbald and Kramer, 1977).

Quantifying this synergistic effect, in terms of the optimum amount of polyunsaturated fat required to increase the dietary energy value of a saturated fat, has received considerable attention. Thus Lewis and Payne (1966) reported that the addition of only 5% of soybean oil relative to tallow (at a total fat inclusion level of 10% of the diet) increased the utilization of the subsequent fat mixture by approximately 16% (*Table 14.11*). Similarly Sibbald (1978) reported higher values for the TME of tallow with successive increases in soybean oil content (giving a total added fat content of 15% of the diet), although the response decreased with higher levels of soybean oil (*Table 14.12*) and Lall and Slinger (1973) reported AME values of mixtures of tallow and rapeseed oil foots (*Table 14.13*) when the level of added fat was 20%. Measurements of the ratio of unsaturated to saturated fatty acids in a mixture has also been used to quantify the degree of synergism. An optimum ratio of between 2.15:1 and 2.62:1 was suggested by Lall and Slinger (1973) and one of at least 2.2:1 by Fuller and Dale (1982) although Muztar, Leeson and Slinger (1981) considered that the proportion of free fatty acids would have a confounding effect.

The relevance of these observations is that the content and structure of the fat in the basal diet employed may influence determined dietary energy values of added fat. Thus Sibbald and Kramer (1978) measured the TME of tallow with poultry and obtained values of 35.4, 44.0, 35.1 and 36.7 MJ/kg when the fat was added at a level of 5% into basal diets based on wheat/soybean meal, maize/soybean meal, wheat/soybean meal/meat meal and wheat/soybean meal/fishmeal respectively. Differences were, it was concluded, attributable to an extent upon both the quantity and fatty acid profile of the basal fat. Similar observations have been reported by Sibbald and Kramer (1980) and Fuller and Dale (1982).

Extensive studies on interactions between added fats and the non-fat component of the basal diet (e.g. Mateos and Sell, 1980a, 1981b) have indicated that the presence of the former may enhance the utilization of the latter. Such an effect has been attributed to added fat reducing overall rate of passage through the gut (Mateos and Sell, 1981b) thus allowing greater time for digestion and absorption. It is a crucial assumption of the substitution method in the evaluation of fats, or of any feedingstuff, that

Table 14.14 AME OF YELLOW GREASE AS INFLUENCED BY NATURE OF BASAL (ON THE ASSUMPTION THAT ANY CHANGE IN DIETARY ENERGY VALUE IS ATTRIBUTABLE TO ADDED FAT). DATA ARE MEAN VALUES OBTAINED AT 3, 6 AND 9% FAT INCLUSION

Basal	AME (MJ/kg)
Maize/soybean meal	36.0
Sucrose[a]	40.9
Glucose	40.8
Fructose	40.6
Glucose and fructose	40.8
Starch	39.9
Maltose	39.8

[a] All carbohydrate added at level of 30% into the basal diet
From Mateos and Sell (1980a)

the basal diet behaves similarly irrespective of the presence of added ingredients. Thus any recorded change in the dietary energy value is attributed to the ingredient under investigation. Such an assumption may not be valid when considering fats, and the variability in AME values (*Table 14.14*) obtained are allegedly a reflection of the differential degree to which added fat enhances the availability of basal components—with those components exhibiting the most rapid transit time being those where the potential for improved utilization is highest following fat addition. Therefore choice of basal diet may be of critical importance in the evaluation of fats. Fuller and Dale (1982) advocated the use of a fat-free basal diet, composed of synthetic ingredients that would be almost completely utilized. Thus there would be no potential for any fat synergism between added fat and basal fat, and a reduction in rate of passage could not improve the digestibility of a basal diet already maximally utilized. Such a procedure would be useful in characterizing the behaviour of fats *per se* but would be of little practical relevance.

The presence of mineral ions in the basal diet has been implicated in reducing the availability of added fat in poultry diets (e.g. Sibbald and Price, 1977) due possibly to the *in vivo* production of insoluble soaps, principally those based upon calcium. As such reactions are likely to proceed more readily if fats are present in the free fatty acid rather than in the esterified form, then it is possible that the free fatty acid content is a determinant of the degree of soap formation. Wiseman and Cole (1983a) demonstrated a trend for pig DE values of a commercial fat blend to decrease with increasing calcium and free fatty acid content (*Table 14.15*).

Finally, it has been suggested that the level of protein in the basal diet may influence availability of added fat. Thus Young, Garrett and Griffiths

Table 14.15 INFLUENCE OF DIETARY LEVELS OF CALCIUM (Ca^{2+}) AND FREE FATTY ACID (FFA) CONTENT ON THE ADE (MJ/kg) OF A FAT BLEND FED TO GROWING PIGS

% Ca^{2+}	0.41	0.74	1.12	1.46
ADE (MJ/kg)	34.6	30.9	26.6	27.8
% FFA	11.6	44.9	85.8	
ADE (MJ/kg)	32.3	30.2	29.1	

Wiseman and Cole (1983a)

288 Assessment of DE and ME of fats for non-ruminants

(1963) demonstrated a definite improvement in absorbability of both lard and tallow when the crude protein of the basal diet was raised from 24% to 28% and 30%, although Sibbald and Slinger (1963) could not detect any fat × basal protein interactions in the determination of AME.

Level of inclusion

The previous section indicated that synergism between fats of different chemical structure could be of importance in the determination of the dietary energy value of a fat when included into a diet containing basal levels of fat. Leeson and Summers (1976) considered the consequences of such an interaction, and concluded that the dietary energy value attributed to a fat will have been dependent upon the level at which it was included in the basal diet, with higher values being recorded at lower levels (*Figure 14.1*). It was therefore assumed that the quantitative contribution of synergism to the dietary energy value of a fat was progressively less at higher inclusion levels. Additionally, the actual chemical nature of the fat in the basal diet will also have had an effect. Sibbald and Kramer (1978, 1980) have indicated that the TME of tallow is, in fact, lower at higher

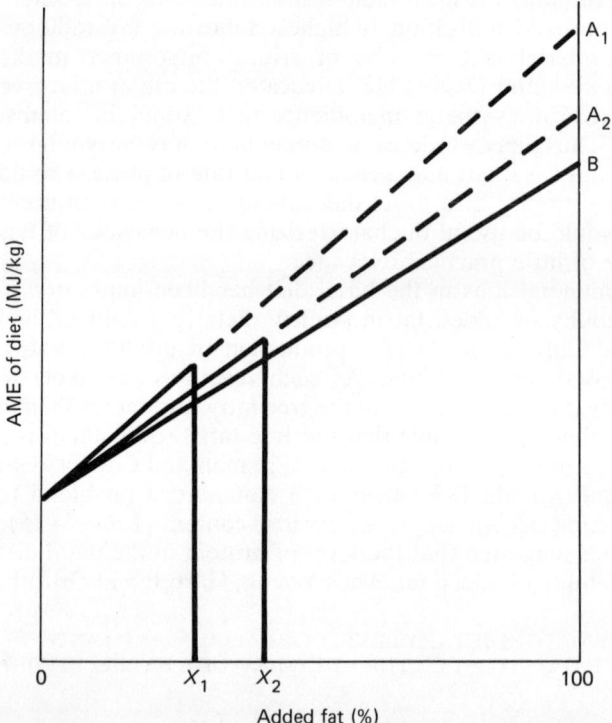

Figure 14.1 Effect of fatty acid synergism on AME value for added fat (determined by extrapolation to 100% fat addition). A_1, A_2 = AME of fat determined at levels X_1, X_2 respectively; B = AME of fat with no synergism. Adapted from Leeson and Summers (1976)

Table 14.16 INFLUENCE OF LEVEL OF ADDITION OF FATS INTO BASAL AND THEIR SUBSEQUENT TME VALUE (MJ/kg)

Fat	Basal	Level of inclusion (%)		
		5	10	15
Tallow	Wheat/soybean meal[a]	35.4	33.6	33.1
	Wheat/soybean meal/fish meal[b]	30.9	26.4	27.8
	Maize/soybean meal[a]	44.0	38.4	33.8
	Wheat/soybean meal/fish meal[b]	36.6	32.3	31.3
	Wheat/soya/meat meal[a]	35.1	32.3	31.6
	Wheat/soybean meal/fish meal[a]	36.7	34.2	32.3
Blend A[c]	Wheat/soybean meal	45.4	37.2	37.2
Blend B[c]	Wheat/soybean meal	42.3	36.9	35.7
Blend C[c]	Wheat/soybean meal	37.1	37.9	40.6
Blend D[c]	Wheat/soybean meal	39.9	38.8	39.8

[a] From Sibbald and Kramer (1978)
[b] From Sibbald and Kramer (1980)
[c] From Halloran and Sibbald (1979)

levels of inclusion although the effects are confounded by both type of basal, with those containing higher levels of unsaturated fatty acids (i.e. based upon maize) resulting in higher values and by type of fat (Halloran and Sibbald, 1979) (*Table 14.16*).

However, studies into the effects of level of inclusion often produce inconclusive results, due possibly both to the very large standard error associated with mean dietary energy values and also to the somewhat random nature of the mean data themselves. Both these problems arise as a consequence of the need to evaluate fats at rather low levels of inclusion. Thus Muztar *et al.* (1981) found no consistent pattern in the AME or TME of fats determined at different levels of inclusion (*Figure 14.2*). It has been suggested (Mateos and Sell, 1980b) that a more appropriate means of fat evaluation would be multi-level assay, where the response in terms of dietary energy value of the complete feed to successive incremental additions of fat would be evaluated. A significant departure from linearity

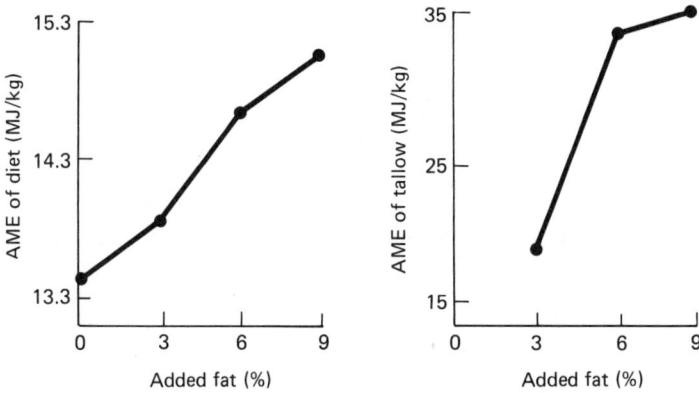

Figure 14.2 Effect of level of inclusion of fat on AME of complete diet, and on AME of added fat determined by difference. From Muztar *et al.* (1981)

in this response would be evidence of interactions between added fat and the basal component (whether fat synergism or interactions between fat and the non-fat component). Such a method has been employed originally by Sibbald and Slinger (1963) who used levels of tallow from 0% to 35% in 5% increments: it has been used subsequently by Mateos and Sell (1980b) with levels from 0% to 15% of yellow grease into a maize/soyabean basal in increments of 3%, and by Mateos and Sell (1981a), who added yellow grease into a sunflower meal/soybean meal basal at levels varying from 0% to 30% in 5% increments. However, although the latter two papers included quadratic regression equations, there was, in fact, no significant departure from linearity in the response outlined above. Similarly, Fuller and Rendon (1979) concluded that level of inclusion of fat had no effect upon its dietary energy value when measured at 5%, 10%, 15% and 20% inclusion levels.

It could be concluded, therefore, that in fact there was no synergism between added fat and basal fat, despite the apparent potential for such a situation occurring in the examples given, with the use of relatively saturated fat sources and basal diets containing a predominance of unsaturated fats.

Conversely, it could be argued that the experimental methodology employed was not sufficiently sensitive to detect any significant departures from linearity. Thus Wiseman and Cole (1983b) evaluated the AME, with broiler chicks of 18 days of age, of a fat blend at 10 levels of inclusion (from 0% to 10% in 1% increments) into both a semi-synthetic fat-free basal, and a commercial basal based upon wheat, maize and soybean meal. There was a significant departure from linearity ($P<0.001$) in the response of dietary AME values to added fat with both basals (*Figure 14.3*). That obtained with the commercial basal could be explained in terms of synergism between added fat and basal fat, which became quantitatively less important at higher levels of inclusion (supporting the observations of Sibbald (1978), and confirming the model of Leeson and Summers (1976)).

In their studies on the influence of added fat on rate of passage of the diet, Mateos, Sell and Eastwood (1982) concluded that, with adult hens, transit time increased linearly when levels of added yellow grease into a commercial basal were raised from 5% to 30% in 5% increments (although the lack of curvilinearity may have been a consequence of experimental insensitivity) and it may be suggested that such an effect would also be apparent with the use of fat-free basals. Thus it could be concluded that the influence of added fat on rate of passage was not responsible for the departure of linearity observed by Wiseman and Cole (1983b). Additionally, the quadratic response obtained with the fat-free basal obviously could not be attributed to fat synergism. It seems likely that the curvilinear response was an illustration of a progressive reduction in the chick's ability to utilize fat at higher inclusion levels.

The effect of level of inclusion was observed by Kussabaiti (1978) who reported an increase in AME of fat-added diets, obtained with 3-week-old chicks, with a reduction in feed intake: this, it was concluded, was due to the bird's finite capacity for fat utilization. The influence of fat structure was also investigated, and it was suggested that the decline in fat utilization with increasing intake was more pronounced with animal-based fats

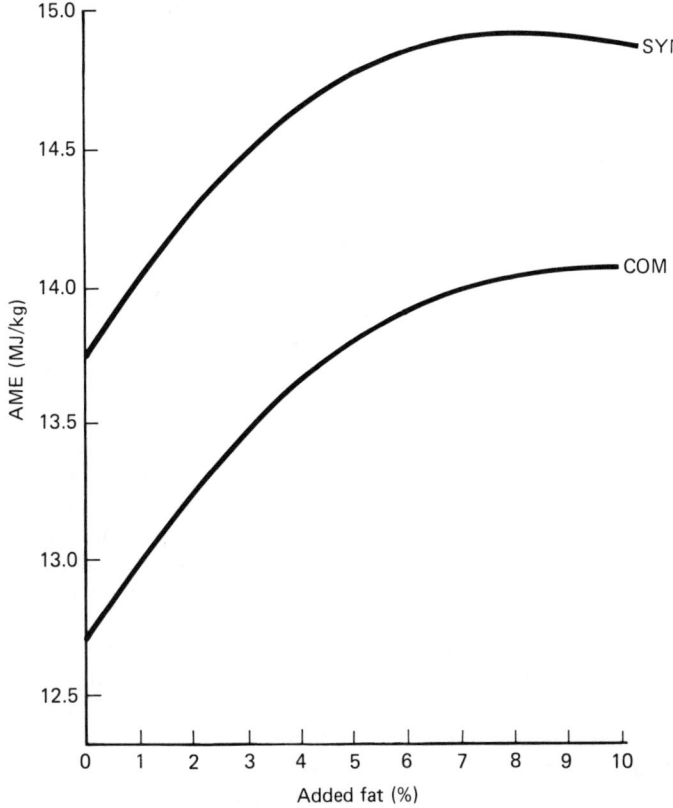

Figure 14.3 Effect of level of inclusion of added fat into a synthetic fat-free (SYN) and commercial (COM) basal on AME values of the resultant diet. Wiseman and Cole 1983b)

containing a greater proportion of palmitic and stearic acids (i.e. those fatty acids that are shown to be poorly utilized by poultry particularly with young chicks). In addition, the effect of age was studied, and it was concluded that the responses outlined were less apparent with older birds. Shannon (1971) had indicated a significant level of intake and free fatty acid interaction (*Table 14.17*). Thus the higher the level of free fatty acids in tallow fed to adult hens, the greater the reduction in AME with increasing levels of intake.

Table 14.17 INFLUENCE OF THE RATIO OF FREE FATTY ACID (FFA) TO TRIGLYCERIDE (TG) AND LEVEL OF FAT INTAKE ON THE AME (MJ/kg) OF TALLOW FED TO POULTRY

FFA:TG	20:80	40:60	60:40	80:20	97:3
AME (MJ/kg)	34.3	32.5	33.1	30.3	30.0
Fat intake (g/d)	3	5	10	20	
AME (MJ/kg)	34.7	31.2	32.3	30.0	

From Shannon (1971)

Interpretation of the multi-level assay data generated by Wiseman and Cole (1983b) in producing dietary energy values of the fat, presents problems. Extrapolation of linear functions to 100% fat inclusion produced AME values for the fat of 27.0 MJ/kg and 27.2 MJ/kg when included in the semi-synthetic and commercial basals respectively. However, because of the significant departure from linearity detected, such an approach is not valid. Extrapolation of quadratic equations to 100% fat inclusion is meaningless. It may be suggested that an appropriate solution is one which solves the quadratic equation at a point where x is equal to the level of fat addition to be used subsequently in practice.

Choice of basal diet in the evaluation of fats may be of importance, as this may affect possible synergistic relationships. However, no statistically significant fat × basal interaction was detected in the data reported by Wiseman and Cole (1983b) and it is therefore likely that the use of a commercial-type basal is acceptable as well as being of greater practical relevance. Because of the interactions between fat structure and level of intake discussed above (Shannon, 1971; Kussabaiti, 1978) it is also important that the experimental techniques employed should allow for the measurement of any such interactions. This may be achieved either by feed restriction, which is possibly of limited practical validity because *ad libitum* feeding systems are more usual, or by using different levels of inclusion of fat.

Recent work at Nottingham (J. Wiseman and D.J.A. Cole, unpublished data) has evaluated a number of fats with broiler chicks using a commercial basal and levels of inclusion ranging from 0% to 15% in 1.5% increments. It was considered that a maximum of 15% was more relevant than one of 10% and that the design employed would be sufficiently sensitive to detect any possible departures from linearity in the response of dietary ME to level of fat. In general, results indicated that responses were dependent upon the nature of the fat evaluated (*Figure 14.4*). Thus no significant departure from linearity was detected for fats such as soybean oil, whereas curvilinearity was observed to a progressively greater degree, the higher the levels of the less readily utilizable fat fractions present (i.e. palmitic and stearic acids, together with free fatty acids). Thus not only does level of inclusion (and hence intake) have an important role in the utilization of a specific fat, but it is also apparent that it may be important in assessing the relative value of a number of fats—differences between two fats may be small at low levels, but may widen at higher ones.

Derivation of actual AME values from response curves produced may be achieved by solving quadratic equations obtained as described previously. It has been suggested that such equations are more valid if the intercept is forced through the determined AME of the basal with no added fat. While this may be statistically unacceptable, it could generate data of more biological relevance. In practice it has been found that there is, in fact, little difference between the two methods of calculation.

Recently, the use of TME has been advocated as the basis for feed evaluation in poultry nutrition. It could be argued that the interactions described involving level of intake in the evaluation of fats remove the major theoretical advantage of TME—that of its independence of level of intake. TME assays are associated with levels of intake far below *ad*

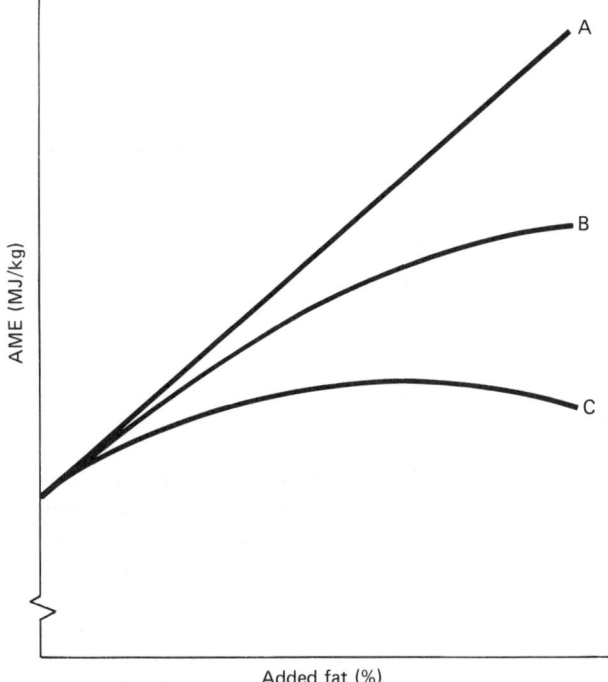

Figure 14.4 Effect of adding fats of high (A), medium (B) and low (C) nutritive value at varying levels of inclusion on AME of resultant diet

libitum, such that any finite capacity for fat digestion, in terms of total amount of fat that can be potentially utilized, may not be reached.

Prediction of dietary energy value of fats

The direct evaluation of fats via metabolism experiments is a lengthy and expensive procedure. Consequently, there has been considerable interest in, and much to be gained from, their indirect assessment through prediction equations based upon their chemical components. The influence of structure (including chain length and degree of saturation of component fatty acids, together with free fatty acid content) on fat utilization has been discussed, and there would therefore appear to be potential for the use of regression equations.

Generally, efforts have been concentrated upon degree of saturation. Thus Sibbald and Kramer (1977) obtained correlations between the TME of a number of fats and, for example, the proportion of saturated fatty acids, the percentage of linoleic acid, or the combined percentages of monoenes, dienes and trienes. Statistical significance was occasionally detected, but no one correlation accounted for more than 44% of the variation in TME values. Halloran and Sibbald (1979) derived prediction equations, based upon iodine value or percentage linoleic acid, which accounted for 52% and 69% respectively of the variability in AME values of 11 animal fats and fat blends. Linoleic acid was also considered by Mateos and Sell (1980b), but they used its level in the entire diet rather than the level in the added fat blend. Lewis (1983) reported a series of experiments that attempted to formulate prediction equations based upon the oleic, linoleic and linolenic acid contents of fats, or upon these fatty acids together with palmitic and stearic acids. Some equations were moderately successful, but there were some anomalies. Thus there was a considerable difference between observed and predicted AME values for marine oil (which, it was accepted, contains a considerable proportion of fatty acids with carbon chain lengths higher than those used in the regression analyses) and, to a lesser extent, for lard. This latter observation has been reported previously (Sibbald and Kramer, 1977) and is attributable, allegedly, to the preponderance of otherwise poorly utilized fatty acids (primarily palmitic) in the 2 position on the glyceride molecule, which render them more readily absorbed. It is for reasons such as these that the compilation of prediction equations has to date proved difficult, and has led to suggestions that it may be necessary to classify fats separately, according to type and origin, for the purposes of formulation of regression equations, and that to extrapolate derived relationships from one type to another is invalid.

In addition to the problems of prediction outlined above, there is of course the interaction between chemical content of a fat (e.g. free fatty acid content) and level of intake. Shannon (1971) in fact produced an equation in which the AME of tallow was related to the percentage free fatty acid content and level of intake. Such an approach, or one considering level of addition, could be potentially useful for other types of fat.

Constituents of fats which do not contribute to subsequent dietary energy value include moisture and unsaponifiable content, which are occasionally used as rough guides to fat quality, particularly with the increasing use of blends from diverse sources which may raise the content of the latter. What has not been established is whether the unsaponifiable content will have an adverse (i.e. negative) influence upon the utilization of the remainder of a fat. Finally, the general assumption that the saponifiable content of a fat is potentially of value itself may be questionable. Thus polymerized fatty acids have been detected in fats, and it is a matter for speculation as to whether they are of any nutritive value. Such a situation may also present problems in measuring the relative proportions of fatty acids in a fat blend. The usual technique, following GLC analysis, is to multiply up the determined levels of fatty acids to the percentage total saponifiable content.

Conclusion

It has been established that in terms of their dietary energy-yielding potential, fats are variable commodities. This may be attributable not only to differences in their chemical structure, but also to the experimental methodology selected for their evaluation. In addition, possible interactions between all the variables concerned confound the problem of assessment. Under practical conditions of feed formulation, however, the need to derive precise dietary energy values may be questionable. Thus a difference of 10% in, for example, AME of a fat will, if included at 5%, make a difference of approximately 1% in the AME of a compound diet. When considering the variability in AME values for the major constituents of such diets, then such a situation may be of limited practical relevance. Conversely, the purchase of fats is frequently on the basis of price per unit of dietary energy value. Reliable and accurate data for such values for fats are thus fundamental to the efficiency of least cost diet formulation.

References

ARTMAN, N.R. (1964). Interactions of fats and fatty acids as energy sources for the chick. *Poultry Science* **43**, 994–1004

BAYLEY, H.S. and LEWIS, D. (1965). The use of fats in pig feeding. II. The digestibility of various fats and fatty acids. *Journal of Agricultural Science* **64**, 373–378

CAREW, L.B., MACHEMER, R.H., SHARP, R.W. and FOSS, D.C. (1972). Fat absorption by the very young chick. *Poultry Science* **51**, 738–742

CARLSON, W.E. and BAYLEY, H.S. (1972). Digestion of fat by young pigs: a study of the amounts of fatty acid in the digestive tract using a fat soluble indicator of absorption. *Journal of Nutrition* **28**, 339–346

DE GROOTE, G. (1974). Utilisation of metabolisable energy. In *Energy Requirements of Poultry* (T.R. Morris and B.M. Freeman, Eds), pp. 113–133. Edinburgh, British Poultry Science Ltd

EUSEBIO, J.A., HAYS, V.W., SPEER, V.C. and McCALL, J.T. (1965). Utilisation of fat by young pigs. *Journal of Animal Science* **24**, 1001–1007

FEDDE, M.R., WAIBEL, P.E. and BURGER, R.E. (1960). Factors affecting the absorbability of certain dietary fats in the chick. *Journal of Nutrition* **70**, 447–452

FREEMAN, C.P., HOLME, D.W. and ANNISON, E.F. (1968). The determination of the true digestibility of interesterified fats in young pigs. *British Journal of Nutrition* **22**, 651–660

FULLER, H.L. and DALE, N.M. (1982). Effect of ratio of basal diet fat to test fat on the true metabolisable energy of the test fat. *Poultry Science* **61**, 914–918

FULLER, H.L. and RENDON, M. (1979). Energetic efficiency of corn oil and poultry fat at different levels in broiler diets. *Poultry Science* **58**, 1234–1238

HALLORAN, H.R. and SIBBALD, I.R. (1979). Metabolisable energy values of fats measured by several procedures. *Poultry Science* **58**, 1299–1307

JANSSEN, W.M.M.A. (1983). Fat evaluation—the Dutch approach. In *Proceedings, National Renderers Association seminar, Coventry* (in press)

JENSEN, L.S., SHUMAIER, G.W. and LATSHAW, J.D. (1970). Extra-calorific effect of dietary fat for developing turkeys as influenced by calorie:protein ratio. *Poultry Science* **49**, 1697–1704

JUST, A., ANDERSON, J.O. and JORGENSEN, H. (1980). The influence of diet composition on the apparent digestibility of crude fat and fatty acids at the terminal ileum and overall in pigs. *Zeitschrift für Tierphysiologie, Tierernährung und Futtermittelkunde* **44**, 82–92

KUSSABAITI, R. (1978). Influence of dietary intake level on the metabolisable energy and the digestibility of lipids in the growing chicken and the adult cockerel. In *Proceedings, 2nd European Symposium on Poultry Nutrition* (C.A. Kan and P.C.M. Simons, Eds), pp. 14–22. Beekbergen, Holland, Worlds Poultry Science Association

LALL, S.P. and SLINGER, S.J. (1973). The metabolisable energy content of rapeseed oil foots and the effect of blending with other fats. *Poultry Science* **52**, 143–151

LEESON, S.J. and SUMMERS, J.D. (1976). Fat ME values: The effect of fatty acid saturation. *Feedstuffs* **48**, (46), 26–27

LESSIRE, M. and LECLERCQ, B. (1982). Metabolisable energy value of fats in chicks and adult cockerels. *Animal Feed Science and Technology* **7**, 365–374

LEWIS, D. (1983). Synergism and related problems in energy prediction of fat. In *Proceedings, National Renderers Association Seminar, Coventry* (in press)

LEWIS, D. and PAYNE, C.G. (1966). Fats and amino acids in broiler rations. 6. Synergistic relationships in fatty acid utilisation. *British Poultry Science* **7**, 209–218

LLOYD, L.E. and CRAMPTON, E.W. (1957). The relation between certain characteristics of fats and oils and their apparent digestibility by young pigs, young guinea pigs and pups. *Journal of Animal Science* **16**, 377–382

MARCH, B. and BIELY, J. (1957). Fat studies in poultry. 6. Utilisation of fats of different melting points. *Poultry Science* **36**, 71–75

MATEOS, G.G. and SELL, J.L. (1980a). Influence of graded levels of fat on utilisation of pure carbohydrate by the laying hen. *Journal of Nutrition* **110**, 1894–1903

MATEOS, G.G. and SELL, J.L. (1980b). True and apparent metabolisable energy value of fat for laying hens: influence of level of use. *Poultry Science* **59**, 369–373

MATEOS, G.G. and SELL, J.L. (1981a). Metabolisable energy of supplemental fat as related to dietary fat level and methods of estimation. *Poultry Science* **60**, 1509–1515

MATEOS, G.G. and SELL, J.L. (1981b). Nature of the extrametabolic effect of supplemental fat used in semipurified diets for laying hens. *Poultry Science* **60**, 1925–1930

MATEOS, G.G., SELL, J.L. and EASTWOOD, J.A. (1982). Rate of food passage (transit time) as influenced by level of supplemental fat. *Poultry Science* **61**, 94–100

MUZTAR, A.J., LEESON, S.J. and SLINGER, S.J. (1981). Effect of blending and level of inclusion on the metabolisable energy of tallow and tower rapeseed soapstocks. *Poultry Science* **60**, 365–372

RENNER, R. and HILL, F.W. (1960). The utilisation of corn oil, lard and tallow by chickens of various ages. *Poultry Science* **39**, 849–854

RENNER, R. and HILL, F.W. (1961a). Factors affecting the absorbability of saturated fatty acids in the chick. *Journal of Nutrition* **74**, 254–258

RENNER, R. and HILL, F.W. (1961b). Utilisation of fatty acids by the chicken. *Journal of Nutrition* **74**, 259–264

SHANNON, D.W.F. (1971). The effect of level of intake and free fatty acid content on the metabolisable energy value and net absorption of tallow by the laying hen. *Journal of Agricultural Science* **76**, 217–221

SIBBALD, I.R. (1978). The true metabolisable energy values of mixtures of tallow with either soybean oil or lard. *Poultry Science* **57**, 473–477

SIBBALD, I.R. and KRAMER, J.K.S. (1977). The true metabolisable energy values of fats and fat mixtures. *Poultry Science* **56**, 2079–2086

SIBBALD, I.R. and KRAMER, J.K.S. (1978). The effect of the basal diet on the true metabolisable energy value of fat. *Poultry Science* **57**, 685–691

SIBBALD, I.R. and KRAMER, J.K.S. (1980). The effect of the basal diet on the utilisation of fat as a source of true metabolisable energy, lipid and fatty acids. *Poultry Science* **59**, 316–324

SIBBALD, I.R. and PRICE, K. (1977). The effects of level of dietary inclusion and of calcium on the true metabolisable energy values of fats. *Poultry Science* **56**, 2070–2078

SIBBALD, I.R. and SLINGER, S.J. (1963). A biological assay for metabolisable energy in poultry feed ingredients together with findings which demonstrate some of the problems associated with the evaluation of fats. *Poultry Science* **42**, 313–325

SIBBALD, I.R., SLINGER, S.J. and ASHTON, G.C. (1961a). The utilisation of a number of fats, fatty materials and mixtures thereof evaluated in terms of metabolisable energy, chick weight gains, and gain feed ratios. *Poultry Science* **40**, 46–61

SIBBALD, I.R., SLINGER, S.J. and ASHTON, G.C. (1961b). Factors affecting the metabolisable energy content of poultry feeds. Variability in the ME values attributed to samples of tallow, and undegummed soybean oil. *Poultry Science* **40**, 303–308

TULLIS, J.B. and WHITTEMORE, C.T. (1980). Digestibility of a fully hydrogenated tallow for growing pigs. *Animal Feed Science and Technology* **5**, 87–91

WHITEHEAD, C.C. and FISHER, C. (1975). The utilisation of various fats by turkeys of different ages. *British Poultry Science* **16**, 481–485

WISEMAN, J. and COLE, D.J.A. (1983a). Interaction between dietary fat, fatty acids and calcium in growing pigs. In *Proceedings, Vth World Conference on Animal Production, Tokyo* (in press)

WISEMAN, J. and COLE, D.J.A. (1983b). Determination of the metabolisable energy content of a commercial fat blend fed to broiler chicks. *World's Poultry Science Journal* **39** (3), 242

YOUNG, R.J. (1961). The energy value of fats and fatty acids for chicks. 1. Metabolisable energy. *Poultry Science* **40**, 1225–1233

YOUNG, R.J. and ARTMAN, N.R. (1961). The energy value of fats and fatty acids for chicks. 2. Evaluated by controlled feed intake. *Poultry Science* **40**, 1653–1661

YOUNG, R.J. and GARRETT, R.L. (1963). Effect of oleic and linoleic acids on the absorption of saturated fatty acids in the chick. *Journal of Nutrition* **81**, 321–329

YOUNG, R.J., GARRETT, R.L. and GRIFFITHS, M. (1963). Factors affecting the absorbability of fatty acid mixtures high in saturated fatty acids. *Poultry Science* **42**, 1146–1154

V

Fats in Animal Feeding Systems

15

THE NUTRIENT DENSITY OF PIG DIETS—ALLOWANCES AND APPETITE

D.J.A. COLE
University of Nottingham School of Agriculture, Sutton Bonington, Loughborough, Leics. LE12 5RD, UK

Introduction

The establishment of precise requirements for different types and levels of production in the pig has commanded considerable interest from research workers. Meeting these requirements is not a simple process and is complicated by the modifying influences of the range of ingredients which has to be used. In combining a number of ingredients the characteristics of the finished diet may be described in many ways. These largely concern the concentration of energy and nutrients within the diet and of particular interest are *nutrient density* and *nutrient balance*. Nutrient density refers to the concentration of energy and nutrients in the diet (i.e. the amounts of energy and nutrients per unit weight of diet), assuming that the balance between energy and nutrients, and the balance between the nutrients themselves, is constant. Although this chapter is largely concerned with a consideration of nutrient density, this is closely linked with *nutrient balance*, which also merits attention. The latter refers to the balance between the various nutrients and energy, and the consequences of varying nutrient balance on production are important.

Of further consequence to nutrient density is the type of feeding system adopted. Confusion has often arisen regarding the definitions of different systems and for convenience the following conventions are suggested:

1. *Ad libitum* feeding refers to the continuous access to a supply of fresh food and water. In the study of voluntary feed intake it is assumed that it refers to *ad libitum* feeding conditions.
2. Restricted feeding is the offering of allowances of feed below *ad libitum* and is synonymous with controlled feeding. It is often based on the liveweight of the animals.
3. Appetite feeding is a form of restricted feeding where appetite is reduced below *ad libitum* by restricting the time allowed for feeding e.g. two thirty-minute periods each day.

Modifying nutrient density

In modifying the nutrient density of the diet, energy concentration is linked to major ingredient changes. Concentration of the diet to produce high nutrient density is inescapably linked to the addition of fat. Thus, the influence of fat needs to be questioned, e.g. in relation to other components of the diet and when included at different levels. This is a major objective of the conference and is dealt with in detail elsewhere in this book.

The vast majority of pig diets are based on cereals which account for some, but not a lot, of the variation in nutrient density. For example, the means of recently quoted digestible energy (DE) values by Wiseman and Cole (1980) were 14.83 MJ/kg DM for barley, 16.09 MJ/kg DM for wheat, 15.87 MJ/kg DM for maize, 12.63 MJ/kg DM for oats and 15.23 MJ/kg DM for rye. It is suggested that with satisfactory preparation and formulation to take account of the differences in energy and nutrient concentration, diets based solely on various individual cereals (barley, wheat, maize and sorghum) and a high protein ingredient can be used to evoke similar responses per unit of energy (Cole, Clent and Luscombe, 1969).

To reduce nutrient density below that resulting from a largely cereal diet generally implies the use of bulky low-energy fibrous ingredients. The influence of crude fibre in the diet has been reviewed on several occasions. It is well established that large quantities of crude fibre in the diet can reduce the digestibility of the whole diet and of individual components. For example, the digestibility of crude protein has been reported to fall by 10–13% (Pond, Lowrey and Maner, 1962), 11% (Lloyd and Crampton, 1955) and 5% (Cole, Duckworth and Holmes, 1967b) when fibre levels were varied over the range 1.5–13.5%. It has been suggested that this reduction could be due to lack of penetration by digestive enzymes or an increase in rate of passage through the gut (Woodman and Evans, 1947), although faster rate of passage of high-fibre diets was not reported by Castle and Castle (1957) or Cole et al., (1967b). It has been suggested that the reduction of protein digestibility can be attributed partly to an increase in metabolic faecal nitrogen (Whiting and Bezeau, 1957).

The imprecise nature of crude fibre is well recognized to cause problems in nutritional considerations which are further complicated by the fact that it is not a nutrient but rather an analytical definition. Efforts to improve measures of crude fibre have included the use of analyses for acid detergent fibre, modified acid detergent fibre and neutral detergent fibre. Digestibility of fibre is generally low and can vary depending on the composition of the fibre. For example, digestibilities of 5, 35 and 51% have been reported for lignin, cellulose and pentosans respectively (Laurentowska, 1959) while digestibility of cellulose has been reported to range from 20% for wood cellulose (Cunningham, Friend and Nicholson, 1962) to an atypical value of 90% for de-lignified root crops (Woodman and Evans, 1947).

Allowances

Response curves of production parameters to individual nutrients are generally characterized by a period of response followed by a plateau or decline in performance. Unless the animal has reached its genetic capacity the inflexion represents the point at which another dietary constituent becomes limiting. Consequently, it is of importance to establish the *nutrient balance* of the diet for different production systems. An example of nutrient balance is inherent in the concept of an ideal protein (Cole, 1979) which has been developed further by the Agricultural Research Council (1981).

The ideal protein was described by Cole (1979) as follows:

> It is likely that the major difference in requirement between pigs growing at different rates and between pigs of different sexes, breeds and liveweights is in the amount of protein that they require, according to their potential for lean deposition. The relative amounts of the different essential amino acids needed for the deposition of 1 g of lean should be the same in each case. Thus, it should be possible to establish an optimum balance of essential amino acids for growth which when supplied with sufficient nitrogen for the synthesis of non-essential amino acids, would constitute the 'ideal protein'. Pigs of different classes (i.e. liveweight, sex, breed, etc) would require different amounts of the ideal protein but the quality of the protein would be the same in each case.

The balance of amino acids used as the basis of an ideal protein is given in *Table 15.1*.

Such an ideal protein has been used as the basis of diets and supplied with 13.65 MJ DE/kg for growing pigs and 13.4 MJ DE/kg for finishing pigs. These diets were used to establish the optimum level of lysine when

Table 15.1 THE AMINO ACID BALANCE USED AS THE BASIS OF THE IDEAL PROTEIN

Lysine	100
Methionine + cystine	50
Threonine	60
Tryptophan	18
Isoleucine	50
Leucine	100
Histidine	33
Phenylaline + tyrosine	100
Valine	70

Based on Cole, 1979

Table 15.2 DIETARY LYSINE REQUIREMENT (% OF DIET) FOR MAXIMUM RESPONSE OF DIFFERENT PRODUCTION CHARACTERISTICS

	Boars	Gilts	Castrated males
Growing pigs (25–55 kg liveweight)			
Daily gain	1.12	1.09	1.01
Food conversion ratio	1.10	1.08	1.03
Lean in ham	1.03	1.00	0.96
Finishing pigs (50–90 kg liveweight)			
Daily gain	0.92	0.83	0.75
Food conversion ratio	0.90	0.84	0.72
Lean in ham	0.93	0.87	0.70

H.T. Yen, D.J.A. Cole and D. Lewis, unpublished data

304 *The nutrient density of pig diets—allowances and appetite*

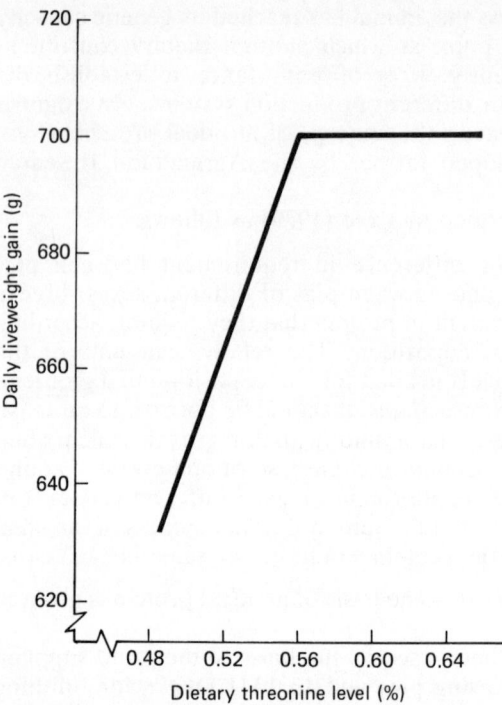

Figure 15.1 Effects of dietary threonine level on daily liveweight gain (after Taylor, Cole and Lewis, 1982). $y = 168.0 + 957.1x$; $y_{max} = 700.4$

associated with a commensurate change in the other components of the ideal protein (*Table 15.2*). The lysine requirements suggested by this approach are considerably higher than earlier estimates which ignored the relationship between individual components of the diet (e.g. ARC, 1967).

The consequences of disrupting nutrient balance are demonstrated in *Figure 15.1* which illustrates that an undersupply of threonine of 0.1% results in a fall in growth rate of 95.7 g/day.

While it is important to establish the nutrient balance of the diet, *nutrient density* is also of consequence. It has been suggested by Cole, Hardy and Lewis (1972) that if high nutrient density diets replace those of lower nutrient density on a weight-for-weight basis, then pig performance is affected in the same way as increasing feed intake under restricted feeding conditions. The outcome is that pigs grow more quickly, utilize their feed less well and produce fatter carcasses. However, when the basis of replacement of diets of low nutrient density by those of high nutrient density is the equivalence of daily intakes of energy and nutrients, such an effect is not found. For example, Bayley and Lewis (1963) reported faster growth rate, better feed utilization and leaner carcasses as nutrient density was increased. In their work, however, part of this improvement must be attributed to the lack of control of the amino acid balance which was better in the higher nutrient density diets because of the greater inclusion of high-quality protein sources, e.g. fishmeal. The higher nutrient density diets also contained increasing proportions of fat which is typical of their formulation. It has been demonstrated that better utilization of such diets might be expected as the heat loss on transformation of feed lipids to body lipids in pigs has been shown to be lower than when carbohydrate is transformed to body lipid (Leroy, 1965) and these aspects are covered elsewhere in this book.

Voluntary feed intake

Although the use of *ad libitum* feeding has been adopted in some situations, a considerable part of pig production in Britain has involved the use of restricted feeding, particularly in the later stages of finishing, to slow down growth rate and improve carcass quality. In most production systems, reduction of growth rate below a maximum is not a desirable feature. Consequently, there is likely to be more interest in *ad libitum* feeding, particularly with the availability of improved genotypes and with the non-castration of males.

Factors affecting level of voluntary feed intake

The amount of food eaten by a pig under conditions of constant access is a compromise between the animal factors (i.e. the requirement of the animal) and the ability of the food to meet those requirements. In this respect it is generally recognized that energy is important in the control of voluntary feed intake.

LIVEWEIGHT

In the growing pig, feed intake is closely related to liveweight. The relationship between energy intake and liveweight suggested by Cole *et al.* (1967b) in Equation 15.1 and by the Agricultural Research Council (1981) in Equation 15.2 show good agreement (*Figure 15.2*).

DE Intake (MJ/day) = $2.405 \, W^{0.675}$ Eq. (15.1)

DE Intake (MJ/day) = $4.7 \, W^{0.51}$ Eq. (15.2)

where W = liveweight (kg).

SEX

There is a large volume of evidence to show that castrated male pigs eat considerably more than gilts (e.g. *Figure 15.3*). The consequence of this greater feed intake is faster growth, poorer feed utilization and fatter carcasses. With the trend to non-castration of boars there has been greater examination of the boar under conditions of *ad libitum* feeding. The boar has a lower voluntary feed intake than the other sexes (*Figure 15.3*) which

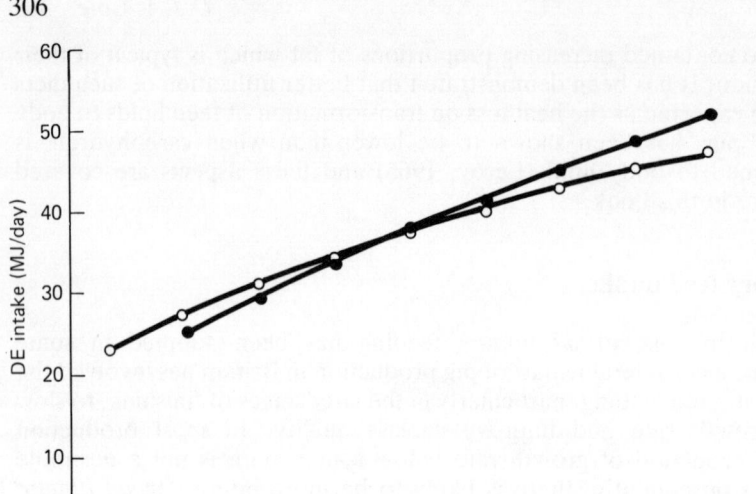

Figure 15.2 The relationship between voluntary energy intake and liveweight.
● DE intake (MJ/day) = $2.405W^{0.675}$ (Cole et al., 1967b); ○ DE intake (MJ/day) = $4.7W^{0.51}$ (ARC, 1981)

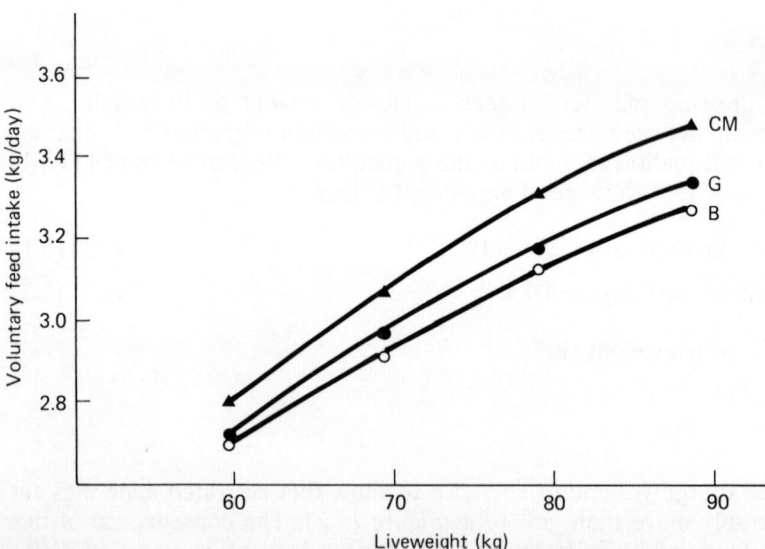

Figure 15.3 The relationship between voluntary feed intake and liveweight for different sexes (CM = castrated males; G = gilts; B = boars), based on the unpublished data of G. Sparkes, D.J.A. Cole and D. Lewis

together with its good feed utilization and carcass quality should hasten the adoption of *ad libitum* feeding in practice. It is likely that the voluntary feed intakes established by the sexes largely reflect their different potentials for tissue growth.

PREVIOUS NUTRITIONAL HISTORY

Feed intake at a particular stage of growth can be influenced considerably by the previous nutritional history of the animal. For example, pigs whose growth over the period 25–50 kg liveweight was restricted to half of that of pigs fed *ad libitum*, ate about 15% more over the period of growth from 60 to 90 kg liveweight (Cole *et al.*, 1968) (*Figure 15.4*). Under *ad libitum*

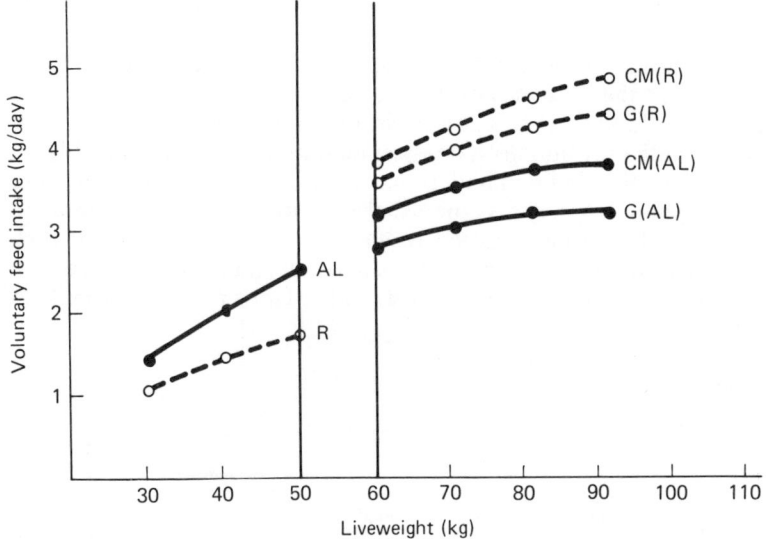

Figure 15.4 The voluntary feed intake (60–90 kg liveweight) of pigs which were fed *ad libitum* (AL) or restricted to a feed intake to give half of the growth rate (R) over the period from 30 to 50 kg liveweight. CM = castrated males; G = gilts. (Based on Cole *et al.*, 1968)

feeding conditions it is likely that any compensatory growth results from a large increase in feed intake rather than any significant improvement in feed utilization.

DIET

Some time ago it was suggested that the pig eats to a constant daily digestible energy intake and that, within limits, it is able to compensate for variation in nutrient density by changes in feed intake. For example, Cole, Duckworth and Holmes (1967a) using diets ranging from 2.57 to 3.30 Mcal DE/kg air dry feed (10.8–13.8 MJ DE/kg) found that feed intake decreased from 3.65 kg DM/day to 2.80 kg DM/day with the highest energy diet. Although there was a 23% reduction in feed intake there was no significant difference for digestible energy intake. It is obvious that such a compensatory mechanism cannot work for indefinite dilutions and that eventually a physical limitation is reached when the pig cannot consume any more food.

The point at which physiological control of feed intake gives way to physical control will be influenced by a number of factors which will include the requirement of the pig and the amount and type of diluent used. One such example of physical limitation has been reported by Cole *et al.* (1968) where pigs were offered a diet containing 2.11 Mcal (8.8 MJ) DE/kg/d air dry diet over the period 59–91 kg liveweight and had an average dry matter intake of 4.05 kg/day with a faecal output of 1.9 kg/day. On the basis of work such as this a model feed intake control has been proposed (*Figure 15.5*).

A further example of compensation with changes in nutrient density of the diet is given in *Figure 15.6* and comes from the same work as *Figure 15.4*. In this work the influence of the various factors affecting feed intake appeared to be additive. This is also illustrated in *Figure 15.7*. However, it should be noted that eventually a physical limitation occurred and with the low-energy diets, pigs that had undergone a period of feed restriction were unable to eat enough to express the usual sex differences in feed intake.

While the pig has been reported to eat less of a high-energy or high nutrient density diet, this does not always result in equal energy intakes on a range of diets. For example, recent work at the University of Nottingham

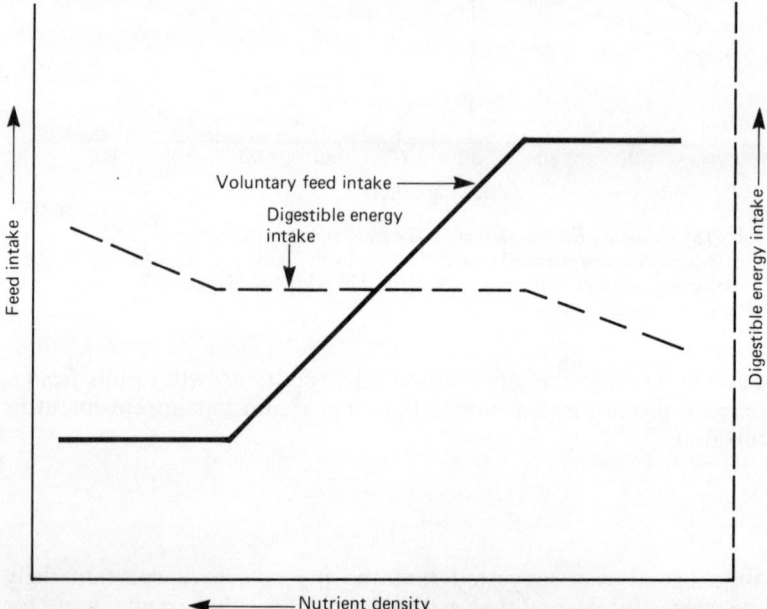

Figure 15.5 Schematic representation of voluntary feed and digestible energy intake under conditions of varying nutrient density (Cole, Hardy and Lewis, 1972)

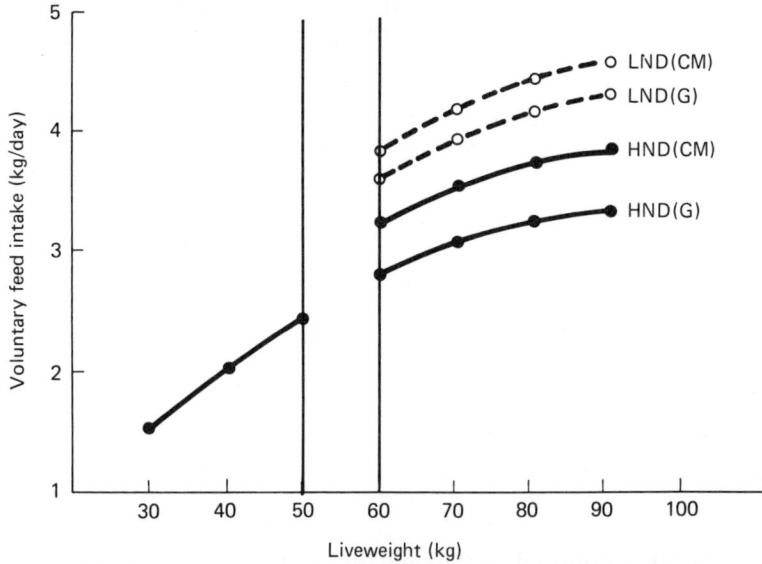

Figure 15.6 The feed intake of pigs fed high nutrient density (HND) diets or low nutrient density (LND) diets over the period 60–90 kg liveweight. CM = castrated males; G = gilts. (From Cole *et al.*, 1968)

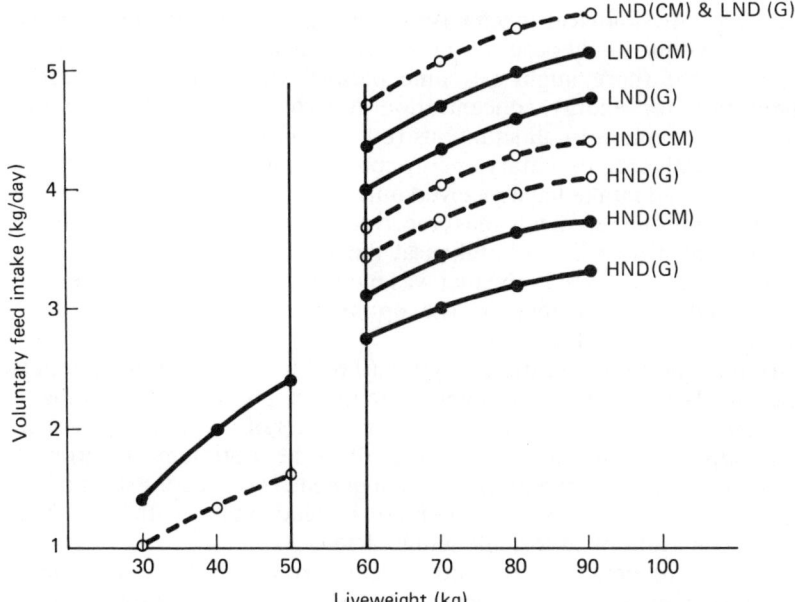

Figure 15.7 The feed intakes of pigs from 60 to 90 kg liveweight. HND = high nutrient density diet; LND = low nutrient density diet; CM = castrated males; G = gilts; ●——● *ad libitum* feeding from 30 to 50 kg liveweight; ○——○ restricted feeding from 50 to 70 kg liveweight. (From Cole *et al.*, 1968)

Table 15.3 THE VOLUNTARY FEED INTAKE OF BOARS AND CASTRATED MALE PIGS GIVEN DIFFERENT LEVELS OF DIETARY ENERGY AND IDEAL PROTEIN

	Ideal protein (%)	7.9	11.3	14.7	18.1		
	Lysine (%)	0.55	0.79	1.03	1.27		
	Dietary energy (MJ ME/kg)	Voluntary feed intake (kg/day)				Mean	Metabolizable energy (MJ/day)
Boars	10.0	1.99	2.08	2.18	1.93	2.04	20.4
	11.4	1.89	2.13	1.78	1.89	1.92	21.9
	12.8	1.85	1.85	1.77	1.87	1.87	23.9
	14.2	1.79	1.80	1.71	1.70	1.75	24.9
	Mean	1.88	1.96	1.86	1.85		
Castrated males	10.0	2.28	2.25	2.31	2.09	2.23	22.3
	11.4	2.13	2.13	2.06	2.02	2.08	23.7
	12.8	2.03	1.93	1.94	1.94	1.96	25.1
	14.2	1.84	1.86	1.99	1.89	1.90	27.0
	Mean	2.07	2.04	2.07	1.94		

G. Sparkes, D.J.A. Cole and D. Lewis, unpublished results

(Sparkes, 1982) has reported energy intakes similar to those obtained by Cole *et al.* (1967a) and Henry (1974). However, when energy concentration or nutrient density were modified, the pigs' response was different (*Table 15.3*) in that although they ate less of the high-energy diets they still had higher energy intakes. This lack of compensation to dietary energy concentration might represent a conflict and compromise between the physical and physiological mechanisms controlling voluntary feed intake although why this should occur in some cases and not others is not clear. It is possible that there might be some parallel with poultry where the compensation for energy concentration reported by Hill and Dansky (1954) is not common to all situations (e.g. see Fisher, Chapter 21).

While the influence of dietary energy concentration and nutrient density on voluntary feed intake have received only little attention in pigs, the role of protein and amino acids has received even less; consequently, a programme of work was undertaken at the University of Nottingham to examine it. When protein in the diet was modified, pigs tended to eat most of what might be described as the optimum protein level (based on knowledge of restricted-fed pigs).

Work on wide ranges of dietary lysine (0.6–3.0%) and methionine plus cystine (0.34–2.0%) has prompted a tentative proposal concerning a model response of feed intake to increasing levels of a single amino acid. The pig appears to increase feed intake up to the 'optimum' for growth, after which there is a sharp decline until a plateau is reached. Presumably, eventually there would be a further fall in feed intake with very high dietary concentrations of a single amino acid.

In work where dietary energy and protein were varied at the same time, the greatest influence on voluntary feed intake was dietary energy (*Table 15.3*).

In conclusion, it can be said that voluntary feed intake is a compromise between the requirements of the animal and the nature of the diet. Neither of these is fixed and, besides being influenced by a number of factors, these factors are themselves subject to change. For example, while breed may influence requirement, the selection pressures of man will alter the genotypes available for production. Similarly, a major change in the inherent characteristics of the animal will result in a change from castration of males to non-castration. At the same time the nature of the ingredients available for incorporation into pig diets may also change. While the relationship between the animal and the diet may be influenced by the physiological and physical controls of feed intake, the relationship between these may also change. For example, little work has been done to examine the influence of liveweight and age on such relationships, yet the information is vitally important to the strategy to be adopted for pig feeding.

References

AGRICULTURAL RESEARCH COUNCIL (1967). *The Nutrient Requirements of Farm Livestock. No. 3. Pigs*. London, Agricultural Research Council

AGRICULTURAL RESEARCH COUNCIL (1981). *The Nutrient Requirements of Pigs*. Slough, Commonwealth Agricultural Bureaux

BAYLEY, H.S. and LEWIS, D. (1963). The use of fat in pig rations. *Journal of Agricultural Science* **61**, 121–125

CASTLE, E.J. and CASTLE, M.E. (1957). Further studies of the rate of passage of food through the alimentary tract of pigs. *Journal of Agricultural Science* **49**, 106–112

COLE, D.J.A. (1979). Amino acid nutrition of the pig. In *Recent Advances in Animal Nutrition—1978* (W. Haresign and D. Lewis, Eds), pp. 59–72. London, Butterworths

COLE, D.J.A., CLENT, E.G. and LUSCOMBE, J.R. (1969). Single cereal diets for bacon pigs. 1. The effect of diets based on barley, wheat, maize meal, flaked maize or sorghum on performance and carcass characteristics. *Animal Production* **10**, 345–357

COLE, D.J.A., DUCKWORTH, J.E. and HOLMES, W. (1967a). Factors affecting voluntary feed intake in pigs. 1. The effect of digestible energy content of the diet on intake of castrated male pigs housed in holding pens and metabolism crates. *Animal Production* **9**, 141–148

COLE, D.J.A., DUCKWORTH, J.E. and HOLMES, W. (1967b). Factors affecting voluntary feed intake in pigs. 2. The effect of two levels of crude fibre in the diet on the intake and performance of fattening pigs. *Animal Production* **9**, 149–154

COLE, D.J.A., HARDY, B. and LEWIS, D. (1972). Nutrient density of pig diets. In *Pig Production* (D.J.A. Cole, Ed.), pp. 243–257. London, Butterworths

COLE, D.J.A., DUCKWORTH, J.E., HOLMES, W. and CUTHBERTSON, A. (1968). Factors affecting voluntary feed intake in pigs. 3. The effect of a period of feed restriction, nutrient density of the diet and sex on intake, performance and carcass characteristics. *Animal Production* **10**, 345–357

CUNNINGHAM, H.M., FRIEND, D.W. and NICHOLSON, J.W.G. (1962). Effect of age, body weight, feed intake and adaptability of pigs on the digestibility and nutritive value of cellulose. *Canadian Journal of Animal Science* **42**, 167–175

HENRY, Y. (1974). Incorporation de proportions variables de matières grasses (huile d'arachide) dans le regime du porc en croissance-finition. II. Influence sur les performances de croissance et la composition corporelle. *Annales de zootechnie* **23**, 171

HILL, F.W. and DANSKY, L.M. (1954). Studies of the energy requirement of chickens. 1. The effect of dietary energy level on growth and food consumption. *Poultry Science* **33**, 143–148

LAURENTOWSKA, C. (1959). *Rocznik nauk rolniczych* **74B**, 567–578

LEROY, A.M. (1965). Analysis of the origin of heat loss of pigs fed normally. In *Energy Metabolism* (K.L. Blaxter, Ed.), pp. 37–48. London and New York, Academic Press

LLOYD, L.E. and CRAMPTON, E.W. (1955). The apparent digestibility of the crude protein of the pig ration as a function of the crude protein and crude fiber content. *Journal of Animal Science* **14**, 693–699

POND, W.G., LOWREY, R.S. and MANER, J.H. (1962). Effect of crude fibre level on ration digestibility and performance in growing fattening swine. *Journal of Animal Science* **21**, 692–696

SPARKES, G.M. (1982). *The Influence of Dietary Protein on Voluntary Feed Intake in Pigs*. PhD thesis, University of Nottingham

TAYLOR, A.J., COLE, D.J.A. and LEWIS, D. (1982). Amino acid requirements of growing pigs. 3. Threonine. *Animal Production* **34**, 1–8

WHITING, F. and BEZEAU, J.M. (1957). The metabolic faecal nitrogen excretion of the pig as influenced by the amount of fibre in the ration and by body weight. *Canadian Journal of Animal Science* **37**, 95–113

WISEMAN, J. and COLE, D.J.A. (1980). Energy evaluation of cereals for pig diets. In *Recent Advances in Animal Nutrition—1979* (W. Haresign and D. Lewis, Eds), pp. 51–67. London, Butterworths

WOODMAN, H.E. and EVANS, R.E. (1947). The nutritive value of fodder cellulose from wheat straw. I. Its digestibility and feeding value when fed to ruminants and pigs. *Journal of Agricultural Science* **37**, 202–210

16

USE OF FATS IN DIETS FOR GROWING PIGS

TIM S. STAHLY
Department of Animal Sciences, University of Kentucky, Lexington, Kentucky, USA

The lipids in feedstuffs commonly consumed by pigs consist mainly of neutral fats, specifically triglycerides. Nutritionally, these fats supply essential fatty acids, facilitate absorption of fat-soluble vitamins and represent a concentrated source of dietary energy-yielding ingredients. Dietary fats also influence diet preference in pigs and the physical properties of feeds, including pellet durability and particle cohesiveness.

Essential fatty acid source

Pigs must consume minimum quantities of fat to meet their physiological needs for essential fatty acids. Essential fatty acid (linoleic acid) levels of 2% and 1% of the digestible energy (DE) have been reported to optimize growth, efficiency of energy utilization and the triene:tetraene index in growing pigs up to 11–14 weeks of age and from 11–14 weeks to 90 kg, respectively (Hill *et al.*, 1961; Leat, 1961; Caster *et al.*, 1962; Christensen, Henckel and Thorbek, 1979). On the basis of recent evidence in man and rats that 50% more linoleic acid may be required to optimize the triene:tetraene ratio, the Agricultural Research Council (ARC, 1981)

Table 16.1 FAT CONTENT OF SOME FEEDSTUFFS[a]

Feedstuff	Composition		
	Fat (%)	Linoleic (%)	% of DE from linoleic
Cereal grains			
Corn	3.9	1.8	4.7
Oats	4.0	1.5	4.9
Wheat	1.7	0.6	1.6
Barley	1.9	0.2	0.6
Protein sources			
Soybean meal	0.8	0.3	0.7
Whey, dried	0.8	—	—

[a] From NRC (1959) and Edwards (1964).

elevated the requirements of linoleic acid to 3.0 and 1.5% of the dietary DE for pigs up to 30 kg and from 30 to 90 kg, respectively.

The pig's estimated need for essential fatty acids will normally be met or exceeded by diets in which the major energy source consists of maize (corn) or oats (*Table 16.1*). However, corn and oats are considerably higher in fat and linoleic acid compared with sorghum grain, barley and wheat (NRC, 1959). Hence, diets based predominantly on barley or wheat and certain milk protein sources may not meet the ARC estimated requirement, particularly in younger animals. In these diets, the addition of 1–1.5% of a fat source containing a high level of linoleic acid may be required to supply the levels of essential fatty acids suggested by the ARC.

Energy source

The primary contribution of fat for pigs is that it serves as a concentrated source of energy-yielding ingredients. The gross energy (GE) content of fats averages 39.3 kJ/g or approximately 2.25 times that of the carbohydrate fraction of the diet.

Digestion and absorption of fat

The proportion of the GE in fat that is digested and absorbed in pigs is directly related to the relative ability of the fat to form micelles with bile salts in the intestinal lumen. Pigs have a high physiological potential for micelle formation except for the initial 2 or 3 weeks postweaning (Scherer, 1973; Kidder and Manner, 1978). The concentrations of fatty acid/ monoglyceride/bile salt micelles in the intestinal lumen of monogastric animals are influenced by the degree of unsaturation and chain length of the fatty acids, the relative concentrations of free versus esterified fatty acids and the positioning of saturated fatty acids on the glycerol molecule.

In pigs, saturated fatty acids alone have a lower micellar formation potential and thus are less efficiently digested than unsaturated fatty acids (Freeman, Holme and Annison, 1968). However, the micellar formation potential and absorption of saturated fatty acids is increased in the presence of unsaturated fatty acids or monoglycerides (Bayley and Lewis, 1965; Carlson and Bayley, 1968; Freeman *et al.*, 1968; Martin, 1977). Therefore, the digestibility of a particular supplemental fat source in pigs is dependent on the fatty acid composition (unsaturated/saturated fatty acid ratio) of the total diet (*Figure 16.1*). The digestibility of fat from diets containing a ratio of unsaturated to saturated (U/S) fatty acids greater than 1.5 is relatively high, averaging 85–92%. The digestibility of fat in diets with a U/S ratio of less than 1.0–1.3 is substantially lower, ranging from 35% to 75%. On this basis, the digestibility of tallow when included as 5.0% of the diet would be expected to be 10–15% higher in pigs fed corn-based diets as compared with barley-based diets in which the total dietary ratio of U/S fatty acids would be approximately 1.5 and 1.0%, respectively (*Table 16.2*). Similarly, the apparent digestibility of hydrogenated tallow in pigs is 5–10% higher when fed in combination with a source of long-chain, unsaturated fatty acids (Martin, 1977).

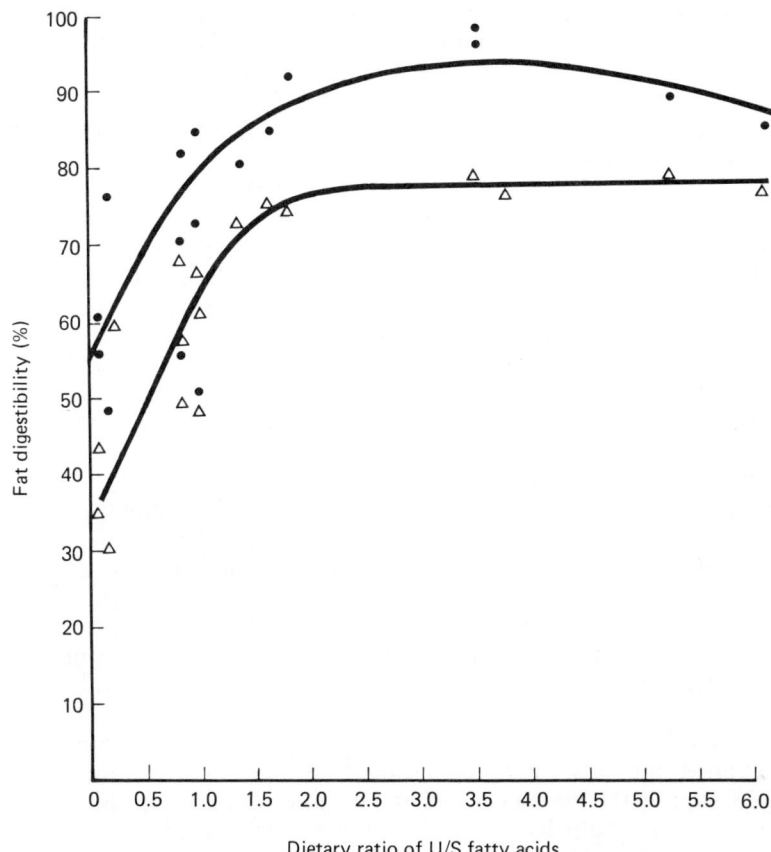

Figure 16.1 Digestibility of fat in pigs as influenced by the ratio of unsaturated to saturated (U/S) fatty acids in the total dietary fat mix. References: Bayley and Lewis, 1965; Carlson and Bayley, 1968; Freeman *et al.*, 1968; Davis and Lewis, 1969; Carlson and Bayley, 1972. Total dietary fat levels ranged from 6.5 to 11.7%. △ Apparent fat digestibility. ● Corrected fat digestibility.

Table 16.2 INFLUENCE OF THE TOTAL DIETARY FATTY ACID PROFILE ON FAT DIGESTION IN PIGS

	Diet composition		*Estimated fat digestibility*[c]
Basal source	*Suppl. fat source*[a]	*Ratio of U/S fatty acids*[b]	
Corn–soy	beef tallow	1.5	85–92
Barley–soy	beef tallow	1.0	70–85
Corn–soy	soy oil	4.8	90–95
Barley–soy	soy oil	4.0	90–95

[a] Supplemental fat source included as 5% of the diet.
[b] Total dietary ratios of unsaturated/saturated (U/S) fatty acids calculated from NRC (1959) and Edwards (1964).
[c] Based on data summarized in *Figure 16.1*.

Short and medium-chain fatty acids (14 carbons or less), although predominantly saturated, readily form micelles. Thus, the apparent digestibility of triglycerides containing a high proportion of medium chain fatty acids in pigs is high (80–95%) regardless of the dietary ratio of U/S fatty acids (Freeman et al., 1968; Hamilton and McDonald, 1969; Frobish et al., 1970; Braude and Newport, 1973).

Long-chain saturated fatty acids are less well absorbed and utilized when present as the free acid versus the ester (Bayley and Lewis, 1965; Hillcoat and Annison, 1974). The digestibility of hydrolysed fats (blends of animal, vegetable or poultry fats) which may contain 20–70% free fatty acids are difficult to evaluate in that the fatty acid profile and the levels of free fatty acids and unsaponified material present in the fat may vary substantially, depending on the composition and proportion of the raw ingredients blended. Apparent fat digestibility values of 60–70% have been reported in young pigs fed a commercial fat blend of hydrolysed vegetable and animal fat containing a high level of free fatty acids (Bayley and Lewis, 1965; Leibbrandt et al., 1975; Martin, 1977).

The positioning of the saturated fatty acids on the glycerol molecule also may influence their digestibility, due to the fact that lipase selectively hydrolyses the bond at the 1 and 3 position of the glycerol moiety. Therefore, fats such as lard, which has palmitic acid predominantly esterified at the 2 position, may be digested slightly more efficiently (2–4%) because of the greater micellar formation potential of a monoglyceride than that of the free acid (Freeman et al., 1968; Davis and Lewis, 1969).

The digestibility of dietary fat is also influenced by dietary components which depress the absorption of nutrients in the small intestine by potentially altering the rate of passage of the digesta or partly insolubilizing the dietary fat. Carbohydrate sources, such as barley straw and potato starch, which depress the absorption of nutrients in the small intestine and stimulate fermentation in the hind gut, have been shown to depress fat digestibility in pigs (Mason and Just, 1976; Just, 1982a). For example, apparent fat digestibility decreases by 1.3–1.5% for each additional 1% crude fibre in the diet (Just, 1982a,b,c). High levels of dietary minerals (Ca, Mg) have also been shown to depress the digestibility of long-chain

Table 16.3 METABOLIZABLE ENERGY VALUE OF DIETARY FATS FOR PIGS[a]

Supplemental fat source[b]	Total dietary U/S ratio[c]	ME (kJ/g)
Tallow	1.3	33.05
Tallow	1.3	32.97
Lard	1.5	33.43
Lard	1.7	32.22
Soybean oil	5.1	30.46
Corn oil	5.4	34.10
Corn oil	5.6	30.79
Average ME of crude fat		32.43

[a] From Tollett, 1961; Diggs et al., 1965; Phillips and Ewan, 1977.
[b] Basal diet consisted largely of a corn-soybean meal mix.
[c] Total dietary ratios of unsaturated/saturated (U/S) fatty acids calculated from NRC (1959) and Edwards (1964).

fatty acids in rats (Cheng, Morehouse and Deuel, 1949) although the magnitude of this relationship in pigs has not been clearly defined.

The amount of digestible energy (DE) in dietary fats also is influenced by the presence of energy diluents in the fat. The normal unsaponifiable material present in fats is digested relatively efficiently, at a level approximating that of triglycerides (Kidder and Manner, 1978). However, moisture and insoluble material in fat act as energy diluents.

The proportion of DE absorbed from the intestinal lumen that is available for cellular metabolism in pigs is similar for fats versus carbohydrates, averaging 94–96% of the DE (Just, 1982d). The metabolizable energy (ME) value of fat is estimated to be 37.0 (39.3 × 0.94) kJ per g of digested fat. In diets containing a ratio of U/S fatty acids greater than 1.3 (assumed to be 85–90% digestible), this equates to a ME value of 31.4 (37 × 0.85) to 33.3 (37 × 0.90) kJ per g of crude fat. These calculations are in good agreement with the values determined experimentally (Tollett, 1961; Diggs *et al.*, 1965; Phillips and Ewan, 1977; *Table 16.3*).

Metabolic utilization of digestible fat

The relative energetic value of fat and carbohydrates for body metabolism is dependent on whether the energy sources are used for body maintenance or tissue synthesis (Chudy and Schiemann, 1969; Nehring and Haenlein, 1973). In animals at maintenance, fat calories are utilized slightly less efficiently than carbohydrates. In growing animals, dietary fat calories appear to be preferentially utilized for body-fat synthesis. The efficiency of utilizing calories from fat and carbohydrates for body-fat synthesis is estimated on a biochemical basis to be 90% and 73%, respectively (Chudy and Schiemann, 1969; Nehring and Haenlein, 1973). The greater efficiency of utilizing fat calories for fatty tissue synthesis is because dietary fat is directly incorporated into body fat, thus minimizing the heat losses normally associated with fat synthesis from acetyl-CoA.

The efficiency of utilization of dietary fat for body-fat synthesis in pigs recently has been estimated to be greater than 90%, possibly because of an inhibitory effect of fat on fermentation and the associated energy losses in the hindgut of the pig (Just, 1982d). If the improved efficiency of energy utilization in pigs fed a fat-supplemented diet is assigned to the dietary fat fraction, the net energy value of fat is estimated to be 43.68 kJ/g of digestible fat (range of 41.96–44.73) or 118% of its ME content (Hillcoat and Annison, 1974; Jordan and Weatherup, 1976; Just, 1982d).

These estimates indicate that pigs fed isocaloric levels of ME would deposit an additional 0.42 g of body fat and 0.47 g of live weight for each additional g of digestible fat consumed (in lieu of soluble carbohydrates; *Table 16.4*). These growth responses of pigs to the isocaloric additions of fat should be achieved only in pigs fed a high level of ME and housed in a thermoneutral environment, because the lower heat production associated with dietary fat utilization would not be energetically advantageous in pigs maintained in a hypothermal environment or fed a submaintenance level of ME.

318 *Use of fats in diets for growing pigs*

Table 16.4 ESTIMATED GROWTH RESPONSES OF PIGS TO DIETARY FAT[a]

Feeding level	Daily gains, g	
	Body fat	Live weight[b]
Isocaloric ME intakes[c]		
Digestible fat, 1 g (37 kJ ME)	1.11	1.23
Digestible carbohydrates, Xg (37 kJ ME)	0.69	0.76
Extra caloric value of fat	+0.42	+0.47
Ad libitum intakes[d]		
Digestible fat, 1% added	+10.4	+11.6

[a] Growing pigs (25–95 kg body weight) assumed to be housed in a thermoneutral environment and fed a grain-based diet.
[b] Adipose tissue assumed to consist of 90% fat and 10% water.
[c] Pigs fed at a level equivalent to 90–100% of appetite.
[d] Fat source included in the diet at the expense of cereal grains.

Dietary fat addition also appears to stimulate the voluntary energy intake of pigs maintained in a warm or hot environment, but not of those housed in a hypothermal environment (Stahly and Cromwell, 1979; Stahly, Cromwell and Overfield, 1981). The voluntary ME intake of pigs housed in a warm environment is increased by 0.2–0.6% for each additional 1% fat added to the diet (Moser and Bitney, 1976; Seerley, McDaniel and McCampbell, 1978; Stahly and Cromwell, 1979; Stahly *et al.*, 1981). The elevated ME intake of pigs fed high-fat diets may be in response to a reduced burden of heat dissipation, particularly in a hyperthermal environment. The greater ME intake of pigs fed fat-supplemented diets may also be partly attributable to a lower proportion of the digestive end-products being glucose rather than a fat effect *per se* (Stephens, 1980).

If the combined effects of dietary fat addition on voluntary ME intake and efficiency of tissue synthesis are considered, pigs maintained in a thermoneutral environment and allowed to consume a diet based on grain–soybean meal *ad libitum* from 25 to 95 kg (average daily feed and ME intake of approximately 2.2 kg and 29.7 MJ, respectively) would be estimated to deposit an additional 10.4 g of body fat and 11.6 g of live weight gain daily over that expected from the ME content of the fat for each addition of 1% fat to the diet, assuming that the fat is 85% digestible (*Table 16.4*).

The influence of environmental temperature on the pig's response to dietary fat additions is shown in *Table 16.5*. In these studies, the dietary inclusion of 5% added fat increased ME intake by 3%, live weight gains by 9%, efficiency of energy utilization by 6% and body fat content by 6%, respectively, in pigs fed in a thermoneutral environment (22.5°C) and by 5, 9, 8 and 7% respectively in animals housed in a hyperthermal (35°C) environment. In contrast, the effects of dietary fat additions on the ME intake (−2%) and the rate (−1%), efficiency (0%) and composition (+4% fat) of growth were minimal in pigs maintained in a hypothermal (10°C) environment. A similar relationship between environmental temperature and dietary fat additions on rate and efficiency of growth has been observed in pigs fed in a warm versus cool seasonal (summer versus winter) or housing (open front versus environmentally regulated) environment (Seerley *et al.*, 1978; Moser, Fritschen and Peo, 1978; Stahly *et al.*, 1981).

As expected, the growth response induced by dietary fat in pigs housed

Table 16.5 THE EFFECT OF DIETARY FAT ADDITIONS ON THE RELATIVE PERFORMANCE OF PIGS HOUSED IN A COLD (10°C), THERMONEUTRAL (22.5°C) OR HOT (35°C) ENVIRONMENT[abc]

Temperature (°C)	10			22.5			35		
Added fat (%)	0	5	Change	0	5	Change	0	5	Change
Daily ME intake	114	112	−2	100	103	+3	72	77	+5
Daily gain	99	98	−1	100	109	+9	66	75	+9
ME intake/gain	116	116	0	100	94	−6	114	106	−8
Carcass backfat	93	97	+4	100	106	+6	85	92	+7

[a] From Stahly and Cromwell, 1979.
[b] Each mean represents 22 individually fed pigs in the 10° and 22.5°C environments and 14 individually fed pigs in the 35°C environment. Diets consisted of a fortified, corn–soybean mix containing 5% corn starch or tallow.
[c] The performance of pigs fed the basal diet and maintained in the 22.5°C environment was considered to be 100%.

Table 16.6 EFFECT OF DIETARY FAT ADDITIONS ON THE COMPOSITION OF GROWTH IN PIGS HOUSED IN COLD, THERMONEUTRAL OR HOT ENVIRONMENTS[ab]

Temperature (°C):	10		22.5		35	
Added fat (%):	0	5	0	5	0	5
ME intake (MJ/day)	30.79	29.54	27.57	26.60	21.57	20.96
Gains (g/day)						
Live weight	785	758	785	804	572	599
Protein	124	122	128	129	104	105
Fat	260	217	213	233	138	186
Energy retained						
Total (MJ of ME/day)	13.39	11.59	11.59	12.43	8.03	9.96
% as fat	77.2	74.4	73.0	74.5	68.3	74.2

[a] From Stahly, 1982.
[b] Each mean represents eight individually fed pigs. Pigs allowed to consume a corn–soybean meal diet containing 5% starch or tallow *ad libitum*.

Table 16.7 EFFECT OF DIETARY FAT ADDITIONS ON ENERGY UTILIZATION IN PIGS HOUSED IN A THERMONEUTRAL ENVIRONMENT[a,b]

Criteria	Added dietary fat (%)			
	0	2.5	5.0	7.5
ME intake (MJ/day)	24.23	24.36	23.96	23.68
Energy utilization (MJ/day)				
Heat production	13.39	13.14	12.91	12.52
Energy retained	10.77	11.26	11.01	11.28
Efficiency of ME utilization				
Heat production (% of ME)	55.3	53.9	53.8	52.8
Energy retained (% of MEp)	64.2	66.0	66.6	68.3

[a] From Hillcoat and Annison, 1974.
[b] Mean values of two fat sources (corn oil and tallow). Dietary fat source substituted for corn starch.

in a thermoneutral environment consists of an increased rate of body fat rather than of lean tissue deposition (Just, 1983; Stahly, 1982; *Table 16.6*). This elevated rate of fat deposition is apparently the result of more energy being available for tissue (fat) synthesis due to less heat being produced from fermentation and nutrient (direct fat deposition) metabolism (Hillcoat and Annison, 1974; Just, 1982d; *Table 16.7*).

In heat-stressed pigs, dietary fat additions partly alleviate the pig's burden of heat dissipation. Dietary fat additions may therefore increase the rate of lean growth as well as that of fat deposition in heat-stressed pigs, because of the stimulatory effect of dietary fat on the pig's voluntary feed intake, and thus on its ingestion of energy and essential amino acids needed for muscle synthesis.

In contrast, the heat generated from nutrient metabolism and fermentation in pigs maintained in a hypothermal environment is relatively efficiently utilized in meeting the maintenance needs of the pig. Thus, the lower level of heat production in pigs fed fat, results in absorbed nutrients being oxidized to maintain body temperature at the expense of fat deposition. In this instance, the efficiency of utilizing absorbed nutrients (ME) for tissue synthesis could actually be depressed by dietary fat additions.

Verification of the nutritional and economic value of fat for pigs

In an attempt to quantify the metabolic and economic value of dietary fat additions for pigs maintained under varying nutritional, environmental and managerial regimens, data reported from the USA were analysed by multiple regression techniques. The data were restricted to those studies involving pigs fed from approximately 21 to 93 kg in which both growth and carcass traits were reported. These comprised 31 trials (involving 1524 pigs) which were conducted in the USA from 1957 to 1982. The basal diets consisted predominantly of cereal grains and soybean meal. A variety of animal fats and vegetable oils were utilized in the studies, although the total dietary ratios of U/S fatty acids were generally greater than 1.3. The ME, fat, crude fibre, protein and lysine content of the diets were calculated from average composition values reported in the literature (NRC, 1979; Ewan, 1979). The nitrogen-free extract (NFE) content of the diets was calculated by difference. Dietary concentrations of ME, fat, crude fibre and protein ranged from 11.38 to 17.28 MJ/kg; 2% to 22%; 1.4% to 7.6% and 9% to 24%, respectively.

An environmental index based on the geographical location, season of the year and type of housing in which the experiment was conducted, was assigned to each trial. The environmental index, which represented the estimated ambient temperature in which the pigs were maintained during their finishing period of growth, ranged from $-5°$ to $28°C$. Approximately 60% of the trials were estimated to have been conducted in a thermoneutral environment (Holmes and Close, 1977).

A growth potential index derived from an estimated rate of muscle deposition (USDA, 1981) also was assigned to the pigs in each trial. Pigs utilized in the experiments conducted in the 1950s and 1960s exhibited a slower estimated rate of muscle deposition than those fed in the 1970s.

A stepwise regression programme utilizing backward elimination (SAS, 1982) was used to identify the linear, quadratic and cross-product components that accounted for a significant ($P<0.10$) proportion of the variability in the pig response to dietary fat additions. The linear components of the independent variables (daily intakes of NFE, fat, fibre, protein (lysine) and the environmental and growth indexes) were forced into the regression equation unless the quadratic and cross-product components of a particular independent variable did not account for a significant proportion of the variability. The regression equations used accounted for 97, 75, 64, 87 and 72% of the variability observed in ME intake, daily live weight gains, ME required per unit of gain, daily muscle deposition and carcass backfat, respectively.

The effect of dietary fat intake on the pig's voluntary ME intake was not influenced by the other independent variables. The effects of dietary fat intake on the growth rate and efficiency of energy utilization in pigs were influenced by the amounts of dietary fat and NFE consumed. The beneficial effect of dietary fat intake on rate and efficiency of growth was reduced as the pig's daily intake of digestible carbohydrate and thus ME were increased. The effect of dietary fat intake on carcass backfat was dependent on the environment in which the pigs were housed, the level of dietary protein consumed and the lean growth potential (index) of the pigs. The dietary addition of fat increased backfat thickness in pigs maintained in a warm environment and fed an adequate dietary protein level, but it had only a marginal effect on the carcass backfat of pigs raised in a cold environment.

The ME intake of the pigs allowed to consume diets based on grain –soybean meal was increased by 38.45 kJ/kg of digested fat consumed or 1.45 kJ over the ME value of the fat *per se*. Fat was assumed to be 85% digestible and was assigned a ME value of 37 kJ/g of digested fat. The relatively small effect of dietary fat on the pig's voluntary ME intake is probably the result of 40% of the studies being conducted in a cold environment. The absence of an interactive effect between the dietary fat intake and environmental index on the pig's voluntary ME intake is probably due to the high variation or inaccuracy in the environmental index assigned to each trial, relative to the small effect of dietary fat on the total ME intake of pigs.

Daily live weight gains of pigs were increased by 1.64–1.72 g per g of digested fat consumed daily in pigs fed a level of ME equivalent to approximately 2.75 times their maintenance (460 kJ ME/kg$^{0.75}$) requirement (*Figure 16.2*). The growth response to dietary fat was reduced to 1.22–1.30 and 0.81–0.89 g per g of digested fat ingested in pigs consuming levels of ME equivalent to 3.25 and 3.75 times maintenance, respectively. The pigs' diminishing growth response to dietary fat additions as their voluntary ME intake increased is probably because the animals which consumed the highest levels of ME were housed in a cold environment. Pigs (25–95 kg) housed in a thermoneutral environment are estimated to consume approximately 3.1–3.5 times their maintenance requirement. On this basis, the live weight gains of pigs in a thermoneutral environment were increased by approximately 1.2 g per g of digested fat consumed or 0.44 g over that expected on the ME value of the fat alone. Using these

322 Use of fats in diets for growing pigs

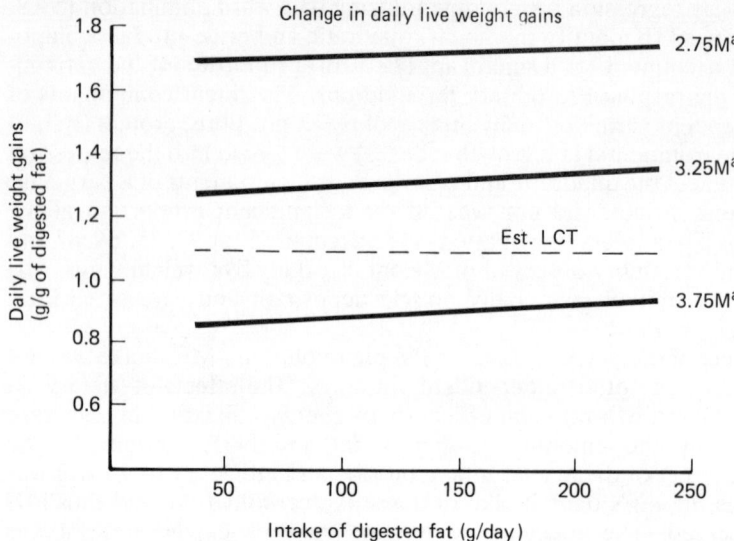

Figure 16.2 Change in daily live weight gain (g) per g of digested fat consumed daily. Est. LCT = estimated lower critical temperature. a: level of feed intake above maintenance level (M).

data, the net energy value of the fat is estimated to be 42.47 kJ/g of digested fat consumed in pigs housed in a thermoneutral environment. This value is slightly less than that previously determined (43.68 kJ) in a limited number of studies conducted under controlled experimental conditions. In contrast, live weight gains of pigs housed in a cold environment (ME intakes of 3.75–4.25 times maintenance) were increased by only 0.7 g per g of digested fat consumed or 0.10 g less than expected on the basis of the ME content of the fat. The net energy value of fat in the cold environment is estimated to be 23.26 kJ per g of digested fat consumed. This low efficiency of utilizing dietary fat calories in the cold environment supports the concept that added dietary fat depresses the level of fermentation and associated heat production in the pig. Reducing heat production in cold-stressed animals in this way would therefore force the pig to oxidize additional quantities of absorbed nutrients, which in turn would reduce the amount of energy available for tissue synthesis.

As expected, the growth response induced by dietary fat in pigs housed in a warm environment largely consisted of an increased level of body-fat deposition (*Figure 16.3*). Carcass backfat thickness was increased by 0.034–0.036 mm for each additional gram of digested fat ingested in pigs maintained in a thermoneutral environment (20°C) and fed an adequate level of dietary protein. The higher level of fat deposition occurred in pigs with the higher growth potential. The failure of carcass backfat to be altered by dietary fat additions in pigs housed in a cold environment is in agreement with previous work (Stahly and Cromwell, 1979; Stahly *et al.*, 1981).

The estimated rate of muscle deposition was increased slightly by dietary fat supplementation (*Figure 16.4*), but this is probably an artefact related

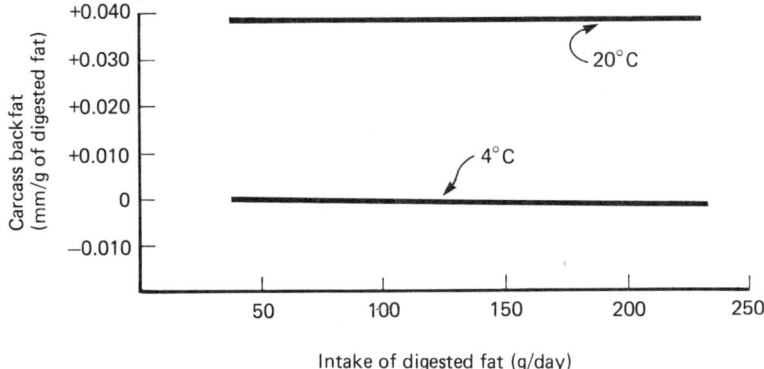

Figure 16.3 Change in carcass backfat* thickness (mm) per g of digested fat consumed daily. *(Adequate daily protein intake).

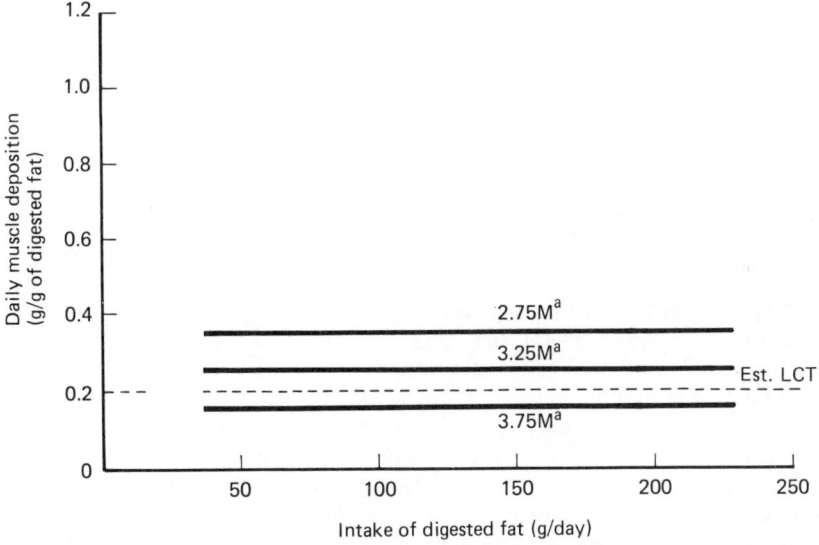

Figure 16.4 Change in daily muscle deposition (g) per g of digested fat consumed daily. a: level of feed intake above maintenance level (M).

to the assumption made in calculating the rate of muscle deposition (USDA, 1981) that the muscle produced contains 10% lipid. As dietary fat additions increase the intramuscular fat content of pork muscles (Smith and Carpenter, 1976; Cromwell et al., 1978), the apparent increase in muscle deposition attributable to dietary fat in pigs maintained in a warm environment (2.75–3.50 maintenance) is probably a reflection of the increased fat content of the muscle. The minimal effect of dietary fat on muscle deposition in pigs housed in the cold environment is probably attributable to a lower level of fat being incorporated in the muscle tissue.

324 Use of fats in diets for growing pigs

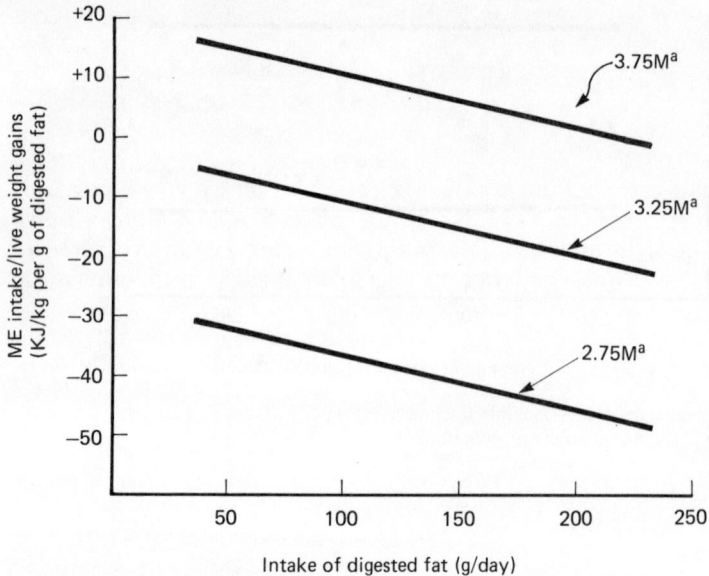

Figure 16.5 Change in ME (kJ) required/kg of gain per g of digested fat consumed daily. a: level of feed intake above maintenance level (M).

The pig's growth response to dietary fat was not altered by the animal's daily intakes of crude protein or fibre. However, the stimulatory effect of dietary fat on fat and muscle deposition was less as the pig's dietary protein intakes increased.

Energy required per unit of weight gain (ME/gain, kJ/kg) was reduced by 32.30–45.48 and 8.54–21.76 kJ per g of digested fat consumed in pigs consuming 2.75 and 3.25 times maintenance, respectively (*Figure 16.5*). In pigs which, it is assumed, were housed in a cold environment (3.75 times maintenance), the energy required per unit of weight gain was actually increased by 1.67–15.06 kJ per g of digested fat consumed. These results support the suggestion that the ingestion of fat calories and the resulting reduction in heat production in cold-stressed animals results in the pig oxidizing alternative nutrient sources at the expense of fat deposition.

Using these data, the energetic value of a kilojoule of metabolizable energy from fat is estimated to be 1.57 times greater than that from

Table 16.8 RELATIVE ENERGETIC VALUE OF DIETARY FAT FOR GROWING PIGS[a]

Nutrient	Environment		
	Hypothermal	Thermoneutral	Hyperthermal
	Energetic value relative to 1 MJ of ME from carbohydrates		
Carbohydrate, digestible	1.00	1.00	1.00
Fat, digestible	0.86	1.57	>1.57
Fat, as % of carbohydrates	86	157	>157

[a] Pigs allowed to consume a grain-based diet *ad libitum* from 25 to 95 kg body weight.

carbohydrates in pigs housed in a thermoneutral environment and fed a level of ME equivalent to 3–3.5 times maintenance (Table 16.8).

The relative economic value of dietary fat may be estimated from the effects of dietary fat additions on the time (days) and MJ of ME required to produce a market-weight pig (25–95 kg body weight) and the backfat thickness of the pig's carcass. On the basis of these three criteria, the effects of dietary fat additions on the cost (non-feed and feed) of producing a market-weight pig and the value of the market animal (carcass) produced, may be determined (Table 16.9). For example, days required to produce 70 kg of live weight gain in pigs maintained in a thermoneutral environment and allowed to consume a grain-based diet *ad libitum* were reduced by approximately 0.057 days for each additional g of digested fat consumed daily in lieu of carbohydrates. ME required to produce 70 kg of live weight gain were reduced by 0.58, 0.79, 1.00, 1.17, 1.34 and 1.51 MJ per g of digested fat consumed in pigs ingesting 38, 77, 116, 154, 193 and 232 g of digested fat daily. The value of dietary fat was partly offset by an increased backfat thickness of 0.035 mm for each additional g of digested fat.

The intramuscular fat content is also increased in pigs fed added dietary fat which may enhance the eating qualities of pork (Table 16.10) from pigs fed dietary fat (Smith and Carpenter, 1976; Cromwell *et al.*, 1978). The firmness and polyunsaturated fat content of the carcass and its susceptibility to oxidative rancidity is related to the proportion of unsaturated fatty

Table 16.9 RELATIVE ECONOMIC VALUE OF DIETARY FAT FOR GROWING PIGS[a]

Criteria	Environment:	Hypothermal		Thermoneutral	
	Level of response:	Average[b]	Range[c]	Average[b]	Range[c]
		Response/g of digestible fat consumed daily[d]			
Days on feed/pig[e]		+0.003	+0.012 to −0.005	−0.057	−0.052 to −0.062
ME required/pig (MJ)[e]		+0.585	+1.05 to +0.12	−1.050	−0.58 to −1.52
Carcass backfat (mm)[f]		−0.002	−0.008 to +0.004	+0.035	+0.029 to +0.041

[a] Pigs allowed to consume a grain-based diet *ad libitum* from 25 to 95 kg body weight.
[b] Average daily intake of 135 g of digestible fat.
[c] Pigs ingesting 38–232 g of digestible fat daily.
[d] Digestible fat consumed in lieu of an isocaloric level of ME from carbohydrates.
[e] Responses to fat independent of daily protein intake
[f] Backfat response in pigs which consume daily intakes of protein (amino acid) that meet or exceed the pig's daily requirement. Ambient temperatures in hypothermal and thermoneutral environments assumed to be 1–7 and 18–24°C, respectively, during the finishing stage of growth

Table 16.10 RELATIONSHIP BETWEEN EATING QUALITY AND COMPOSITION OF PORK[a]

Consumer satisfaction[b]	Proportion of longissimus muscle (%)		
	Fat	Protein	Moisture
6.0–8.0	9.1	19.4	70.1
5.1–6.0	4.8	20.2	73.5
4.0–5.0	4.0	20.8	73.7

[a] From Smith and Carpenter (1976).
[b] Consumer overall satisfaction scores: extremely desirable = 9; extremely undesirable = 1.

acids in the dietary fat mix, but the economic significance of these effects in terms of consumer acceptance or demand has been minimal for pork products from pigs fed low levels of supplemental fat (Smith and Carpenter, 1976; Lewis et al., 1976; Hartman et al., 1983).

Quality control considerations for dietary fat sources

As with all feed ingredients, quality control specifications must be established and adhered to in order to maximize the nutritional value of the supplemental fat sources used and to ensure the wholesomeness of the final feed mixture. The quality control specifications recommended for feed fats are outlined by Hathaway (1975) and Fuller (1975). Field reports suggest that diets containing high levels of free fatty acids are less palatable to pigs. Blended fats containing a high level of free fatty acids have been reported to be slightly less palatable than those which consist predominantly of triglycerides (Leibbrant et al., 1975). The presence of odours or off-flavours in fat may also influence diet preference in pigs in addition to potentially altering the palatability of pork end-products for the human consumer. Peroxide formation, which is an intermediate step in oxidative rancidity, has not been shown to be closely related to pig performance (Carpenter, 1968); however, the inclusion of an antioxidant and the testing of the stability of feed fats (AOM test) is recommended.

Dietary formulation considerations

The ingestion of long-chain fatty acids *per se* does not depress (and may enhance) protein digestibility (Just, 1982d; Sauer et al., 1980; *Table 16.11*). Similarly, protein retention in pigs is not altered by dietary fat additions (Hillcoat and Annison, 1974; Stahly, 1982), with the possible exception of medium-chain fatty acids (Jordan and Weatherup, 1976; Ehrensvard et al., 1976; DeWilde, 1981). Therefore the pig's daily requirement for essential amino acids is not detrimentally influenced by the ingestion of long-chain

Table 16.11 EFFECT OF DIETARY FAT ON THE APPARENT ILEAL AND FECAL DIGESTIBILITY OF CRUDE PROTEIN AND AMINO ACIDS IN PIGS[(ab)]

Site:	Ileal			Faecal		
Dietary fat (%):	4.5	17.0	26.8	4.5	17.0	26.8
	Apparent digestion coefficients (%)					
Crude protein	74	73	76	82	81	85
Amino acids						
Lysine	83	84	85	83	84	87
Tryptophan	77	79	81	89	89	92
Threonine	69	70	73	84	85	88
Methionine	83	82	84	83	85	86
Average	78	79	81	87	88	90

[(a)]From Sauer et al., 1980.
[(b)]Pigs (50–70 kg body weight) were fed isonitrogenous diets at the rate of 1.42 kg of dry matter daily.

fats. However, because the pig tends to eat to meet its energy need, the concentrations of amino acids in fat-supplemented diets containing marginal levels of protein need to be increased in proportion to changes in the energy density of the diet.

The effect of a dietary addition of a highly digestible fat on mineral absorption in pigs also appears to be minimal (Jordan and Weatherup, 1978). Therefore, dietary mineral concentrations which allow the pig to consume the daily mineral intakes recommended by the ARC (1981) and NRC (1979) would appear to be adequate.

High levels of dietary fibre are potentially antagonistic to the beneficial effects of fat on efficiency of energy utilization, in that dietary fibre depresses nutrient absorption in the small intestine and stimulates fermentation in the hind gut of pigs (Just, 1982c,d). The degree to which the energetic value of dietary fat and fibre for tissue synthesis are influenced by their respective dietary concentrations is not known.

Summary

The nutritional value of fat as an energy source for pigs is influenced largely by the digestibility of the dietary fat mix (U/S ratio), the quantity of ME and fat consumed daily by the pigs and the environmental temperature in which the pigs are housed. The relative economic value of fat as a dietary energy source is determined largely by the relative cost of a unit of ME from fat versus carbohydrates, the environment in which the pigs are housed and the magnitude of the financial incentive for the production of lean carcasses. In the production of commercial feeds, the effect of dietary fat on diet palatability and pellet durability must also be assessed.

Acknowledgements

Appreciation is expressed to Mr Bryan Rudolph and Professor Richard Anderson for their assistance in statistically analysing portions of the data included in this chapter and Ms Virginia Moore for her efforts in typing the manuscript.

The investigation reported in this paper (No. 83-5-59) is in connection with a project of the Kentucky Agricultural Experiment Station and is published with approval of the Director.

References

ARC (1981). *The Nutrient Requirements of the Pig.* London, Commonwealth Agricultural Bureaux–Agricultural Research Council
BAYLEY, H.S. and LEWIS, D. (1965). The use of fats in pig feeding. II. The digestibility of various fats and fatty acids. *Journal of Agricultural Science* **64**, 373–378
BRAUDE, R. and NEWPORT, M.J. (1973). Artificial rearing of pigs. 4. The

replacement of butterfat in a whole milk diet by either beef tallow, coconut oil or soya-bean oil. *British Journal of Nutrition* **29**, 447–455

CARLSON, W.E. and BAYLEY, H.S. (1968). Utilization of fat by young pigs: fatty acid composition of ingesta in different regions of the digestive tract and apparent and corrected digestibilities of corn oil, lard and tallow. *Canadian Journal of Animal Science* **48**, 315–322

CARLSON, W.E. and BAYLEY, H.S. (1972). Digestion of fat by young pigs: a study of the amounts of fatty acid in the digestive tract using a fat-soluble indicator of absorption. *British Journal of Nutrition* **28**, 339–346

CARPENTER, K.J. (1968). Possible adverse effects of oxidized fat in feeds. In *Proceedings, Nutrition Conference for Feed Manufacturers* (H. Swan and D. Lewis, Eds), pp. 54–71. Nottingham, England, J.A. Churchill

CASTER, W.O., AHM, P., HILL, E.G., MOHRHAUER, H. and HOLMA, R.T. (1962). Determination of linoleate requirement of swine by a new method of estimating nutritional requirement. *Journal of Nutrition* **78**, 147

CHENG, A.L.S., MOREHOUSE, M.G. and DEUEL, HARRY J., Jr (1949). The effect of the level of dietary calcium and phosphorus on the digestibility of fatty acids, simple triglycerides and some natural and hydrogenated fats. *Journal of Nutrition* **37**, 237–250

CHRISTENSEN, K., HENCKEL, S. and THORBEK, G. (1979). Effect of different dietary levels of linoleic acid on energy metabolism and mitochondrial activity in young pigs. In *Eighth Symposium on Energy Metabolism* (L.E. Mount, Ed.), pp. 107–110. EAAP Publication 26. London, Butterworths

CHUDY, A. and SCHIEMANN, R. (1969). Utilization of dietary fat for maintenance and fat deposition in model studies with rats. In *Energy Metabolism of Farm Animals* (K.L. Blaxter, J. Kielanowski and G. Thorbek, Eds), pp. 161–170. EAAP Publication 12. Newcastle upon Tyne, Oriel Press Ltd

CROMWELL, G.L., HAYS, V.W., TRUJILLO-FIGUEROA, V. and KEMP, J.D. (1978). Effects of dietary protein and energy levels for growing-finishing swine on performance, muscle composition and eating quality of pork. *Journal of Animal Science* **47**, 505–513

DAVIS, R.H. and LEWIS, D. (1969). The digestibility of fats differing in glyceride structure and their effects on growth performance and carcass composition of bacon pigs. *Journal of Agricultural Science* **72**, 217–222

DEWILDE, R.O. (1981). De energiewaarede van voedervetten bij mestvarkens. 1981. *Landbouwtijdschrift* **34**, 249–258

DIGGS, B.G., BECKER, D.E., JENSEN, A.H. and NORTON, H.W. (1965). Energy value of various feedstuffs for the young pig. *Journal of Animal Science* **24**, 555–558

EDWARDS, H.M., Jr (1964). Fatty acid composition of feedstuffs. *Georgia Agricultural Experimental Station Technical Bulletin 36*. Athens, University of Georgia

EHRENSVARD, U., BERSCHAUER, F., MENKE, K.H., RODGDAKIS, E. and STURM, G. (1976). Effect of ration composition on plasma insulin and partition of energy for protein and fat synthesis in the growing pig. In *Energy Metabolism of Farm Animals* (M. Vermorel, Ed.), pp. 85–88. EAAP Publication 19. Institut National de la Recherche Agronomique

EWAN, R.C. (1979). *Energy Value of Food Ingredients*. Ames, Iowa State University
FREEMAN, C.P., HOLME, D.W. and ANNISON, E.F. (1968). The determination of the true digestibilities of interesterified fats in young pigs. *British Journal of Nutrition* **22**, 651–660
FROBISH, L.T., HAYS, V.W., SPEER, V.C. and EWAN, R.C. (1970). Effect of fat source and level on utilization of fat by young pigs. *Journal of Animal Science* **30**, 197–202
FULLER, H.L. (1975). Effect of processing animal and vegetable fats on their stability and nutritional value. In *Effect of Processing on the Nutritional Value of Feeds*, pp. 131–141. Washington, DC, National Academy of Sciences
HAMILTON, R.M. and McDONALD, B.E. (1969). Effect of dietary fat source on the apparent digestibility of fat and the composition of fecal lipids of the young pig. *Journal of Nutrition* **97**, 33–41
HARTMAN, A.D., COSTELLO, W.J., WAHLSTROM, R.C. and LIBAL, G.W. (1983). Effect of various levels of sunflower seeds on carcass characteristics of swine. *Journal of Animal Science* **57** (Suppl. 1), 79
HATHAWAY, H.D. (1975). Quality control of fats for animal feed. In *Proceedings of the Georgia Nutrition Conference*, pp. 47–55. Athens, Georgia, University of Georgia
HILL, E.G., WARMANEN, E.L., SILBERNICK, C.L. and HOLMAN, R.T. (1961). Essential fatty acid nutrition in swine. 1. Linoleate requirement estimated from triene:tetraene ratio of tissue lipids. *Journal of Nutrition* **74**, 335–341
HILLCOAT, J.B. and ANNISON, E.F. (1974). The efficiency of utilization of diets containing maize oil, tallow and tallow acid oil in the pig. *Energy Metabolism of Farm Animals* (K.H. Menke, H.J. Lanze and J.R. Reichl, Eds), pp. 141–143. EAAP, Publication 14. Stuttgart, B.R.D., Universität Hohenheim
HOLMES, C.W. and CLOSE, W.H. (1977). The influence of climatic variables on energy metabolism and associated aspects of productivity in the pig. In *Nutrition and Climatic Environment* (W. Haresign, H. Swan and D. Lewis, Eds), pp. 51–73. London, Butterworths
JORDAN, J.W. and WEATHERUP, S.T.C. (1976). The effect of fat and carbohydrate and their ration in the diet on energy and protein utilization by early weaned pigs. In *Energy Metabolism of Farm Animals* (M. Vermorel, Ed.), pp. 150–152. Beaumont, France, Institut National de la Recherche Agronomique. EAAP Publication 19
JORDAN, J.W. and WEATHERUP, S.T.C. (1978). The effect of dietary fat levels on the absorption and retention of calcium, phosphorus and magnesium by early weaned pigs. *Records of Agricultural Research* **26**, 29–34
JUST, A. (1982a). The influence of ground barley straw on the net energy value of diets for growth in pigs. *Livestock Production Science* **9**, 717–729
JUST, A. (1982b). The net energy value of balanced diets for growth in pigs. *Livestock Production Science* **8**, 541–555
JUST, A. (1982c). The influence of crude fiber from cereals on the net energy value of diets for growth in pigs. *Livestock Production Science* **9**, 569–580

JUST, A. (1982d). The net value of crude fat for growing pigs. *Livestock Production Science* **9**, 501–509

JUST, A. (1983). Nutritional manipulation and interpretation of body composition in growing swine. *Journal of Animal Science* (in press)

KIDDER, D.E. and MANNER, M.J. (1978). *Digestion in the pig*. Bristol, England, Scientechnica

LEAT, W.M.F. (1961). Studies on pigs reared on diets low in tocopherol and essential fatty acids. *British Journal of Nutrition* **15**, 259–270

LEIBBRANDT, V.D., HAYS, V.W., EWAN, R.C. and SPEER, V.C. (1975). Effect of fat on performance of baby and growing pigs. *Journal of Animal Science* **40**, 1081–1085

LEWIS, P.K., NOLAND, P.R., BROWN, C.J. and HECK, M.C. (1976). *Effects of Feeding Rancid Fat in the Diet on the Carcass and Storage Quality of Pork. Bulletin 818.* Fayetteville, University of Arkansas

MARTIN, C.K. (1977). *Effect of Dietary Fat Source on Performance and Carcass Characteristics of Growing-Finishing Swine*. MS thesis, University of Kentucky, Lexington

MASON, V.C. and JUST, A. (1976). Bacterial activity in the hindgut of pigs. 1. Its influence on the apparent digestibility of dietary energy and fat. *Zeitschrift für Tierphysiologie, Tierernährung und Futtermittelkunde* **36**, 301–310

MOSER, B.D. and BITNEY, L. (1976). G-F swine diets. In *Nebraska Swine Report. E.C.* pp. 76–219

MOSER, B.D., FRITSCHEN, R.D. and PEO, E.R., Jr (1978). High energy diets fed to G-F swine in different housing facilities. *Journal of Animal Science* **47** (Suppl. 1), 6

NRC (1959). *Joint US–Canadian Tables of Feed Composition. National Academy of Sciences Publication 659*. Washington, DC, National Academy of Sciences

NRC (1979). *Nutrient Requirements of Domestic Animals. No. 2. Nutrient Requirements of Pigs*, eighth revised edition. Washington, DC, National Academy of Sciences

NEHRING, K. and HAENLEIN, G.F.W. (1973). Feed evaluation and ration calculation based on net energy. *Journal of Animal Science* **36**, 949–964

PHILLIPS, B.C. and EWAN, R.C. (1977). Utilization of milo and soybean oil by young swine. *Journal of Animal Science* **44**, 990–997

SAS (1982). *SAS Users Guide* (A.A. Ray, Ed.). Cary, NC, SAS Institute Inc.

SAUER, W.C., JUST, A., JORGENSEN, H.H., FEKADU, M. and EGGUM, B.O. (1980). The influence of diet composition on the apparent digestibility of crude protein and amino acids at the terminal ileum and overall in pigs. *Acta agriculturae scandinavica* **30**, 449–459

SCHERER, C.W. (1973). *Effects of Dietary Fat and Age on Performance, Digestibility and Pancreas Lipase Activity in Pigs*. PhD dissertation, University of Kentucky, Lexington

SEERLEY, R.W., McDANIEL, M.C. and McCAMPBELL, H.C. (1978). Environmental influence on utilization of energy in swine diets. *Journal of Animal Science* **47**, 427–434

SMITH, G.C. and CARPENTER, Z.L. (1976). Eating quality of meat animal products and their fat content. In *Fat Content and Composition of*

Animal Products (National Research Council, Eds), pp. 147–182. Washington, DC, National Academy of Sciences

STAHLY, T.S. (1982). The energetics of growth in pigs as influenced by environmental temperature and diet composition. In *Proceedings, Maryland Nutrition Conference* (J.D. DeBarthe, Ed.), pp. 58–64. Washington, DC, University of Maryland

STAHLY, T.S. and CROMWELL, G.L. (1979). Effect of environmental temperature and dietary fat supplementation on the performance and carcass characteristics of growing and finishing swine. *Journal of Animal Science* **49**, 1478–1488

STAHLY, T.S., CROMWELL, G.L. and OVERFIELD, J.R. (1981). Interactive effects of season of year and dietary fat supplementation, lysine source and lysine level on the performance of swine. *Journal of Animal Science* **53**, 1269–1277

STEPHENS, D.E. (1980). The effects of alimentary infusions of glucose, amino acids, or neutral fat on meal size in hungry pigs. *Journal of Physiology* **299**, 453–463

TOLLETT, J. (1961). *The Available Energy Content of Feedstuffs for Swine*. PhD dissertation, University of Illinois, Champaign

USDA (1981). *Guidelines for Uniform Swine Improvement Programs*. National Swine Improvement Federation and US Department of Agriculture (SEA, Program Aid 1157). Washington, DC, US Printing Office

THE USE OF FAT IN SOW DIETS

R.W. SEERLEY
Department of Animal and Dairy Science, University of Georgia, Athens, Georgia 30602, USA

The neonatal pig

Incidence and causes of mortality

Pre-weaning mortality is a major source of loss to the pig industry. The period of highest death loss of newborn pigs is the first 4 days after parturition, with approximately 70% of all pre-weaning deaths occurring during this period (Pomeroy, 1960). Although newborn pigs appear mature at birth, there is a 10–30% mortality between birth and weaning (ARS, 1965; Leman *et al.*, 1972).

The evidence available suggests that baby pigs have low energy reserves at birth, experience rapid body heat losses after birth and are subject to stresses of being small at birth, born in cool facilities, and encounter littermate competition for an inadequate supply of poorer-quality milk. Newborn pigs with a lower birth weight than 900 g have a poor chance of survival (Pomeroy, 1960; Kernkamp, 1965; Sharpe, 1966). These smaller pigs have larger body surface areas and, therefore, lose body heat more rapidly than larger littermates (Stanton and Carroll, 1974). In comparison with larger pigs, smaller pigs have certain anatomical differences that are suggestive of morphological immaturity. It has been observed that small pigs have a lower percentage of protein and DNA in some organs, larger brains, hearts, spleens and stomachs, smaller and less mature bones and less highly developed brains chemically than larger littermates (Widdowson, 1971; Adams, 1971; Dickerson, Meret and Widdowson, 1971). The immaturity of these animals indicates that the maturation process of these neonates is not complete and special care must be taken to protect the animal from extensive extra-uterine stress (Noland and Johnson, 1972).

Piglets require a warm and dry environment for survival, especially during the first few days after birth. They must adjust to a new energy budget at birth. They are susceptible to cold, and as a direct effect, chilling may irreversibly lower body temperature of neonatal pigs (Curtis, 1970). In litters farrowed during cold weather, chilling reportedly caused about a quarter of the piglet deaths during the first postnatal day but less than 5%

of those on the second and third days. After 5 days of age, no deaths were attributed directly to chilling (Bauman, Kadlec and Powlen, 1966). Indirectly, chilling may weaken defensive responses to certain critical challenges (Curtis, Heidemeich and Harrington, 1967).

Thermostability

Little is known about specific mechanisms of thermoregulation in pigs. Wild piglets are confronted with, and adapt better to, more rigorous external environments than domestic piglets. Studies of the adaptive mechanisms of these pigs have contributed to the understanding of thermoregulation in domestic piglets.

Wild piglets are more thermostable, as judged by a cold test, than domestic piglets, and the effect cannot be explained by surface to mass ratio since wild piglets weighed less than domestic piglets (Foley *et al.*, 1971). Resistance to cold stress increases by 3 days of age in both strains of pigs (Curtis *et al.*, 1967). This corresponds well to improved survivability after the immediate postpartum period. Pelage insulation is the major component of total thermal insulation in most animals (Bianca, 1968). At 6 hours of age, pelage removal is more deleterious to cold resistance in wild than in domestic piglets (Foley *et al.*, 1971). Neonatal wild pigs have greater average pelage weight density and pelage population density than domestic piglets. Both traits relate directly to insulation (Hansen *et al.*, 1972). In fact, pelage is so slight in domestic piglets that its removal does not greatly alter thermostability. Mount (1964) estimated that only 15% of total thermal insulation is contributed by pelage in the domestic pig.

Oxygen consumption rates during cold stress were higher in wild than domestic piglets. In domestic pigs, this rate was higher at 54 hours than at 6 hours of age. In contrast, metabolic response to cold by wild piglets was similar at both times and probably accounted in part for the greater thermostability of the wild piglet. Removal of pelage caused no change in oxygen consumption rate during cold stress in wild or domestic species, suggesting that this rate was at or near its maximum (Foley *et al.*, 1971). The advantage of the wild pig seemed to be attributable partly to extra pelage and partly to the more metabolic reaction to cold. The ability to enhance this metabolic response to cold with age appeared to be important to thermoregulation in the domestic pig.

Other types of physiological immaturity undoubtedly contribute to poor survival rates. The pig has no brown fat; this further complicates its ability to maintain body temperature (Manners and McCrea, 1963). The small amount of carcass fat of 1–2% at birth, limits the lipid available as a substrate for oxidation. Consequently, the piglet is dependent almost exclusively on carbohydrate metabolism for subsistence since even if metabolic pathways for the utilization of lipid are present, the substrate supply is limited. For the weak pig, these metabolic defects may be disastrous (Mersmann, 1974).

Carbohydrate metabolism

Graham, Sampson and Hester (1941) described a fatal condition in newborn pigs characterized by acute hypoglycaemia. This condition can be readily produced in newborn pigs by fasting (Sampson, Hester and Graham, 1942) and is related to the age of the newborn pig. Hanawalt and Sampson (1947) showed that pigs subjected to fasting 12–24 h after farrowing were highly susceptible to the development of hypoglycaemia, whereas those which were allowed to nurse for 6 days were less susceptible. The mechanism for this increased tolerance to fasting is unknown. Impaired glycolysis does not account for the susceptibility to fasting since loss of hepatic glycogen occurs rapidly during a fast (McCance and Widdowson, 1959).

The susceptibility of the newborn pig to hypoglycaemia places a major emphasis on carbohydrate energy storage. There is little glycogen in the liver early in gestation, but at later stages it increases rapidly and reaches higher than adult concentrations by the end of gestation. The time of increased glycogen storage appears to be related to gestation length. In those species with long gestation periods, accumulation begins early and proceeds at a steady rate. In species where the gestation period is short, such as the pig, the rise occurs later in gestation and proceeds at a rapid rate until term (Shelley, 1961). Traces of glycogen are present in the pig liver as early as day 44 of gestation but glycogen storage does not increase appreciably until day 86. Similarly, carcass glycogen begins to rise sharply after day 67 of gestation. The carcass contributes about 90% of the total body glycogen throughout gestation. By term, liver glycogen supplies most of the remaining 10% of total glycogen. Other tissues such as the heart and lungs contribute to glycogen reserves during gestation: however, the importance of heart and lung tissue glycogen decreases with gestational age, while liver glycogen increases (Okai *et al.*, 1978).

The liver stores of glycogen fall rapidly after birth (12–18 h), whereas in skeletal muscle the rate of decrease after birth is less rapid (36–48 h). The mobilization of glycogen is reflected by an initial rise in blood glucose levels during the first 6 hours of fasting after which it is poorly maintained (Goodwin, 1957; Swiatek *et al.*, 1968; Gent *et al.*, 1970). The reliance on carbohydrate stores as the major metabolic fuel immediately after birth is common to most species; however, in the neonatal pig, it is even more important since depot lipid levels are so low. The dependence of the pig on glycolysis, the major pathway for glucose utilization, is indicated by high plasma lactate levels at birth (Pettigrew, Zimmerman and Ewan, 1971).

Several factors are involved in the maintenance of blood glucose levels. Gluconeogenesis is important to this and is mainly an activity of the liver. Gluconeogenic capacity is deficient in fetal animals from most species (Ballard, 1970; Walker, 1971). It therefore seems likely that there is an inadequate response of gluconeogenesis in the neonatal pig. The mechanisms for an impaired response may include defects or lack of enzymes, poor hormonal response or inadequate substrates for gluconeogenesis (Swiatek *et al.*, 1968).

The postnatal period is associated with rapid changes in certain hepatic enzymes in some species. The fact that the 4-day-old pig does not develop hypoglycaemia during fasting, and that even the 6-hour-fed newborn animal exhibits increased tolerance to starvation, indicates that enzyme defects are not the cause of reduced gluconeogenesis. Liver slices from 1-day-old pigs exhibit the capacity to convert certain gluconeogenic substrates into glucose, while livers from fasted animals have significantly increased gluconeogenic capacity (Swiatek, 1971). Further, substantial activity of most hepatic gluconeogenic enzymes is present at birth, and these activities are increased by ageing and fasting (Swiatek *et al.*, 1970). As considerable enzyme activities are present at birth, primary regulation of gluconeogenesis is probably by some means other than enzyme synthesis.

Several hormones are known to influence gluconeogenesis. During starvation, peripheral glucose uptake by the insulin-dependent skeletal muscle, liver and adipose tissues is severely curtailed. This preserves available glucose for tissues with an absolute requirement for sugar. The newborn pig responds in a normal fashion to starvation by decreasing insulin and increasing growth hormone secretion. It is not felt that impaired gluconeogenesis can be attributed to inappropriate response of these hormones (Swiatek *et al.*, 1968).

Lipid metabolism

During starvation, the most striking difference between newborn and older pigs, other than hypoglycaemia, is the plasma free fatty acid (FFA) level. Investigations with other species (sheep, rat, man) have provided information about the physiological significance of plasma FFA in energy metabolism during starvation. Within the first 2 hours of life of unfed lambs and human infants, there is a very marked rise in plasma FFA as a result of increased lipid mobilization (Van Duyne and Havel, 1959; Persson and Gentz, 1966). This increase in FFA level is sixfold compared with only a twofold increase in the neonatal pig.

In the pig at birth, total body fat represents 1% of the carcass; however, much of this fat may be present as structural fat and not available for mobilization. Further, the concentration of plasma glycerol and β-hydroxybutyrate, which are indicators of lipid mobilization and utilization, are low compared with those in human infants. This indicates that the total amount of lipid mobilized during a fast is small and is also compatible with the fact that the unfed pig is very susceptible to starvation. The hormone-responsive lipolytic system appears to be functional in newborn adipose tissue because there is at least some increase in plasma FFA (twofold) (Swiatek *et al.*, 1968). The failure to maintain an elevated plasma FFA level seems to indicate a depletion of available mobilizable fat stores rather than a deficiency in a hormone-responsive system.

When pigs are allowed to nurse, the plasma level of FFA and glycerol increases rapidly after birth. From 24 hours to 4 weeks of life there is a progressive rise of plasma FFA, glycerol and glucose that coincides with accumulation of body fat. Concurrently, the pig develops an increased ability to tolerate starvation (Goodwin, 1957; Swiatek *et al.*, 1968).

Low plasma FFA in the neonatal pig could affect gluconeogenesis. Mammals are not capable of the net synthesis of carbohydrate from fat; however, evidence shows that increased FFA load presented to the liver causes a rapid increase in glucose production and release (Herrera *et al.*,

1966). In studies with rats, Ferre et al. (1978) found that fasting hypoglycaemia was partly reversed by giving triglycerides (soybean oil) or an injection of gluconeogenic substrates (lactate, pyruvate, glycerol, alanine, glutamate, serine) and completely reversed when these two treatments were combined. The rise in blood glucose was not secondary to decreased glucose utilization. Their data suggest that lack of gluconeogenic substrate was not the primary factor limiting the activity of gluconeogenesis in the neonate during a fast. Rather, FFA availability and, therefore, hepatic fatty acid oxidation, are essential to provide the energy required as ATP, NADH and acetyl CoA for the maintenance of the high rate of hepatic gluconeogenesis required for glucose homeostasis in the newborn.

In contrast to these results, the infusion of oleate into 1-day-old piglets had virtually no effect on the conversion of lactate and alanine to glucose by liver slices (Beiber et al., 1979). It is possible that deficient oxidation of this fat occurred, as there are relatively few liver mitochondria in the pig at birth. Further, the source of fat, oleate, may have affected these results. Wolfe, Maxwell and Nelson (1978) reported that the neonatal pig oxidized palmitate at a faster rate than stearate or myristate, while it has been shown in rats that fatty acid oxidation rate was greatest with shorter chain length and polyunsaturated fatty acids (Carroll, 1958; Lynn and Brown, 1959). Miller et al. (1971b) reported that laurate was oxidized significantly more rapidly than palmitic, oleic or linoleic acids by 1-day-old pigs.

This evidence suggests that certain fatty acids may be more efficient gluconeogenic stimulators than others. This may be due to the ability of the pig to oxidize these fats more readily, thus supplying the energy required for maintenance of hepatic gluconeogenesis. There is other evidence that modern lean swine are less able to utilize fat in diets than are obese-type pigs (Kveragas, 1982). In this study the evidence was that obese pigs increased their weight on a percentage basis faster than lean pigs when fed diets with either 5% or 25% dry matter from lipids, and blood parameters provided evidence of superior lipid utilization.

Adding lipids to sow diets

There is evidence that additional lipids in the diets of sows during late gestation and/or lactation cause changes in sow milk production and composition and possibly changes in the baby pig. Seerley et al. (1974) initially examined the value of feeding lipids to sows for a 5-day period prior to parturition. A basal diet was fed plus supplemental energy in the form of corn starch or corn oil. Sows receiving corn oil had a 90% survival rate for piglets at 21 days of age. This was higher, but not significantly so, than the 80.3% or 75.2% survival from feeding the basal or cornstarch-supplemented diets, respectively.

The results led to interest from many researchers. Much information has been published and, understandably, a certain amount of controversy exists in the literature. Reviews on the subject (Pettigrew, 1979, 1981; Moser and Lewis, 1980; Seerley, 1981; Coffey, 1981) indicate that improved survival can be achieved by feeding fat to sows prior to parturition. Differences in results stem from variations in design of experiments, litter sizes, length of feeding supplemental fat, amount of fat fed and initial survival rates of particular herds. In some cases, lack of improvement in survival has been due to feeding low levels of lipids, feeding lipids only after parturition, and a small number of sows per treatment. Herds with an average survival rate of 80% have a greater chance of showing a response than do herds with a 90% survival rate (Seerley, 1981). Pettigrew (1979) summarized data from 39 sources and suggested a 4.1% increase in survival rate when the herd survival rate was relatively low (<80%). This led to the conclusion that fat supplementation will be of little use if the herd survival rate is already above 85% and that added fat in sow diets protects the piglets from an adverse environment.

The timing of feeding fat affects the results. Experiments have been designed where treatments are imposed in late gestation only, in lactation only or both. Most reports from the gestation only group produced improvement in piglet survival rate. When fat was added after farrowing, little improvement in survival was noted. Feeding dietary lipids to sows requires adjustments to absorb and transport them to the mammary glands. The fat is then available as increased milk fat to prevent or reverse an energy deficiency and thus improve survival. The lack of response in 'lactation only' feeding results from the failure of these events to occur in time to prevent mortality during the first 3 days of lactation (Pettigrew, 1979).

Most experiments have used fat levels between 7.5% and 15%. This is the range that has given the greatest response in survival. Moser and Lewis (1980) concluded that a minimum of 7.5% fat should be added, while Pettigrew (1979) suggested that at least 1 kg of fat should be fed to the sow before farrowing.

It was stated earlier that birth weight seems to be a critical factor in survival. A more consistent response to lipid feeding has been the improved survival of smaller pigs in the litter. The addition of corn oil or animal fat to sow diets during gestation and lactation improved survival among pigs in light and medium weight ranges (<900 g and 900–1135 g). These results agree with other reports of higher survival rates for piglets weighing less than 1000 g at birth when lipids were added to the diet (Seerley *et al.*, 1974; Boyd *et al.*, 1978b).

Milk production

The susceptibility of the newborn pig to starvation makes the manipulation of sow's milk composition of interest in regard to improved survival. The gross composition of sow's colostrum and milk has been examined by several authors (Braude *et al.*, 1947; Perrin, 1955; Lodge, 1959). In general, crude protein is highest at birth; lactose increases from colostrum to milk; total solids and non-fat solids decrease slightly as lactation progresses; and total lipids are lower in colostrum, but increase in early lactation and then decline to weaning.

Research by Willett and Maruyama (1946) indicated that fat in sow diets could increase the fat content in milk. The fatty acid composition of colostrum fat contains lower percentages of myristic and palmitoleic acids, while the C18 fatty acid content is relatively high compared with that in milk. A high level of linoleic acid in colostrum fat compared with milk fat is the most marked difference in any single fatty acid (de Man and Bowland, 1963; Duncan and Garton, 1966).

The fatty acid pattern in milk is readily altered by the fat source included in the sow diet. Feeding tallow increases the percentage of oleic acid; lard increases levels of oleic and other unsaturated fatty acids; and corn oil increases the level of linoleic acid (Salmon-Legagneur, 1965; Trollerz and Lindberg, 1965; Miller, Conrad and Harrington, 1971a). There has been interest in the linoleic acid content of sow's milk. Colostrum contains relatively high amounts of this essential fatty acid even when there are low amounts of this fatty acid in feed. Feeding corn oil or soybean oil increases the percentage of linoleic acid in both colostral and milk fat. However, increased levels of linoleic acid in sow's feed, and therefore in plasma, do not enhance the incorporation of the essential fatty acid arachidonate, of which linoleic acid is a precursor, into the sow's milk (Kruse et al., 1977). The plasma of sows and piglets is also readily altered by the fat source in the diet, and composition of colostrum and milk lipids reflect changes in the composition of the plasma lipids of sows. Apparently, lactating sows incorporate fatty acids of dietary origin into milk fat preferentially to those from body storage (Witter and Rock, 1970).

Along with these specific effects on fatty acid content of milk, an increase in total lipid is a consistent response to added fat in sow diets. A variable which may influence this response is the length of time of feeding fat. Feeding fat during late gestation or lactation has been examined: feeding during gestation alone increases fat in colostrum, while feeding during lactation alone increases fat in subsequent milk. Adding fat during both periods increases the fat content both of colostrum and of milk (Pettigrew, 1979).

Milk production was increased approximately 30% by the addition of lipids to sow diets in a recent study in our laboratory (Coffey, 1981). Isocaloric diets provided 10% dietary fat from day 109 postcoitum to farrowing or continued in lactation; these were compared with a control group of sows. The feeding of fat prior to farrowing only, or on through lactation, were equally effective in increasing milk yield on day 14 of lactation. Increased milk yield due to the feeding of lipids to sows has been reported by Kruse et al. (1977), Pettigrew (1978) and Boyd (1979); the percentage increase ranged from 8% to 18% in those studies (*Table 17.1*). Average pig and litter weights at birth, or on day 7, 14 and 21 of lactation, did not differ between treatments. Boyd (1979) reported a non-significant

Table 17.1 MILK YIELDS OF SOWS (kg/day)

Reference	Control sows	Fat-fed sows	Increase (%)
Kruse et al., 1977[a]	4.60	5.32	16
Pettigrew, 1978[a]	3.82	4.48	18
Boyd, 1979[a]	8.73	9.45	8
Coffey and Seerley, 1981	5.45	7.18	31
			$\bar{x}$ 18

[a] Cited by Moser and Lewis, 1980

trend towards heavier weaning weight as a result of lipid feeding. Lewis, Speer and Haught (1978) reported a relatively low correlation between milk yield and pig weight gains, suggesting that pig weight gain alone is not a good indication of sow milk yield. The extra energy available to pigs may be metabolized by increased activity behaviour.

The changes in milk composition and yield caused by fat treatment may be important with regard to improved survival. The increased lipids and energy at day 3 may be particularly important because piglet glycogen stores are depleted by 72 hours after birth in fed and fasted pigs (Seerley and Poole, 1974). The value of the extra milk produced should be substantial for the survival of smaller piglets in the litter and piglets in a marginal energy balance.

Piglet composition

The energy reserves in newborn pigs have been relatively difficult to increase by dietary manipulation. In our first report (Seerley et al., 1974) we suggested the possibility of improving the piglet's energy status, and thus its prospects of survival, by feeding fat to the sow immediately prior to farrowing. Increased energy storage as carcass fat by feeding lipids to sows is not a consistent response. Several studies have shown non-significant increases in body fat of newborns (Okai, Aherne and Hardin, 1977; Boyd et al., 1978a; Seerley, Snyder and McCampbell, 1981). The length of feeding supplemental fat to sows prior to farrowing has varied in studies reviewed here; the maximum time has been 15 days. Energy storage in the fetus occurs rapidly and late in gestation, but this time frame may not be long enough to display the potential for increasing carcass fat in the piglet at birth. Boyd et al. (1978a) reported increased concentration of plasma FFA in the dam on day 110 of gestation from feeding tallow. It has been shown repeatedly that fatty acids are transported from the mother to the fetus in several species (Van Duyne et al., 1960; Van Duyne, Havel and Felts, 1962; Hershfield and Nemeth, 1968). In addition, piglet-carcass fatty acid percentages are influenced by the sow's diet, reflecting the lipid source (Seerley, Griffin and McCampbell, 1978).

The major form of energy storage in the newborn pig is glycogen. Body reserves of glycogen are also difficult to affect consistently by short-term manipulation of energy in sow diets (Okai et al., 1977, 1978; Seerley et al., 1978). Seerley et al. (1974) reported that piglets from dams fed corn starch or corn oil had more glycogen per unit of weight in the longissimus muscle than those from the lower-energy control group. Both groups had more glycogen in the liver.

Tallow added to the sow's diet at 20% produced a slight increase in pig liver glycogen at birth, and this difference was maintained throughout the 12 hours after birth. Corn starch fed at the same energy intake as tallow maintained a high glycogen level at 6 hours but not at 12 hours after birth (Boyd et al., 1978a). This suggests that feeding fat to sows slows the depletion of liver glycogen during the first 12 hours after birth.

Several authors (Seerley et al., 1974; Cast et al., 1977; Boyd, 1979) found that pigs from sows fed diets with added fat have higher blood glucose concentrations at birth than those of pigs from control sows. However, Friend (1974) reported exactly the opposite. Blood glucose concentration may be indicative of persistent liver glycogen concentration and reflect other benefits of increased energy storage. Boyd et al. (1978a) tested this hypothesis in smaller, less competitive pigs, which were defined as those weighing between 680 g and 1090 g at birth. Pigs from dams fed tallow had higher blood glucose concentrations at birth, 6 hours and 24 hours of age compared with lower-energy controls. Boyd et al. considered that this finding was attributable to: (1) greater concentration of liver fat at birth; (2) a larger portion of the energy being contributed by fat via the milk; or (3) possible greater concentrations of carcass lipid at birth, or a combination of these factors.

Thermostability is another indication of energy status in the newborn pig. Piglets from sows fed fat were more thermostable at 54 hours of age than were piglets from sows fed isocaloric amounts of corn starch (Seerley et al., 1974). The difference was not apparent at 6 hours of age.

A study recently reported from our laboratory (Coffey et al., 1982b) provided evidence that longer-term feeding of fat to sows in gestation has significant benefits for pigs. Levels of animal fat and corn starch were fed to provide three levels of energy (22.91, 25.85 or 28.84 MJ/sow/day) in late gestation. Results of the trial are shown in *Tables 17.2, 17.3* and *17.4*. Differences in feeding fat at lower to higher levels for 5 days compared with 35 days were clearly illustrated in the data collected. The levels were representative of 5%, 10% or 15% added fat. In the 5-day feeding trial there was no significant treatment difference in piglet-blood free fatty acids (FFA) at birth or after fasting. Blood glucose levels were similar at birth and after a 24 h fast; however, after fasting for 48 hours, pigs whose dams received the highest level of added animal fat had higher ($P<0.05$) blood glucose than those whose dams received 0.39 or 0.58 kg/day of corn starch. Glycogen was increased ($P<0.05$) at birth in pigs whose dams received 0.27 kg/day fat compared to piglets from sows fed corn starch. Pigs whose dams received 0.27 or 0.58 kg/day fat or corn starch, respectively, maintained more ($P<0.05$) liver glycogen after 24 hours. In addition, feeding 0.27 kg/day of animal fat increased ($P<0.05$) pig-blood β-hydroxybutyrate in comparison with the two higher levels of corn starch at birth, increased ($P<0.05$) the ability of the liver to oxidize palmitate to β-hydroxybutyrate *in vitro*, and increased ($P<0.05$) colostrum total lipids with a non-significant increase in energy.

Feeding the dietary treatments for the last trimester of pregnancy (trial 2) did not have any effect on piglet blood FFA level. At birth, blood glucose was increased ($P<0.05$) and the level was higher after the 48 h fast for treatments with added fat. Pigs from sows fed fat in trial 2 had more ($P<0.05$) liver glycogen at birth than those from sows fed corn starch. β-Hydroxybutyrate production in the piglets was not different at birth but was slightly elevated after a 24 h fast and increased ($P<0.05$) after a 48 h fast, as a result of addition of fat to sow diets. Palmitate and ^{14}C glucose utilization *in vitro* by livers from pigs fasted for 24 hours was increased ($P<0.05$) as a result of lipid feeding.

Sows fed 0.18 kg/day or 0.27 kg/day of animal fat had more piglets ($P<0.05$) born live when compared with all treatments except the highest

Table 17.2 EFFECT OF AMOUNT AND SOURCE OF ENERGY FED TO SOWS FOR 5 OR 35 d ON PIG BLOOD METABOLITES AT BIRTH AND AFTER 24 h OR 48 h FAST

Item	Age (h)	Animal fat (kg/d)			Corn starch (kg/d)			SE
		0.09	0.18	0.27	0.20	0.39	0.58	
Trial 1 (5-day feeding)								
Free fatty acids (μEq palmitate/litre)[a]	0	148.1	144.3	229.1	228.8	150.8	180.9	14.8
	24	437.1	296.2	340.9	349.2	317.2	213.5	15.2
	48	249.8	297.8	261.4	298.2	213.1	214.4	17.4
Glucose (mg/100 ml)[a]	0	45.2	52.8	56.6	56.9	50.8	65.0	3.0
	24	46.7	52.4	41.5	32.2	51.1	48.1	2.8
	48	25.0[de]	23.2[de]	28.9[d]	23.8[de]	17.3[e]	15.6[e]	1.1
β-hydroxybutyrate (mmol/litre)	24	0.38[de]	0.38[de]	0.48[d]	0.44[de]	0.35[e]	0.36[e]	0.02
	24	0.53	0.57	0.57	0.65	0.51	0.54	0.02
	48	0.61	0.66	0.66	0.76	0.65	0.71	0.02
Trial 2 (35-day feeding)								
Free fatty acids (μEq palmitate/litre)	0	321.1	223.7	275.0	216.5	257.8	238.5	12.4
	24	370.2	427.6	308.4	295.4	328.9	423.9	16.3
	48	403.7	340.9	448.3	387.2	374.7	336.2	19.3
Glucose (mg/100 ml)[b]	0	106.2[d]	72.0[d]	80.9[d]	63.3[e]	65.8[e]	59.4[e]	4.2
	24	57.6[e]	71.3[de]	69.7[de]	64.8[e]	51.3[e]	89.7[d]	3.2
	48	39.9	43.4	45.4	20.3	30.3	28.5	4.5
β-hydroxybutyrate (mmol/100 mg)[c]	0	0.55	0.54	0.47	0.61	0.46	0.57	0.03
	24	0.61[de]	0.67[de]	0.63[d]	0.54[de]	0.56[de]	0.48[e]	0.03
	48	0.86[de]	0.70[e]	0.97[d]	0.65[e]	0.58[e]	0.64[e]	0.05

[a] Energy source means for fat or starch were 51.2, 57.6; 50.4, 44.0; 26.1, 18.8 mg/100 ml at age 0, 24 or 48 h respectively. The 48 h means were different ($P<0.05$).

[b] Energy source means for fat or starch were 85.38, 63.06; 66.81, 68.98; 42.09, 27.35 mg/100 ml at age 0, 24 or 48 h respectively. The 0 h and 48 h means were different ($P<0.05$). Energy source × age interaction was significant.

[c] Energy source means for fat or starch were 0.52, 0.56; 0.64, 0.56; 0.79, 0.62 nmol/litre at ages 0, 24 or 48 h, respectively. The 48 h means were different ($P<0.05$). Energy source × age interaction was significant.

[d,e] Means in the same row with different superscripts differ ($P<0.05$).

Table 17.3 EFFECT OF AMOUNT AND SOURCE OF ENERGY FED TO SOWS FOR 5 OR 35 d ON PIG LIVERS AND *IN VITRO* LIVER METABOLITES AT BIRTH AND AFTER A 24 h OR 48 h FAST

Item	Age (h)	Animal fat (kg/d)			Corn starch (kg/d)			Energy source		SE
		0.09	0.18	0.27	0.20	0.39	0.58	Animal fat	Corn starch	
Trial 1 (5-day feeding)										
β-hydroxybutyrate (produced *in vitro*) (nmol/100 mg)	24	0.073[d]	0.089[cd]	0.100[c]	0.086[cd]	0.088[cd]	0.085[d]	0.087	0.086	0.0026
[14]C palmitate converted to CO_2 (*in vitro*) (nmol/100 mg)	24	2.07	2.54	2.72	3.56	2.26	2.47	2.44	2.76	0.144
Glycogen (mg/g)[a]	0	437.2[cd]	393.2[cd]	486.7[c]	378.8[de]	381.8[de]	307.4[e]	435.6[c]	354.4[d]	4.3
	24	11.0[e]	104.1[cd]	176.5[c]	7.0[e]	82.6[de]	136.1[cd]	96.7	80.1	4.1
	48	4.1	11.0	4.0	3.3	3.7	3.3	6.4	3.5	0.34
Trial 2 (35-day feeding)										
β-hydroxybutyrate (produced *in vitro*) (nmol/100 mg)	24	0.099[cd]	0.122[c]	0.128[c]	0.109[cd]	0.101[cd]	0.084[d]	0.115[c]	0.096[d]	0.0056
[14]C glucose converted to CO_2 (*in vitro*) (nmol/100 mg)	24	57.9[c]	50.0[c]	58.7[c]	43.9[d]	43.0[d]	40.1[d]	55.3[c]	42.5[d]	3.81
Glycogen (mg/g)[b]	0	497.2	480.0	459.8	437.2	431.5	423.8	477.7[c]	430.8[d]	3.3
	24	6.2[d]	18.0[d]	137.0[c]	27.8[d]	21.5[d]	124.1[c]	51.0	65.8	3.5
	48	3.0	3.2	3.2	4.4	4.0	3.5	3.2	3.8	0.12

[a] Energy source means for fat or starch were 175.5[c], 145.99[d] mg/g, respectively. The energy source × age and treatment × age interactions were significant.
[b] Energy level means for low, medium and high energy were 161.1[c], 160.5[c] and 191.7[d] mg/g, respectively. The treatment × age interaction was significant.
[c,d,e] Means in the same row with different superscripts differ ($P<0.05$).

Table 17.4 EFFECT OF AMOUNT AND SOURCE OF ENERGY FED TO SOWS ON COLOSTRUM AND LITTER CHARACTERISTICS

Item	Animal fat (kg/d)			Corn starch (kg/d)			Energy source		
	0.09	0.18	0.27	0.20	0.39	0.58	Animal fat	Corn starch	SE
Trial 1 (5-day feeding)									
No. born live	12.4	12.2	11.2	11.6	11.0	11.9	11.9	11.4	0.009
No. born dead	0.6	0.6	0.6	1.0	0.4	0.6	0.60	0.67	0.004
Birth weight (kg)	1.27	1.32	1.31	1.13	1.36	1.30	1.29	1.27	0.003
Colostrum (cal/g)	1307	1397	1584	1241	1437	1448	1429	1375	10.29
Colostrum total lipids (%)	3.87[e]	5.73[e]	9.26[d]	3.87[e]	3.74[e]	3.81[e]	6.07[d]	3.80[e]	0.100
Trial 2 (35-day feeding)									
No. born live[a]	11.2[e]	13.2[d]	13.4[d]	12.4[e]	11.8[e]	12.6[de]	12.7	12.2	0.009
No. born dead[b]	1.0[e]	0.2[d]	0.2[d]	1.6[e]	1.2[e]	0.4[d]	0.47[d]	1.07[e]	0.006
Birth weight (kg)[c]	1.17[e]	1.34[d]	1.30[d]	1.16[e]	1.26[e]	1.24[e]	1.27[d]	1.22[e]	0.002
Colostrum (cal/g)	1424	1562	1470	1250	1299	1261	1490[d]	1271[e]	9.45
Colostrum total lipids (%)	6.78	8.42	7.52	5.48	6.52	5.48	7.63	5.83	0.083

[a] Energy level means for low, medium and high were 11.8[d], 12.5[d] and 13.0[e], respectively.
[b] Energy level means for low, medium and high were 1.3[d], 0.7[e] and 0.3[e], respectively.
[c] Energy level means for low, medium and high were 1.17[d], 1.30[e] and 1.27[e], respectively.
[d,e] Means with different superscripts differ ($P<0.05$)

level of corn starch. The number of pigs born dead was lower ($P<0.05$) for these treatment groups. Birth weights were heavier ($P<0.05$) for pigs farrowed by sows receiving the two highest levels of fat. The average energy and total lipid content of colostrum from sows fed fat was higher ($P<0.05$) than the average of those fed starch.

The significance of the data is that feeding fat to sows affects the metabolism in newborn pigs as well as the sow and these changes can be extremely beneficial with respect to the health and well-being of both the sow and the pigs. Changes in metabolism are partly attributable to other supportive energy metabolism pathways. Studies in energy metabolism pathways will be useful to improve energy efficiency in sows, as they are with dairy cows to avoid substrate cycling (Madsen and Miller, 1982).

Sow conditioning and recycling

The current NRC (1979) recommendation to provide 23.72 MJ of ME per day of gestation appears to be nearly correct. However, the data from feeding lipids and increased feed intake in late gestation are evidence that the requirement is approximately 29.29 MJ/day in late gestation. The S-145 committee on sow reproduction (Cromwell et al., 1982) reported that an increase in feed intake from 1.82 kg to 3.2 kg from day 90 postcoitum to farrowing increased birth weights and survivability of pigs. Lipid feeding to sows in late gestation offers advantages (on an isocaloric basis) for sows and pigs to utilize more effectively other metabolic pathways for improved performance.

The energy requirement in lactation is related to body size and condition, number of pigs nursing, milking ability and other factors such as ambient temperature. On the basis of the research findings of Lodge (1969), Vermedahl et al. (1969), Hitchcock et al. (1971), Libal and Wahlstrom (1975) and Stahly, Cromwell and Simpson (1981) the small increase in weaning weight does not justify maximum caloric intake by lactating sows. An increase in milk yield has been achieved by feeding 8.37 MJ DE per day (O'Grady et al., 1973) or lipids (Pettigrew, 1979; Coffey et al., 1982b); however, the lipid feeding in lactation has not increased weaning weights (Coffey, 1981; Stahly et al., 1981). Apparently, the extra milk is needed for maintenance and activity.

In studies to determine the effect of reproduction on sow's body composition, Whittemore, Franklin and Pearce (1980) reported an increase in body weight by sows in subsequent parities, but body fat may decrease. Extreme losses in body fat can diminish reproductive efficiency (Elsley, 1968) which occurs in sows producing large amounts of milk and those nursing large litters.

The use of fat in late gestation and lactation diets by US swine producers caused an additional unexpected response. Some producers reported a shortened interval between weaning and oestrus. Subsequent research (Reese et al., 1980) illustrated delayed oestrus by sows fed 33.47 MJ daily in lactation versus sows fed 66.94 MJ: therefore, lower caloric intake decreases body conditioning and causes delayed oestrus. Research by Cox and Britt (1982) and Cox et al. (1983) showed that the percentage of first-parity gilts in oestrus within 7 days after weaning is less than that of second-parity sows and the percentage in oestrus in both groups is lower in summer than in winter months. The feeding of a 10% fat lactation diet increased the percentage in oestrus from 34% to 59% in summer months, but the percentage in the fat-fed group tended to be less in winter months (74% versus 82%). Crawford (1982) reported an improvement in survival (86% versus 74%) by feeding fat to sows in late gestation. The greatest difference in survival occurred in the hot summer months during a heat stress period when only 54% of control pigs survived, versus 82% of the pigs from fat-fed sows. These data provided by Cox et al. (1983) and Crawford (1982), as well as data with growing pigs (Coffey et al., 1982a) give evidence that the lower heat increment of lipids provides an additional benefit. In the past, there has been a tendency to add bulk and decrease energy in lactation diets in warmer months, but a better procedure is to add fat to maintain energy intake while permitting feed intake to decrease slightly.

Summary

The objective of sow feeding is to maintain the animals in moderate condition by controlling calorie intake. A reasonable guideline for an average sow is to provide 25.10, 29.29 and 66.94 MJ/day during gestation, late gestation and lactation, respectively. Energy intake in lactation is more variable: therefore, more feeding management is needed to maintain conditioning in sows. Sources of energy should be considered for maximum productivity. The feeding of lipids in late gestation and lactation provides opportunities for increased quantity and quality of milk and improved conditioning, rebreeding and longevity of sows in the breeding herd. High-producing sows gain the most from lipid feeding. Piglets may have increased energy reserves and more blood glucose at birth, improved energy metabolism and increased survival, especially of smaller, energy-marginal pigs. A few of these parameters can be duplicated by feeding good carbohydrate diets such as a corn–soybean meal diet, excellent husbandry and expensive facilities, but not all parameters can be duplicated, especially on average hog farms. Some parameters are unique to lipid feeding. If only a single parameter is improved in a small way, the more expensive fat may not be necessary, but generally multiple benefits are achieved to justify lipids in a sow-feeding programme. Furthermore, the findings through research in the 1970s are useful to show weaknesses in sow feeding: improved feeding through future research will improve sow productivity and permit maximum genetic expression for reproduction.

References

ADAMS, P.H. (1971). Intra-uterine growth retardation in the pig. II. Development of the skeleton. *Biology of the Neonate* **19**, 341–346

ARS (1965). *Losses in Agriculture. Handbook no. 29*. Washington, DC, ARS/USDA

BALLARD, F.J. (1970). Carbohydrates. In *Physiology of the Perinatal Period, Vol. 1* (U. Stave, Ed.), pp. 417–440. New York, Appleton-Century-Crofts

BAUMAN, R.H., KADLEC, J.E. and POWLEN, P.A. (1966). Some factors affecting death loss in baby pigs. *Purdue University Agricultural Experimental Station Bull. 810*

BIANCA, W. (1968). Thermoregulation. In *Adaptation of Domestic Animals* (E.S.E. Hafey, Ed.), pp. 97–118. Philadelphia, Lea and Febiger

BIEBER, L.L., HELMITH, T., DOLONSKI, E.A., OLGOARD, M.K., CHOI, Y. and BELANGER, L.L. (1979). Gluconeogenesis in neonatal piglet liver. *Journal of Animal Science* **49**, 250–257

BOYD, R.D. (1979). *Glucose Homeostasis and Fatty Acid Utilization in the Neonatal Pig*. PhD thesis, University of Nebraska, Lincoln

BOYD, R.D., MOSER, B.D., PEO, E.R., Jr and CUNNINGHAM, P.J. (1978a). Effect of energy source prior to parturition and during lactation on tissue lipid, liver glycogen and plasma levels of some metabolites in the newborn pig. *Journal of Animal Science* **47**, 874–882

BOYD, R.D., MOSER, B.D., PEO, E.R. Jr and CUNNINGHAM, P.J. (1978b). Effect of energy source prior to parturition and during lactation on piglet survival and growth and on milk lipids. *Journal of Animal Science* **47**, 883–892

BRAUDE, R., COATES, M.E., HENRY, K.M., KON, S.K., ROWLAND, S.J., THOMPSON, S.Y. and WALTER, D.M. (1947). A study of the composition of sows' milk. *British Journal of Nutrition* **1**, 64–70

CARROLL, K.K. (1958). Digestibility of individual fatty acids in the rat. *Journal of Nutrition* **64**, 399–410

CAST, W.R., MOSER, B.D., PEO, E.R., Jr and CUNNINGHAM, P.J. (1977). Fat, choline and thyroprotein additions to the diet of lactating swine. *Journal of Animal Science* **45** (Suppl. 1), 80 (Abstr.)

COFFEY, M.T. (1981). *Energy Homeostasis and Utilization in the Neonatal Pig as Affected by Duration of Feeding and Source of Additional Energy in Sow Diets*. PhD dissertation, University of Georgia

COFFEY, M.T., SEERLEY, R.W., FUNDERBURKE, D.W. and McCAMBPELL, H.C. (1982a). Effect of heat increment and level of dietary energy and environmental temperature on the performance of growing-finishing swine. *Journal of Animal Science* **54**, 95–105

COFFEY, M.T., SEERLEY, R.W., MARTIN, R.J. and MABRY, J.W. (1982b). Effect of level, source and duration of feeding supplemental energy in sow diets on metabolic and hormonal traits related to energy utilization in the baby pig. *Journal of Animal Science* **55**, 329–336

COX, N.M. and BRITT, J.H. (1982). Feeding for rebreeding. *Pig American* June, pp. 26–28

COX, N.M., BRITT, J.H., ARMSTRONG, W.D. and ALHERSEN, H.D. (1983). Effect of feeding fat and altering weaning schedule on rebreeding in primiparous sows. *Journal of Animal Science* **56**, 21–29

CRAWFORD, J.F. (1982). *The Effect of Dietary Lipids and Sucrose on the Survivability of Baby Pigs*. MS thesis, University of Georgia

CROMWELL, G.L., PRINCE, T.J., COMBS, G.E., MAXWELL, C.V., KNABE, D.A. and ORR, D.E. (1982). [S-145 Committee on Nutrition Systems for Swine to Increase Reproductive Efficiency.] Effects of additional feed during late gestation on reproductive performance of sows—a cooperative study. *Journal of Animal Science* ASAS Abstract No. 371, p. 268

CURTIS, S.E. (1970). Environmental thermoregulatory interactions and neonatal piglet survival. *Journal of Animal Science* **31**, 576–587

CURTIS, S.E., HEIDEMEICH, C.J. and HARRINGTON, R.B. (1967). Age dependent changes of thermostability in neonatal pig. *American Journal of Veterinary Research* **28**, 1887–1890

DE MAN, J.M. and BOWLAND, J.P. (1963). Fatty acid composition of sow's colostrum, milk and body fat as determined by gasliquid chromatography. *Journal of Dairy Research* **30**, 339–343

DICKERSON, J.W.T., MERET, A. and WIDDOWSON, E.M. (1971). Intra-uterine growth retardation in the pig. III. The chemical structure of the brain. *Biology of the Neonate* **19**, 354–360

DUNCAN, W.R.H. and GARTON, G.A. (1966). The component fatty acids of the colostral fat and milk fat of the sow. *Journal of Dairy Research* **33**, 255–259

ELSLEY, F.W.H. (1968). The influence of feeding level upon the reproductive performance of pregnant sows. *Veterinary Record* **83**, 93–97

FERRE, P., PEGARIER, J.P., MARLISS, E.B. and GIRARD, J.R. (1978). Influence of exogenous fat and gluconeogenic substrates on glucose homeostasis in the newborn rat. *American Journal of Physiology* **234**, E129–E136

FOLEY, C.W., SEERLEY, R.W., HANSEN, W.J. and CURTIS, S.E. (1971). Thermoregulatory responses to cold environment by neonatal wild and domestic piglets. *Journal of Animal Science* **32**, 926–929

FRIEND, D.W. (1974). Effect on performance of pigs from birth to market weight of adding fat to the lactation diet of their dams. *Journal of Animal Science* **29**, 1073–1081

GENTZ, J., BENGTSSON, G., HAKKARAINEN, J., HELLSTROM, R. and PERSSON, B. (1970). Metabolic effects of starvation during neonatal period in the piglet. *American Journal of Physiology* **218**, 662–668

GOODWIN, R.F.W. (1957). The relationship between the concentration of blood sugar and some vital body functions in the newborn pig. *Journal of Physiology* **136**, 208–217

GRAHAM, R., SAMPSON, J. and HESTER, H.R. (1941). I. Acute hypoglycemia in newly born pigs (so-called baby pig disease). *Proceedings of the Society for Experimental Biology* **47**, 338–339

HANAWALT, W.M. and SAMPSON, J. (1947). Studies on baby pig mortality. V. Relationship between age and time of onset of acute hypoglycemia in fasting newborn pigs. *American Journal of Veterinary Research* **8**, 235–239

HANSEN, W.J., FOLEY, C.W., SEERLEY, R.W. and CURTIS, S.E. (1972). Pelage traits in neonatal wild, domestic, and 'crossbred' piglets. *Journal of Animal Science* **34**, 100–102

HERRERA, M.G., KAMM, D., RUDERMAN, N. and CAHILL, G.F., Jr (1966). Non-hormonal factors in the control of gluconeogenesis. *Advances in Enzyme Research* **4**, 225–235

HERSHFIELD, M.S. and NEMETH, A.M. (1968). Placental transport of free palmitic and linoleic acids in the guinea pig. *Journal of Lipid Research* **9**, 460–468

HITCHCOCK, J.P., SHERRITT, G.W., GOBBLE, J.L. and HAZLETT, V.E. (1971).

Effect of lactation feeding level of the sow on performance and subsequent reproduction. *Journal of Animal Science* **33**, 30–34

KERNKAMP, H.C.H. (1965). Birth and death statistics on pigs of preweaning age. *Journal of the American Veterinary Medical Association* **146**, 337–340

KRUSE, P.E., DANIELSON, V., NEILSON, H.E. and CHRISTENSEN, K. (1977). The influence of different dietary levels of linoleic acid on reproductive performance and fatty acid composition of milk fat and plasma lipids in pigs. *Acta agriculturae scandinavica* **27**, 289–296

KVERAGAS, C.L. (1982). *The Utilization of Dietary Lipids by Obese Ossabaw and Lean Domestic Baby Pigs*. MS thesis, University of Georgia

LEMAN, A.D., KNUDSON, C., RODEFFER, H.E. and MUELLER, A.G. (1972). Reproductive performance of swine on 76 Illinois farms. *Journal of the American Veterinary Medical Association* **161**, 1248–1250

LEWIS, A.J., SPEER, V.C. and HAUGHT, D.G. (1978). Relationship between yield and composition of sow's milk and weight gains of nursing pigs. *Journal of Animal Science* **47**, 634–638

LIBAL, G.W. and WAHLSTROM, R.C. (1975). Effect of level of feeding during lactation on sow and pig performance. *Journal of Animal Science* **41**, 1542–1545

LODGE, G.A. (1959). The energy requirements of lactating sows and the influence of level of food intake upon milk production and reproductive performance. *Journal of Agricultural Science* **53**, 177–191

LODGE, G.A. (1969). The effects of pattern of feed distribution during the reproductive cycle on the performance of sows. *Animal Production* **11**, 133–143

LYNN, W.S. and BROWN, R.H. (1959). Oxidation and activation of the unsaturated fatty acids. *Archives of Biochemistry and Biophysics* **81**, 353–362

McCANCE, R.W. and WIDDOWSON, E.M. (1959). The effect of lowering the ambient temperature on the metabolism of the newborn pig. *Journal of Physiology* **147**, 124–134

MADSEN, F.C. and MILLER, J.K. (1982). Avoiding substrate cycling in dairy cattle. *Feedstuffs* **54**, 73–74

MANNERS, M.J. and McCREA, M.R. (1963). Changes in the chemical composition of sow-reared piglets during the first month of life. *British Journal of Nutrition* **17**, 495–513

MERSMANN, H.J. (1974). Metabolic patterns in the neonatal swine. *Journal of Animal Science* **38**, 1022–1030

MILLER, G.M., CONRAD, J.H. and HARRINGTON, R.B. (1971a). Effect of dietary unsaturated fatty acids and stage of lactation in milk composition and adipose tissue in swine. *Journal of Animal Science* **32**, 79–83

MILLER, G.M., CONRAD, J.H., KEENAN, T.W. and FEATHERSTON, W.R. (1971b). Fatty acid oxidation in young pigs. *Journal of Nutrition* **101**, 1343–1349

MOSER, B.D. and LEWIS, A.J. (1980). Adding fat to sow diets. *Feedstuffs* **52**, 36–38

MOUNT, L.E. (1964). The thermal insulation of the newborn pig. *Journal of Physiology* **168**, 698–765

NOLAND, P.R. and JOHNSON, Z. (1972). Influence of birth weight of pigs on their mortality and performance. *Arkansas Farm Research* **21**, 7–9

NRC (1979). *Nutritional Requirements of Domestic Animals, No. 2: Nutrient Requirements of Swine*, 8th revised edn. Washington, DC, National Academy of Sciences–Nutritional Research Council

O'GRADY, J.F., ELSLEY, F.W.H., MacPHERSON, R.M. and McDONALD, I. (1973). The response of lactating sows and their litters to different energy allowances. I. Milk yield and composition, reproductive performance of sows and growth rate of litters. *Animal Production* **17**, 65–74

OKAI, D.B., AHERNE, F.X. and HARDIN, R.T. (1977). Effects of sow nutrition in late gestation on the body composition and survival of the neonatal pig. *Canadian Journal of Animal Science* **57**, 439–448

OKAI, D.B., WYLLIE, D., AHERNE, F.X. and EWAN, R.C. (1978). Glycogen reserves in the fetal and newborn pig. *Journal of Animal Science* **46**, 391–401

PERRIN, D.R. (1955). The chemical composition of the colostrum and milk of the sow. *Journal of Dairy Research* **22**, 103–107

PERSSON, B. and GENTZ, J. (1966). The pattern of blood lipids, glycerol and ketone bodies during the neonatal period, infancy and childhood. *Acta paediatrica scandinavica* **55**, 353–362

PETTIGREW, J.E. (1978). Fat in gestation and lactation diets for sow. *Invited paper presented at the 71st Annual Meeting, American Society for Animal Science, Tucson, AZ*

PETTIGREW, J.E., Jr (1981). Supplemental dietary fat for peripartal sows: A review. *Journal of Animal Science* **53**, 107–117

PETTIGREW, J.E., ZIMMERMAN, D.R. and EWAN, R.C. (1971). Plasma carbohydrate levels in the neonatal pig. *Journal of Animal Science* **32**, 895–899

POMEROY, R.W. (1960). Infertility and neonatal mortality in the sow. III. Neonatal mortality and foetal development. *Journal of Agricultural Science* **54**, 31–56

REESE, D.E., MOSER, D.B., PEO, E.R., LEWIS, A.J., ZIMMERMAN, D.R., KINDER, J.E. and JOHNSON, R.K. (1980). Influence of dietary energy intake during lactation on the interval to first postweaning estrus in swine. *Journal of Animal Science* **51** (Suppl. 1), 217

SALMON-LEGAGNEUR, E. (1965). *Quelques Aspects des Relations Nutritionelles entre la Gestation et la Lactation chez la Truie*. Doctoral thesis, Institut National de la Recherche Agronomique

SAMPSON, H., HESTER, J.R. and GRAHAM, R. (1942). Studies on baby pig mortality. II. Further observations on acute hypoglycemia in newly born pigs (so-called baby pig disease). *American Journal of Veterinary Research* **100**, 33–37

SEERLEY, R.W. (1981). High fat rations for sows can benefit piglets. *Feedstuffs* **53**, 34–35

SEERLEY, R.W. and POOLE, D.R. (1974). Effect of prolonged fasting on carcass composition and glucose of neonatal swine. *Journal of Nutrition* **104**, 210–217

SEERLEY, R.W., GRIFFIN, F.M. and McCAMPBELL, H.C. (1978). Effects of sows' dietary energy on source of sows' milk and piglet carcass composition. *Journal of Animal Science* **46**, 1009–1017

SEERLEY, R.W., SNYDER, R.A. and McCAMPBELL, H.C. (1981). The influence of sow dietary lipids and choline on piglet survival, milk and carcass composition. *Journal of Animal Science* **52**, 542–550

SEERLEY, R.W., PACE, T.A., FOLEY, C.W. and SCARTH, R.D. (1974). Effect of energy intake prior to parturition on milk lipids and survival rate, thermostability and carcass composition of piglets. *Journal of Animal Science* **38**, 64–70

SHARPE, H.B.A. (1966). Pre-weaning mortality in a herd of large white pigs. *British Veterinary Journal* **122**, 99–111

SHELLEY, H.J. (1961). Glycogen reserves and their changes at birth and in anoxia. *British Medical Bulletin* **17**, 137–143

STAHLY, T.S., CROMWELL, G.L. and SIMPSON, W.S. (1981). Effects of level and source of supplemental fat in the lactation diet of sows on the performance of pigs from birth to market weight. *Journal of Animal Science* **51**, 352–360

STANTON, H.C. and CARROLL, J.K. (1974). Potential mechanisms responsible for prenatal and perinatal mortality or low viability of swine. *Journal of Animal Science* **38**, 1037–1044

SWIATEK, K.R. (1971). Development of gluconeogenesis in pig liver slices. *Biochimica et biophysica acta* **252**, 274–279

SWIATEK, K.R., CHAO, K.L., CHAO, H.L., CORNBLATH, M. and TILDON, J.T. (1970). Enzymatic adaptations in newborn pig liver. *Biochimica et biophysica acta* **222**, 145–154

SWIATEK, K.R., KIPNIS, D.M., MASON, G., CHAO, K.L. and CORNBLATH, M. (1968). Starvation and hypoglycemia in newborn pigs. *American Journal of Physiology* **214**, 400–405

TROLLERZ, G. and LINDBERG, P. (1965). Influence of dietary fat and short time starvation on the composition of sow-milk fat. *Acta veterinaria scandinavica* **6**, 118–134

VAN DUYNE, C.M. and HAVEL, R.J. (1959). Plasma unesterified fatty acid concentration in foetal and neonatal life. *Proceedings of the Society for Experimental Biology and Medicine* **120**, 599–602

VAN DUYNE, C.M., HAVEL, R.J. and FELTS, J.M. (1962). Placental transfer of palmitic acid-1-C^{14} in rabbits. *American Journal of Obstetrics and Gynecology* **84**, 1069–1074

VAN DUYNE, C.M., PARKER, R.H., HAVEL, R.J. and HOLM, L.W. (1960). Free fatty acid metabolism in fetal and newborn sheep. *American Journal of Physiology* **199**, 987–990

VERMEDAHL, L.D., MEADE, R.J., HANKE, H.E. and RUST, J.W. (1969). Effects of energy intake of the dam on reproductive performance, development of the offspring and carcass characteristics. *Journal of Animal Science* **28**, 465–472

WALKER, D.G. (1971). Development of enzymes for carbohydrate metabolism. In *The Biochemistry of Development* (P. Benson, Ed.), pp. 77–95. Philadelphia, J.B. Lippincott Co.

WIDDOWSON, E.M. (1971). Intra-uterine growth retardation in the pig. I. Organ size and cellular development at birth and after growth to maturity. *Biology of the Neonate* **19**, 329–340

WILLETT, E.L. and MARUYAMA, C. (1946). The effect of intake of garbage fat upon fat content of sows' milk. *Journal of Animal Science* **5**, 365–370

WITTER, R.C. and ROCK, J.A.F. (1970). The influence of the amount and nature of dietary fat on milk fat composition in the sow. *British Journal of Nutrition* **24**, 749–760

WHITTEMORE, C.T., FRANKLIN, M.R. and PEARCE, B.S. (1980). Fat changes in breeding sows. *Animal Production* **31**, 183–190

WOLFE, R.G., MAXWELL, C.U. and NELSON, E.C. (1978). Effect of age and dietary fat level on fatty acid oxidation in the neonatal pig. *Journal of Nutrition* **108**, 1621–1634

SUPPLEMENTAL FATS AND ENERGY DENSITY IN PIG DIETS*

ROBERT H. WILSON
Wandalup Farms, Mandurah, Western Australia
and
JAMES E. PETTIGREW, Jr
Department of Animal Science and the Swine Center, University of Minnesota, St Paul, MN, USA

The presentation by Drs Cole, Stahly and Steerley stimulated considerable discussion, and it was thought useful to include a report of it in the final proceedings.

On farm trial work reported from Western Australia, sows were fed 8 MJ DE/day extra energy as tallow for 10 days *pre-partum*. This resulted in a 34% reduction in stillborn piglets from 4.24% for control-fed sows and 2.80% for sows fed fat supplement. More live piglets at birth were produced from sows fed the extra fat than from the control group (10.22 and 9.64 piglets born alive per litter, respectively). The average litter size at weaning in the treatment group was 0.56 pigs greater than in the controls.

Sows fed the supplementary fat in the trial were also in better body condition at weaning than the control sows, as determined by the body weight and backfat measurement. The average weaning to first service interval was 5.7 d for the treatment group and 6.8 d for the control group.

Experience of the pig farming industry in the United Kingdom was discussed. Feeding a fat supplement before farrowing was reported to give varying degrees of success under practical farm situations. However, the trial work discussed showed evidence of improved piglet survival, reduced stillborns and greater weaning weights of piglets from sows fed supplementary fat. The improvement in piglet survival to weaning was reported to be 2–4%.

The fat supplements used commercially in the UK also provide a supplement of vitamin C, which it was suggested was important in reducing the number of stillbirths. The Western Australian work also provided 1 g/day of vitamin C to the treatment sows for 10 days *pre-partum*. There was no discussion, however, on the possible mechanisms involved or the role that vitamin C may have in reducing stillbirths.

Fat supplementation of the sows' diet during late gestation modifies the sows' metabolism to produce a semi-ketotic state, and this increases the amount of energy from colostral fat available to the piglets. This was

*Synopsis of discussion by the members of the Easter School following the presentations by Drs Cole, Stahly and Steerley.

reported in the paper by Dr Seerley and also from the United Kingdom industry trial results. However, it may be possible to provide supplemental fat directly to the piglets rather than involving the sow as an intermediate step. Research workers at the University of Minnesota have investigated this possibility in a study with over 1800 piglets.

Half of the piglets in each litter were given small (2 ml) doses of corn oil orally by syringe four times during the first 2 days *post-partum*. The overall survival rate in this study was very high, about 89%, and was not improved by the fat doses. The work with supplemental fat in sows' diet during late gestation suggests that it is very difficult to improve a survival rate as high as 89% by provision of additional dietary energy to the piglets. There was no improvement in the survival rate of any of several birthweight classes of piglets examined in this study. However, piglets below average in birthweight that were destined to die, tended to live longer if dosed with fat than if not dosed. While this is not of economic benefit, it does indicate that the pigs utilized the oral oil to some extent. This observation suggests that this approach should be pursued in herds with lower survival rates, perhaps with larger doses of fat.

In general discussion of the formal papers presented, there were comments on the value of the slight increase in body fat content at birth in piglets born to sows supplemented with fat in late gestation. Although this increase is a very small percentage of the total body fat content, it is a much larger percentage of the fat available for utilization as an energy source. This is because much of the body fat in the piglet at birth is phospholipid needed as a membrane component or in lipid transport, and thus not available for use by the piglet as an energy source. However, even though the value of this increase in body fat is greater than would at first appear, the total amount of additional energy available to the piglets from this source as a result of supplementing the sows' diet with fat is much less than the total amount of additional energy from colostral fat.

There was discussion of whether dietary fat in late gestation simply supplies additional fat calories to the sow, or whether it meets a specific essential fatty acid requirement. If it is the latter, then the source of fat would be important, and the zinc status of the animal would also be of interest. While it is not possible with present data to answer this question directly, the fact that some studies have shown a response to supplemental tallow suggests that it is not simply a matter of meeting essential fatty acid requirements.

A very interesting aspect of the subject is the adequacy of enzyme co-factors involved in the utilization of the supplemental fat. Perhaps when fat is added to the diet, the amounts of some, or all of, the B vitamins should also be increased. It is interesting to note that all three studies of supplemental fat reported earlier in this discussion involved the inclusion of vitamins as well as supplemental fat in the treatments that yielded a positive response, and that two of these studies used B vitamins.

The medium-chain fatty acids in colostrum may be of specific importance to the piglets as sources of ketone bodies, and thus of energy for the central nervous system. A question not clearly answered in this symposium or in this discussion is the effect of the supplemental fat in the sows' diet on the amount of medium-chain fatty acids in the colostrum and milk.

However, it is well demonstrated that when fat is added to the diet, the fatty acid composition of the milk mimics that of the dietary fat source.

The participants were reminded that early work by Elsley and others agreed with the French work quoted by Dr Seerley in showing that an increase in energy intake during gestation increases the birth weight of the piglets, but that this effect is very small if dietary energy intake is at normal or higher levels.

As described by Dr Seerley, a major reason for including fat in the lactation diet of sows is to minimize weight loss during lactation, and thus to alleviate or eliminate the reproductive problems that result from sows becoming too thin. Field studies in the United Kingdom and Western Australia have shown that sows lose considerable amounts of body fat during their first lactation and that this mobilization of body reserves is associated with increased returns to service following weaning. There was disagreement among the participants at the conference on this point. Some suggested that since many sows will not consume an adequate amount of feed in lactation, it may be helpful to include fat in the diet as a means of increasing the energy intake. Others suggested that voluntary feed intake in the lactating sow will be adequate if the sow is not fed excess energy during gestation.

Both the Western Australia farm trial and field trials in the United Kingdom demonstrated that feeding supplementary fat to sows before farrowing decreased the average number of days from weaning to service by about 1 day. Researchers from the University of Western Australia have shown that a high feed intake during lactation is required to reduce backfat loss and to minimize the period of post-weaning anoestrus in first-litter sows. These research workers have also shown that increasing both the energy and protein intakes of sows during their first lactation reduced the weaning-to-mating interval from 16.1 to 10.5 days.

The effect of supplemental fat in the sows' diet on piglet weaning weights seems inconsistent, since in some cases it is increased and in others it stays the same. However, total litter weaning weight is usually greater even if the average piglet weaning weight is not, because the number of pigs weaned is usually increased.

On the subject of nutrient density in pig diets, researchers at Denmark's National Institute of Animal Science Research in Pigs and Horses have investigated the influence of dietary energy concentration on the efficiency of utilization of metabolizable energy. As the metabolizable energy content of the diet was decreased by the addition of fibre, the net energy as a percentage of metabolizable energy also decreased. Expressed another way, as the percentage of the digested energy disappearing in the hindgut increased, the efficiency of utilization of the metabolizable energy (net energy as percentage of metabolizable energy) decreased.

18

USE OF FATS IN DIETS FOR LACTATING DAIRY COWS

DONALD L. PALMQUIST
Ohio Agricultural Research and Development Center, The Ohio State University, Wooster 44691, USA

The effect of dietary fat on metabolism and milk synthesis in the dairy cow was reviewed in depth recently by Storry (1981). Other reviews on the general topic of lipid nutrition and metabolism in ruminants are by Bauman and Davis (1975), Leat and Harrison (1975), Scott and Cook (1975), Palmquist (1976), Emery (1980), Palmquist and Jenkins (1980), Bell (1981), Christie (1980), Harfoot (1981), Moore and Christie (1981), Noble (1981) and Vernon (1981).

The classic studies of supplemental fat in dairy concentrates by Maynard and associates at Cornell University, reported from 1929 to 1944 (summarized by Maynard, Loosli and McCay, 1941; Loosli, Maynard and Lucas, 1944) are instructive in any review of the use of fat in diets of dairy cows. In tests of high- vs low-fat concentrate mixes ranging from 0.7% to 7.5% ether extract fed to a total of 185 cows, a small but consistent advantage in milk production was observed for the high-fat concentrates. Increased milk energy value exceeded the calculated energy value of the fat which led Maynard and associates to conclude the supplemental fat improved the energetic efficiency of milk production (Loosli, Maynard and Lucas, 1944).

Dietary fat and rumen fermentation

In many studies, dietary fat has caused milk-fat depression, which may be related to decreased fibre digestion in the rumen (Lucas and Loosli, 1944; Brooks *et al.*, 1954; Ward *et al.*, 1957; White *et al.*, 1958; Davison and Woods, 1963; Czerkawski, Blaxter and Wainman, 1966; El Hag and Miller, 1972; Devendra and Lewis, 1974a; Kowalczyk *et al.*, 1977; Palmquist and Jenkins, 1982; Ikwuegbu and Sutton, 1982) and to a narrower acetate:propionate ratio of rumen contents (Steele and Moore, 1968; Nicholson and Sutton, 1971; Rohr, Daenicke and Oslage, 1978; Ikwuegbu and Sutton, 1982). Beitz and Davis (1964) observed no change in the rumen volatile fatty acid (VFA) ratio even though 225 g of cod-liver oil fed once daily caused milk-fat depression.

The mechanism by which fat affects rumen fermentation is not clearly established. Devendra and Lewis (1974a) summarized four theories to explain depressed fibre digestion: (1) physical coating of the fibre by fat; (2) modified rumen microbial population due to toxic effects of fat; (3) inhibition of microbial activity from surface-active effects of fatty acids on cell membranes; (4) reduced cation availability from formation of insoluble complexes with long-chain fatty acids (LCFA). Devendra and Lewis (1974a) favoured a theory of physical coating of feed fibre with LCFA. Conversely, El Hag and Miller (1972) concluded that depression in digestibility results from inhibitory effects of LCFA on microbial growth, as shown by Galbraith *et al.* (1971). White *et al.* (1958) suggested microbial inhibition based on the observation that recovery of cellulose digestion did not occur until 17 days after removal of fat from the diet. Further, Ørskov, Hine and Grubb (1978) showed that physical coating of dried grass with tallow did not inhibit digestion of fibre in nylon bags suspended in the rumen of sheep fed normal diets, but digestion was inhibited in sheep fed tallow. Growth of the cellulolytic species *Butyrivibrio fibrisolvens*, *Ruminococcus albus* and *Ruminococcus flavefaciens* was inhibited by oleic acid in the presence of a soluble substrate (cellobiose) (Maczulak, Dehority and Palmquist, 1981). Finally, Harfoot *et al.* (1974) showed that fatty acids in rumen contents preferentially bind to food particles, competing with bacteria for adsorption. They proposed that such a mechanism would *decrease* inhibitory effects of LCFA on rumen micro-organisms. Purified cellulose (Solka Floc) added to pure culture incubations reversed effects of added LCFA on microbial growth (Maczulak *et al.*, 1981). Harfoot *et al.* (1974) observed that adsorption of LCFA on feed particles decreased as unsaturation increased, a possible factor in the greater toxicity observed with these fatty acids. Triglycerides were adsorbed to bacteria less strongly than fatty acids, but were adsorbed to food particles. Apparently no studies have demonstrated preferential binding of LCFA to different feedstuffs.

Lucerne ash reversed inhibition of cellulose digestion caused by added maize oil *in vitro* (Brooks *et al.*, 1954) and in sheep fed a semi-purified diet containing ground maize cobs as the sole roughage. White *et al.* (1958) showed that lucerne ash could be replaced by calcium and they concluded that for optimal cellulose utilization the ruminal requirement for calcium was increased by supplemental fat.

The most detailed studies of LCFA–cation interactions have been by Galbraith, Miller and associates (Galbraith *et al.*, 1971; El Hag and Miller, 1972; Galbraith and Miller, 1973). Gram-positive bacteria were inhibited by LCFA, whereas little or no inhibition was observed with Gram-negative organisms (Galbraith *et al.*, 1971; Henderson, 1973). A more variable response was observed by Maczulak *et al.* (1981). Elaidic acid (Galbraith *et al.*, 1971) and vaccenic acid (Maczulak *et al.*, 1981) were less inhibitory than oleic acid. Henderson (1973) concluded that minimum inhibition of Gram-positive organisms by LCFA was consistent with higher propionate usually observed in rumen content of fat-fed animals.

Among several alkaline earth metals and iron tested for ability to reverse inhibitory effects of fat on digestion *in vitro*, only calcium gave consistent results (El Hag and Miller, 1972). Davison and Woods (1963) also failed to show a beneficial response to magnesium carbonate *in vivo*. The effect of adding increasing amounts of linolenic acid to fat-extracted malt distiller's

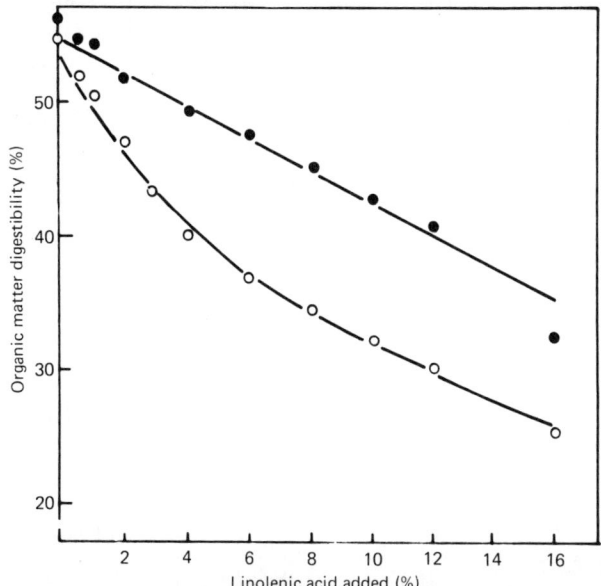

Figure 18.1 Effect of adding increasing amounts of linolenic acid to fat-extracted malt distillers grains on *in vitro* organic matter digestibility (OMD); –●–, OMD in the presence of 0.1 mmol Ca; –○–, OMD in the absence of added Ca. (Courtesy of El Hag and Miller, 1972)

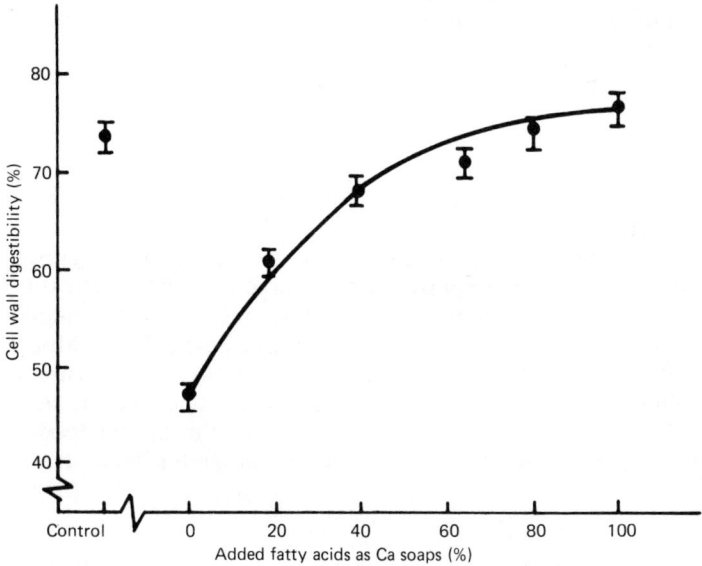

Figure 18.2 Effect of increasing proportion of calcium soaps on cell wall digestibility *in vitro*. Control: 30% concentrate:70% timothy hay; Experimental: 25% concentrate: 65% timothy hay:10% soya oil fatty acids. (Courtesy of Palmquist and Jenkins, 1982)

grains in the presence and absence of calcium *in vitro* is shown in *Figure 18.1* (El Hag and Miller, 1972). Even though digestibility was higher in the presence of calcium, there was a progressive decrease in organic matter digestibility with increasing fatty acid. Thus, calcium was not capable of *completely* reversing the inhibitory effects of linolenic acid, probably because the fatty acid did not react rapidly and completely in the aqueous medium (Jenkins and Palmquist, 1982). Compare the effect of increasing the proportion of fatty acid present as the *preformed* calcium soap on cell wall digestibility (*Figure 18.2*). These data show that the insoluble soaps of fatty acids are not inhibitory; whether a free carboxyl group is required for

Table 18.1 EFFECT OF TALLOW FATTY ACIDS OR CALCIUM SOAPS ON NUTRIENT DIGESTIBILITY IN TWO STEERS

Parameter	Diet		
	Control	Fatty acid	Calcium soap
Dry matter intake (kg/day)	5.9	4.8 (81)[a]	4.9 (83)
True digestibility in rumen (%):			
Dry matter	58.4	47.4 (81)	56.0 (96)
Acid detergent fibre	54.4	27.3 (50)	56.1 (103)
Total digestibility (%):			
Dry matter	71.6	71.7 (100)	69.8 (97)
Acid detergent fibre	56.0	51.3 (92)	53.4 (95)

[a] Values in parentheses = % of control. (Courtesy of Palmquist and Jenkins, 1982)

inhibition is not determined. In duodenally cannulated steers, rumen dry matter and fibre digestibilities were normal when the diet contained 5% calcium soap of tallow, whereas the same amount of tallow fatty acids reduced rumen fibre digestibility from 59% to 27% (*Table 18.1*; Palmquist and Jenkins, 1982).

Role of biohydrogenation

The significance of biohydrogenation in total ecology of the rumen is uncertain; however, biohydrogenation most certainly relieves the toxic effects of dietary polyunsaturated LCFA on rumen micro-organisms (Harfoot, 1981). Biohydrogenation was more complete, with accumulation of less *trans* monoenoic acids, when linoleic acid was infused intraruminally as triacylglycerol, compared with infusion of the free acid (Noble, Moore and Harfoot, 1974). Differences in the chemical nature of the precursors (esterified vs unesterified) which determine completeness and isomeric composition of biohydrogenation products, have significance in feeding practice, in particular the use of free fats or full-fat oil seeds (Steele, Noble and Moore, 1971).

Effects of fat on digestibility

In sheep, dietary fat usually depresses fibre digestibility (Sundstøl, 1974) (*Figure 18.3*), whereas this may not be true in lactating cows (Palmquist

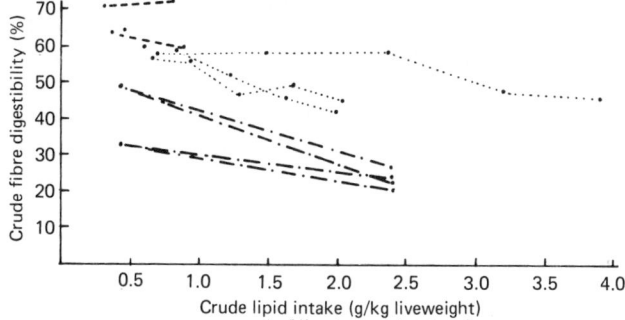

Figure 18.3 Effect of animal fat supplementation on digestibility of crude fibre by wethers in several studies. (After van der Honing et al., 1981)

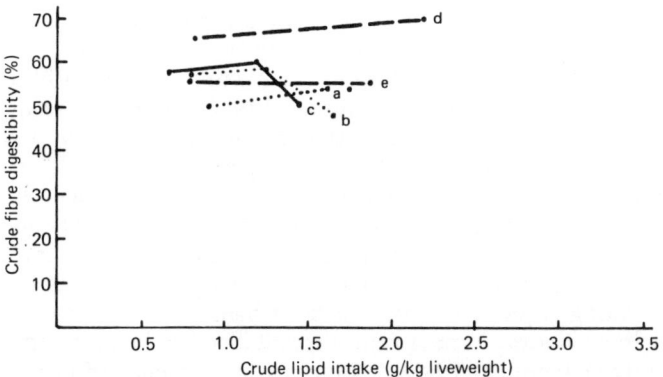

Figure 18.4 Effect of fat supplementation on digestibility of fibre by dairy cows. a = soya oil; b = coconut fat; c,d,e = animal tallow. (After van der Honing et al., 1981)

and Conrad, 1978, 1980; van der Honing et al., 1981) (*Figure 18.4*). In studies with unprotected fats, all fats increased ether extract digestibility (Rohr et al., 1978; Palmquist and Conrad, 1978, 1980; van der Honing et al., 1981). Apparent digestibility of nitrogen was increased by blended animal–vegetable fat (Palmquist and Conrad, 1978, 1980), and by coconut-palm oil (50:50), soya oil and animal tallow (Rohr et al., 1978). Fibre digestibility was unaffected by dietary fat (Palmquist and Conrad, 1978, 1980; van der Honing et al., 1981), or depressed by coconut-palm oil or tallow (Rohr et al., 1978). Bines et al. (1978) reported reduced digestibility of fibre when incompletely protected tallow was fed. Whole cottonseed fed up to 25% of the total diet (7% ether extract) increased digestibilities of lipid, nitrogen and energy, with no depressing effect on digestibilities of fibre, calcium, phosphorus, or magnesium (Smith et al., 1981).

There is a limit to the quantity of fat which may be absorbed by the cow (Palmquist and Conrad, 1978). Although ruminants are less affected than non-ruminants by the degree of fatty acid saturation (Andrews and Lewis, 1970a,b), highly saturated fats are less well digested (Macleod and

Buchanan-Smith, 1972). Bile acids are definitely required for absorption of fatty acid (Heath and Morris, 1963; Harrison and Leat, 1972); equally important is the need for phospholipid, pancreatic lipase, and pancreatic phospholipase (Smith and Lough, 1976; Lough and Smith, 1976). Whether any or all of these are limiting with high fat feeding is unknown. Bines *et al.* (1978) concluded that cows could efficiently absorb up to 1.4 kg exogenous (protected) fatty acid per day. Wrenn *et al.* (1976) observed sharp increases in faecal fat when protected lipid intake increased from 0.9 to 1.7 kg/day. Digestibility of unprotected lipid decreased when intake increased from 774 to 1360 g/day in Jersey cows (Palmquist and Conrad, 1978).

Site of digestion

Dietary oils tend to shift the site of digestion of organic matter from the rumen to the hindgut (*Table 18.1*), at least at relatively low levels of intake (Devendra and Lewis, 1974b; Sutton, 1980; Palmquist and Jenkins, 1982; Ikwuegbu and Sutton, 1982). The effect is due mainly to inhibition of fibre digestion in the rumen (Ikwuegbu and Sutton, 1982). Tamminga *et al.* (1982) found no inhibition of rumen fermentation when tallow was supplemented at 7% or 12% of the diet of lactating cows.

Absence of inhibitory effects of fat on total fibre digestibility in lactating cows (*Figure 18.4*) was attributed to high feed intake and rate of passage (Palmquist and Conrad, 1978) or to high feed intake and high concentrate feeding (van der Honing *et al.*, 1981), which led to low fibre digestion coefficients in lactating cows. The importance of high fibre and calcium intake on maintaining normal rumen fibre digestibility has been emphasized more recently (Palmquist, 1981a). Alternatively, absence of effects on total digestibility could be due to compensatory effects of hindgut fermentation (Palmquist and Jenkins, 1982). High calcium, especially limestone, should enhance fermentation in the hindgut.

Dietary fat and energy metabolism

Supplemental dietary fat may be beneficial in at least three ways: (1) when feed capacity limits energy intake and production, high energy density of fat may permit greater energy consumption; (2) at any level of feed intake, increasing fat intake to achieve an optimum ratio with other nutrients may increase efficiency of energy utilization (Kronfeld, 1976, 1982; Macleod, Yu and Schaeffer, 1977; Brumby *et al.*, 1978); (3) when excessive amounts of cereal grains are fed to maximize digestible energy intake, fat may be substituted for starch, increasing ration forage/concentrate ratio, normalizing rumen fermentation and correcting milk fat percentage (Palmquist and Conrad, 1978).

Dry-matter intake must be maintained if supplemental fat is to increase energy intake. Some studies suggest that dietary fat reduces forage consumption (Orth, Kauffman and Rohr, 1966; Bines *et al.*, 1978), whereas others report no effect (Palmquist and Conrad, 1978; Kronfeld *et al.*, 1980; van der Honing *et al.*, 1981; Østergaard *et al.*, 1981). Inhibition of fibre digestibility could reduce forage intake: thus a more appropriate test of effects of fat on energy metabolism may be with protected lipids. Bines et al. (1978) fed high-concentrate (75% of feed dry matter) diets containing 600–1400 g/day of protected tallow. As lipid intake increased, both hay and concentrate consumption decreased, the effect being moderate during weeks 1–6 of lactation, but dry-matter intake decreased markedly at the highest lipid intake during weeks 7–13, so that total energy consumed was reduced at the higher levels of lipid supplement. They concluded that reduced intake was caused by free fatty acid inhibition of rumen digestibility, due to incomplete protection of the supplement. Conversely, Smith, Dunkley and Franke (1978) concluded that total energy absorbed, rather than gut fill, limited dry-matter intake in cows fed protected tallow at approximately 12% of dry matter fed, even though body-weight loss continued during the first month of lactation. In other studies, protected oil seeds (Yang, Baldwin and Russell, 1978) or protected tallow (Sharma, Ingalls and McKirdy, 1978; Wrenn et al., 1978) did not reduce forage intake. Macleod et al. (1977) observed decreased concentrate intake and decreased forage intake when protected tallow was 8% or 13% of the concentrate. However, the cows were allowed concentrate for only two 1 h periods each day, with *ad libitum* forage for 22 hours.

Supplementation of unprotected lipid caused rate of eating to be decreased, especially size and length of the initial meal (Heinrichs, Palmquist and Conrad, 1982; de Visser, Tamminga and van Gils, 1982); however, compensation was achieved through increased size and number of spontaneous meals, so that total dry-matter consumption was not changed. The effect may be extra-ruminal, as Sharma et al. (1978) reported slower eating with an apparently well-protected tallow supplement. These studies seem to suggest that the possibility of increasing energy intake of lactating cows by feeding supplemental fat is limited, at best, to increases of 4–5% of total energy intake (Palmquist and Conrad, 1980; van der Honing et al., 1981).

Efficiency of energy utilization

Direct transfer of LCFA from diet to milk is inherently more efficient than *de novo* synthesis from carbohydrate or volatile fatty acids by the animal (Kronfeld, 1976, 1982; Baldwin et al., 1980). However, this transfer is limited, as few if any milk-fat C18 LCFA are synthesized *de novo* in the normal case (Palmquist et al., 1969), so that increased transfer of diet LCFA to milk fat must be at the expense of short- and medium-chain fatty acid synthesis (Storry, Hall and Johnson, 1973; Mattos and Palmquist, 1974; Smith, Dunkley and Franke, 1978; Christie, 1981), and/or by greater total secretion of milk fat. In either case, milk fatty acid composition is changed to a lower percentage of short- and medium-chain fatty acids and a higher percentage of LCFA. As milk fat probably has an obligatory content of short-chain fatty acids to maintain liquid melting point at body temperature (Moore and Christie, 1981), total milk-fat secretion may be limited by short-chain fatty acid content. It is interesting that most consistent increases in milk-fat percentage are reported when protected polyunsaturated fatty acids are fed (McDonald and Scott, 1977), although moderate increases with protected tallow have been reported (Smith *et al.*, 1978; Bines *et al.*, 1978; Wrenn *et al.*, 1978). Possibly the high linoleic acid content of such milk contributes physical characteristics usually provided by short-chain fatty acids which are required for normal secretion.

Estimates of diet LCFA transfer to milk fat vary widely, and are dependent on such factors as quantity of fat in the diet, level of milk-fat production and the metabolic state of adipose tissue (Christie, 1981; Storry, 1981). Although diet fat increases blood lipid concentration, these changes have been difficult to relate to milk-fat production (Palmquist, 1976). Baldwin *et al.* (1980) have shown that triacylglycerol uptake by mammary tissue is a curvilinear function of arterial concentration, best described by Michaelis–Menten kinetics. Input–output estimates for protected polyunsaturated fats range from 14% to 42% (Yang *et al.*, 1978; Storry, 1981). In addition to the factors listed above, estimates for protected polyunsaturated fats are influenced by underprotection from rumen biohydrogenation and by overprotection which decreases absorption. Further, only about one-half of the absorbed linoleic acid is in lipoprotein triacylglycerol, available for mammary uptake (Palmquist and Mattos, 1978). Banks, Clapperton and Ferrie (1976a) used input–output to estimate that approximately 50% of dietary C18 fatty acids were transferred to milk, whereas several literature estimates ranged from 30% to 40%. However, the high-fat diets used by Banks *et al.* (1976a) provided only about 550 g dietary fat/cow/day, low enough to provide a relatively high estimate of transfer according to the curvilinear relationships described by Baldwin *et al.* (1980). Palmquist and Mattos (1978) used re-injection techniques with ^{14}C-labelled chylomicrons to estimate that 76% of gut-synthesized lipoprotein triglyceride was taken up by the mammary gland of cows consuming 400–500 g/day LCFA. Further, the proportion of milk fat contributed by diet LCFA increased exponentially as total milk-fat production increased, suggesting some limitation of *de novo* synthesis.

Kronfeld (1976) calculated that milk production might be maximally efficient when the diet provided 16% of metabolizable energy as LCFA, which draws support from the experiments of Brumby *et al.* (1978), who concluded from regression analysis that maximum efficiency of energy utilization occurred when fatty acid from protected tallow was 12–16% of the digested energy (*Figure 18.5*). The lower value (12%) was in weeks 2–6 of lactation, whereas the higher value was optimal at weeks 7–13, suggesting that mobilized adipose fatty acids contributed to metabolizable energy in early lactation. Significant increases in efficiency of utilization of gross (10.1%) and of metabolizable energy (4.75%) were reported by van der Honing *et al.* (1981) when 7% tallow was added to the concentrate (7.9% crude fat in feed dry matter; about 16–20% of metabolizable energy as LCFA).

The energy cost of ATP generation from acetate is about 10% greater than from LCFA (Baldwin *et al.*, 1980). As acetate accounted for 23% of the total carbon dioxide produced in lactating goats, and LCFA provided less than 5% (Annison *et al.*, 1967), it seems that there is room to increase oxidative as well as synthetic energetic efficiency by increasing LCFA intake.

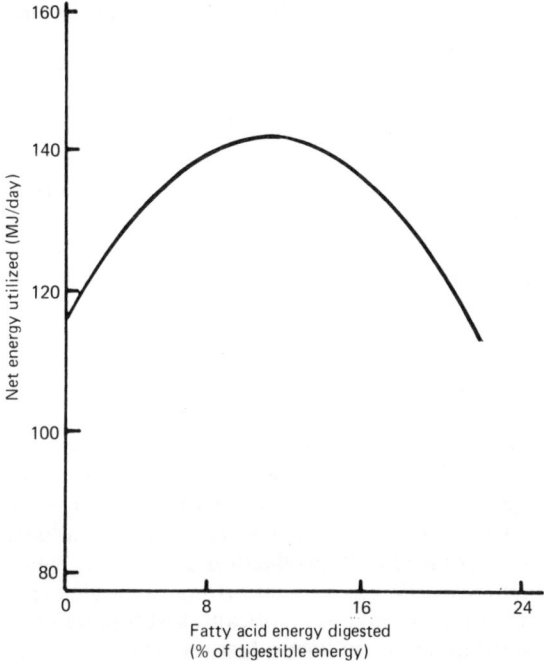

Figure 18.5 Relationship between net energy utilized (net energy in milk + net energy in liveweight gain + maintenance requirement) and the fatty acid energy digested (% DE) corrected for endogenous losses. (After Brumby *et al.*, 1978, courtesy of Cambridge University Press)

The utilization (partition) of dietary LCFA may be for milk-fat synthesis, oxidation or deposition as adipose. Partition between milk and adipose is influenced by endocrine status which is dependent on stage of lactation and plane of nutrition (Bines and Hart, 1982). High-concentrate diets cause energy to be partitioned away from milk and into adipose (Tyrrell, 1980), which may be mediated by insulin (Yang and Baldwin, 1973; Trenkle, 1981; Bines and Hart, 1982) in a 'glucogenic response' (Van Soest, 1963). Substituting dietary fat for starch reversed milk-fat depression and decreased body-weight gain (Palmquist and Conrad, 1978), presumably by normalizing rumen fermentation and minimizing the glucogenic response; however, an effect of diet LCFA on insulin and growth hormone is also possible, as diet fat decreased the plasma insulin concentration and induced insulin resistance (Palmquist and Moser, 1981), and increased the growth hormone/insulin ratio in early lactation (Palmquist, 1981b). Adipose sensitivity to lipolytic stimuli is greatly increased and re-esterification is minimal or absent in early lactation (Metz and van den Bergh, 1977). As growth hormone is well documented to mobilize adipose LCFA, directing them to milk fat (Peel *et al.*, 1981; Baumann, Eisemann and Currie, 1982; Bines and Hart, 1982), diet fat may promote its own partition to milk fat, at least in early lactation. The growth hormone/insulin ratio was decreased by dietary fat in the second half of lactation (Palmquist, 1981b).

One might anticipate that supplemental fat in early lactation should minimize body-weight loss (Macleod, Yu and Schaeffer, 1977). Cows fed protected lipid in early lactation were reported to have lower circulating ketones (Bines *et al.*, 1978; Kronfeld *et al.*, 1980), which Kronfeld (1982) attributed to displacement of mobilized adipose (plasma non-esterified fatty acids) by dietary fat (plasma lipoprotein triglycerides). In contrast to these observations, Østergaard *et al.* (1981) and Rijpkema and de Visser (1982) found that the energy of supplemental fat was converted to milk with no decrease in body-weight loss. The latter observation would be consistent with endocrine status orientated to adipose mobilization in early lactation (homeorhesis; Bauman and Currie, 1981).

Dietary fat and milk production

The variability of feeding trials for milk production is such that increases of 5% are often not significant with less than 10 cows per group (Palmquist and Conrad, 1978). In the extensive and comprehensive studies of Maynard and associates (Maynard *et al.*, 1941; Loosli *et al.*, 1944), adding 3–4% fat to the concentrates increased milk production by 2–10%, which often was not statistically significant. When the data from all studies (185 cows) were pooled, the effect of fat was highly significant. Østergaard *et al.* (1981) summarized the results of many milk-production trials in which cows were fed added fat (*Figure 18.6*) to demonstrate clearly a curvilinear

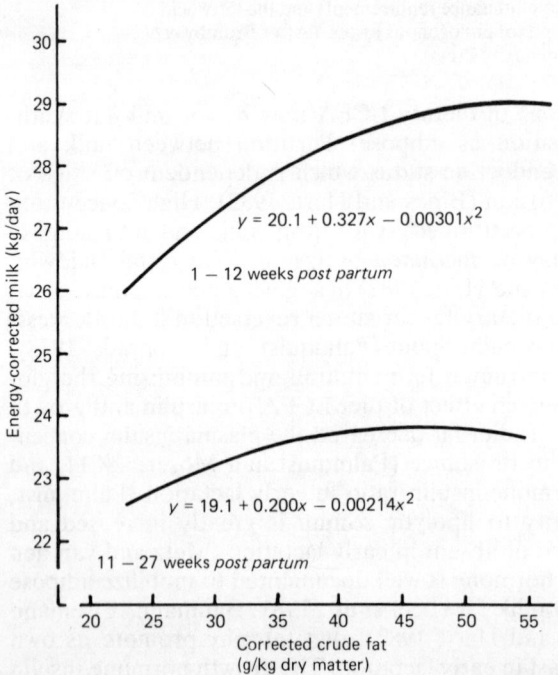

Figure 18.6 Quantitative relationships between the content of corrected crude fat in the ration and milk yield. (After Østergaard *et al.*, 1981)

Table 18.2 LEAST SQUARES MEANS OF MAIN EFFECTS FOR TRAITS MEASURED ON FIRST- AND SECOND-LACTATION ANIMALS IN THE FIRST 100 d OF LACTATION

	First-lactation Ration		Second-lactation Ration	
	Control	5% fat	Control	5% fat
Milk yield (kg)	2454	2525	3051[a]	3336[b]
Fat content (%)	3.6[a]	3.8[b]	3.7	3.6
Fat-corrected milk yield (kg)	2310[a]	2455[b]	2936[a]	3125[b]

[a,b] Means within lactation with dissimilar superscripts differ ($P<0.01$). (After Mattias et al., 1982)

Table 18.3 SUBSTITUTION OF FAT BLEND FOR CONCENTRATES: COMPOSITION OF THE TOTAL DIET DRY MATTER

Factor	Diet[a]			
	1	2	3	4
Crude protein (%)	13.6	15.9	13.5	16.3
Acid detergent fibre (%)	17.4	19.3	22.8	21.3
Ether extract (%)	3.30	2.88	5.90	6.80
Gross energy (MJ/kg)	19.4	18.9	19.9	20.0
Calcium (%)	0.43	0.64	0.65	0.65
Magnesium (%)	0.17	0.19	0.20	0.20
Grain in diet (%)[b]	60	60	46	46

[a] Diets 1 and 2: 1 part dehydrated lucerne pellets, 2 parts grain concentrate, 3 parts maize silage (wet basis). Diets 3 and 4: 1 part dehydrated lucerne pellets, 1 part grain concentrate, 3 parts maize silage (wet basis).
[b] Assumes 40% of maize silage dry matter is grain.
(After Palmquist and Conrad, 1978)

Table 18.4 SUBSTITUTION OF FAT BLEND FOR CONCENTRATES: FEED INTAKE, MILK PRODUCTION AND WEIGHT CHANGE

Factor	Diet[a]			
	1	2	3	4
Dry matter intake (kg/day)	19.3	19.3	19.6	20.2
Milk (kg/day)	29.4	29.6	28.6	30.3
FCM (kg/day)	23.7[b]	24.2[b]	26.3[bc]	27.7[c]
Milk fat (kg/day)	0.80[b]	0.83[b]	1.01[c]	1.10[c]
Milk protein (kg/day)	0.98	1.01	0.97	1.02
Milk fat (%)	2.71[b]	2.71[b]	3.42[c]	3.44[c]
Milk protein (%)	3.29	3.38	3.32	3.23
Body weight change (kg/day)	1.19[bc]	1.48[b]	0.72[bc]	0.33[c]

[a] See Table 18.3 for diet composition.
[bc] Means on the same line with different superscripts are different ($P<0.05$). (After Palmquist and Conrad, 1978)

Table 18.5 SUBSTITUTION OF FAT BLEND FOR CONCENTRATES: RUMEN VOLATILE FATTY ACIDS

Acid	Diet[a]			
	1	2	3	4
Acetic (molar %)	57.8[b]	59.4[bc]	61.2[bc]	62.2[c]
Propionic (molar %)	30.6[b]	29.2[bc]	26.4[bc]	24.9[c]
Butyric (molar %)	11.5	11.4	12.4	12.9
Ac:Pr	1.89	2.03	2.32	2.50

[a] See *Table 18.3* for diet composition.
[bc] Means on the same line with different superscripts are different ($P<0.05$).
(From Palmquist and Conrad, 1978)

response to fat which was greater for high-yielding cows. In a recent report, Mattias *et al.* (1982) fed concentrates containing 5% added tallow to first- and second-lactation Holstein cows for the first 100 days of lactation. Both first- and second-lactation cows fed fat produced more fat-corrected milk, but heifers responded only with a higher milk-fat percentage, while older cows had an equal milk fat percentage but produced more milk (*Table 18.2*). Interactions of genetic potential and fat supplementation were evident.

An alternative system for feeding fat addresses the problem of the low milk fat syndrome caused by feeding excess highly fermentable carbohydrate (Davis and Brown, 1970). By substituting blended animal–vegetable fat for concentrates, Palmquist and Conrad (1978) increased the forage/concentrate ratio while maintaining energy intake (*Table 18.3*): milk production was maintained, the milk fat percentage was normalized, and body-weight gain was minimized (*Table 18.4*). They postulated that the effect was mediated by normalization of rumen fermentation, as evidenced by rumen volatile fatty acid ratios (*Table 18.5*). Steele *et al.* (1971) reported similar positive production responses when ground whole soyabeans isocalorically replaced starch in the ration.

Milk fat production and composition

Increasing dietary LCFA increases their secretion in milk and inhibits *de novo* synthesis of short- and medium-chain fatty acids (except butyric) in mammary tissue. Usually the predominant increase is in oleic acid because of mammary desaturase activity on the stearic acid formed by biohydrogenation in the rumen. Banks *et al.* (1980) have used this principle to improve the low-temperature spreadability of butter (i.e. to produce milk fat of high oleic:palmitic ratio). Dietary fatty acids which are not modified in the rumen, either naturally, as for example lauric acid (Rindsig and Schultz, 1974), or by protection from biohydrogenation (Yang *et al.*, 1978) are transferred unchanged to milk fat, increasing their relative proportion. Storry *et al.* (1973) showed that total milk-fat production depends on a balance between transfer of dietary LCFA to milk fat and *de novo* synthesis in the mammary gland. Whether specific fatty acids have unusual effects on the secretion of total milk fat remains largely unresolved. For example, highly polyunsaturated LCFA of fish oils may reduce mammary uptake of LCFA from plasma lipoproteins by inhibition of lipoprotein lipase (Brumby, Storry and Sutton, 1972; Storry *et al.*, 1974c) although this concept is challenged by Christie (1981); it has been postulated that *trans* unsaturated acids inhibit milk-fat synthesis in an undefined manner (Selner and Schultz, 1980); the possibility that total unsaturation could limit synthesis by influencing melting point is intriguing. With regard to the last point, intravenous infusion of sterculic acid, an inhibitor of stearoyl CoA desaturase, into lactating animals reduced milk-fat secretion (Bickerstaffe and Johnson, 1972; Cook *et al.*, 1976). Milk-fat production is most consistently increased by feeding protected fats, particularly unsaturates, which increase plasma triglyceride concentration and mammary uptake (Gooden and Lascelles, 1973).

Milk protein concentration and production

Dietary fat decreases the content of protein in milk (*Figure 18.7*), although not in all cases (Palmquist and Conrad, 1978, 1980). The effect is apparently specific for casein synthesis (Dunkley, Smith and Franke, 1977; Storry *et al.*, 1974b; Schaar, Ahrné and Palmquist, unpublished work), implying specific effects on mammary gland metabolism. Palmquist and Moser (1981) postulated mediation by insulin, as insulin resistance was observed when protected soya oil was fed (*Figure 18.8*).

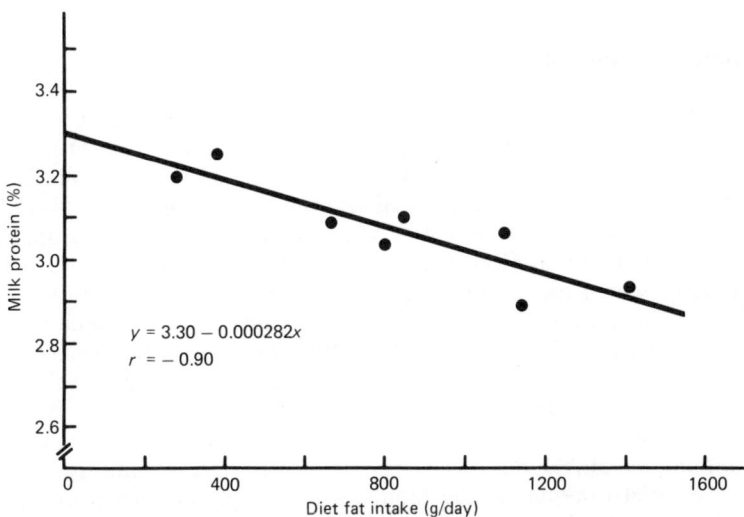

Figure 18.7 Relation between diet fat intake and milk protein concentration. (Data from Bines *et al.*, 1978)

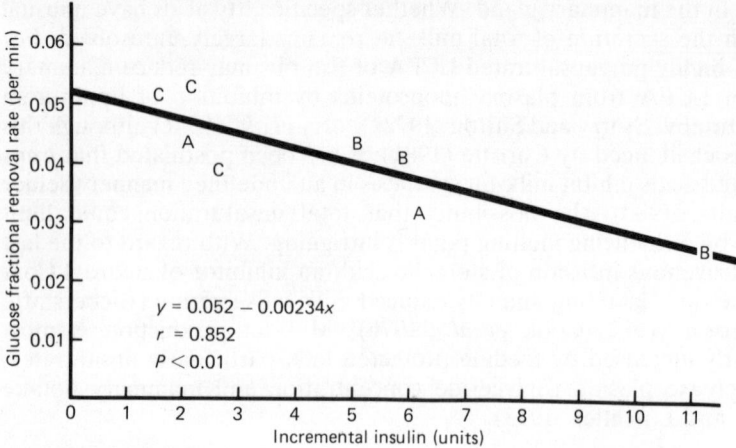

Figure 18.8 Glucose fractional removal rate vs insulin secretion after intravenous glucose load. A, B = high fat diets; C = control diet. (From Palmquist and Moser, 1981)

Other effects of diet fat on milk composition

No consistent effects of dietary fat on lactose content have been documented, as would be expected from its osmotic relationship to milk secretion. Dietary fat increases the citrate content of milk, which may be related to its function as a precursor of NADPH for fatty acid synthesis (Faulkner and Peaker, 1982).

Feed formulation and management

An important consideration for successful feeding of unprotected LCFA is to maximize forage intake. This principle is based on observation of fatty acid adsorption to feed particles (Harfoot, Noble and Moore, 1973; Harfoot *et al.*, 1974), maintenance of saliva production (Orth, Kauffman and Rohr, 1966) for normal rumen function and pH, which maximize biohydrogenation potential (Latham, Storry and Sharpe, 1972), and on feeding trials in which more milk and fat were produced when high-fat diets contained high forage (Brown, Stull and Stott, 1962; Hutjens and Schultz, 1971; Palmquist and Conrad, 1978, 1980). Of equal importance is the provision of relatively saturated fats, as for example, animal tallow (Brown *et al.*, 1962; Steele and Moore, 1968; Macleod and Wood, 1972; Banks *et al.*, 1976b; van der Honing *et al.*, 1981). Positive responses to blended animal–vegetable fats have also been reported (Palmquist and Conrad, 1978, 1980). Whether the blends have unique characteristics is uncertain, but they tend to cause less inhibition of fibre digestion *in vitro* than either completely esterified or unesterified fatty acids (Jenkins and Palmquist, unpublished work).

Alternatively, vegetable oils may be fed successfully as the unextracted oil in the whole seed, as shown for cottonseed (Anderson *et al.*, 1979; Smith *et al.*, 1981) and sunflower seed (Rafalowski and Park, 1982). Similar benefits may be obtained from crushed or extruded whole sunflower seed (McGuffey and Schingoethe, 1982), crushed whole soyabeans (Larson and Schultz, 1970; Hutjens and Schultz, 1971; Steele, Noble and Moore, 1971) and extruded whole soyabeans (Smith *et al.*, 1980; Mielke and Schingoethe, 1981). Feeding of raw soyabeans is limited to 10% of feed dry matter by the presence of trypsin inhibitor (Palmquist and Conrad, 1971). Limited data suggest that crushed rapeseed may be used as a fat supplement (Christensen, Cochran and Steacy, 1978); crushed rapeseed maintained normal milk and fat production in straw-based diets (Frank, 1979a,b, 1980).

Storry *et al.* (1974a) fed high levels of protected tallow to cows with the low milk fat syndrome induced by restricted roughage feeding. Nearly complete recovery of normal milk-fat production was achieved by transfer of about 20% of the supplemental tallow fatty acids directly to milk fat. Partial recovery of the milk-fat percentage was observed when high-grain diets were supplemented with 5% tallow or blended animal–vegetable fat (Palmquist and Conrad, 1980), but complete recovery occurred only when the grain content of the diet was reduced.

The energy-corrected milk production of cows supplemented with protected and unprotected lipids has been summarized by Østergaard *et al.* (1981) and Danfaer (1981); data are in *Figure 18.9*. It is evident that cows respond to higher levels of protected than of unprotected LCFA. Despite the relatively consistent response in production by cows fed protected lipid, this material has not become commercially viable because of the high cost of manufacture, difficulty in achieving consistent quality control, and lack of governmental approval to include formaldehyde-treated products in diets of lactating animals. The calcium soaps of LCFA are insoluble in

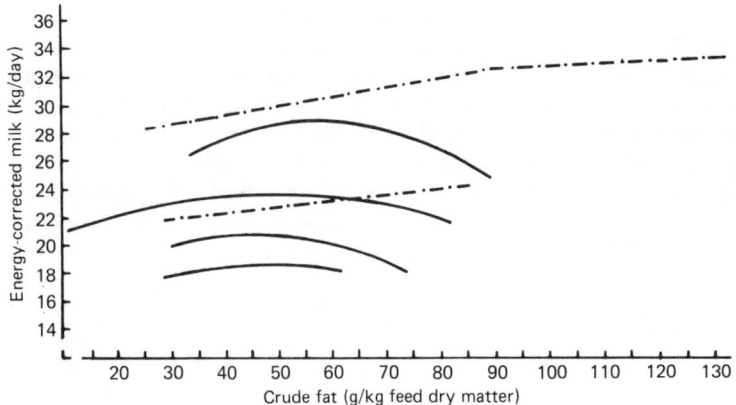

Figure 18.9 Effects of increasing amounts of animal fat (protected –•– and unprotected ──) on milk yield. Individual curves are composed from data from different experiments and represent cows of equal production potential. (After Østergaard *et al.*, 1981, and Danfaer, 1981)

the rumen and have been suggested to be a 'new generation of protected lipids' (Palmquist and Jenkins, 1982). Although no long-term milk-production data are available, calcium soaps of LCFA are being produced commercially as a dairy feed supplement.

The physical and biological characteristics of saturated fatty acids in the rumen (high melting point, low microbial inhibition) are the basis of a commercial venture in feed fat formulation. A 'dairy fat prill' containing approximately 45% palmitic, 45% stearic and 10% oleic as unesterified fatty acids was reported to increase milk production and to maintain the milk-fat percentage (Clapperton and Steele, 1982).

In feeding unprotected LCFA to lactating cows, care must be taken to provide adequate calcium (Palmquist and Conrad, 1980) and magnesium (Sundstøl, 1974; Bines et al., 1978) in the diet. Calcium increases the digestibility of all diet components (Palmquist and Conrad, 1980) and increases rate and extent of calcium soap formation in the rumen (Jenkins and Palmquist, 1982). The specific quantitative requirement of calcium in high-fat diets has not been determined; 0.9–1.0% of the diet dry matter has been suggested to feed compounders (Palmquist, 1981c). Hypomagnesaemia has been reported with feeding of high-fat diets (Sundstøl, 1974; Bines et al., 1978). The mechanism by which this occurs and the precise level required in fat feeding have not been established.

Handling fat products

By-product feed fats are difficult to manage in feed-compounding systems, because most are solid at ambient temperatures, and must be melted and pumped for storage, transport and mixing (Wilson, 1968). Further, their inclusion in compounded feeds is usually limited to 3–6%, as higher amounts cause bridging of meals in bulk storage and reduce strength of pellets. These problems have, for a long time, encouraged manufacturers to seek methods to produce 'dry' fats. Dry fat products include formaldehyde-treated, protein-coated protected fats, 'prills' of saturated fatty acids, calcium soaps, and fats adsorbed on to various carriers. Carriers range in adsorptive capability from about 40% to 65%, and since most are various types of earth, increase considerably the cost of shipping and dilute the energy value of the feed supplement. Some nutritive carriers, as for example, ground maize cobs and 'bee's wing' of maize cobs, have been used with moderate success. The lack of suitable dry fat sources generally limits the use of fat by-products to feed compounders.

Summary

Evidence from feeding trials indicates that fat may be used to increase milk and milk-fat production, and efficiency of production. However, many challenges remain for biochemists, nutritionists and dairymen. These include increased understanding of effects of fat on rumen fermentation, e.g. fibre digestibility, microbial protein synthesis, LCFA biohydrogenation and feed intake. Much remains to be learned of the effects of increased fat intake on endocrinology and energy utilization by the cow and on the synthesis of milk protein. More information on the use of fat in different feeding systems is needed to improve the predictability of animal response to dietary fat. Continued development of dry fat products should help to solve problems of fat effects on rumen fermentation, aid in development of practical fat-feeding systems and decrease labour and investment required by feed compounders to include fat in their programmes.

Acknowledgements

Funds for the Ohio research reported herein were provided by the Ohio Agricultural Experiment Station, Wooster; the Fats and Proteins Research Foundation, Des Plaines, IL; and Jacob Stern and Sons, Inc., Jenkintown, PA. Appreciation is due to Drs T.C. Jenkins and D.S. Kronfeld for reviewing the manuscript, and to Gloria Solze for typing the manuscript.

References

ANDERSON, M.J., ADAMS, D.C., LAMB, R.C. and WALTERS, J.L. (1979). Feeding whole cottonseed to lactating dairy cows. *Journal of Dairy Science* **62**, 1098–1103

ANDREWS, R.J. and LEWIS, D. (1970a). The utilization of dietary fats by ruminants. I. The digestibility of some commercially available fats. *Journal of Agricultural Science* **75**, 47–53

ANDREWS, R.J. and LEWIS, D. (1970b). The utilization of dietary fats by ruminants. II. The effect of fatty acid chain length and unsaturation on digestibility. *Journal of Agricultural Science* **75**, 55–60

ANNISON, E.F., LINZELL, J.L., FAZAKERLEY, S. and NICHOLS, B.W. (1967). The oxidation and utilization of palmitate, stearate, oleate and acetate by the mammary gland of the fed goat in relation to their overall metabolism, and the role of plasma phospholipids and neutral lipids in milk-fat synthesis. *Biochemical Journal* **102**, 637–647

BALDWIN, R.L., SMITH, N.E., TAYLOR, J. and SHARP, M. (1980). Manipulating metabolic parameters to improve growth rate and milk secretion. *Journal of Animal Science* **51**, 1416–1428

BANKS, W., CLAPPERTON, J.L. and FERRIE, MORAG E. (1976a). Effect of feeding fat to dairy cows receiving a fat-deficient basal diet. II. Fatty acid composition of the milk fat. *Journal of Dairy Research* **43**, 219–227

BANKS, W., CLAPPERTON, J.L., FERRIE, MORAG E. and WILSON, AGNES G. (1976b). Effect of feeding fat to dairy cows receiving a fat-deficient basal diet. I. Milk yield and composition. *Journal of Dairy Research* **43**, 213–218

BANKS, W., CLAPPERTON, J.L., KELLY, M.E., WILSON, A.G. and CRAWFORD, R.J.M. (1980). The yield, fatty acid composition and physical properties of milk fat obtained by feeding soya oil to dairy cows. *Journal of the Science of Food and Agriculture* **31**, 368–374

BAUMAN, D.E. and CURRIE, W.B. (1981). Partitioning of nutrients during pregnancy and lactation: a review of mechanisms involving homeostasis and homeorhesis. *Journal of Dairy Science* **63**, 1514–1529

BAUMAN, D.E. and DAVIS, C.L. (1975). Regulation of lipid metabolism. In *Digestion and Metabolism in the Ruminant* (I.W. McDonald and A.C.I. Warner, Eds), pp. 496–509. Armidale, NSW, New England Publishing Unit

BAUMAN, D.E., EISEMANN, J.H. and CURRIE, W.B. (1982). Hormonal effects on partitioning of nutrients for tissue growth: role of growth hormone and prolactin. *Federation Proceedings* **41**, 2538–2544

BEITZ, D.C. and DAVIS, C.L. (1964). Relationship of certain milk fat depressing diets to changes in the proportions of the volatile fatty acids produced in the rumen. *Journal of Dairy Science* **47**, 1213–1216

BELL, A.W. (1981). Lipid metabolism in liver and selected tissues and in the whole body of ruminant animals. *Progress in Lipid Research* **18**, 117–164

BICKERSTAFFE, R. and JOHNSON, A.R. (1972). The effect of intravenous infusions of sterculic acid on milk fat synthesis. *British Journal of Nutrition* **27**, 561–570

BINES, J.A. and HART, I.C. (1982). Metabolic limits to milk production, especially roles of growth hormone and insulin. *Journal of Dairy Science* **65**, 1375–1389

BINES, J.A., BRUMBY, P.E., STORRY, J.E., FULFORD, R.J. and BRAITHWAITE, G.D. (1978). The effect of protected lipids on nutrient intakes, blood and rumen metabolites and milk secretion in dairy cows during early lactation. *Journal of Agricultural Science* **91**, 135–150

BROOKS, C.C., GARNER, G.B., GEHRKE, C.W., MUHRER, M.E. and PFANDER, W.H. (1954). The effect of added fat on the digestion of cellulose and protein by ovine rumen microorganisms. *Journal of Animal Science* **13**, 758–764

BROWN, W.H., STULL, J.W. and STOTT, G.H. (1962). Fatty acid composition of milk. I. Effect of roughage and dietary fat. *Journal of Dairy Science* **45**, 191–196

BRUMBY, P.E., STORRY, J.E. and SUTTON, J.D. (1972). Metabolism of cod-liver oil in relation to milk fat secretion. *Journal of Dairy Research* **39**, 167–183

BRUMBY, P.E., STORRY, J.E., BINES, J.A. and FULFORD, R.J. (1978). Utilization of energy for maintenance and production in dairy cows given protected tallow during early lactation. *Journal of Agricultural Science* **91**, 151–159

CHRISTENSEN, D.A., COCHRANE, M. and STEACY, G. (1978). Utilization of protected and unprotected rapeseed by lactating dairy cows. In *Proceedings, 5th International Rapeseed Conference, Malmö*, Vol. 2, pp. 217–219. Malmö, Conference Organizing Committee

CHRISTIE, W.W. (1980). The effects of diet and other factors on the lipid composition of ruminant tissues and milk. *Progress in Lipid Research* **17**, 245–278

CHRISTIE, W.W., Ed. (1981). *Lipid Metabolism in Ruminant Animals*. New York, Pergamon Press
CLAPPERTON, J.L. and STEELE, W. (1982). Different forms of fat in the diet of dairy cows. *Proceedings of the Nutrition Society* **41**, 136A
COOK, L.J., SCOTT, T.W., MILLS, S.C., FOGERTY, A.C. and JOHNSON, A.R. (1976). Effects of protected cyclopropene fatty acids on the composition of ruminant milk fat. *Lipids* **11**, 705–711
CZERKAWSKI, J.W., BLAXTER, K.L. and WAINMAN, F.W. (1966). The effect of linseed oil and of linseed oil fatty acids incorporated in the diet on the metabolism of sheep. *British Journal of Nutrition* **20**, 485–494
DANFAER, A. (1981). The effect of dietary fat on milk production. In *Fats in Feeds and Feeding* (R. Marcuse, Ed.), pp. 84–91. Gothenburg, Scandinavian Forum for Lipid Research and Technology
DAVIS, C.L. and BROWN, R.E. (1970). Low milk fat syndrome. In *Physiology of Digestion and Metabolism in the Ruminant* (A.T. Phillipson, Ed.), pp. 545–565. Newcastle upon Tyne, Oriel Press
DAVISON, K.L. and WOODS, W. (1963). Effect of calcium and magnesium upon digestibility of a ration containing corn oil by lambs. *Journal of Animal Science* **22**, 27–29
DEVENDRA, C. and LEWIS, D. (1974a). The interaction between dietary lipids and fibre in the sheep. 2. Digestibility studies. *Animal Production* **19**, 67–76
DEVENDRA, C. and LEWIS, D. (1974b). The interaction between dietary lipids and fibre in the sheep. IV. Duodenal studies. *Malaysian Agricultural Research* **3**, 228–241
DE VISSER, H., TAMMINGA, S. and VAN GILS, L.G.M. (1982). Further studies on the effect of fat supplementation of concentrates fed to lactating dairy cows. 1. Effect on feed intake, feed intake pattern and milk production and composition. *Netherlands Journal of Agricultural Science* **30**, 347–352
DUNKLEY, W.L., SMITH, N.E. and FRANKE, A.A. (1977). Effects of feeding protected tallow on composition of milk and milk fat. *Journal of Dairy Science* **60**, 1863–1869
EL HAG, G.A. and MILLER, T.B. (1972). Evaluation of whisky distillery by-products. VI. The reduction in digestibility of malt distiller's grains by fatty acids and the interaction with calcium and other reversal agents. *Journal of the Science of Food and Agriculture* **23**, 247–258
EMERY, R.S. (1980). Mobilization, turnover and disposition of adipose tissue lipids. In *Digestive Physiology and Metabolism in Ruminants* (Y. Ruckebush and P. Thivend, Eds), pp. 541–558. Lancaster, England, MTP Press, Ltd
FAULKNER, A. and PEAKER, M. (1982). Reviews of the progress of dairy science: secretion of citrate into milk. *Journal of Dairy Research* **49**, 159–169
FRANK, B. (1979a). *Fatty Rape Products for Dairy Cows. 1. Rapeseed in a Straw-based Diet. Report 70*. Uppsala, Swedish University for Agricultural Science, Department of Animal Husbandry
FRANK, B. (1979b). *Fatty Rape Products for Dairy Cows. 2. Rapeseed and Rapeseed Expeller cake in Straw-based diets. Report 71*. Uppsala,

Swedish University for Agricultural Science, Department of Animal Husbandry

FRANK, B. (1980). *Fatty Rape Products for Dairy Cows. 3. Rapeseed in Different Rations. Report 75.* Uppsala, Swedish University for Agricultural Science, Department of Animal Husbandry

GALBRAITH, H. and MILLER, T.B. (1973). Effect of metal cations and pH on the antibacterial activity and uptake of long chain fatty acids. *Journal of Applied Bacteriology* **36**, 635–646

GALBRAITH, H., MILLER, T.B., PATON, A.M. and THOMPSON, J.K. (1971). Antibacterial activity of long chain fatty acids and the reversal with calcium, magnesium, ergocalciferol and cholesterol. *Journal of Applied Bacteriology* **34**, 803–813

GOODEN, J.M. and LASCELLES, A.K. (1973). Effect of feeding protected lipid on the uptake of precursors of milk fat by the bovine mammary gland. *Australian Journal of Biological Sciences* **26**, 1201–1210

HARFOOT, C.G. (1981). Lipid metabolism in the rumen. In *Lipid Metabolism in Ruminant Animals* (W.W. Christie, Ed.), pp. 21–55. New York, Pergamon Press

HARFOOT, C.G., NOBLE, R.C. and MOORE, J.H. (1973). Food particles as a site for biohydrogenation of unsaturated fatty acids in the rumen. *Biochemical Journal* **132**, 829–832

HARFOOT, C.G., CROUCHMAN, M.L., NOBLE, R.C. and MOORE, J.H. (1974). Competition between food particles and rumen bacteria in the uptake of long-chain fatty acids and triglycerides. *Journal of Applied Bacteriology* **37**, 633–641

HARRISON, F.A. and LEAT, W.M.F. (1972). Absorption of palmitic, stearic and oleic acids in the sheep in the presence or absence of bile and/or pancreatic juice. *Journal of Physiology* **225**, 565–576

HEATH, T.J. and MORRIS, B. (1963). The role of bile and pancreatic juice in the absorption of fat in ewes and lambs. *British Journal of Nutrition* **17**, 465–474

HEINRICHS, A.J., PALMQUIST, D.L. and CONRAD, H.R. (1982). Feed intake patterns of cows fed high fat grain mixtures. *Journal of Dairy Science* **65**, 1325–1328

HENDERSON, C. (1973). The effects of fatty acids on pure cultures of rumen bacteria. *Journal of Agricultural Science* **81**, 107–112

HUTJENS, M.F. and SCHULTZ, L.H. (1971). Addition of soybeans or methionine analog to high-concentrate rations for dairy cows. *Journal of Dairy Science* **54**, 1637–1644

IKWUEGBU, O.A. and SUTTON, J.D. (1982). The effect of varying the amount of linseed oil supplementation on rumen metabolism in sheep. *British Journal of Nutrition* **48**, 365–375

JENKINS, T.C. and PALMQUIST, D.L. (1982). Effect of added fat and calcium on in vitro formation of insoluble fatty acid soaps and cell wall digestibility. *Journal of Animal Science* **55**, 957–963

KOWALCZYK, J., ØRSKOV, E.R., ROBINSON, J.J. and STEWART, C.S. (1977). Effect of fat supplementation on voluntary food intake and rumen metabolism in sheep. *British Journal of Nutrition* **37**, 251–257

KRONFELD, D.S. (1976). The potential importance of the proportions of glucogenic, lipogenic and aminogenic nutrients in regard to the health and productivity of dairy cows. *Advances in Animal Physiology and Animal Nutrition* **7**, 5–26

KRONFELD, D.S. (1982). Major metabolic determinants of milk volume, mammary efficiency, and spontaneous ketosis in dairy cows. *Journal of Dairy Science* **65**, 2204–2212

KRONFELD, D.S., DONOGHUE, S., NAYLOR, J.M., JOHNSON, K. and BRADLEY, C.A. (1980). Metabolic effects of feeding protected tallow to dairy cows. *Journal of Dairy Science* **63**, 545–552

LARSON, S.A. and SCHULTZ, L.H. (1970). Effects of soybeans compared to soybean oil and meal in the ration of dairy cows. *Journal of Dairy Science* **53**, 1233–1240

LATHAM, M.J., STORRY, J.E. and SHARP, M.E. (1972). Effect of low roughage diets on the microflora and lipid metabolism in the rumen. *Applied Microbiology* **24**, 871–877

LEAT, W.M.F. and HARRISON, F.A. (1975). Digestion, absorption and transport of lipids in the sheep. In *Digestion and Metabolism in the Ruminant* (I.W. McDonald and A.C.I. Warner, Eds;, pp. 481–495. Armidale, NSW, University of New England Publishing Unit

LOOSLI, J.K., MAYNARD, L.A. and LUCAS, H.L. (1944). IV. Further studies of the influence of different levels of fat intake upon milk secretion. *Cornell University Agricultural Experimental Station Memoir 265*, 3–32

LOUGH, A.K. and SMITH, A. (1976). Influence of the products of phospholipolysis of phosphatidylcholine on micellar solubilization of fatty acids in the presence of bile salts. *British Journal of Nutrition* **35**, 89–96

LUCAS, H.L. and LOOSLI, J.K. (1944). The effect of fat upon the digestion of nutrients by dairy cows. *Journal of Animal Science* **3**, 3–11

McDONALD, I.W. and SCOTT, T.W. (1977). Foods of ruminant origin with elevated content of polyunsaturated fatty acids. *World Review of Nutrition and Dietetics* **26**, 144–207

McGUFFEY, R.K. and SCHINGOETHE, D.J. (1982). Whole sunflower seeds for high producing dairy cows. *Journal of Dairy Science* **65**, 1479–1483

MACLEOD, G.K. and BUCHANAN-SMITH, J.G. (1972). Digestibility of hydrogenated tallow, saturated fatty acids and soybean oil-supplemented diets by sheep. *Journal of Animal Science* **35**, 890–895

MACLEOD, G.K. and WOOD, A.S. (1972). Influence of amount and degree of saturation of dietary fat on yield and quality of milk. *Journal of Dairy Science* **55**, 439–445

MACLEOD, G.K., YU, Y. and SCHAEFFER, L.R. (1977). Feeding value of protected animal tallow for high yielding dairy cows. *Journal of Dairy Science* **60**, 726–738

MACZULAK, A.E., DEHORITY, B.A. and PALMQUIST, D.L. (1981). Effects of long-chain fatty acids on growth of rumen bacteria. *Applied and Environmental Microbiology* **42**, 856–862

MATTIAS, J.E., RUEGSEGGER, G.J., SCHULTZ, L.H. and TYLER, W.J. (1982). Effect of feeding animal fat to dairy cows in early lactation. *Journal of Dairy Science* **65** (Suppl. 1), 151

MATTOS, W. and PALMQUIST, D.L. (1974). Increased polyunsaturated fatty acid yields in milk of cows fed protected fat. *Journal of Dairy Science* **57**, 1050–1054

MAYNARD, L.A., LOOSLI, J.K. and McCAY, C.M. (1941). III. Further studies of the influence of different levels of fat intake upon milk secretion. *Cornell University Agricultural Experimental Station Bulletin* **753**, 1–18

METZ, S.H.M. and VAN DEN BERGH, S.G. (1977). Regulation of fat mobilization in adipose tissue of dairy cows in the period around parturition. *Netherlands Journal of Agricultural Science* **25**, 198–211

MIELKE, C.D. and SCHINGOETHE, D.J. (1981). Heat-treated soybeans for lactating cows. *Journal of Dairy Science* **64**, 1579–1585

MOORE, J.H. and CHRISTIE, W.W. (1981). Lipid metabolism in the mammary gland of ruminant animals. In *Lipid Metabolism in Ruminant Animals* (W.W. Christie, Ed.), pp. 227–277. New York, Pergamon Press

NICHOLSON, J.W.G. and SUTTON, J.D. (1971). Some effects of unsaturated oils given to dairy cows with rations of different roughage content. *Journal of Dairy Research* **38**, 363–372

NOBLE, R.C. (1981). Digestion, absorption and transport of lipids in ruminant animals. In *Lipid Metabolism in Ruminant Animals* (W.W. Christie, Ed.), pp. 57–93. New York, Pergamon Press

NOBLE, R.C., MOORE, J.H. and HARFOOT, C.G. (1974). Observations on the pattern of biohydrogenation of esterified and unesterified linoleic acid in the rumen. *British Journal of Nutrition* **31**, 99–108

ØRSKOV, E.R., HINE, R.S. and GRUBB, D.A. (1978). The effect of urea on digestion and voluntary intake by sheep of diets supplemented with fat. *Animal Production* **27**, 241–245

ORTH, A., KAUFFMAN, W. and ROHR, K. (1966). Beitrag zur Frage des Einflusses höherer und verschiedenartiger Fettgaben auf die Leistung von Milchkühen und die verdauungsvorgänge im Pansen. *Zeitschrift für Tierphysiologie, Tierernährung und Futtermittelkunde* **21**, 83–96

ØSTERGAARD, V., DANFAER, A., DANGAARD, J., HINDHEDE, J. and THYSEN, I. (1981). *The Effect of Dietary Lipids on Milk Production in Dairy Cows*. Copenhagen, Beretning Fra Statens Husdyrbrugs Forsøg 508

PALMQUIST, D.L. (1976). A kinetic concept of lipid transport in ruminants. A review. *Journal of Dairy Science* **59**, 355–363

PALMQUIST, D.L. (1981a). Fat as an energy source in lactation diets. In *Proceedings, Georgia Nutrition Conference*, pp. 148–160. University of Georgia

PALMQUIST, D.L. (1981b). Metabolite, insulin and growth hormone concentrations in blood plasma of cows fed high fat diets for entire lactations. *Journal of Dairy Science* **64** (Suppl. 1), 159

PALMQUIST, D.L. (1981c). Use of fat in formulating dairy rations. In *Proceedings, Maryland Nutrition Conference* (J.H. Vandersall, Ed.), pp. 60–64. University of Maryland

PALMQUIST, D.L. and CONRAD, H.R. (1971). High levels of raw soybeans for dairy cows. *Proceedings of the American Society of Animal Science* **33**, 295

PALMQUIST, D.L. and CONRAD, H.R. (1978). High fat rations for dairy cows. Effects on feed intake, milk and fat production, and plasma metabolites. *Journal of Dairy Science* **61**, 890–901

PALMQUIST, D.L. and CONRAD, H.R. (1980). High fat rations for dairy cows. Tallow and hydrolyzed blended fat at two intakes. *Journal of Dairy Science* **63**, 391–395

PALMQUIST, D.L. and JENKINS, T.C. (1980). Fat in lactation rations: Review. *Journal of Dairy Science* **63**, 1–14

PALMQUIST, D.L. and JENKINS, T.C. (1982). Calcium soaps as a fat supplement in dairy cattle feeding. In *Proceedings, XIIth World Congress on Diseases of Cattle, Amsterdam*, pp. 477–481. Dutch Section of the World Association for Buiatrics, Utrecht, The Netherlands

PALMQUIST, D.L. and MATTOS, W. (1978). Turnover of lipoproteins and transfer to milk fat of dietary (1-Carbon-14) linoleic acid in lactating cows. *Journal of Dairy Science* **61**, 561–565

PALMQUIST, D.L. and MOSER, E.A. (1981). Dietary fat effects on blood insulin, glucose utilization, and milk protein content of lactating cows. *Journal of Dairy Science* **64**, 1664–1670

PALMQUIST, D.L., DAVIS, C.L., BROWN, R.E. and SACHAN, D.S. (1969). Availability and metabolism of various substrates in ruminants. V. Entry rate into the body and incorporation into milk fat of D(−) β-hydroxybutyrate. *Journal of Dairy Science* **52**, 633–638

PEEL, C.J., BAUMAN, D.E., GOREWIT, R.C. and SNIFFEN, C.J. (1981). Effect of exogenous growth hormone on lactational performance in high yielding dairy cows. *Journal of Nutrition* **111**, 1662–1671

RAFALOWSKI, W. and PARK, C.S. (1982). Whole sunflower seed as a fat supplement for lactating cows. *Journal of Dairy Science* **65**, 1484–1492

RIJPKEMA, Y.S. and DE VISSER, H. (1982). Rundvet in krachtvoer voor melkkoeien. *Bedrijfsontwikkeling* **13**, 39–45

RINDSIG, R.B. and SCHULTZ, L.H. (1974). Effect of feeding lauric acid to lactating cows on milk composition, rumen fermentation, and blood lipids. *Journal of Dairy Science* **57**, 1414–1418

ROHR, K., DAENICKE, R. and OSLAGE, H.J. (1978). Untersuchungen über den Einfluss verschiedener Fettbeimischungen zum Futter auf Stoffwechsel und Leistung von Milchkühen. *Landbauforschung Volkenrode* **28**, 139–150

SCOTT, T.W. and COOK, L.J. (1975). Effect of dietary fat on lipid metabolism in ruminants. In *Digestion and Metabolism in the Ruminant* (I.W. McDonald and A.C.I. Warner, Eds), pp. 510–532. Armidale, NSW, University of New England Publishing Unit

SELNER, D.R. and SCHULTZ, L.H. (1980). Effects of feeding oleic acid or hydrogenated vegetable oils to lactating cows. *Journal of Dairy Science* **63**, 1235–1241

SHARMA, H.R., INGALLS, J.R. and McKIRDY, J.A. (1978). Replacing barley with protected tallow in ration of lactating Holstein cows. *Journal of Dairy Science* **61**, 574–583

SMITH, A. and LOUGH, A.K. (1976). Micellar solubilization of fatty acids in aqueous media containing bile salts and phospholipids. *British Journal of Nutrition* **35**, 77–87

SMITH, N.E., DUNKLEY, W.L. and FRANKE, A.A. (1978). Effects of feeding protected tallow to dairy cows in early lactation. *Journal of Dairy Science* **61**, 747–756

SMITH, N.E., COLLAR, L.S., BATH, D.L., DUNKLEY, W.L. and FRANKE, A.A. (1980). Whole cottonseed and extruded soybean for cows in early lactation. *Journal of Dairy Science* **63** (Suppl. 1), 153–154

SMITH, N.E., COLLAR, L.S., BATH, D.L., DUNKLEY, W.L. and FRANKE, A.A. (1981). Digestibility and effects of whole cottonseed fed to lactating cows. *Journal of Dairy Science* **64**, 2209–2215

STEELE, W. and MOORE, J.H. (1968). The effects of a series of saturated fatty acids in the diet on milk-fat secretion in the cow. *Journal of Dairy Research* **35**, 361–370

STEELE, W., NOBLE, R.C. and MOORE, J.H. (1971). The effects of 2 methods of incorporating soybean oil into the diet on milk yield and composition in the cow. *Journal of Dairy Research* **38**, 43–48

STORRY, J.E. (1981). The effect of dietary fat on milk composition. In *Recent Advances in Animal Nutrition—1981* (W. Haresign, Ed.), pp. 3–33. London, Butterworths

STORRY, J.E., HALL, A.J. and JOHNSON, V.W. (1973). The effects of increasing amounts of dietary tallow on milk-fat secretion in the cow. *Journal of Dairy Research* **40**, 293–299

STORRY, J.E., BRUMBY, P.E., HALL, A.J. and JOHNSON, V.W. (1974a). Responses in rumen fermentation and milk-fat secretion in cows receiving low-roughage diets supplemented with protected tallow. *Journal of Dairy Research* **41**, 165–173

STORRY, J.E., BRUMBY, P.E., HALL, A.J. and JOHNSON, V.W. (1974b). Response of the lactating cow to different methods of incorporating casein and coconut oil in the diet. *Journal of Dairy Science* **57**, 61–67

STORRY, J.E., BRUMBY, P.E., HALL, A.J. and TUCKLEY, B. (1974c). Effects of free and protected forms of codliver oil on milk fat secretion in the dairy cow. *Journal of Dairy Science* **57**, 1046–1049

SUNDSTØL, F. (1974). *Hydrogenated Marine Fat as Feed Supplement: Department of Animal Nutrition Report No. 159*, Agricultural University of Norway

SUTTON, J.D. (1980). Digestion and end-product formation in the rumen from production rations. In *Digestive Physiology and Metabolism in Ruminants* (Y. Ruckebusch and P. Thivend, Eds), pp. 271–290. Lancaster, England, MTP Press, Ltd

TAMMINGA, S., VAN VUUREN, A.M., VAN DER KOELEN, C.J., KHATTAB, H.M. and VAN GILS, L.G.M. (1983). Further studies on the effect of fat supplementation of concentrates fed to lactating dairy cows. 3. Effect on rumen fermentation and site of digestion of dietary components. *Netherlands Journal of Agricultural Science* **31**, 249–258

TRENKLE, A. (1981). Endocrine regulation of energy metabolism in ruminants. *Federation Proceedings* **40**, 2536–2541

TYRRELL, H.F. (1980). Limits to milk production efficiency by the dairy cow. *Journal of Animal Science* **51**, 1441–1447

VAN DER HONING, Y., WIEMAN, B.J., STEG, A. and VAN DONSELAAR, B. (1981). The effect of fat supplementation of concentrates on digestion and utilization of energy by productive dairy cows. *Netherlands Journal of Agricultural Science* **29**, 79–92

VAN SOEST, P.J. (1963). Ruminant fat metabolism with particular reference to factors affecting low milk fat and feed efficiency. A review. *Journal of Dairy Science* **46**, 204–216

VERNON, R.G. (1981). Lipid metabolism in the adipose tissue of ruminant animals. In *Lipid Metabolism in Ruminant Animals* (W.W. Christie, Ed.), pp. 179–362. New York, Pergamon Press

WARD, J.K., TEFFT, C.W., SIRNY, R.J., EDWARDS, H.N. and TILLMAN, A.D. (1957). Further studies concerning the effect of alfalfa ash upon the utilization of low-quality roughages by ruminant animals. *Journal of Animal Science* **16**, 633–641

WHITE, T.W., GRAINGER, R.B., BAKER, F.H. and STROUD, J.W. (1958). Effect of supplemental fat on digestion and the ruminal calcium requirement of sheep. *Journal of Animal Science* **17**, 797–803

WILSON, J.D. (1968). Mill problems of fat inclusion. In *Proceedings, 2nd Nutrition Conference for Feed Manufacturers* (H. Swan and D. Lewis, Eds), pp. 75–86. London, J. & A. Churchill, Ltd

WRENN, T.R., WEYANT, J.R., WOOD, D.L., BITMAN, J., RAWLINGS, R.M. and LYON, K.E. (1976). Increasing polyunsaturation of milk fats by feeding formaldehyde-protected sunflower–soybean supplement. *Journal of Dairy Science* **59**, 627–635

WRENN, T.R., BITMAN, J., WATERMAN, R.A., WEYANT, J.R., WOOD, D.L., STROZINSKI, L.L. and HOOVEN, N.W., Jr (1978). Feeding protected and unprotected tallow to lactating cows. *Journal of Dairy Science* **61**, 49–58

YANG, Y.T. and BALDWIN, R.L. (1973). Lipolysis in isolated cow adipose cells. *Journal of Dairy Science* **56**, 366–374

YANG, Y.T., BALDWIN, R.L. and RUSSELL, J. (1978). Effects of long supplementation with lipids on lactating dairy cows. *Journal of Dairy Science* **61**, 180–188

19

THE USE OF FAT IN DOG AND CAT DIETS

P.T. KENDALL
Animal Studies Centre, Freeby Lane, Waltham-on-the-Wolds, Leicestershire LE14 4RT, UK

The functions of dietary fat for dogs and cats are similar to those of other mammals. Fat serves as a concentrated energy source, provides EFA, is a carrier of fat-soluble vitamins and conveys palatability and texture to food. There are very few data on *in vivo* energy values of different fats in dog and cat diets. Most fats appear to be well absorbed by both species, with high apparent digestibilities which rise hyperbolically with fat intake. Average true digestibilities of fat for commercial foods were estimated to be 98% and 83% in dogs and cats, respectively. Dogs appear to have a significantly greater ability than cats to digest and absorb fats in most diets, but the differences are not explained by a higher metabolic faecal fat loss in cats. In general, however, both species can tolerate very wide ranges in dietary fat level (5–66%), provided that EFA and other nutrient requirements are met.

The EFA metabolism of dogs conforms with the normal mammalian pattern and while precise quantitative needs have not been established, 1% dietary linoleic acid in the diet dry matter (DM)—i.e. 2% dietary metabolizable energy (ME)—appears to be adequate for all life stages. In contrast, the EFA metabolism of cats is remarkable because of an apparent inability to convert linoleic to γ-linolenic acid, explained by the absence or low activity of the Δ^6-desaturase enzyme. Alternative pathways for the synthesis of PUFA, such as arachidonate, have been postulated and contradictory clinical and biochemical evidence has been presented when cats receive vegetable oil as the only dietary EFA. Thus, the qualitative nature of the feline EFA requirement remains confused but in practice it is prudent to include at least 2% of the dietary ME as linoleic acid and some PUFA of animal origin.

Fat-soluble vitamin needs of dogs and cats have been the subject of limited study. Cats appear to be unable to convert carotenes into materials with vitamin A activity and require dietary supplies of synthetic vitamin A or animal tissues rich in vitamin A. In contrast, vitamin A metabolism of dogs conforms to the orthodox mammalian pattern. The best published estimates of fat-soluble vitamin needs of both species are reviewed.

Considerable emphasis is placed on diet palatability within the pet-food industry and the contribution of fats is believed to be important. However, data on the physico-chemical properties and method of incorporation of fat on palatability of dog and cat diets have rarely been published.

Introduction

Before considering the use of fat in dog and cat diets it is important to understand how the nutrition of these species differs from that of the food animals discussed previously. Companion animals such as the dog and cat are rarely kept for purely economic reasons, and there has not been the same incentive, nor the same funding, for research compared with that for food animals. In addition, because there is no real economic value attributable to particular rates of growth or reproductive efficiency, it is difficult to define the criteria necessary to assess the optimum level of nutrient input. The objectives of feeding companion animals are essentially maintenance of health, freedom from disease, and longevity; in effect, maximum long-term physical and mental fitness. These objectives, though eminently laudable, are difficult to measure. In practice such factors have combined to limit the expansion of nutritional knowledge of dogs and cats to an extent where definition of requirements for some nutrients has scarcely passed the qualitative stage. What has, however, emerged during the last decade or so is increasing evidence that cats differ from the normal mammalian pattern of nutrition and metabolism in several areas (Morris and Rogers, 1982). At least two of these, vitamin A metabolism and essential fatty acid (EFA) needs, have direct relevance to the role of fat in cat diets and will be discussed later.

The functions of fat in the diet of carnivores are similar to those for other mammalian species. Dietary fat serves as a concentrated source of energy, provides EFA, is a carrier of fat-soluble vitamins, and lends palatability and desirable texture to dog and cat food. While there are some data for most components of the role of dietary fat in dog and cat diets, considerable research and development emphasis within the pet food industry has been placed on the link between fat and food palatability. Pet food manufacturers have been historically reluctant to publish work in this area because, unlike diets for food animals, palatability (or, rather, owner's perception of animal-food palatability) is widely believed to be a sensitive barometer of a product's likely commercial success. A product which is unpalatable and refused by the animal makes little headway in the market place. On the other hand, data on EFA and vitamin needs have been published and thereby shared by those interested in dog and cat nutrition. Nutritional requirements as opposed to palatability are seen as an area of commercial neutrality by most responsible manufacturers, because it is clearly in the interests of all connected with the welfare of dogs and cats that nutritional status of populations is optimized. Therefore, there is little capital to be made by the 'nutritional auction' approach to product marketing which encourages the promotion of specific nutrients or their levels by manufacturers. The remainder of this chapter reviews what is known concerning fat in the diet of dogs and cats and how this knowledge is applied to the practical manufacture of commercial pet foods.

Energy value and digestibility of fats

There appear to be few *in vivo* data for the energy values of specific fats in dog and cat diets. In the absence of this information the NRC (1974, 1978) recommended that dietary fat should be assigned an energetic value of 37 kJ/g (9 kcal/g) based on the Atwater system. Such an energetic value assumes apparent fat digestibility of 96% (Harris, 1966) but apparent fat digestibilities of dog and cat diets measured *in vivo* generally show this to be an overestimate, especially for commercial foods (Kendall, Holme and Smith, 1982a; Kendall, Smith and Holme, 1982b). The published data on fat digestibility are outlined in *Table 19.1*. When measuring apparent fat digestibility of dog and cat diets the analytical method used to measure fat in food and faeces is very important. Unless ether extraction is preceded by acid hydrolysis, triglycerides will not be completely released and estimates of fat content in samples may be 50–100% too low (Budde, 1952; Hoffman, 1953). There also appears to be a need to use a chloroform–methanol mixture to obtain full fat extraction of phospholipids in certain animal products. Thus the preferred method of fat analysis for dog and cat digestibility studies probably involves the acid hydrolysis technique (Cox and Pearson, 1962) with chloroform–methanol mixture as solvent.

The data in *Table 19.1* indicate that dogs and cats generally digest and absorb fats efficiently, with some difference between laboratories which may be explained by method of fat analysis. The exceptions are the data for highly saturated beef tri-stearin which appears to be very poorly digested by cats. Mean apparent fat digestibility of a semi-purified diet containing 25% beef tri-stearin and 1% vegetable oil was 29% and 91% in cat and dogs, respectively. The data for cats agree with the observation of Janssen (1983) who reported an apparent fat digestibility of 50.5% for a commercial fat product based on stearic acid in mature cockerels.

Cats seem to be generally less efficient fat digesters than dogs (Kendall, Holme and Smith, 1982c) when apparent fat digestibilities are compared under the same conditions. *Table 19.2* outlines mean apparent fat digestibilities of eight foods evaluated in both dogs and cats. Mean apparent fat digestibility (92.2%) in dogs significantly exceeded that for cats (75.6%); in addition there was a significant ($P<0.001$) interaction between food and species. There are several possible reasons for this difference but a higher metabolic faecal fat loss for cats does not appear to be responsible; estimates of endogenous faecal fat losses in dogs and cats are given later. Factors possibly contributing to lower apparent fat digestibilities in cats include real differences between species in digestive ability, modulating effect of level of fat intake and the influence of dietary fibre on faecal fat excretion. Some support for the latter hypothesis stems from comparative fat digestibilities for dry cat food (in *Table 19.2*) of about 96% and 56%, respectively in dogs and cats. The dry cat food was probably highest in fibre of those studied, whereas the between-species difference in fat digestibility for the lowest-fibre food, fresh mince, was small. The low fat digestibility of the first sample of canned dog food measured in cats may partly reflect the low food and fat intake on this diet. Clearly, level of fat intake has an important influence on apparent fat digestibility measured in monogastric carnivores. In both dogs and cats, apparent fat digestibility

Table 19.1 APPARENT DIGESTIBILITIES OF FAT IN THE DIETS OF DOGS AND CATS

Diet	Number of samples	Fat sources	Dietary level[a]	Mean (± SEM)[b] apparent digestibility (%)	Range	Reference
Dogs						
Mixed	1	Protein concentrate	–	84	79–95	James and McCay (1950)
Sledge dog	1	Cereals, wheat germ	33	97	–	Orr (1965)
	1	Pemican (beef base)	45	87	–	ibid
	1	Nutrican (whale meat base)	66	88	–	ibid
	1	Seal				
Canned	42	Mixed animal	5–45	88 ± 1.2[c]	62–96	Kendall, Holme and Smith (1982a)
Semi-moist	24	Beef tallow/protein concentrate	7–22	91 ± 0.6[c]	80–98	ibid
Dry complete	12	Beef tallow/cereal/protein concentrate	4–17	84 ± 1.2[c]	74–89	Kendall (unpublished)
Mixer biscuit	11	Beef tallow/cereal	9–13	72 ± 2.3[c]	61–86	ibid
Semi-purified	1	Vegetable oil	39	91 ± 1.4		ibid
Cats						
Canned	1	Mixed animal	36	91 ± 0.8	80–94	Lewis, Boulay and Chow (1979)
Canned	43	Mixed animal	12–49	79 ± 1.2[c]	52–92	Kendall, Holme and Smith (1982a)
Dry	1	Not given	14	83	–	Thrall and Miller (1976)
Dry	1	Not given	9	77		
Dry	28	Cereal/tallow/protein concentrate	6–17	79[c]	52–92	Kendall, Smith and Holme (1982b)
Semi-purified	1	Butter	25	97	–	Kane, Morris and Rogers (1981)
Semi-purified	1	Lard	25	99	–	ibid
Semi-purified	1	Unbleached tallow	25	98	–	ibid
Semi-purified	1	Bleached tallow	25	98	–	ibid
Semi-purified	1	Yellow grease	10–50	95	–	ibid
Semi-purified	1	Chicken fat	25	99	–	ibid
Semi-purified	1	Beef and mutton	35	99 ± 0.01	–	Morris, Trudell and Pencovic (1977)
Semi-purified	1	Beef tri-stearin	30	29 ± 1.4[c]	8–56	Kendall (unpublished)
Semi-purified		Vegetable oil				

(a) Approximate percentage level in diet dry matter
(b) Standard error of the mean
(c) Based on acidified ether extract determinations.

Table 19.2 MEAN[a] FAT LEVELS AND APPARENT FAT DIGESTIBILITY PERCENTAGES FOR EIGHT FOODS MEASURED IN DOGS AND CATS (AFTER KENDALL, HOLME AND SMITH, 1982c)

Food	Fat level (% fat dry matter)	Apparent fat digestibility (%)	
		Dogs	Cats
Dry cat food	9.7	95.6	56.2
Canned cat	21.0	90.8	78.0
Canned dog	25.0	89.8	51.6
Canned dog	25.1	94.6	81.8
Canned cat	35.9	92.6	84.9
Semi-purified diet	36.9	88.0	79.2
Semi-purified diet	37.4	88.2	76.5
Fresh mince	46.1	97.7	95.7
Mean	–	92.2	75.6 ($P<0.001$)
Standard error of the mean		1.5	

[a] Each value is a mean for six adult Beagles or domestic short-haired cats fed to estimated maintenance needs

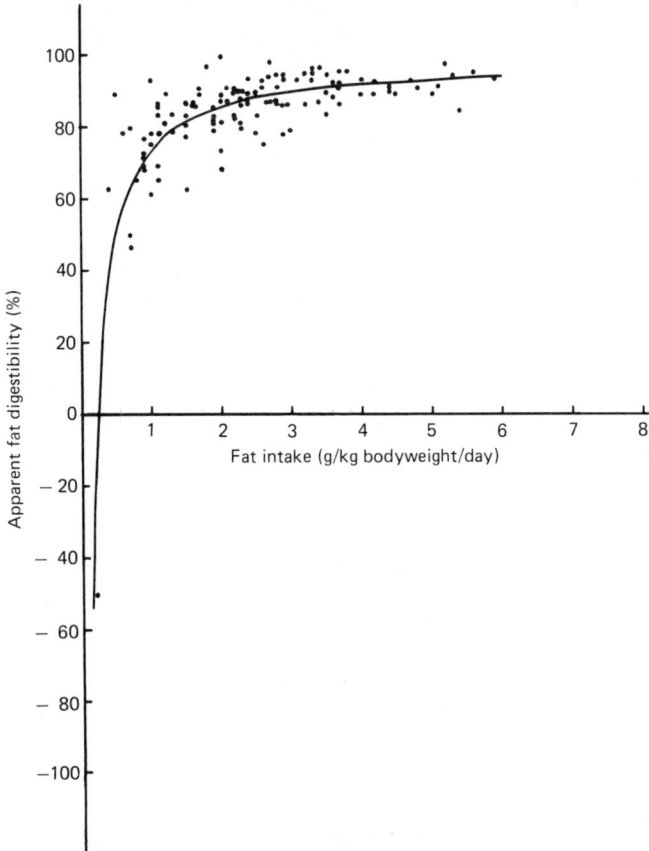

Figure 19.1 Relationship between apparent fat digestibility (%) and fat intake for 122 dog digestibility trials

388 The use of fat in dog and cat diets

rises hyperbolically with fat intake (Kendall *et al.*, 1982b) to levels which asymptotically approach the true digestibility. *Figure 19.1* shows the relationship between apparent fat digestibility percentage and daily fat intake (x, g/kg bodyweight) for 122, 14-day dog digestibility trials carried out with panels of six adult Beagles fed commercial products. This relationship could best be described by the following significant ($P<0.001$) hyperbolic regression:

Apparent fat digestibility (%) = $97.7 - 23.7/x$;

$R^2 = 0.72$; Residual standard deviation (RSD) = 7.6% Eq. (19.1)

Mean ($\pm$ SE) apparent fat digestibility for the 122 trials was $83.4 \pm 1.3\%$ with a mean daily fat intake of 2.3 ± 0.14 g/kg bodyweight. The true digestibility of fat corresponds to the constant term in the hyperbolic equation which gives an average value of 97.7% for dogs. Metabolic faecal fat can also be estimated as the intercept on the x-axis corresponding to zero apparent digestibility. This produces a value of 242 mg metabolic faecal fat/kg bodyweight/day.

Figure 19.2 outlines the same relationship between apparent fat digestibility percentage and fat intake for 97 digestibility trials, each conducted

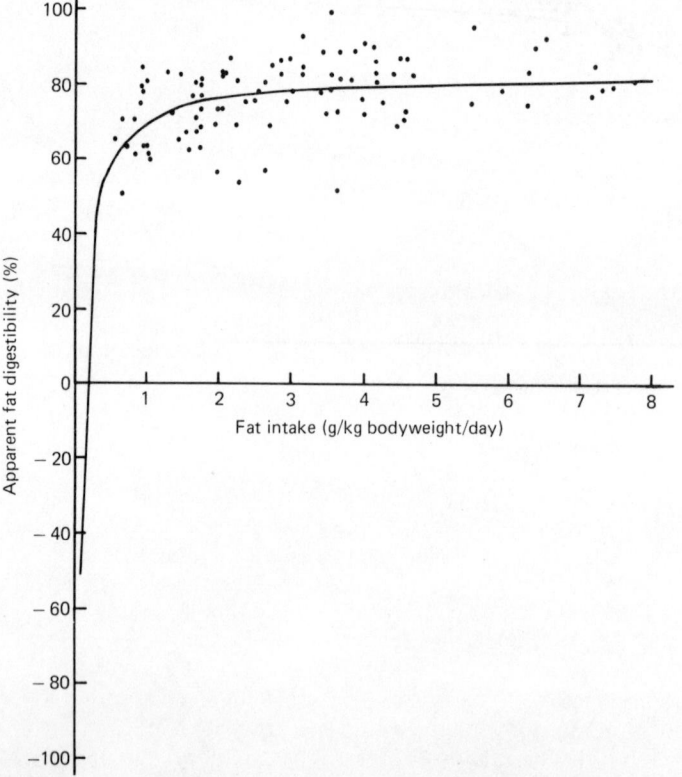

Figure 19.2 Relationship between apparent fat digestibility (%) and fat intake for 97 cat digestibility trials

with six adult cats fed commercial foods. The following significant ($P<0.001$) hyperbolic regression equation was generated:

Apparent fat digestibility (%) = $82.6 - 12.7/x$;

$R^2 = 0.21$; RSD = 8.5% Eq. (19.2)

The mean (± SE) apparent fat digestibility was 76.6 ± 1.0% over the 97 trials for a mean daily intake of 3.0 ± 0.18 g/kg bodyweight. Thus, cats consumed more fat on average than dogs, but had significantly ($P<0.001$) lower apparent fat digestibility. The true fat digestibility in cats at 82.6% was also considerably below the value for dogs. In contrast, a lower metabolic faecal fat value of about 150 mg/kg bodyweight/day was estimated for cats compared with 242 mg/kg bodyweight/day for dogs. Thus the lower fat digestibilities measured in cats are not explained here by higher metabolic faecal fat losses, although there is a shortage of data for low fat intakes, especially with cats. In conclusion, the lower fat digestibilities for cats relative to dogs seem most likely to be explained by enzymatic fat digestive ability or absorptive capacity.

Dietary fat levels

The ability of dogs and cats to digest fat is rarely exceeded in normal individuals. Extremely wide variations in fat intake appear generally to be acceptable, if EFA and other nutrient requirements are met. Siedler and Schweigert (1952) fed a diet containing about 4% ether extract (dry basis) to growing puppies while Orr (1965) fed seal meat (skin, blubber and lean meat) containing 66% ether extract (dry basis) to adult dogs. Both groups of dogs appeared normal. Morgan (1935, 1940) fed diets containing 10–24% fat for two years without producing harmful effects. Similar satisfactory performance has been observed in puppies and breeding dogs when dietary fat level ranged between 5% and 40% (Linton, 1934; Ivy, 1936; Axelrod, Gullberg and Morgan, 1951).

Relatively high dietary fat levels from both animal and plant sources, have been used traditionally in feeding trials with cats. Purified diets composed of 25–30% fat are commonly used which appear to be more palatable than those with low fat content (Greaves, 1965). Scott (1966) found satisfactory performance in kittens fed diets with 22% fat. Thus it appears that both dogs and cats can tolerate a very wide range of dietary fat provided that the nutritional balance of the diet is maintained. The NRC (1974, 1978) recommended at least 5% and 9% fat in the dry matter of dog and cat diets, respectively, but higher levels are typically observed in commercial diets, as outlined in *Table 19.3*.

In general, commercially prepared foods for dogs and cats are higher in fat and digestible energy (DE) compared with diets for food animals. Canned meats tend to be higher in fat than commercially prepared packeted products, which reflects the fat profile of the raw ingredients rather than any particular nutritional objective. The highest fat and DE levels are formulated into products specially designed for growing puppies

Table 19.3 TYPICAL RANGES FOR FAT LEVELS IN COMMERCIAL DOG AND CAT DIETS[a]

Food type	Fat (% DM)	Digestible energy (MJ/kg DM)
Dry		
Complete for dogs	5–19	14–16
Complete for cats	5–12	14–16
Mixer for dogs	5–8	13–16
Semi-moist		
Dog	9–12	15–17
Cat	15–25	16–20
Canned		
Meaty chunks in jelly—dogs	20–30	18–22
Meat and cereal—dogs	5–10	14–16
Meaty food for puppies	32	22
Meaty chunks—cats	15–30	16–20
Fish/meat/cereal—cats	5–8	14–16
Meaty food for kittens	33	22
Canned meaty dog food and biscuit mixture (3:1 by weight)	16	18

[a]There are many brands of foods available and their nutrient content varies widely. The tabulated values are indicative of the general values found.

or kittens to help ensure satisfactory growth and development within the limits of voluntary appetite of the animal.

Dry dog and cat foods tend to have lower fat and DE levels more in line with those found in pig and broiler diets. Dry dog foods are usually extruded, expanded, kibbles or loose mixtures containing ingredients similar to diets for monogastric food animals (Holme, 1982). As such they are higher in carbohydrate and fibre than most canned foods.

Fat malabsorption/maldigestion syndromes

Although most dogs and cats digest fats with high efficiency, problems associated with fat digestion and absorption are not infrequent in some dogs. German Shepherds (Alsatians) show a relatively high incidence of pancreatic exocrine insufficiency (Darke, 1979) which may be hereditary; males are more commonly affected, usually young adults about 1–3 years old. Pancreatic exocrine insufficiency is characterized by the production of foul-smelling fatty faeces (steatorrhoea) and the dietary management usually involves a reduction in dietary fat level, vegetable oil supplementation and pancreatic enzyme replacement therapy.

Essential fatty acids

Although dietary fat performs several functions, its physiological need is the reflection of a requirement for certain essential polyunsaturated fatty acids that it contains. The role of EFA in the diet of the dog and cat has recently been reviewed (Rivers and Frankel, 1980b; McLean, 1981), yet there is considerable confusion as to even qualitative needs, particularly in the case of cats.

Essential fatty acid needs of dogs

Much of the work published on the role of EFA in dog nutrition originated from Hansen, Wiese and colleagues. If the diet is very low in fat or if the fat is completely saturated, skin lesions appear (Hansen, Beck and Wiese,

Table 19.4 A COMPARISON OF EFA DEFICIENCY IN RATS, DOGS AND CATS (COURTESY OF RIVERS, 1982)

	Rat (HCO or fat-free diet)	Dog (fat-free diet)	Cat (HCO diet)	Cat (SSO or SBOL diet)
Gross clinical picture				
Reduced growth	+	+	+	−
Failure of catch-up growth	0	0	+	0
Adults emaciated	+	+	+	+
Coat 'roughened' or staring	+	+	+	+
Dandruff or scaly skin	+	+	+	+
Greasy coat	0	0	+	+
Alopecia	+	+	−	−
Diminished skin pigmentation	+	−	−	−
Caudal necrosis	+	−	−	−
Failure of wound healing	+	0	+	+
Non-bacterial ear discharge	−	+	−	−
Increased wax in ears	0	0	+	0
Increased susceptibility to infections	+	+	+	0
Increased susceptibility to radiation	+	0	0	0
Muscles feel flaccid	+	0	+	+
Testes feel flaccid or fail to develop	+	0	+	+
Oestrus irregular or absent	+	0	+	+
Male animals generally refuse to mate	+	0	+	+
Pregnant females resorb frequently	+	0	+	+
Stillbirths and neonatal deaths frequent	+	0	+	+
Pathology and histology				
Skin: hyperplasia and hyperkeratosis	+	+	+	+
Liver: fatty infiltration and degeneration	+	0	+	+
Testes: degeneration	+	0	0	0
Kidney damage	+	−	−	−
Anaemia	0	−	−	−
Elevated white cell count	+	0	+	+
Physiology and biochemistry				
Elevated BMR	+	0	0	+
Elevated rate of skin water loss	+	0	+	−
Notching of QRS complex of ECG	+	0	−	−
20:3w9 produced	+	+	trace	trace
Increased serum cholesterol and triglyceride	+	+	0	0
5,11,14,-20:3 produced	trace	0	trace	+
Factors affecting deficiency				
Saturated fat in diet exacerbates	+	0	+	0
Low humidity exacerbates	+	+	0	0
Fast growth exacerbates	+	+	+	+

+, Effect occurs; −, effect does not occur; 0, no data

1948; Hansen, Sinclair and Wiese, 1954; Hansen and Wiese, 1951; Wiese et al., 1965, 1966) which can be prevented or cured by linoleic acid, γ-linolenic acid or arachidonic acid. Because these fatty acids are interconvertible within the tissues (Steinberg et al., 1956), if any one of the three is present in adequate amounts the EFA requirement will be met. Normally γ-linolenic acid and arachidonic acid are not major components of natural fats: hence, the effectiveness of dietary fat in preventing and curing an EFA deficiency is usually related to its linoleic acid content. The role and requirements of γ-linolenic acid in the dog are not known and no significant investigations have been carried out (McLean, 1981).

There is a marked lack of quantitative data on EFA requirements of dogs. The minimum linoleic acid requirement has not been precisely defined. The pathological and biochemical changes in the skin (*Table 19.4*) produced by an EFA deficiency can be reversed in growing puppies when 2–6% of the ME requirement is provided by linoleic or arachidonic acid (Hansen and Wiese, 1951; Wiese et al., 1966). In contrast, 1% of the ME requirement as linoleic acid or about 0.4% in the diet dry matter does not appear to be adequate for growing puppies (Wiese et al., 1966). Rate of growth and ME intake also influences the development of EFA deficiency (Wiese, Hansen and Coon, 1962); energy restriction and slow growth rate apparently protects against EFA deficiency. The NRC (1974) recommended that 1% linoleic acid in the diet dry matter should meet EFA needs of dogs for all life stages.

In summary it may be concluded that the scheme elucidated for PUFA interconversion in the rat is a reasonable qualitative explanation of the pathways existing in the dog (Rivers and Frankel, 1980b) but more information is needed about rate of transformations to establish minimum quantitative EFA requirements for different physiological states.

Essential fatty acid needs of cats

Until 1975 no EFA requirement had been demonstrated for the cat, and where estimates of requirement were made, these were based upon analogy with the rat. Since then a considerable amount of research has been carried out by at least three independent groups of workers, which has revealed a complex and unusual pattern to EFA metabolism in the cat. Rivers and colleagues were first to explore EFA needs of cats and published papers (Rivers, Sinclair and Crawford, 1975; Rivers, Hassam and Alderson, 1976a; Rivers et al., 1976b; Hassam, Rivers and Crawford, 1977; Frankel and Rivers, 1978; Rivers et al., 1979; Rivers and Frankel, 1980a,b; Rivers, 1982) which argued that an EFA deficiency syndrome could be induced in the cat by feeding an EFA-free diet based on hydrogenated coconut oil (HCO) as a fat source. Of perhaps even greater significance were their remarkable findings that cats fed semi-purified diets containing safflower seed oil (SSO; n-3 deficient) and a mixture of soybean oil and linseed oil (SBOL) providing n-3 and n-6 EFA, also showed clinical signs of EFA deficiency after 12–18 months on the diets. The deficiency syndromes have been separately characterized in *Table 19.4* and compared with typical signs of EFA deficiency in rats and dogs fed a HCO or fat-free diet. Rivers (1982) states that, although the signs of EFA deficiency were most pronounced in cats fed HCO diets, this did not permit the disease to be differentiated between the dietary groups and accords with the observation in rodents that HCO feeding induces a worse deficiency state than feeding of fat-free diets (Rivers and Frankel, 1980a). The evidence from this work suggests that an EFA deficiency exists in the cat which resembles the disease in other laboratory animals or man, with the exception that it can be induced on diets containing vegetable oils that are potent EFA sources for other species.

Rivers et al. (1975) postulated that this apparent inability to metabolize linoleic acid was due to the absence of Δ^8- and Δ^6-desaturase enzymes (*Figure 19.3*), and a failure to find significant amounts of di-homo-γ-linolenic acid in tissues of cats fed linoleic acid supported this hypothesis. From further feeding and biochemical studies (Hassam et al., 1977a; Frankel and Rivers, 1978) they concluded that the Δ^5-desaturase, as well as the Δ^8- and Δ^6-desaturases were not operative in the cat. The nutritional significance of these observations would make it essential for the cat to receive a diet containing fat of animal origin, in order to provide a source of the long-chain PUFA metabolites derived from linoleic and linolenic acids (McLean, 1981).

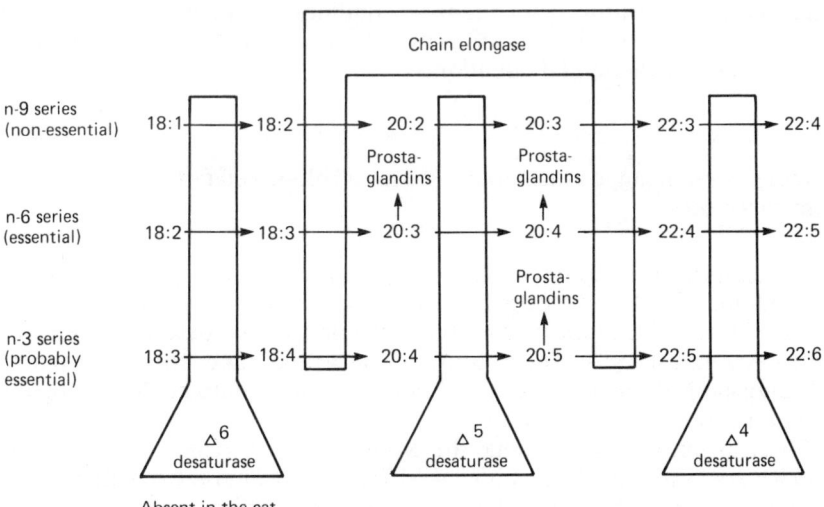

Figure 19.3 The predominant pathways of polyunsaturated fatty acid metabolism in mammals (courtesy of Rivers and Frankel, 1980)

Subsequent studies by Sinclair, McLean and Monger (1979) using isotopically labelled fatty acids have confirmed the absence of the Δ^6-desaturase enzyme in the cat but γ-linolenate was converted to arachidonic acid, indicating the presence of the Δ^5-desaturase enzyme. In further experiments (Sinclair et al., 1980; Rivers, 1982) the fatty acid ($\Delta^{5,8,11}$-eicosatrienoic acid) normally associated with EFA deficiency was identified from cats fed diets containing only saturated fatty acids whilst $\Delta^{5,11,14}$-eicosatrienoic acid was identified from animals fed diets rich in linoleic acid. McLean (1981) concluded that since the synthesis of these two fatty acids requires the presence of a Δ^5-desaturase together with either a Δ^6- or Δ^8-desaturase, this means that the conversion of linoleic acid to arachidonic must also be possible in the cat. Data from feeding studies have been presented to support this hypothesis (McLean, 1981; MacDonald, Rogers and Morris, 1981a,b). These groups of workers appear to have evidence that 5% safflower seed oil in the diet could be nutritionally adequate as the only EFA source for periods of at least 3 years, during which time satisfactory maintenance, growth and reproduction has been claimed. In addition, signs of EFA deficiency produced by removing safflower seed oil from the diet could be largely reversed by including safflower seed oil (Kuchel, 1979).

Thus, there are now doubts as to the correctness of the original interpretation and conclusions made by Rivers and colleagues concerning EFA metabolism and nutrition of the cat. It is generally agreed that tissue levels of n-6 EFA can be dramatically reduced by the feeding of vegetable oil diets, compared with meat-based diets containing fat of animal origin. Whether this reflects adaptation within the physiological range or a chronic pathological condition has not been effectively answered. McLean (1981) reports that cats fed vegetable oil diets maintain relatively stable tissue arachidonic acid levels after 3 years, which suggests a residual production rate. This differs little from Frankel's findings (Frankel, 1980) that red blood arachidonic acid levels are slowly declining on SSO diets, according to the double exponential equation:

$$p = 22.3e^{-0.176t} + 2.6e^{-0.00433t} \qquad \text{Eq. (19.3)}$$

where p = percentage arachidonic acid in red blood cell fatty acids and t = time in weeks

This relationship (*Figure 19.4*) predicts an extremely slow decline in arachidonic acid levels of approximately 0.3% per annum from about 3 years of age. To reduce levels to a probable limit of detection with the GLC system (0.05% of the fatty acids) would take about 18 years. Therefore it is difficult to see how Frankel's model could be differentiated from a residual production rate.

It is also generally agreed that the Δ^6-desaturase enzyme is absent in the cat (Rivers *et al.*, 1975; Sinclair *et al.*, 1979) but an alternative pathway for arachidonic acid production has been proposed which involves Δ^5- and Δ^8-desaturases (Sinclair *et al.*, 1979). However, whether these transformations are sufficient to maintain optimum EFA status of cats is still not clear (Rivers and Frankel, 1981). A summary of current status of the metabolism of n-6 series EFA by the cat is shown in *Figure 19.5*. The soundest practical advice for formulation of cat diets is that linoleic acid levels should exceed 1% in the dry matter (NRC, 1978) and it would seem prudent to include some PUFA, such as found in meat and fish products in cat diets (McLean, 1981).

The whole topic of n-3 EFA requirements of the cat has not yet been studied and the cat could make an attractive model to ascertain the role of n-3 EFA in mammals in general.

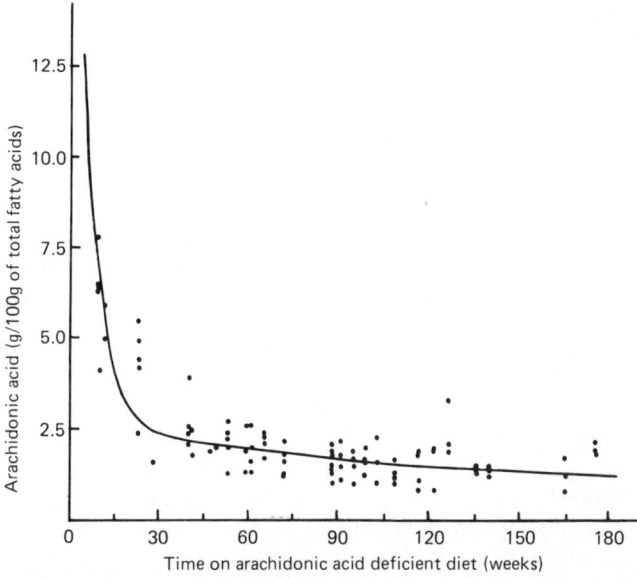

Figure 19.4 Decrease in the level of arachidonic acid in red blood cell ethanolamine phosphoglycerides of cats given a linoleic acid supplemented, but arachidonic acid deficient diet (after Frankel, 1980)

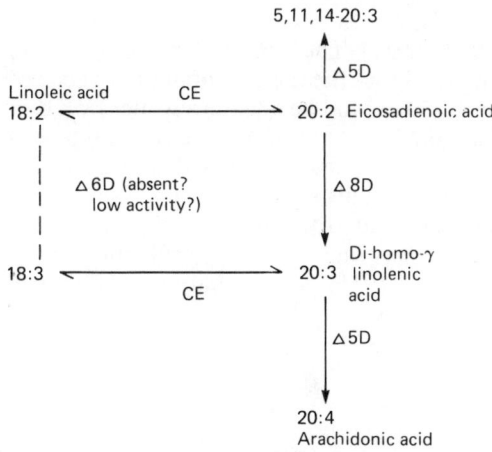

Figure 19.5 A summary of current views of the metabolism of n-6 series EFA by the cat (after Rivers, 1982). CE = chain elongations; D = desaturase enzyme

In summary, the recent statement of Rivers (1982) appears apt when considering feline EFA requirements. He states that knowledge of requirements has gone from a passive acceptance of the equivalence of the cat with the rat, through a simple view of the uniqueness of the cat, into a final state of confusion. What has been established, as so often happens in feline nutrition, is that the cat is not a rat; how and why it is different has yet to be determined.

Fat-soluble vitamins

Precise quantitative requirements have not been established for each vitamin in different physiological states. The most comprehensive reviews of requirements are those of the NRC (1974, 1978).

Vitamin A

The vitamin A needs of dogs can be partly met by metabolic conversion of carotenes to vitamin A (Bradfield and Smith, 1938) whereas cats require preformed dietary vitamin A because they lack the ability effectively to convert β-carotene to vitamin A (Ahmad, 1931; Rea and Drummond, 1932; Gershoff et al., 1957). The vitamin A needs of adult dogs and growing puppies can be met by 110 IU and 220 IU/kg bodyweight (Bradfield and Smith, 1938) and the NRC (1974) recommends a dietary vitamin A concentration of 5000 IU/kg dry matter as adequate for all life stages.

Kidney vitamin A levels in cats appear to be higher than do vitamin A levels in kidneys of other species (Lowe, Morton and Vernon, 1957; Moore, Sharman and Scott, 1963); nevertheless, liver vitamin A appears to constitute the primary reserve and liver concentrations appear to be more closely related to dietary intake than do kidney concentrations. Liver, cod-liver oil, retinyl acetate or palmitate are satisfactory sources of vitamin A for the cat. Within the range 1–26% dietary fat, vitamin A absorption as estimated by serum vitamin A levels, appears to be positively correlated with dietary fat level (Gershoff et al., 1957). The NRC (1978), in reviewing the published literature, recommended that the dietary vitamin A needs of kittens and cats in the maintenance state could be adequately met by a diet supplying 10000 IU vitamin/kg dry matter. Vitamin A toxicity is a relatively frequent problem in cats and is usually associated with the exclusive feeding of liver-rich diet (Seawright, English and Gartner, 1967). The naturally occurring disease can be duplicated within 10 weeks by feeding about 30 μg vitamin A per gram bodyweight. This approximates to a dietary vitamin A concentration of 5×10^6 IU/kg dry matter or 500 times the NRC (1978) recommended allowance.

Vitamin D

Vitamin D requirements are dependent on dietary concentrations of calcium and phosphorus, the dietary calcium-to-phosphorus ratio, physiological state and perhaps breed and sex. Vitamin D requirements of puppies have been investigated by several groups of workers (Kozelka, Hart and Bohstedt, 1933; Arnold and Elvehjem, 1939; Michaud and Elvehjem, 1944). The NRC (1974) concluded that when the dietary calcium-to-phosphorus ratio is 1.2, daily vitamin D requirements should be met by 11 IU/kg bodyweight for adult maintenance and 22 IU/kg bodyweight for growing puppies. It was stated that these amounts would be provided by a dietary concentration of 500 IU vitamin D per kg dry matter.

Vitamin D requirements of cats have not been widely studied. Gershoff *et al.* (1957) found that 250 IU of cholecalciferol given orally, twice a week, prevented the development of rickets in kittens from 3–6 months of age to 21 months of age. Rivers *et al.* (1979) were unable to induce signs of vitamin D deficiency in adult cats fed a diet deficient in vitamin D for a year without access to sunlight. The NRC (1978) concluded that the vitamin D needs of cats could be met by diets supplying 1000 IU/kg dry matter.

Vitamin E

Most of the literature on vitamin E requirements of dogs was published before the dietary relationship between vitamin E and selenium had been established. Consequently the NRC (1974) found the data inadequate to set a species-specific vitamin E requirement for dogs. On the basis of research with other species, and assuming a dry diet containing 1% linoleic acid and 0.1 mg/kg selenium, they therefore recommended that vitamin E needs of dogs could be met by diets supplying 50 IU/kg dry matter. With diets containing high levels of PUFA it was suggested that the DL-α-tocopheryl:PUFA ratio be maintained at least at 0.5.

The need for vitamin E in the diet of cats is markedly influenced by dietary composition. Several reports have noted an association of steatitis (yellow fat disease) with consumption of fish-based diets, particularly red tuna (Cordy, 1954; Coffin and Holzworth, 1954; Munsen *et al.*, 1958; Griffiths, Thornton and Willson, 1960). The NRC (1978) vitamin E requirement of 80 IU/kg dry matter is largely based on the work of Gershoff and Norkin (1962) which shows that supplemental vitamin E at levels of 34–68 IU/kg diet prevented lesions of vitamin E deficiency in kittens.

Vitamin K

The NRC (1974, 1978) found no evidence for a vitamin K requirement in normal dogs and cats, as simple vitamin K deficiency states have not been described in either species.

Fat and diet palatability

Over the past few years an increasing number of pet food manufacturers have resorted to a variety of 'palatability enhancers' to improve animal acceptance of their products (Zorich, 1982). These include animal digests and sprays which rely heavily on fat as a key ingredient. However, published data on the influence of fat on the palatability of dog and cat diets are sparse, mainly because this information is considered to be of great commercial value by pet food manufacturers. In one study (Kendall, Blaza and Smith, unpublished work), individually housed cats received eight experimental dry cat foods containing 3.1–7.0% fat *ad libitum*. Food

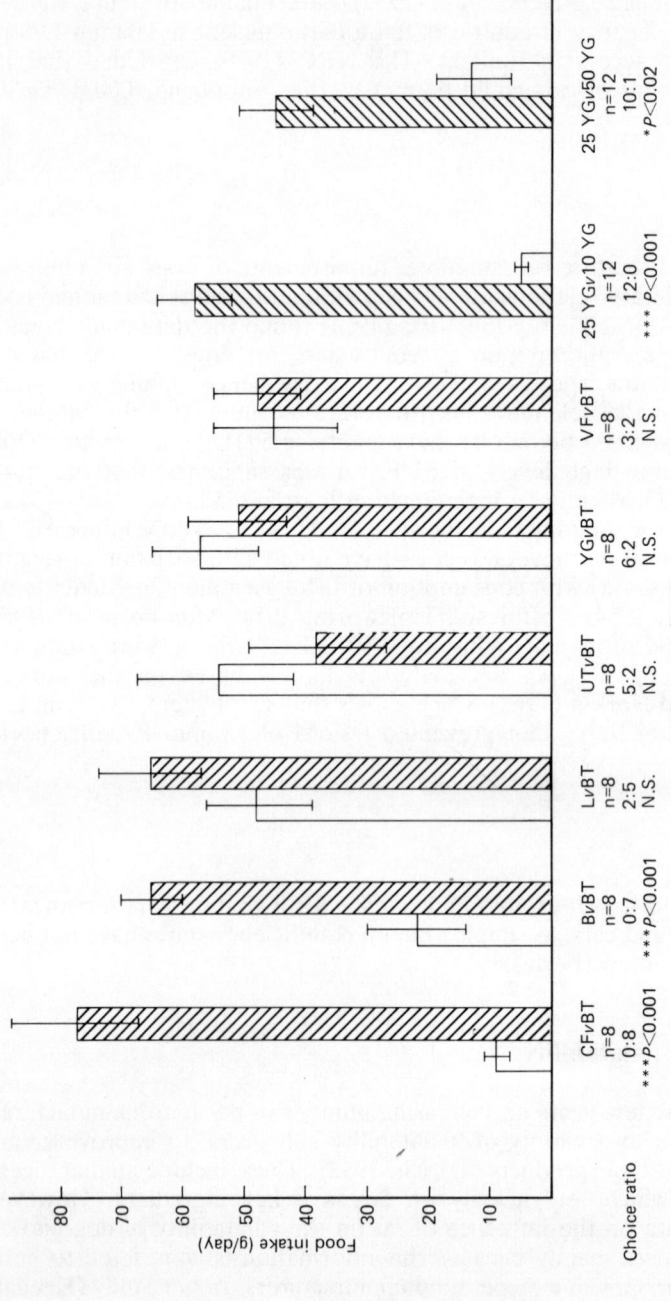

Figure 19.6 Acceptance of cat diets made with various sources and levels of fat (courtesy of Kane, Morris and Rogers, 1981). BT, bleached tallow; CF, chicken fat; YG, yellow grease; L, lard; B, butter; UT, unbleached tallow; VF, hydrogenated vegetable fat

Table 19.5 RELATIONSHIP BETWEEN COMPOSITION (% AS IS) OF DRY CAT FOOD AND VOLUNTARY FOOD INTAKE

Food	Crude protein (CP)	Acid ether extract (AEE)	Nitrogen-free extract (NFE)	Mean (N = 6) voluntary food intake (g/cat)
1	48.6	6.9	21.3	77
2	37.8	3.1	38.8	72
3	32.9	3.8	44.7	68
4	44.2	4.8	28.8	68
5	52.2	7.0	23.3	80
6	39.3	6.7	29.9	73
7	32.7	5.8	38.7	67
8	45.3	5.3	25.7	68
Pooled SEM[a]	–	–	–	6.6

Voluntary Food Intake I (g/cat/day) = –13.7 + 1.12 CP + 2.35 AEE + 0.83 NFE ($R^2 = 0.72$, $P<0.01$)

[a] Standard error of treatment mean

composition and mean daily voluntary intakes for six individually housed adult cats are shown in *Table 19.5*. Mean daily food intakes (*I*) ranged from 67 g to 80 g/cat and were related to food percentage crude protein (*CP*), acid ether extract (*AEE*) and nitrogen-free extract (*NFE*) by the equation:

$$I = -13.7 + 1.12\ CP + 2.35\ AEE + 0.83\ NFE\ (R^2 = 0.72,\ P<0.01,\ RSD = 3.0\,g) \qquad \text{Eq. (19.4)}$$

Thus the regression coefficient for fat (*AEE*) has a weighting on voluntary intake at least double that for other major organic nutrients. This tends to support the widely held belief that cats prefer diets with higher fat levels, especially when food fat level is low at below 10%.

In other studies (Kane, Morris and Rogers, 1981) the palatability of purified diets, each containing a 25% level of one of seven fat sources, was compared by two-choice preference test with bleached tallow as a control. The other fat sources examined were unbleached tallow, lard, chicken fat, yellow grease, butter and vegetable fat. All eight cats used in the study preferred ($P<0.01$) diets made with bleached tallow to those with chicken fat or butter. Five of the eight cats preferred bleached tallow to lard and five of the eight cats preferred unbleached tallow to bleached tallow. Cats showed no preference between vegetable fat and bleached tallow. These results are summarized in *Figure 19.6* and show that cats exhibit marked preferences for diets based on certain fats. In a further experiment the effect of level of dietary fat on acceptance was examined (Kane *et al.*, 1981). Cats preferred a purified diet with 25% yellow grease over diets made with 10% or 50% yellow grease, which indicates that the response to fat level is not linear over the entire range, but that an optimum exists. This optimum may be a result of enhancement of flavour or consistency.

Fat can be incorporated into dry and semi-moist foods either as a spray or in the core biscuit. Very few published data are available on the relative effectiveness of the different application routes or the physico-chemical properties of the fats used.

Conclusions

The uses of fat in the diet of carnivores are similar to those for other mammals and include meeting obligatory physiological needs for fats and fat-soluble nutrients and enhancing food palatability with fat or fat-containing mixtures. Recommended (NRC, 1974, 1978) allowances for fats and fat-related nutrients have been summarized in *Table 19.6*. Little emphasis has to date been placed on the energetic value of fats in dog and cat diets, compared with diets for agricultural animals; this mainly reflects the different feeding objectives for the species. In contrast, more emphasis is placed within the pet food industry on the ability of fat to improve food palatability than is perhaps the case with manufacturers of foods for agricultural animals.

Table 19.6 RECOMMENDED NUTRIENT ALLOWANCES FOR FAT AND FAT-RELATED SUBSTANCES PER KILOGRAM DRY MATTER (DM) IN DOG AND CAT DIETS SUPPLYING ABOUT 17 MJ/kg DM

	Unit	Dogs[a]	Cats[b]
Fat	g	50	90
Linoleic acid	g	10	10
Arachidonic acid	g	–	1
Vitamin A	IU	5000[c]	10000
Vitamin D	IU	500[d]	1000
Vitamin E	IU	50[e]	80

[a] NRC (1974)
[b] NRC (1978)
[c] Equivalent to 1.5 mg of all-*trans* retinol
[d] Equivalent to 12.5 µg of cholecalciferol
[e] Equivalent to 50 µg of DL-x-tocopheryl acetate

References

AHMAD, B. (1931). The fate of carotene after absorption in the animal. *Biochemical Journal* **25**, 1195–1204

ARNOLD, A. and ELVEHJEM, C.A. (1939). Nutritional requirements of dogs. *Journal of the American Veterinary Medical Association* **95**, 187–194

AXELROD, H.E., GULLBERG, M.G. and MORGAN, A.F. (1951). Carbohydrate metabolism in riboflavin deficient dogs. *American Journal of Physiology* **165**, 604–619

BRADFIELD, D. and SMITH, M.C. (1938). The ability of the dog to utilize vitamin A from plant and animal sources. *American Journal of Physiology* **124**, 168–173

BUDDE, E.F. (1952). The determination of fat in baked biscuit type of dog foods. *Journal of the Association of Official Agricultural Chemists* **35**, 799–805

COFFIN, D.L. and HOLZWORTH, J. (1954). "Yellow fat" in two laboratory cats: acid-fast pigmentation associated with a fish-bone ration. *Cornell Veterinarian* **44**, 63–71

CORDY, D.R. (1954). Experimental production of steatitis (yellow fat disease) in kittens fed a commercial canned cat food and prevention of the condition by vitamin E. *Cornell Veterinarian* **44**, 310–318

COX, H.E. and PEARSON, D. (1962). General methods—fat. In *Chemical Analysis of Foods*, pp. 26–28. London, Churchill

DARKE, P.G.G. (1979). Pancreatic dysfunction. In *Diarrhoea in the Dog* (E.T.B. Edney, Ed.), pp. 27–28. Waltham Symposium No. 1. Waltham, Leicestershire, Pedigree Petfoods

FRANKEL, T.L. (1980). *Essential Fatty Acid Deficiency in the Cat* (Felis catus L.). PhD thesis, University of Cambridge, UK

FRANKEL, T.L. and RIVERS, J.P.W. (1978). The nutritional and metabolic impact of γ-linolenic acid (18:3w6) on cats deprived of animal lipid. *British Journal of Nutrition* **39**, 227–231

GERSHOFF, S.N. and NORKIN, S.A. (1962). Vitamin E deficiency in cats. *Journal of Nutrition* **77**, 303–308

GERSHOFF, S.N., ANDRUS, S.B., HEGSTED, D.M. and LENTINI, E.A. (1957). Vitamin A deficiency in cats. *Laboratory Investigation* **6**, 227–240

GREAVES, J.P. (1965). Protein and calorie requirements of the feline. In *Canine and Feline Nutritional Requirements* (O. Graham-Jones, Ed.), pp. 33–45. London, Pergamon Press

GRIFFITHS, R.C., THORNTON, G.W. and WILLSON, J.E. (1960). Pancreatitis ("yellow fat") in cats. *Journal of the American Veterinary Medical Association* **137**, 126–130

HANSEN, A.E. and WIESE, H.F. (1951). Fat in the diet in relation to nutrition of the dog. I. Characteristic appearance and gross changes of animals fed diets with and without fat. *Texas Reports on Biology and Medicine* **9**, 491–515

HANSEN, A.E., BECK and WIESE, H.F. (1948). Susceptibility to infection manifested by dogs on a low fat diet. *Federation Proceedings* **7**, 289

HANSEN, A.E., SINCLAIR, J.G. and WIESE, H.F. (1954). Sequence of histological changes in skin of dogs in relation to dietary fat. *Journal of Nutrition* **52**, 541–554

HARRIS, L.E. (1966). *Biological Energy Interrelationships and Glossary of Energy Terms*. Washington, DC, National Academy of Sciences

HASSAM, A.G., RIVERS, J.P.W. and CRAWFORD, M.A. (1977a). The failure of the cat to desaturate linoleic acid: its nutritional implications. *Nutrition and Metabolism* **21**, 321–328

HOFFMAN, H.H. (1953). Report on crude fat in baked dog food. *Journal of the Association of Official Agricultural Chemists* **36**, 208–211

HOLME, D.W. (1982). Foods for dogs and cats. In *Dog and Cat Nutrition* (A.T.B. Edney, Ed.), pp. 33–45. Oxford, Pergamon Press

IVY, A.C. (1936). Starch digestion in the dog. *Veterinary Medicine* **31**, 145

JAMES, W.T. and McCAY (1950). A study of food intake, activity and digestive efficiency in different type foods. *American Journal of Veterinary Research* **11**, 412–413

JANSSEN, W.M.M.A. (1983). Fat evaluation, the Dutch approach. In *Proceedings, National Renderers Association Seminar, Coventry* (in press)

KANE, E., MORRIS, J.G. and ROGERS, Q.R. (1981). Acceptability and digestibility by adult cats of diets made with various sources and levels of fat. *Journal of Animal Science* **53**, 1516–1523

KENDALL, P.T., HOLME, D.W. and SMITH, P.M. (1982a). Methods of prediction of the digestible energy content of dog foods from gross energy value, proximate analysis and digestible nutrient content. *Journal of the Science of Food and Agriculture* **33**, 823–831

KENDALL, P.T., HOLME, D.W. and SMITH, P.M. (1982c). Comparative evaluation of net digestive and absorptive efficiency in dogs and cats fed a variety of contrasting diets. *Journal of Small Animal Practice* **23**, 577–587

KENDALL, P.T., SMITH, P.M. and HOLME, D.W. (1982b). Factors affecting digestibility and in-vivo energy content of cat foods. *Journal of Small Animal Practice* **23**, 538–554

KOZELKA, F.L., HART, E.B. and BOHSTEDT, G. (1933). Growth, reproduction, and lactation in the absence of the parathyroid glands. *Journal of Biological Chemistry* **100**, 715–729

KUCHEL, T.R. (1979). *Studies on the Refeeding of Essential Fatty Acid Deficient Cats with Linoleic acid.* MVS thesis, University of Melbourne, Victoria, Australia

LEWIS, L.D., BOULAY, J.P. and CHOW, F.H.C. (1979). Fat excretion and assimilation by the cat. *Feline Practice* **9**, 46–49

LINTON, R.G. (1934). Canine nutrition. *12th International Veterinary Congress, NY 3*, pp. 488–502

LOWE, J.A., MORTON, R.A. and VERNON, J. (1957). Unsaponifiable constituents of kidney in various species. *Biochemical Journal* **67**, 228–234

MacDONALD, M.L., ROGERS, Q.R. and MORRIS, J.G. (1981a). Influence of dietary fat and essential fatty acids on growth in kittens. *Federation Proceedings* **40**, 882, Abst. 3704

MacDONALD, M.L., ROGERS, Q.R. and MORRIS, J.G. (1981b). Clinical signs of essential fatty acid deficiency in the cat. In *Nutrition in Health and Disease and International Development. Symposia from the XII International Congress of Nutrition* (A.E. Harper and G.K. Davis, Eds). New York, NY, Alan R. Liss

McLEAN, J.G. (1981). Essential fatty acids in the dog and cat. In *Veterinary Annual* (C.S.G. Grunsell and F.W.G. Hill, Eds), 21st edn, pp. 167–170. Bristol, Scientechnica

MICHAUD, L. and ELVEHJEM, C.A. (1944). The nutritional requirements of dogs. *North American Veterinarian* **25**, 657–666

MOORE, T.I., SHARMAN, M. and SCOTT, P.P. (1963). Vitamin A in the kidney of the cat. *Research in Veterinary Science* **4**, 397–403

MORGAN, A.F. (1935). The food needs of dogs. *Veterinary Journal* **91**, 204–210

MORGAN, A.F. (1940). Deficiencies and fallacies in canine diet. *North American Veterinarian* **21**, 476–486

MORRIS, J.G. and ROGERS, Q.R. (1982). Metabolic basis for some of the nutritional peculiarities of the cat. *Journal of Small Animal Practice* **23**, 599–613

MORRIS, J.G., TRUDELL, J. and PENCOVIC, T. (1977). Carbohydrate digestion by the domestic cat (*Felis catus*). *British Journal of Nutrition* **37**, 365–373

MUNSON, T.O., HALZWORTH, E., SMALL, E., WITZEL, JONES, T.C. and LUGIN-

BUHL. H. (1958). Steatitis ('yellow fat') in cats fed canned red tuna. *Journal of the American Veterinary Medical Association* **133**, 563-568
NATIONAL RESEARCH COUNCIL (1974). *Nutrient Requirements of Dogs, No. 8*. Washington, DC, National Academy of Sciences
NATIONAL RESEARCH COUNCIL (1978). *Nutrient Requirements of Cats, No. 13*. Washington, DC, National Academy of Sciences
ORR, N.W.M. (1965). The food requirements of Antarctic sledge dogs. In *Canine and Feline Nutritional Requirements* (O. Graham-Jones, Ed.), pp. 101-112. London, Pergamon Press
REA, J. and DRUMMOND, J.C. (1932). Formation of vitamin A from carotene in the animal organism. *Zeitschrift für Vitamin und Hormone Ferment-Forschung* **1**, 177-183
RIVERS, J.P.W. (1982). Essential fatty acids in cats. *Journal of Small Animal Practice* **23**, 563-576
RIVERS, J.P.W. and FRANKEL, T.L. (1980a). Fat in the diet of cats and dogs. In *Nutrition of the Dog and Cat* (R.S. Anderson, Ed.), pp. 67-99. Oxford, Pergamon Press
RIVERS, J.P.W. and FRANKEL, T.L. (1980b). Essential fatty acids in feline nutrition. In *Proceedings of the 3rd Kal Kan Symposium* (R.L. Wyatt, Ed.), pp. 48-52. Vernon, California, Kal Kan Foods
RIVERS, J.P.W. and FRANKEL, T.L. (1981). The production of 5,8,11-eicosatrienoic acid (20:3ω9) in the EFA-deficient cat. *Proceedings of the Nutrition Society* **40**, 117A
RIVERS, J.P.W., HASSAM, A.G. and ALDERSON, C. (1976a). The absence of Δ^6-desaturase activity in the cat. *Proceedings of the Nutrition Society* **35**, 67A
RIVERS, J.P.W., SINCLAIR, A.J. and CRAWFORD, M.A. (1975). Inability of the cat to desaturate essential fatty acids. *Nature* **258**, 171-173
RIVERS, J.P.W., FRANKEL, T.L., JUTTLA, S. and HAY, A.W.M. (1979). Vitamin D in the nutrition of the cat. *Proceedings of the Nutrition Society* **38**, 36A
RIVERS, J.P.W., SINCLAIR, A.J., MOORE, D.P. and CRAWFORD, M.A. (1976b). The abnormal metabolism of essential fatty acids in the cat. *Proceedings of the Nutrition Society* **35**, 66A
SCOTT, P.P. (1966). Nutrition. In *Diseases of the Cat* (G.T. Wilkinson, Ed.), pp. 1-31. London, Pergamon Press
SEAWRIGHT, A.A., ENGLISH, P.B. and GARTNER, R.J.W. (1967). Hypervitaminosis A and deforming cervical spondylosis of the cat. *Journal of Comparative Pathology* **77**, 29-35
SIEDLER, A.J. and SCHWEIGERT, B.S. (1952). Effect of the level of fat in the diet on the growth performance of dogs. *Journal of Nutrition* **48**, 81-90
SINCLAIR, A.J., McLEAN, J.G. and MONGER, E.A. (1979). Metabolism of linoleic acid in the cat. *Lipids* **14**, 932-936
SINCLAIR, A.J., McLEAN, J.G. and MONGER, E.A. (1980). Polyunsaturated fatty acid metabolism in the cat. *Proceedings of the Australian Biochemical Society* **13**, 128
STEINBERG, G., SLATON, W.J., Jr, HOWTON, D.R. and MEAD, J.P. (1956). Metabolism of essential fatty acids. IV. Incorporation of linoleate into arachidonic acid. *Journal of Biological Chemistry* **220**, 257-264

THRALL, B.E. and MILLER, L.G. (1976). Water turnover in cats fed dry rations. *Feline Practice* **6**, 10–17

WEISE, H.F., HANSEN, A.E. and COON, E. (1962). Influence of high and low caloric intakes on fat deficiency of dogs. *Journal of Nutrition* **76**, 73–81

WIESE, H.F., BENNETT, M.J., COON, E. and YAMANAKA, W. (1965). Lipid metabolism of puppies as affected by kind and amount of fat and dietary carbohydrate. *Journal of Nutrition* **86**, 271–279

WIESE, H.F., YAMANAKA, W., COON, E. and BARBER, S. (1966). Skin lipids of puppies as affected by kind and amount of dietary fat. *Journal of Nutrition* **89**, 113–122

ZORICH, C.L. (1982). *Animal Digests, Petfood Industry* March/April, pp. 6–7

VI

Carcase Considerations

20

FAT DEPOSITION AND THE QUALITY OF FAT TISSUE IN MEAT ANIMALS

J.D. WOOD
Animal Physiology Division, ARC Meat Research Institute, Langford, Bristol, UK

Introduction

As far as most consumers are concerned, meat should contain only a small amount of fat. Too much fat discourages the purchase of meat and is commonly removed either before cooking or during the meal. But most consumers want some fat, partly because the concept of the ideal cut of meat includes fat, at least in Britain, and also because a small amount of fat is required for optimum eating quality. This is because fat confers the characteristic species flavour on meat through a complex interaction between components of fat and lean and also because it prevents drying out during cooking. The actual amount of fat required for these purposes is currently the subject of research and recent findings suggest that, as far as flavour is concerned, it is very small indeed (Mottram and Edwards, 1983). This could explain the observation that above a minimum fat level there is only a small, although positive, correlation between the fat content of meat and its eating quality (Rhodes, 1973).

However, the quality of the fat tissue in terms of firmness and appearance depends to some extent on the quantity of fat and some meat traders have expressed concern that, as overall fatness is reduced through changes in breeding and feeding on the farm, fat quality will decline. This chapter examines these interrelations and the effects on fat quality of production factors designed to decrease fatness or to increase the efficiency of meat production.

Definition of fat quality

In this chapter, fat quality will be taken to be largely concerned with visual and textural aspects of the carcass fat tissues. Good quality will imply firm white fat in pigs and firm creamy-white fat in cattle and sheep. This consensus definition is derived from butchery and cookery manuals in which words used to describe poor quality fat include soft, oily, wet, grey and floppy. An additional aspect of quality relates to the flavour of fat when eaten, especially the presence of unusual flavours.

Quality problems can affect all the fat tissues in the body but the carcass tissues (subcutaneous and intermuscular) are most significant when meat is sold fresh. Subcutaneous fat has been most widely studied.

Deposition of fat in cattle, sheep and pigs

It is well known that during growth to commercial slaughter weights there is a marked change in the composition of the body with the proportion of fat increasing, bone decreasing and lean remaining fairly constant. In terms of the growth coefficient, b, in the linear allometric model

$$\log y = \log a + b \log x$$
when x = tissue weight
y = live weight

values of approximately 0.7 for bone, 1.0 for lean and above 1.5 for fat are commonly found (Butler-Hogg and Wood, 1982; Butler-Hogg, 1984; Wood et al., 1983b). All three species are slaughtered at around 200–300 g total fat per kg live weight with a similar value for carcass fat in relation to carcass weight.

Fat tissue accumulates in the same parts of the body, termed depots, in cattle, sheep and pigs but the proportions in the major depots differ between species, as shown in *Table 20.1* in which the comparison is made at approximately the same proportion of total body fat. Pigs have much more subcutaneous and less abdominal fat than cattle and sheep. Despite these differences, the growth of the fat depots in relation to total body fat is quite similar in the three species. Average standard errors for b were 0.02, 0.04 and 0.03 in the cattle, sheep and pigs respectively. In pigs, perirenal fat seems to be particularly fast-growing and mesenteric fat slow-growing, although omental fat grows rapidly in all species. In the case of the sheep, the value of b for perirenal fat was 1.08 when growth from 21 to 65 kg was considered (the ruminant stage).

Table 20.1 PARTITION OF BODY FAT BETWEEN DEPOTS IN CATTLE, SHEEP AND PIGS AT SIMILAR FAT CONTENT (APPROX. 200 g/kg LIVE WEIGHT); AND RELATIVE GROWTH (b) OF FAT DEPOTS (y) ON TOTAL FAT (x) USING MODEL $\log y = \log a + b \log x$

	Cattle[a]		Sheep[b]		Pigs[c]	
	Total fat (g/kg)	b	Total fat (g/kg)	b	Total fat (g/kg)	b
Carcass						
Subcutaneous	240	1.17	433	1.15	680	1.04
Intermuscular	364	0.89	327	0.92	215	0.91
Abdomen						
Perirenal–retroperitoneal	172	1.06	95	0.86	61	1.21
Mesenteric	127	1.03	36	0.89	38	0.65
Omental	97	1.18	109	1.18	7	1.04

[a] Friesian steers 41–460 kg live weight, 1.6–90.0 kg total fat (Butler-Hogg and Wood, 1982)
[b] Southdown ewes and wethers 3–65 kg live weight, 0.1–34.9 kg total fat (Butler-Hogg, 1984)
[c] Large White gilts and hogs 15–120 kg live weight, 2.0–39.0 kg total fat (Wood et al., 1983b)

Fat also accumulates between muscle fibres as intramuscular (marbling) fat and forms approximately 100 g/kg of total dissectible fat although it can be assessed only by solvent extraction. Intramuscular fat is most obvious in muscles from animals which are fat overall but this does not mean that it is late-developing since it had slower relative growth (lower b value) than subcutaneous, intermuscular and 'internal cavity' fat in sheep (Broad and Davies, 1981a,b) and pigs (Davies and Pryor, 1977).

In cattle there are important breed differences in the proportions of body fat in the various depots. Breeds noted for beef production, such as Herefords, have more subcutaneous and less abdominal fat at the same weight of total fat than breeds noted for milk production, such as Friesians (Truscott, Wood and MacFie, 1983). Jerseys have an even lower proportion of subcutaneous fat and a higher proportion of abdominal fat than Friesians (Butler-Hogg and Wood, 1982). Breed effects in sheep and pigs are smaller than those in cattle (Wood *et al.*, 1983b).

Just as fat accumulates in the various depots in a predictable way during growth so it accumulates at a faster rate in some parts of the depots than others. Results in *Figure 20.1* show the relative growth of regions of subcutaneous fat in the cattle and pigs described in *Table 20.1* (data from the sheep were not strictly comparable because fewer regions were

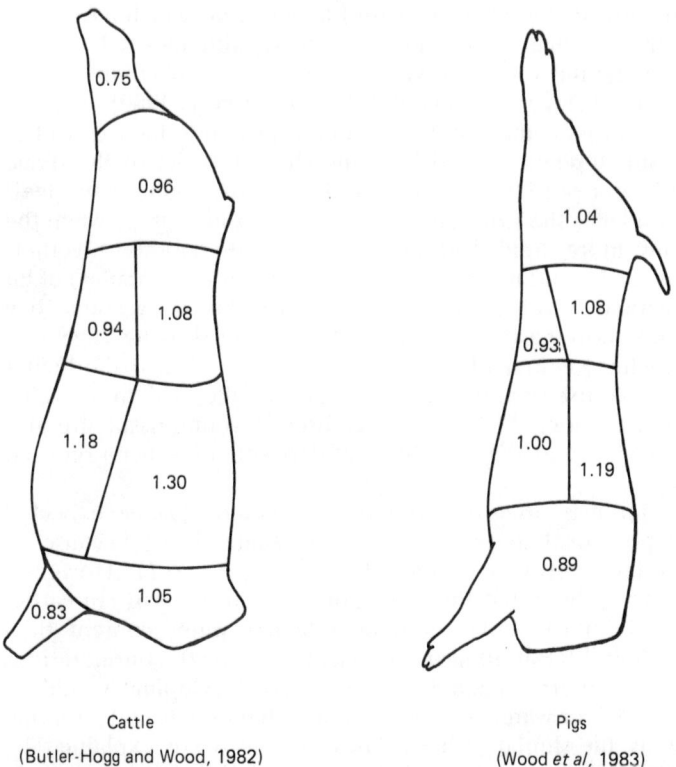

Cattle
(Butler-Hogg and Wood, 1982)

Pigs
(Wood *et al*, 1983)

Figure 20.1 Relative growth (b) of regions of subcutaneous fat (y) on total subcutaneous fat (x) in cattle and pigs using model $\log y = \log a + b \log x$

410 *Fat deposition and the quality of fat tissue in meat animals*

studied). In both cattle and pigs, relative growth was highest in the dorsal part of the forequarter (foreloin in pigs, crop in cattle) and, in general, ranking of regions was similar. More regions were studied in cattle where relative growth was lowest in the shin and leg.

Factors affecting fat quality during growth and development

Gross chemical and histological composition of tissues

Changes in the relative proportions of lipid (the major constituent), water and connective tissue occur during growth and have important effects on fat quality. Young fat tissue contains a high proportion of water and connective tissue and a low proportion of lipid contained in small cells (*Figure 20.2*). As the animal gets older and dietary energy is diverted more and more into fat growth, cell size increases; consequently the proportion of lipid increases and those of connective tissue and water decrease. These changes occur in a uniform way and have been used to predict overall carcass composition (Aberle, Etherton and Allen, 1977). In subcutaneous fat the number of fat cells also increases during growth (Wood, Enser and Restall, 1978; Broad, Davies and Tan, 1981; Truscott, Wood and Denny, 1983). This is mainly attributable to recent filling of cells which have been present from an early age (Leat and Cox, 1980), although it has been shown that the potential for cell division is present in subcutaneous fat from cattle of normal slaughter weights (Plaas and Cryer, 1980).

Such changes in gross composition would be expected to have considerable effects on the appearance and handling characteristics of the tissue although there have been few definitive studies. Thus, young tissue feels relatively wet and lacks the firmness which comes in older tissue when the fat cells contain more lipid and are packed more closely together. Separation between fat and muscle and between the groups (lobules) of fat cells within the tissue, which are themselves surrounded by connective tissue, also occurs more easily in young underdeveloped tissue. The grey (as opposed to white) colour of young tissue is partly due to the higher concentration of connective tissue which lowers the whiteness value (MacDougall and Disney, 1967). These features of young tissue are also seen in animals of the same age in which fat deposition has been reduced by underfeeding.

Consistent differences in gross composition would also be expected between fat depots or between parts of the same depot because of differences in relative growth (*Table 20.1* and *Figure 20.1*). However, results in *Table 20.2* show that this does not necessarily occur. In cattle, subcutaneous and intermuscular fat had a higher water content than perirenal and omental fat at all ages whereas from growth considerations alone (*Table 20.1*), intermuscular fat, being early-developing, would be expected to have a lower water content than the other depots and perirenal and subcutaneous fat similar values. These results were explained by Truscott (1980) as indicating that the abdominal depots grew mainly through an increase in the size of the cells visible at 6 months or before, whereas the carcass depots grew partly through an increase in the number

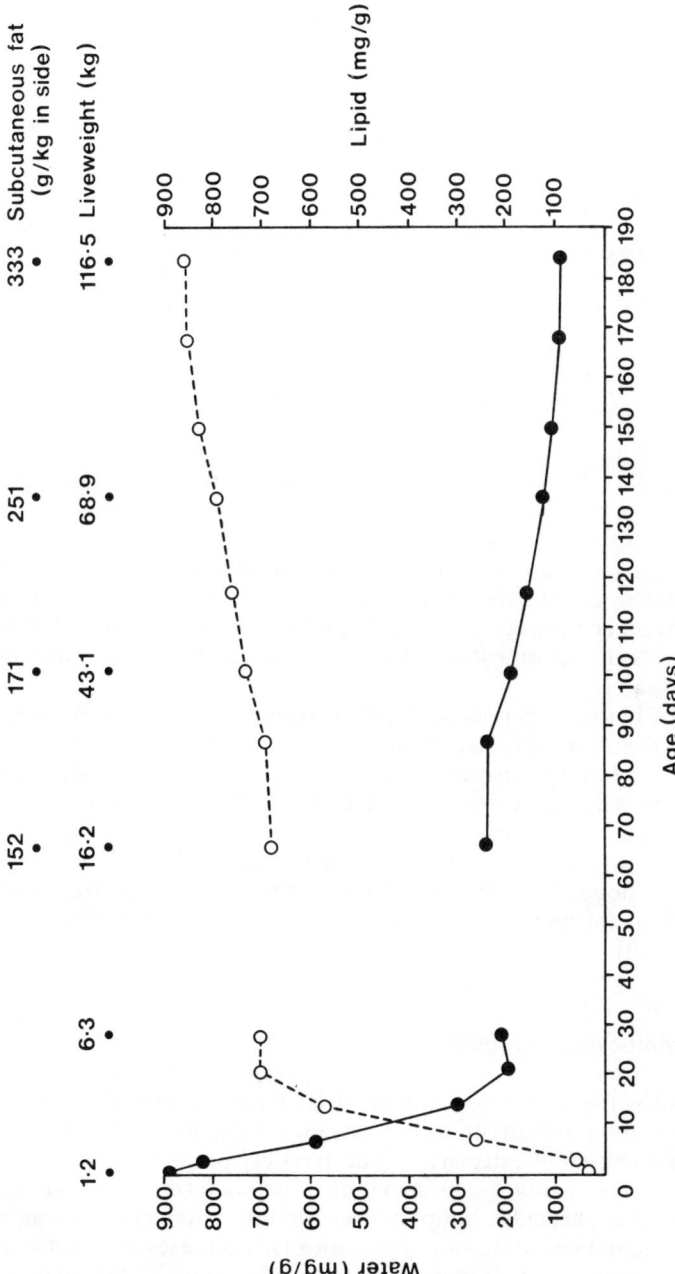

Figure 20.2 Water and total lipid concentration of outer layer of subcutaneous fat in castrated male and female Large White pigs. Samples from shoulder (days 0–28) and last rib region (days 66–184). Values are means of two pigs (days 0–28) (Moody et al., 1978) or four pigs (days 66–184) (Wood et al., 1983b). From 66 days a pelleted diet containing 190 g crude protein, 35 g fat and 13.0 MJ ME/kg was fed

Table 20.2 WATER CONTENT OF FAT TISSUE (mg/g ± SE) IN HEREFORD STEERS OF THREE AGES AFTER *AD LIBITUM* CONSUMPTION OF A COMPLETE PELLETED DIET

	Age (months)		
	6	13	20
Carcass			
Subcutaneous–Rump	205	133	90
	11	28	10
Midloin	341	109	104
	42	18	17
12th rib	260	103	85
	47	19	5
Brisket	459	296	215
	37	3	21
Intermuscular–prescapular	195	220	97
	25	12	12
Abdomen			
Perirenal	92	59	44
	19	1	3
Omental	197	111	64
	6	8	4

Number of steers was 4, 2 and 15 at 6, 13 and 20 months respectively. Truscott (1980)

of cells. Within subcutaneous fat there was also no clear association with growth rate when all four sites were considered (see *Figure 20.1*) although the large difference between brisket and rump could be interpreted in this way. In all sites the relationship between the concentrations of water and lipid was the same.

Differences between depots in chemical composition do not appear to modify the conclusions on relative growth based on dissection information alone (*Table 20.1* and *Figure 20.1*). This even applies to intramuscular fat which contains a higher proportion of phospholipid (approximately 250 mg/g lipid) than the dissected fat depots ($<$ 50 mg/g lipid). In sheep, phospholipid had a lower relative growth rate than triglyceride but comparison of depots on the basis of the storage component triglyceride still resulted in intramuscular fat having the lowest value (Broad and Davies, 1981a,b).

Fatty acid composition of lipid

Because lipid is the major constituent of fat tissue in animals at normal slaughter weights, most attention in work on fat quality has been paid to lipid and its component fatty acids. The physical property of fatty acids which most affects quality is melting point, which determines the firmness of the tissue at a particular temperature. Melting point increases as the carbon chain lengthens and, more dramatically, decreases as unsaturated linkages are introduced. For example, stearic acid melts at 69°C, oleic acid at 13.4°C or 16.3°C depending on the polymorphic form, linoleic acid at –5°C and linolenic acid at –11°C. The composite fatty acids from meat animal fat tissues melt between 25°C and 50°C (Deuel, 1951).

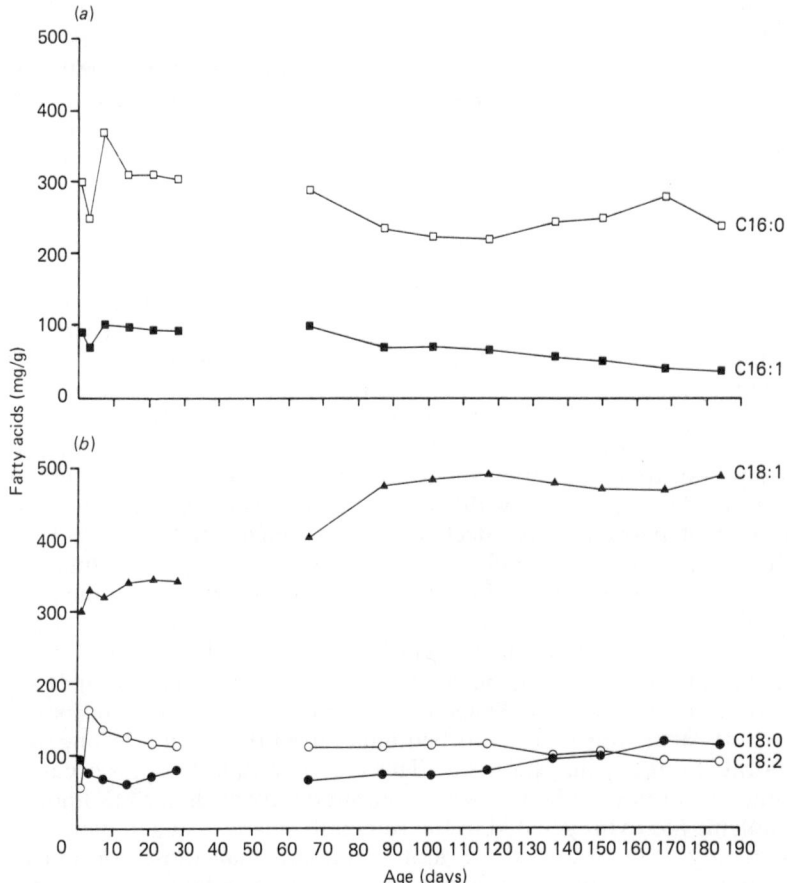

Figure 20.3 Concentrations (mg/g fatty acids) of the C16 (a) and C18 (b) fatty acids in total lipid from outer layer of subcutaneous fat in Large White pigs. The animals and samples are those described in *Figure 20.2* (Moody *et al.*, 1978; Wood *et al.*, 1983b). Up to day 28 the piglets were suckling

The colour of fat tissue at a particular temperature is partly determined by the extent to which the lipid has solidified and other constituents such as connective tissue, capillaries containing blood and carotenoid pigments can be observed. Fat tissue in which the lipid has not fully solidified appears relatively grey or yellow for this reason.

Fatty acid composition changes during growth and fattening (*Figure 20.3*). In pigs fed low-fat diets (i.e. < 40 g/kg) the two C16 and three C18 fatty acids constitute approximately 980 mg/g of total fatty acids and changes in their concentrations have the greatest effect on melting point. The only other fatty acid whose concentration exceeds 10 mg/g is myristic which reached 17 mg/g at 184 days in the pigs described in *Figure 20.3*. In these typical data, the concentrations of both C16 fatty acids fell between birth and 184 days of age but palmitoleic more so than palmitic, leading to an increase in saturation during the period. The concentrations of both stearic and oleic acid increased and there was no overall change in their ratio. The concentration of linoleic acid increased markedly from birth to 3 days of age, an effect of high levels of linoleic acid in colostrum (180–200 mg/g, Duncan and Garton, 1966) and its preferential deposition, but declined from then until 28 days and from 28 until 184 days of age. This is due firstly, to a fall in the concentration of linoleic acid in sows' milk as lactation progresses and secondly, to the increasing importance of *de novo* fatty acid synthesis in total fat deposition, linoleic acid being entirely derived from the diet. As a result the C18 fatty acids also increased in saturation during the period and the combined changes in C16 and C18 fatty acid concentrations led to an increase in the melting point of lipid and firmness of fat tissue. This can be roughly predicted from the concentration of stearic acid alone (Wood *et al.*, 1978); in the pigs in *Figure 20.3*, predicted melting (clarification) point in the outer layer of subcutaneous fat was 37°C and 43°C at 28 and 184 days of age respectively.

The overwhelming effect of triglyceride fatty acids on the fatty acid composition of total lipid after about 1 week of age was illustrated in the pigs described in *Figure 20.3* by the fact that arachidonic acid, an important constituent of phospholipids, declined from 38 mg/g total fatty acids at birth to 18 mg/g at 3 days and 5 mg/g at 7 days. Thereafter it remained below 5 mg/g (Moody *et al.*, 1978). Other workers have found that the fatty acid composition of the total lipids and triglycerides from pig backfat are indistinguishable (Scott, Cornelius and Mersmann, 1981).

In pigs and other monogastric animals the fatty acid composition of fat tissue triglycerides can be changed by increasing the concentration of dietary fatty acids which are absorbed intact from the small intestine and incorporated directly into fat tissue. This incorporation does not occur to the same extent in ruminants since unsaturated fatty acids such as linoleic and linolenic are hydrogenated in the rumen to stearic acid and oleic acid and the fatty acids absorbed are more saturated than those consumed. More *trans*-unsaturated, odd-chain and branched-chain fatty acids are also absorbed from the ruminant gut and incorporated into fat tissue (Christie, 1978). Nevertheless, the C16 and C18 *cis*-fatty acids make up > 900 mg/g of total fatty acids and, as in pigs, have the biggest effect on the melting point of the lipid. As with pigs, the fatty acid with a concentration most highly correlated with lipid melting point and fat firmness is stearic (L'Estrange and Mulvihill, 1975; Busboom *et al.*, 1981) although the concentration of stearic acid and hence the melting point is higher in ruminants. This causes beef and lamb fat to be hard when cold and gives the sticky taste to cold lamb fat which many people find unattractive. Unlike pigs, most studies with cattle and sheep, particularly cattle, have shown that the concentration of saturated fatty acids in fat tissue lipid decreases with age and body fat content and that of the unsaturated fatty acids increases. Moulton and Trowbridge (1909) had indirect evidence for this and it has been found more recently in biopsy studies by Link *et al.* (1970) and Leat (1975). Leat (1975) found that the ratios of the concentrations of palmitic to palmitoleic, stearic to oleic and stearic to linoleic acids in subcutaneous fat increased from 3 to 12 months of age in Jersey steers and fell thereafter. The concentrations of stearic acid and oleic acid were 158 mg/g and 427 mg/g fatty acids respectively at 9 months and 42 mg/g and 494 mg/g at

Figure 20.4 Lipid extracted from midloin subcutaneous fat from an old fat steer (a) and a young lean heifer (b) showing the effects of age and fat content on the melting point of lipid. Details are given in *Table 20.3*

Table 20.3 EFFECT OF AGE AND FAT CONTENT ON THE MELTING POINT AND FATTY ACID COMPOSITION OF SUBCUTANEOUS LIPID

Animal	Age (yr, approx.)	Carcass weight (kg)	Composition of side (g/kg)		Melting (clarification) point (°C)	Fatty acids (mg/g)		
			Lean	Fat		16:0	18:0	18:1
Fat steer	11	550	493	373	25.0	140	27	564
Young heifer	1	122	674	137	43.5	267	147	415

24 months of age, these changes being apparently closely related to the deposition of subcutaneous fat. An extreme example of the effect of age and fat content on the melting point and fatty acid composition of subcutaneous lipid is shown in *Figure 20.4* and *Table 20.3*. The steer is grossly over-fat by commercial standards. Evidence that the same effects occur in sheep was presented by Bensadoun and Reid (1965).

As with water, lipid and connective tissue, the concentrations of individual fatty acids vary with the site of deposition in the body in a way that is only partly related to the amount or rate of fat deposition in the different sites. In general there is a progressive increase in saturation from peripheral (subcutaneous) through intermuscular and intramuscular to deep body sites in cattle, sheep and pigs (Leat, 1977, *Table 20.4*; Duncan

Table 20.4 FATTY ACID COMPOSITION (mg/g) OF LIPID FROM FOUR BODY SITES IN ABERDEEN ANGUS AND FRIESIAN STEERS AND HEIFERS AGED 16–22 MONTHS FED A HAY DIET. ONE ANIMAL OF EACH SEX FROM EACH BREED = FOUR ANIMALS IN EACH MEAN

Fatty acid	Body site			
	Subcutaneous[a]	Intramuscular[b]	Intermuscular[c]	Perirenal
14:0	33	30	35	45
16:0	260	316	312	336
9-16:1	94	43	41	20
18:0	82	189	224	252
9-18:1	447	366	322	282
9,12-18:2	21	12	11	10

[a]Brisket. [b]m. longissimus, 10th rib. [c]Adjacent to intramuscular sample. Leat (1977)

and Garton, 1967; Christie, Jenkinson and Moore, 1972). In sheep these differences between perirenal and subcutaneous fat had developed by 1 year of age although they were not present at birth (Garton and Duncan, 1969). Temperature differences between the sites are partly responsible for the differences in fatty acid composition, since a low environmental temperature reduced the melting point of subcutaneous lipid and increased its iodine number in sheep (Marchello, Cramer and Miller, 1967) and produced more unsaturated fat in pigs (MacGrath *et al.*, 1968). However, temperature is not the only controlling factor and others, such as the preferential deposition of absorbed fatty acids in internal sites in ruminants, are also important (Duncan and Garton, 1967).

Fat quality in practical meat production

Cattle and sheep

In general, cattle and sheep have firmer fat than pigs and the fatty acid composition of body fat is less susceptible to change by dietary means. Nevertheless, some of the changes in production methods that are used to reduce carcass fatness and increase the efficiency of meat production have consequences for fat quality. These can be conveniently grouped under the headings diet, breed and sex.

DIETARY EFFECTS

The effect of wide variations in energy intake on the composition of fat tissue was demonstrated by Vickery (1977) in a study of dry cows differing in 'condition' at slaughter. The water content of brisket subcutaneous fat increased from 187 mg/g in 'fat' to 782 mg/g in 'emaciated' animals. Clearly, what fat tissue there was in the emaciated animals was little more than connective tissue. There were changes in fatty acid composition as condition changed from fat to emaciated, i.e. the concentration of stearic acid increased and that of oleic acid decreased, which is in agreement with the reports on fatty acid changes during growth. However, these changes and their combined effects on fat quality were much smaller than those in gross chemical composition.

Table 20.5 FATTY ACID COMPOSITION (mg/g TOTAL FATTY ACIDS) OF SUBCUTANEOUS FAT IN CATTLE AND SHEEP FED FORAGE OR CONCENTRATE DIETS

Animal	Feed	Fatty acid						
		14:0	16:0	9-16:1	18:0	9-18:1	9,12-18:2	9,12,15-18:3
Cattle[a]	Hay	33	260	94	82	447	21	11
	Barley	34	239**	106	69	487**	15	5
Sheep[b]	Pasture	32	208	45	259*	367	66	24*
	Drylot	36	223	43	201	378	88*	15
Sheep[c]	Pasture	35	289**	47	187***	428	20	—
	Purified	37	253	69*	137	476*	24	—

[a] Brisket fat from Aberdeen Angus and Friesian steers and heifers aged 16–24 months. Four animals in each mean. Leat (1977).
[b] Dorsal lumbar fat from crossbred ewe, wether and ram lambs 32–50 kg live weight fed either blue grass/clover pasture (16) or 130 g/kg protein concentrate diet (16). Kemp et al. (1981).
[c] Inguinal fat from four sheep fed on pasture and 30 fed diets composed of casein, glucose and starch. Tove and Matrone (1962).

In cattle and sheep the ratio of forage to concentrates in the diet affects the fatty acid composition of body fat and sometimes has significant effects on fat quality. In both species, increasing the proportion of concentrates leads to an increase in propionate production in the rumen, which results in a shift towards synthesis of more unsaturated fatty acids. This is greater than that expected from the increased total fat synthesis which occurs in animals fed high-energy diets (Roberts, 1966). Some results are given in *Table 20.5*. No adverse comments on fat quality were made in the study on cattle by Leat (1977) (1st reference) or on sheep by Kemp et al. (1981) (reference 2) but in the study of Tove and Matrone (1962) (reference 3), extracted subcutaneous lipid from sheep fed purified diets was liquid at room temperature and that from pasture-fed controls was solid, even though the differences in fatty acid composition were not greater than in the other examples. Other work suggests that this was due to high levels of C14–C16 branched-chain fatty acids with particularly low melting points which are synthesized in sheep when the soluble carbohydrate content of the diet is increased and excessive amounts of propionate (> 300 mmol/mol, Ørskov, Duncan and Carnie, 1975) are produced in the rumen

(Duncan, Ørskov and Garton, 1972; Garton, Hovell and Duncan, 1972). The production of branched-chain fatty acids and soft fat are increased when grains are thoroughly processed before feeding, e.g. by rolling, but are reduced when the grain is fed loose. For example, feeding the barley whole as part of a loose mixture ensured that the concentration of branched-chain fatty acids remained below 40 mg/g and that firm fat was produced. However, pelleting the barley whole or rolled increased the concentration of branched-chain fatty acids to > 50 mg/g and unacceptably soft fat resulted (Ørskov, Fraser and Gordon, 1974; Duncan *et al.*, 1974). Interestingly, cattle deposit fewer branched-chain fatty acids when they are fed the same processed grain diets as sheep (Duncan and Garton, 1978), which explains why complaints of soft oily fat are not levelled at carcasses produced using the barley beef system.

In intensively fed ruminants, increasing the level of dietary fat up to a certain level has attractions because it increases energy utilization (Black, 1971). The subsequent effect on depot fatty acids seems unpredictable, however. Some workers have shown that the mainly unsaturated fatty acids (e.g. linoleic) present in fat sources such as soyabean oil or sunflower seed oil are completely hydrogenated in the rumen, leading to increased concentrations of stearic acid in depot fat with little change in the concentration of linoleic (Tove and Mochrie, 1963). However, Gibney and L'Estrange (1975) found that the concentration of linoleic acid in perirenal fat increased from about 20 mg/g to 70 mg/g when lamb diets were supplemented with sunflower oil (58 g/kg), causing a decline in melting point from 46.7°C to 43.3°C. Clearly, considerable amounts of dietary fat escaped ruminal hydrogenation in this case.

The advantages of increasing the energy density of diets without reducing digestibility can be achieved if the dietary fat is protected from hydrogenation in the rumen. This has been achieved by feeding the fat supplement in a water suspension, thereby activating the oesophageal groove reflex (Kowalczyk *et al.*, 1977) and, in some experiments, by protecting the lipid with formaldehyde-treated protein (Scott, Cook and Mills, 1971). This has a dramatic effect on the concentration of linoleic acid in fat tissue. In one experiment with steers (Cook *et al.*, 1972), the concentration of linoleic acid in perirenal fat increased from 50 mg/g to 350 mg/g after 8 weeks' consumption of a diet containing 200 g/kg protected sunflower oil and had reached 200 mg/g after only 2 weeks. Incorporation into internal fat depots was greater than that into subcutaneous fat, supporting the view that there is preferential uptake of absorbed fatty acids in the internal depots of ruminants.

Apart from the possibility of increasing growth efficiency, 'protection' of dietary lipid was developed to produce more unsaturated milk and meat products which would be beneficial for people with cardiac disorders. Although, apparently, most people find meat with high levels of linoleic acid acceptable, there have been reports of 'oily', 'sweet' or 'bland' tastes, resulting from the oxidative breakdown of linoleic acid during cooking (McDonald and Scott, 1977). This also occurs if aerobic storage at too high temperatures (> −20°C) is prolonged (Bremner *et al.*, 1976). There have been few reports on the firmness of the fat produced but, since concentrations of linoleic acid > 150 mg/g are associated with soft fat in pigs, this would presumably also occur in the beef and lamb products. However, in one study in lamb in which the concentration of linoleic acid in meat reached 170 mg/g, the meat was found to be entirely satsifactory and 'it did not have the slightly sticky fat taste in the palate' which is normally associated with saturated lamb fat (Astrup and Nedkvitne, 1975). Clearly, other fat sources such as tallow and mixed fats can be 'protected' to regulate the fatty acid composition and quality characteristics of meat fat.

Other work has shown that meat from animals fed predominantly grain or concentrate diets, containing higher concentrations of oleic and lower concentrations of stearic acid than those fed forage diets, has a more desirable flavour (Rhodes, 1971; Westerling and Hedrick, 1979; Kemp *et al.*, 1981). Westerling and Hedrick (1979) found a high positive correlation between flavour score and the concentration of oleic acid, which suggested a specific effect on flavour. However, as oleic acid increases with fat content, this may have been an effect of fatness *per se* since some studies (but by no means all) suggest an effect of fatness on flavour. On the other hand, other workers have reported low and variable correlations between fatty acid composition and eating quality characteristics (Dryden and Marchello, 1970).

GENETIC EFFECTS

Breed affects the composition and quality of fat tissue mainly through its effect on total fat content (Leat, 1977). At the same body weight, late-maturing breeds have a lower concentration of fat than early-maturing breeds and the fat tissue itself has a higher concentration of water and a lower concentration of lipid in smaller cells. There are also more unfilled cells (Truscott *et al.*, 1983). At the same age, these differences between breeds are considerably reduced. Complaints about sloppy fat in extremely lean breeds, as with animals fed low-energy diets, are more likely to be due to these gross chemical effects than to changes in fatty acid composition since the leaner breeds have more saturated lipid.

Results of a study by Pyle *et al.* (1977) show the importance of total fatness in explaining differences in fatty acid composition between cattle breeds. Eighty steer carcasses from 13 sire breed crosses on Angus cows were examined. Overall, there were high correlations between the concentration of subcutaneous fat in the hindquarter and the concentrations of palmitoleic ($r = 0.48$) and stearic acid ($r = -0.46$) in brisket subcutaneous fatty acids. What differences remained after correction for differences in fatness must have been very small since the total range in oleic acid concentration was only 502–545 mg/g. The lowest concentration of stearic acid was 75 mg/g and the highest ratio of palmitoleic to stearic, an indication of firmness, was 1.44, lower than that found by Leat (1975) in Jersey steers over 15 months of age which were particularly fat.

Differences in total fat content were also responsible for the differences in fatty acid composition found between double-muscled and control Charolais steers by Bailey *et al.* (1982). Concentrations of stearic acid were higher and linoleic acid lower in three fat depots from the double-muscled animals because they were leaner. The implication is that where reported differences in fatty acid composition between breeds or selection lines are small, differences in body fat content will also be small (Rumsey et al., 1972; Sumida et al., 1972 in cattle and Boyland, Berger and Allen, 1976; Ch'ang, Evans and Hood, 1980 in sheep).

Fat with a pronounced yellow colour is undesirable although it is often found in old cattle and those fed grass as opposed to grain (Forrest, 1981). There are also breed effects, with Jerseys and Guernseys having yellower fat than other breeds. This is caused by a higher concentration of carotene in body fat resulting from excessive absorption from the intestine or a reduced conversion to vitamin A (Morgan and Everitt, 1969). In a 1969 review, the problem was considered to be serious in New Zealand because of the high proportion of Jersey cows in the national herd from which much beef is derived (Morgan and Everitt, 1969); replacement of Jerseys with Friesians was suggested as a solution. Similarly in sheep, xanthophylls are present in fat tissue at very low concentrations but there have been reports from Ireland and Iceland of breeds and strains in which the concentration is very high, due to the presence of a recessive gene (Patterson, 1965; Kirton et al., 1975).

SEX EFFECTS

Castration of bulls and rams increases fat deposition and so there is great interest on the part of farmers in producing entire males, if suitable management systems can be found.

Bulls have wetter fat than steers (*Table 20.6*), even at the same age. These differences are greater than those observed in cattle breeds of the same age (Truscott, 1980) and might be a sex effect independent of fatness although the differences in carcass fat content are also large. Poorly developed fat in excessively lean bulls sometimes leads to allegations of low carcass quality (MacDougall and Rhodes, 1972) and this will be more marked in countries such as Britain with traditional 'straight-line' butchery

Table 20.6 LEAN AND FAT CONTENT OF SIDE AND CHEMICAL COMPOSITION OF LEAN AND FAT TISSUE IN SEVEN PAIRS OF TWIN CALVES. ONE MEMBER OF THE PAIR WAS LEFT ENTIRE AND THE OTHER CASTRATED AT 6 WEEKS OF AGE. THE ANIMALS WERE FED *AD LIBITUM* ON A COMPLETE DIET AND SLAUGHTERED AT 400 DAYS OF AGE

	Bulls	Steers	Significance
Live weight (kg)	409	379	*
Lean in side[a] (g/kg)	657	601	***
Water	743	734	*
Lipid	25	42	***
Protein	217	207	**
Fat in side[b] (g/kg)	178	238	**
Water	274	223	***
Lipid	612	688	***
Protein	112	86	**

[a] Minced lean from whole side.
[b] Minced subcutaneous and intermuscular fat from whole side.
Fisher, Wood and Tas (unpublished)

methods which rely on firm fat 'setting-up' the muscle in joints than, for example, in France or Germany where fat is completely removed from muscles before sale. This may partly explain the greater popularity of bulls for meat in these countries than in Britain.

Differences in fatty acid composition between the sexes in cattle have usually been found to be small, and those that there are can be explained by differences in overall fat content (Roberts, 1966; Hood and Allen, 1971; Thrall and Cramer, 1971). This does not appear to be the case in sheep, however, where a specific problem of soft oily fat is found in ram lambs fed high-concentrate diets. For example, Crouse et al. (1972) found that ram lambs were leaner than wethers but had soft fat; at the same concentration of lipid in the carcass rams had approximately 20 mg/g less stearic acid, an unexpected result. Lower concentrations of stearic acid in subcutaneous fat from rams than in that from wethers were also found by Tichenor et al. (1970), Kemp et al. (1981) and Busboom et al. (1981). Ørskov, Duncan and Carnie (1975) found that feeding whole barley (as opposed to rolling it) prevented soft fat in wether but not ram lambs; this was not associated with differences in rumen propionate production.

Pigs

Attempts to include high levels of fat in pig diets in efforts to reduce costs and increase feed conversion efficiency have to take into account the effect on carcass fat quality because dietary fatty acids, apart from being used as an energy source, are also incorporated unchanged into body fat. As in cattle and sheep, breed and sex also affect fat quality and it is important to know whether these effects are independent of the amount of body fat and whether there are interactions with diet.

In many European countries there has been a great reduction in the fat content of pig carcasses in recent years. In Britain, for instance, fat thickness at the last rib (P_2) has fallen by 5 mm over the last 10 years. However this has led to allegations of poor fat quality by some meat traders and it has become important to define lower fat levels commensurate with good carcass and eating quality.

DIETARY EFFECTS

Studies in the United States between 1920 and 1930 showed the dramatic effects of the fatty acid composition of dietary fat on the fatty acid composition and quality of body fat in the pig (e.g. Ellis and Isbell, 1926a,b). These studies were undertaken to solve the problem of soft and oily pork fat in pigs which had grazed (hogged-off) or otherwise eaten large quantities of peanuts or soyabeans (*Table 20.7*). High levels of linoleic acid in the oils from these plants were implicated and the linoleic acid content of the diet has since been regarded as critical in determining carcass fat

Table 20.7 THE INFLUENCE OF THE AMOUNT AND TYPE OF DIETARY FAT ON THE QUALITY OF BACKFAT IN PIGS[a]

Main ingredient in diet	Dietary fat		Iodine No.	Melting point (°C)	Backfat	Fatty acids (mg/g)	
	In diet (g/kg)	Iodine No.			Grade[b]	18:1	18:2
Brewers rice	8	100	54.7	37.3	Hard	561	19
Corn	43	126	60.8	39.1	Hard	495	82
Peanut meal:corn meal (1:2)	50	94	72.6	28.7	Soft	526	130
Rice polish	97	100	76.9	28.6	Soft	541	137
Soyabeans + corn	approx. 95	approx. 100	78.3	27.6	Soft	441	200
Peanuts grazed or self-fed	331	94	89.6	19.4	Oily	550	203
Soyabeans grazed	175	128	93.2	26.0	Oily	391	306

[a] 6–20 hogs on each diet slaughtered at 73–108 kg live weight.
[b] Chilled carcasses graded hard, medium hard, medium soft, soft, oily.
Ellis and Isbell, 1926a,b

quality. Conclusions from the earlier and subsequent studies on dietary effects, concentrating on those involving linoleic acid, can be summarized:

1. Linoleic acid has little effect on the melting point of fat tissue triglycerides when its concentration remains below about 150 mg/g: at these low levels there are insufficient triglyceride molecules containing both oleic and linoleic acid, which are liquid at room temperature, to affect the melting point of the whole mixture. Above this value, sufficient molecules contain both unsaturated fatty acids to have a significant effect. At concentrations of 100–150 mg/g linoleic acid, which commonly occur in pigs of average fat content fed conventional diets (i.e. < 40 g/kg dietary fat) (Wood and Enser, 1982), melting point is determined by stearic acid (Wood et al., 1978) or the 'monoene:saturated ratio'—the ratio of the concentrations of the C16 and C18 mono-unsaturated to saturated fatty acids (Lea, Swoboda and Gatherum, 1970).
2. Values of > 150 mg linoleic acid/g fatty acids in body fat are commonly observed when the dietary fat concentration exceeds approximately 40 g/kg, giving a value for linoleic acid of about 16 mg/g in diets containing plant oils.
3. The concentration of linoleic acid in body fat is directly proportional to the amount consumed (*Figure 20.5*) or, more specifically, the amount consumed which is available for fat deposition (Brooks, 1971). In the work described in *Figure 20.5*, diets were formulated to a particular protein content: when this was derived from roasted soyabeans the level of dietary fat and the consequent effect on carcass quality was

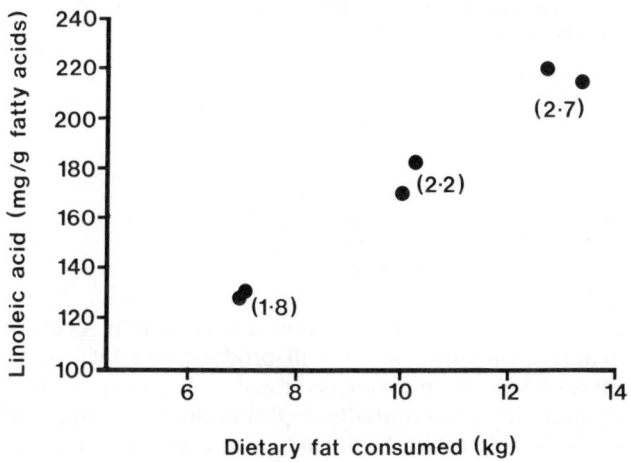

Figure 20.5 Effect of dietary fat consumption on linoleic acid content of backfat. Fat firmness scores in brackets: 1, firm; 2, moderately firm; 3, moderately soft. Each point is the average for seven hogs weighing 93 kg live weight. From 23 kg the pigs were fed on diets with 140 or 160 g/kg crude protein containing low (33 g/kg), high (70 g/kg) or intermediate (51 g/kg) levels of fat provided respectively by soyabean meal, roasted soyabeans or a combination of both. Linoleic acid provided approximately 500 g/kg of the dietary fat (Hanson et al., 1970)

greater than when it was derived from soyabean meal, which has a much lower fat content. Work by Brooks (1971) showed that the fatty acid composition of intramuscular lipid was less affected by diet than that of subcutaneous lipid. This could be due to the faster growth of subcutaneous fat (*Table 20.1*).

4. The same relationships between the consumption of linoleic acid and its concentration in body fat are apparently not maintained when feed intake and the rate of fat deposition are reduced to low levels. Under these conditions the concentration in body fat increases relative to that in the diet because, whereas about 400 g/kg of the dietary fatty acid continues to be deposited, there is a marked reduction of *de novo* fatty acid synthesis (Hilditch, Lea and Pedelty, 1939). On low fat diets, several workers have observed a close inverse relationship between the concentration of linoleic acid and the rate of deposition of fat (*Figure 20.6*). In the experiment shown, the diet contained 190 g crude

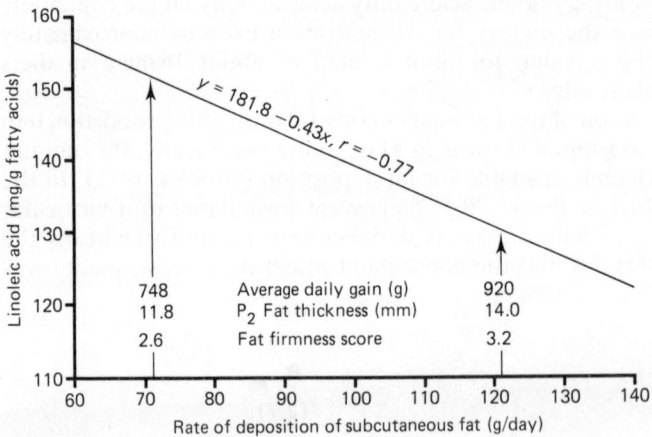

Figure 20.6 Relationship between the rate of growth of subcutaneous fat and the concentration of linoleic acid in backfat from last rib region in 32 boar pigs. Sixteen pigs were fed from 27 to 88 kg live weight at a high level of intake and 16 at a low level. Means for each group are also shown. Fat firmness was subjectively scored 1 (very soft) to 5 (very hard). (Wood and Enser, 1982)

protein, 35 g fat and 13.6 MJ DE/kg diet and it is clear that levels of linoleic acid bordering on those which will produce soft fat can be achieved through relatively slight underfeeding of such a 'normal' diet.

5. Linoleic acid is apparently preferentially deposited in that, compared with other fatty acids which are both dietary constituents and the products of *de novo* synthesis, its concentration in body fat increases more rapidly in relation to that in the diet (Dahl and Persson, 1965; Brooks, 1971). This may be due to inhibition by linoleic acid of synthetic and desaturase enzymes responsible for the production of the other fatty acids, and also to the high digestibility of linoleic acid.

6. To obtain energy-dense diets without incurring the carcass quality disadvantages of high levels of linoleic acid in backfat, various saturated fat sources, e.g. hydrogenated marine fat (Gjefsen and Lyso,

1979) and hydrogenated tallow (Tullis and Whittemore, 1980), have been suggested. The disadvantage of completely saturated fats is their low digestibility, although inclusion of some unsaturated fat in a predominantly saturated mix, especially in a 'natural' form, appears to overcome this (Kidder and Manners, 1978).

Reducing the level of feed intake, apart from leading to an increase in the concentration of linoleic acid in body fat, also affects its gross chemical composition, as in cattle and sheep. Results in *Table 20.8* show that when low-level feeding reduced daily growth by 0.18 (18%) in gilts, the concentration of linoleic acid in both layers of backfat increased, although not markedly, and there were large changes in water and lipid content which may have caused the tendency towards lower fat-quality scores in the underfed group.

Table 20.8 EFFECT OF REDUCING THE LEVEL OF FEED INTAKE BETWEEN 20 AND 68 kg LIVE WEIGHT ON CARCASS COMPOSITION AND FAT QUALITY IN GILTS. DIET CONTAINED 35 g FAT AND 13.0 MJ DE/kg; ELEVEN GILTS WERE FED AT HIGH LEVEL AND TEN AT LOW LEVEL

	Level of intake		Significance of difference
	High	Low	
Average daily gain (g)	667	547	***
P_2 fat thickness (mm)	17.0	13.3	***
Lean in side (g/kg)	547	598	***
Outer fat layer[a]			
Water (g/kg)	112	154	***
Lipid (g/kg)	861	803	***
Linoleic acid (mg/g fatty acids)	97	125	***
Inner fat layer			
Water (g/kg)	120	196	***
Lipid (g/kg)	841	750	***
Linoleic acid (mg/g fatty acids)	90	117	***
Fat firmness score[b]	3.6	3.1	NS
Fat whiteness score[c]	3.7	3.1	NS

[a] Samples from last rib. [b] Scores 1 (very soft) to 5 (very hard). [c] Scores 1 (very grey) to 5 (very white).
Wood (unpublished data)

The storage life of meat is limited to a great extent by the oxidative breakdown of unsaturated fatty acids, leading to the production of peroxides and eventually to rancidity (Enser, 1974). Linoleic acid is particularly susceptible, twelve times more so than oleic acid (Enser, 1974) and so adequate temperature control and packaging of pigmeat with soft fat is critical. If rancidity is avoided there is no effect of fatty acid composition on eating quality. When pigs were fed on diets with different types of added fat, so that the linoleic acid content of backfat varied from 137 mg/g to 367 mg/g fatty acids (the meat being stored at −30°C before use), there were no treatment effects on taste panel scores for tenderness, flavour, juiciness or overall acceptability of meat (Koch *et al.*, 1968). Peroxides can also accumulate in body fat during growth if very unsaturated animal fats (e.g. marine oils) are fed and the animal is deficient in vitamin E (Agricultural Research Council, 1981). This results in yellow fat which is unattractive and leads to downgrading of the carcass (Kirby, 1981).

Finally, it has been observed that addition of copper to diets at a level sufficient to promote growth, i.e. 250 ppm, produces soft fat. Moore et al. (1969) found that the melting point of lipid was reduced by 10°C and in another study, 0.8 (80%) of pigs receiving 250 ppm copper had soft fat compared with 0.05 (5%) of a similar group receiving no copper (Elliot and Bowland, 1968). The effect is due to an increase in the ratio of oleic to stearic acid, possibly caused by activation by copper of desaturase enzymes (Thompson, Allen and Meade, 1973) and also to a change in the structure of triglycerides (Christie and Moore, 1969). The effect on fatty acid composition seems greater than that expected from the reduction in fat content which probably results when copper is fed. Since most commercial pig diets in Britain include copper as a growth promoter, its effects on fat quality have presumably to be tolerated, although whether these are significant at the level of 125 ppm recently suggested as a maximum in the EEC have not been established.

GENETIC EFFECTS

In a study of soft fat in carcasses from several breeds and crosses, Lea et al. (1970) commented that 'there has been a feeling in the (British pig) industry that pig fats tend to be softer than they were pre-war' and the implication was that this was due to genetic changes. They found that the more the Hampshire breed was represented in a cross, the more likely the carcasses were to have soft fat because of an increase in the monoene:saturated ratio. This ratio was thought to be independent of dietary influences and possibly of body fat content. Villegas et al. (1973) also found a higher monoene:saturated ratio in Hampshire pigs (*Table 20.9*) but in this case the concentration of linoleic acid was also higher in Hampshires, especially on the higher fat diet (diet 1) and would have had an effect on melting

Table 20.9 EFFECT OF BREED ON FATTY ACID COMPOSITION OF SUBCUTANEOUS FAT (10–12th RIB) IN PIGS

Fatty acids (mg/g)	Breed and diet[a]							
	Yorkshire		Crossbred[b]		Hampshire		Duroc	
	1	2	1	2	1	2	1	2
14:0	13[d]	15[d]	13[d]	13[d]	12[e]	14[d]	14[d]	16[c]
16:0	241[de]	256[e]	242[de]	243[d]	211[c]	218[c]	237[d]	275[f]
9-16:1	17[c]	23[d]	20[cd]	30[e]	24[d]	39[f]	19[cd]	29[e]
18:0	137[d]	139[d]	122[cd]	142[e]	108[c]	107[c]	125[d]	130[d]
9-18:1	323[c]	424[f]	382[cd]	422[ef]	384[cd]	447[f]	377[cd]	398[de]
9,12-18:2	180[e]	107[c]	182[e]	113[c]	229[f]	138[d]	185[e]	114[c]
9,12,15-18:3	20[e]	7[c]	20[e]	8[c]	24[f]	8[c]	17[d]	8[c]
Monoene:saturated ratio	0.90	1.13	1.10	1.17	1.28	1.50	1.09	1.05

[a] Four castrated males in each breed × diet group, fed *ad libitum* from 31 to 102 kg live weight. Diet 1 contained approx. 220 g/kg roasted soyabeans and 62 g/kg fat; Diet 2 contained approx. 180 g/kg soyabean meal and 35 g/kg fat.
[b] Duroc × Hampshire × Yorkshire
Means in same line with different superscripts are significantly different ($P<0.05$)
Villegas et al., 1973.

Table 20.10 SELECTION EFFECTS ON FATTY ACID COMPOSITION (mg/g) OF SUBCUTANEOUS FAT IN PIGS

Fatty acid	Reference 1			Reference 2		Reference 3	
	Obese	Contemporary	Lean	Control	Selected	Line 4	Line 6
14:0	13[a]	11[b]	9[c]	14	14	17[a]	14[b]
16:0	237[a]	220[b]	213[b]	287	285	266[a]	239[b]
9-16:1	24[a]	20[b]	24[a]	24	25	52	56
18:0	118	117	103	154	153	110	104
9-18:1	394[a]	367[b]	379[b]	433[a]	415[b]	465[a]	480[b]
9,12-18:2	167[a]	216[b]	230[c]	93[a]	107[b]	91[a]	106[b]
Monoene:saturated ratio	1.17[a]	1.15[a]	1.28[b]	1.04	1.00	1.37	1.45
Melting (clarification) point (°C)				47.5	47.3		

Reference 1. Obese and lean pigs were from selection lines in Duroc and Yorkshire breeds in which backfat thickness in obese line was double that in lean. Contemporary pigs were Duroc × Large White. Six gilts in each group fed *ad libitum* on a diet containing 750 g/kg corn and 22 g/kg soyabean meal to 6 months of age. Samples from inner layer at shoulder. Scott, Cornelius and Mersmann, 1981.

Reference 2. Large Whites selected for low backfat thickness for seven generations had 14 mm fat over *m. longissimus* at last rib cf. 18 mm in randomly-mated controls. Seventy-seven selected and 80 control boars fed *ad libitum* on a diet containing 30 g/kg fat of which linoleic acid was 400 g/kg fatty acids. Samples from inner layer at last rib. Wood *et al.*, 1978.

Reference 3. The fattest (line 4) and leanest (line 6) of 7 selection lines examined. Line 4 were randomly-mated control Lacombes, line 6 were Poland China × Lacombe. Thirteen gilts and 13 hogs in each group fed *ad libitum* on a standard diet to 90 kg live weight. Samples from belly at last rib. Martin *et al.*, 1972.

Means with different superscripts are significantly different ($P<0.05$).

point (this was not measured). Other work suggests that these results were due to lower body fat levels in the Hampshires because the general finding has been for leaner breeds or strains to have higher concentrations of linoleic acid and sometimes, although not always, higher monoene:saturated ratios. Results in *Table 20.10* are taken from three selection experiments. The biggest differences in fatty acid composition were found between lean and obese Duroc × Yorkshire pigs differing markedly in fat content (reference 1). Linoleic acid differed most between the lines. Much smaller differences were observed between selected and control Large Whites by Wood *et al.* (1978) (reference 2) and in this case the monoene:saturated ratio was slightly lower in selected pigs and there was no difference in melting point between the lines. In both references 1 and 2 (but not in reference 3) the concentration of oleic acid was lower in the lean lines and it has been suggested that the pig can regulate melting point within certain limits by reducing the concentration of oleic acid in response to an increase in linoleic.

As indicated by the results in *Table 20.10*, the fatty acid with a concentration most affected by the amount of fat deposited and showing the highest correlation with indices of fatness between and within breeds, is linoleic acid (Martin *et al.*, 1972; Wood *et al.*, 1978). As with changes in the level of feeding, it seems that the relationship between the amount of linoleic acid consumed and that deposited is similar between genotypes, so that dilution of linoleic acid in the synthesized fatty acids is a measure of the amount of synthesized fat.

The consequences of these results are that genetically lean pigs will have relatively unsaturated and soft fat. Although the genetic effect appears to be small and much less than the range of dietary effects, some workers have said that it is significant. Vold (1976) for instance, found that raw sausages which included backfat from a lean line of Norwegian Landrace pigs developed oxidative rancidity more quickly than those made with fat from a fat line and higher levels of linoleic acid in the lean line were implicated. It is also apparent that when diets high in linoleic acid are fed, the consequences for fat quality are greater in lean pigs, as shown by the results for Hampshires in *Table 20.9* and for the lean Duroc × Yorkshires in *Table 20.10*. There have been few detailed studies of the interactions between genetics and diet as far as fat quality is concerned, but this seems to be an important omission. It should also be remembered that the type of diet consumed early in life, e.g. following weaning, could be critical because it will most affect the composition of the early developing outer fat layer which is more unsaturated anyway (*Table 20.8*).

Genetically leaner pigs also have a higher concentration of water and a lower concentration of lipid in fat tissue but results found by Wood *et al.* (1983b) were too small to have affected fat quality.

SEX EFFECTS

As in cattle and sheep, castration of male pigs increases fat deposition and reduces the efficiency of conversion of feed into meat. Farmers from many countries wishing to produce lean meat efficiently have therefore turned to boars when contractual arrangements with meat companies have made this possible. The meat trade has had several objections to boars, foremost of which is 'boar taint' (androstenone), but soft and floppy fat has figured prominently in the list (Wood and Enser, 1982).

Several workers have observed that lipid extracted from boar fat tissue is slightly more unsaturated than that from castrates when comparisons are made at the same carcass weight, and boars are leaner (Malmfors, Lundström and Hansson, 1978; Smithard, Smith and Ellis, 1980; Wood and Enser, 1982). The major differences are similar to those noted for lean breeds, i.e. boars have a higher concentration of linoleic acid and a slightly higher monoene:saturated ratio than castrates but, as with breeds, the castration effect appears to be small. Wood and Enser (1982) showed that the differences in linoleic acid content were explained by differences in the rate of subcutaneous fat deposition. This work also found large differences in the water and lipid content of backfat between boars and castrates but here the differences were still apparent when comparisons were made at the same proportion of subcutaneous fat in the carcass. Boars also had higher concentrations of fat-free dry matter (suggesting more connective tissue) in backfat. However, there was no evidence that these differences in gross composition had a specific effect on fat quality as determined by an experienced assessor, since firmness and whiteness scores were roughly in line with the concentrations of linoleic acid (Wood and Enser, 1982).

Subsequent work examined fat quality in boars and castrates in more detail (Wood et al., 1983a). Using an Instron materials testing machine to obtain an objective assessment of fat firmness, it was concluded that, below 9 mm P_2, boars had slightly firmer fat than castrates; this is possibly a reflection of the greater concentration of connective tissue. Above this, castrates had firmer fat but in both cases the castration effect was small. In addition, it was found that the force required to separate the muscle and fat layers in the last rib region, which is an important aspect of bacon quality, was related to the thickness of backfat and there was no inherent difference between boars and castrates.

Together, these results show that boars will have slightly lower fat quality scores than castrates in most practical situations because they are leaner. At the same fat thickness the differences are negligible. The implication is that boars from lean strains should not be underfed (as the results in *Figure 20.6* show) and, as with breeds, attention should be paid to the linoleic acid content of the diet.

Conclusions

A large part of the variation in fat quality which occurs in commercial meat production is caused by variation in the amount of fat tissue. In cattle, sheep and pigs, fat tissue from excessively lean carcasses has higher concentrations of connective tissue and water and appears wet and floppy. In pigs, the lipid present in lean carcasses is relatively unsaturated, further softening the tissue, but this does not occur in cattle and sheep where soft fat arising from an increase in the concentration of unsaturated fatty acids occurs only in excessively fat animals. Fat-quality problems are always greatest in subcutaneous fat, which has higher concentrations of water and unsaturated acids than the other depots.

Fat-quality differences between breeds and sexes are also largely caused by differences in the amount of fat tissue except in rams, which have more unsaturated lipid and softer fat than expected. Diet has a more marked effect on fat quality than breed or sex, especially in pigs, in which the potentially undesirable consequences of combining lean breeds, low levels of energy intake, boars and diets high in linoleic acid should be avoided. There are also some important dietary effects in cattle and sheep, for instance the production of soft fat in lambs fed processed grain diets. More dramatic dietary effects are observed when unsaturated fatty acids are protected from hydrogenation in the rumen.

There is no convincing evidence that fat quality declines to unacceptably low levels in lean carcasses except when these are excessively lean. In the future, attention will continue to be paid to reducing overall fat levels in meat, although lower limits will also need to be defined for the sake of good meat quality.

References

ABERLE, E.D., ETHERTON, T.D. and ALLEN, C.E. (1977). Prediction of pork carcass composition using subcutaneous adipose tissue moisture or lipid concentration. *Journal of Animal Science* **45**, 449–456

AGRICULTURAL RESEARCH COUNCIL (1981). *The Nutrient Requirements of Pigs*. Slough, Commonwealth Agricultural Bureaux

ASTRUP, H.N. and NEDKVITNE, J.J. (1975). The improved production of unsaturated lamb meat. *Acta agriculturae scandinavica* **25**, 49–52

BAILEY, A.J., ENSER, M.B., DRANSFIELD, E., RESTALL, D.J. and AVERY, N.C. (1982). In *Muscle Hypertrophy of Genetic Origin and Its Use to Improve Beef Production* (J.W.B. King and F. Menissier, Eds), pp. 178–202. The Hague, Netherlands, Martinus Nijhoff

BENSADOUN, A. and REID, J.T. (1965). Effect of physical form, composition and level of intake of diet on the fatty acid composition of the sheep carcass. *Journal of Nutrition* **87**, 239–244

BLACK, J.L. (1971). A theoretical consideration of the effect of preventing rumen fermentation on the efficiency of utilization of dietary energy and protein in lambs. *British Journal of Nutrition* **25**, 31–55

BOYLAN, W.J., BERGER, Y.M. and ALLEN, C.E. (1976). Fatty acid composition of Finnsheep crossbred lamb carcasses. *Journal of Animal Science* **42**, 1421–1426

BREMNER, H.A., FORD, A.L. MacFARLANE, J.J., RATCLIFFE, D. and RUSSELL, N.T. (1976). Meat with high linoleic acid content: Oxidative changes during frozen storage. *Journal of Food Science* **41**, 757–761

BROAD, T.E. and DAVIES, A.S. (1981a). Pre- and postnatal study of the carcass growth of sheep. 1. Growth of dissectable fat and its chemical components. *Animal Production* **31**, 63–71

BROAD, T.E. and DAVIES, A.S. (1981b). Pre- and postnatal study of the carcass growth of sheep. 3. Growth of dissectable and chemical components of muscle and changes in the muscle:bone ratio. *Animal Production* **32**, 234–243

BROAD, T.E., DAVIES, A.S. and TAN, G.Y. (1981). Pre- and postnatal study of the carcass growth of sheep. 2. The cellular growth of adipose tissue. *Animal Production* **31**, 73–79

BROOKS, C.C. (1971). Fatty acid composition of pork lipids as affected by basal diet, fat source and fat level. *Journal of Animal Science* **33**, 1224–1231

BUSBOOM, J.R., MILLER, G.J., FIELD, R.A., CROUSE, J.D., RILEY, M.L., NELMS, G.E. and FERRELL, C.L. (1981). Characteristics of fat from heavy ram and wether lambs. *Journal of Animal Science* **52**, 83–92

BUTLER-HOGG, B.W.B. (1984). Fat partition in Clun and Southdown sheep: body composition and growth of the fat depots. *Animal Production* (in press)

BUTLER-HOGG, B.W.B. and WOOD, J.D. (1982). The partition of body fat in British Friesian and Jersey steers. *Animal Production* **35**, 253–262

CH'ANG, T.S., EVANS, R. and HOOD, R.L. (1980). Sire effect on fatty acid composition of ovine adipose tissue. *Journal of Animal Science* **51**, 1314–1320

CHRISTIE, W.W. (1978). The composition, structure and function of lipids in the tissues of ruminant animals. *Progress in Lipid Research* **17**, 111–205

CHRISTIE, W.W. and MOORE, J.H. (1969). The effect of dietary copper on the structure and physical properties of adipose tissue triglycerides in pigs. *Lipids* **4**, 345–349

CHRISTIE, W.W., JENKINSON, D.M. and MOORE, J.H. (1972). Variation in lipid composition through the skin and subcutaneous adipose tissue of pigs. *Journal of the Science of Food and Agriculture* **23**, 1125–1129

COOK, L.J., SCOTT, T.W., FAICHNEY, G.J. and DAVIES, H.L. (1972). Fatty acid interrelationships in plasma, liver, muscle and adipose tissues of cattle fed sunflower oil protected from ruminal hydrogenation. *Lipids* **7**, 83–89

CROUSE, J.D., KEMP, J.D., FOX, J.D., ELY, D.G. and MOODY, W.G. (1972). Effects of castration on ovine neutral and phospholipid deposition. *Journal of Animal Science* **34**, 388–392

DAHL, O. and PERSSON, K. (1965). Properties of animal depot fat in relation to dietary fat. *Journal of the Science of Food and Agriculture* **16**, 452–455

DAVIES, A.S. and PRYOR, W.J. (1977). Growth changes in the distribution of dissectable and intermuscular fat in pigs. *Journal of Agricultural Science* **89**, 257–266

DEUEL, H.J. (1951). *The Lipids: their Chemistry and Biochemistry 1. Chemistry*. New York, Interscience Publishers Inc.

DRYDEN, F.D. and MARCHELLO, J.A. (1970). Influence of total lipid and fatty acid composition upon the palatability of three bovine muscles. *Journal of Animal Science* **31**, 36–41

DUNCAN, W.R.H. and GARTON, G.A. (1966). The component fatty acids of the colostral fat and milk fat of the sow. *Journal of Dairy Research* **33**, 255–259

DUNCAN, W.R.H. and GARTON, G.A. (1967). The fatty acid composition and intramuscular structure of triglycerides derived from different sites in the body of the sheep. *Journal of the Science of Food and Agriculture* **18**, 99–102

DUNCAN, W.R.H. and GARTON, G.A. (1978). Differences in the proportions of branched-chain fatty acids in subcutaneous triacylglycerols of barley-fed ruminants. *British Journal of Nutrition* **40**, 29–33

DUNCAN, W.R.H., ØRSKOV, E.R. and GARTON, G.A. (1972). Fatty acid composition of triglycerides of lambs fed on barley-based diets. *Proceedings of the Nutrition Society* **31**, 19A–20A

DUNCAN, W.R.H., ØRSKOV, E.R., FRASER, C. and GARTON, G.A. (1974). Effect of processing of dietary barley and of supplementary cobalt and cyanocobalamin on the fatty acid composition of lamb triglycerides, with reference to branched-chain components. *British Journal of Nutrition* **32**, 71–75

ELLIS, N.R. and ISBELL, H.S. (1926a). Soft pork studies. 2. The influence of the character of the ration upon the composition of the body fat of hogs. *Journal of Biological Chemistry* **69**, 219–238

ELLIS, N.R. and ISBELL, H.S. (1926b). Soft pork studies. 3. The effect of food fat upon body fat, as shown by the separation of the individual fatty acids of the body fat. *Journal of Biological Chemistry* **69**, 239–248

ELLIOT, J.I. and BOWLAND, J.P. (1968). Effects of dietary copper sulphate on the fatty acid composition of porcine depot fats. *Journal of Animal Science* **27**, 956–960

ENSER, M. (1974). Factors affecting the development of oxidative rancidity in frozen meat. In *Proceedings, Meat Research Institute Symposium No. 3. Meat Freezing—Why and How* (C.L. Cutting, Ed.), pp. 11.1–11.5. Langford, Bristol, Meat Research Institute

FORREST, R.J. (1981). Effect of high concentrate feeding on the carcass quality and fat coloration of grass-reared steers. *Canadian Journal of Animal Science* **61**, 575–580

GARTON, G.A. and DUNCAN, W.R.H. (1969). Composition of adipose tissue triglycerides of neonatal and year-old lambs. *Journal of the Science of Food and Agriculture* **20**, 39–42

GARTON, G.A., HOVELL, F.D.DeB. and DUNCAN, W.R.H. (1972). Effect of dietary propionate on the fatty acid composition of lamb triglycerides. *Proceedings of the Nutrition Society* **31**, 20A–22A

GIBNEY, M.J. and L'ESTRANGE, J.L. (1975). Effects of dietary unsaturated fat and of protein source on melting point and fatty acid composition of lamb fat. *Journal of Agricultural Science* **84**, 291–296

GJEFSEN, T. and LYSO, A. (1979). Hydrogenated marine fat with high content of free fatty acids in feed mixtures for growing–finishing pigs. *Acta agriculturae scandinavica* **29**, 65–70

HANSON, L.E., ALLEN, C.E., MEADE, R.J., RUST, J.W. and MILLER, K.P. (1970). Cooked soybeans for swine and effects on carcass composition. *Feedstuffs* **42**, 16–18

HILDITCH, T.P., LEA, C.H. and PEDELTY, W.H. (1939). The influence of low and high planes of nutrition on the composition and synthesis of fat in the pig. *Biochemical Journal* **33**, 493–504

HOOD, R.L. and ALLEN, C.E. (1971). Influence of sex and postmortem aging on intramuscular and subcutaneous bovine lipids. *Journal of Food Science* **36**, 786–790

KEMP, J.D., MAHYUDDIN, M., ELY, D.G., FOX, J.D. and MOODY, W.G. (1981). Effect of feeding systems, slaughter weight and sex on organoleptic properties and fatty acid composition of lamb. *Journal of Animal Science* **51**, 321–330

KIDDER, D.E. and MANNERS, M.J. (1978). *Digestion in the Pig*, pp. 180–186. Bristol, Scientechnica

KIRBY, P.S. (1981). Steatitis in fattening pigs. *Veterinary Record* **109**, 385

KIRTON, A.H., CRANE, B., PATERSON, D.J. and CLARE, N.T. (1975). Yellow fat in lamb caused by carotenoid pigmentation. *New Zealand Journal of Agricultural Research* **18**, 267–272

KOCH, D.E., PEARSON, A.M., MAGEE, W.T., HOEFER, J.A. and SCHWEIGERT, B.S. (1968). Effect of diet on the fatty acid composition of pork fat. *Journal of Animal Science* **27**, 360–365

KOWALCZYK, J., ØRSKOV, E.R., ROBINSON, J.J. and STEWART, C.S. (1977). Effect of fat supplementation on voluntary food intake and rumen metabolism in sheep. *British Journal of Nutrition* **37**, 251–257

LEA, C.H., SWOBODA, P.A.T. and GATHERUM, D.P. (1970). A chemical study of soft fat in cross-bred pigs. *Journal of Agricultural Science* **74**, 279–289

LEAT, W.M.F. (1975). Fatty acid composition of adipose tissue of Jersey cattle during growth and development. *Journal of Agricultural Science* **85**, 551–558

LEAT, W.M.F. (1977). Depot fatty acids of Aberdeen Angus and Friesian cattle reared on hay and barley diets. *Journal of Agricultural Science* **89**, 575–582

LEAT, W.M.F. and COX, R.W. (1980). Fundamental aspects of adipose tissue growth. In *Growth in Animals* (T.L.J. Lawrence, Ed.), pp. 137–174. London, Butterworths

L'ESTRANGE, J.L. and MULVIHILL, T.A. (1975). A survey of fat characteristics of lamb with particular reference to the soft fat condition in intensively fed lambs. *Journal of Agricultural Science* **84**, 281–290

LINK, B.A., BRAY, R.W., CASSENS, R.G. and KAUFFMAN, R.G. (1970). Fatty acid composition of bovine subcutaneous adipose tissue lipids during growth. *Journal of Animal Science* **30**, 722–725

McDONALD, I.W. and SCOTT, T.W. (1977). Foods of ruminant origin with elevated content of polyunsaturated fatty acids. *World Reviews of Nutrition and Dietetics* **26**, 144–207

MacDOUGALL, D.B. and DISNEY, J.G. (1967). Quality characteristics of pork with specific reference to Pietrain, Pietrain × Landrace and Landrace pigs at different weights. *Journal of Food Technology* **2**, 285–297

MacDOUGALL, D.B. and RHODES, D.N. (1972). Characteristics of the appearance of meat. III. Studies on the colour of meat from young bulls. *Journal of the Science of Food and Agriculture* **22**, 637–647

MacGRATH, W.S., VANDER NOOT, G.W., GILBREATH, R.L. and FISHER, H. (1968). Influence of environmental temperature and dietary fat on backfat composition of swine. *Journal of Nutrition* **96**, 461–466

MALMFORS, B., LUNDSTRÖM, K. and HANSSON, I. (1978). Fatty acid composition of porcine backfat and muscle lipids as affected by sex, weight and anatomical location. *Swedish Journal of Agricultural Research* **8**, 25–38

MARCHELLO, J.A., CRAMER, D.A. and MILLER, L.G. (1967). Effects of ambient temperature on certain ovine fat characteristics. *Journal of Animal Science* **26**, 294–297

MARTIN, A.H., FREDEEN, H.T., WEISS, G.M. and CARSON, R.B. (1972). Distribution and composition of porcine carcass fat. *Journal of Animal Science* **35**, 534–541

MOODY, W.G., ENSER, M.B., WOOD, J.D., RESTALL, D.J. and LISTER, D. (1978). Composition of fat and muscle development in Pietrain and Large White piglets. *Journal of Animal Science* **46**, 618–632

MOORE, J.H., CHRISTIE, W.W., BRAUDE, R. and MITCHELL, K.G. (1969). The effect of dietary copper on the fatty acid composition and physical properties of pig adipose tissues. *British Journal of Nutrition* **23**, 281–287

MORGAN, J.H.L. and EVERITT, G.C. (1969). Yellow fat colour in cattle. *New Zealand Journal of Agricultural Science* **4**, 10–18

MOTTRAM, D.S. and EDWARDS, R.A. (1983). The role of triglycerides and phospholipids in the aroma of cooked meat. *Journal of the Science of Food and Agriculture* **34**, 517–522

MOULTON, C.R. and TROWBRIDGE, P.F. (1909). Composition of the fat of beef animals on different planes of nutrition. *Journal of Industrial Engineering Chemistry* **1**, 761–768

ØRSKOV, E.R., DUNCAN, W.R.H. and CARNIE, C.A. (1975). Cereal processing and food utilization by sheep. 3. The effect of replacing whole barley by whole oats on food utilization and firmness and composition of subcutaneous fat. *Animal Production* **21**, 51–58

ØRSKOV, E.R., FRASER, C. and GORDON, J.G. (1974). Effect of processing of cereals on rumen fermentation, digestibility, rumination time, and firmness of subcutaneous fat in lambs. *British Journal of Nutrition* **32**, 59–69

PATTERSON, D.S.P. (1965). The association between depot fat mobilization and the presence of xanthophyll in the plasma of normal sheep. *Journal of Agricultural Science* **64**, 273–278

PLAAS, H.A.K. and CRYER, A. (1980). The isolation and characterisation of a proposed adipocyte precursor cell type from bovine subcutaneous white adipose tissue. *Journal of Developmental Physiology* **2**, 275–289

PYLE, C.A., BASS, J.J., DUGANZICH, D.M. and PAYNE, E. (1977). The fatty acid composition of the subcutaneous brisket fat from steers obtained from Angus cows mated to 13 different beef breeds. *Journal of Agricultural Science* **89**, 571–574

RHODES, D.N. (1971). A comparison of the quality of meat from lambs reared intensively indoors and conventionally on grass. *Journal of the Science of Food and Agriculture* **22**, 667–668

RHODES, D.N. (1973). *Fatness of Beef and Eating Quality: Meat Research Institute Memorandum No. 15*. Langford, Bristol, Meat Research Institute

ROBERTS, W.K. (1966). Effects of diet, degree of fatness, and sex upon fatty acid composition of cattle tissues. *Canadian Journal of Animal Science* **46**, 181–190

RUMSEY, T.S., OLTJEN, R.R., BOVARD, K.P. and PRIODE, B.M. (1972). Influence of widely diverse finishing regimes and breeding on depot fat composition in beef cattle. *Journal of Animal Science* **35**, 1069–1075

SCOTT, R.A., CORNELIUS, S.G. and MERSMANN, H.J. (1981). Fatty acid composition of adipose tissue from lean and obese swine. *Journal of Animal Science* **53**, 977–981

SCOTT, T.W., COOK, L.J. and MILLS, S.C. (1971). Protection of dietary polyunsaturated fatty acids against microbial hydrogenation in ruminants. *Journal of the American Oil Chemists Society* **48**, 358–364

SMITHARD, R.R., SMITH, W.C. and ELLIS, M. (1980). A note on the fatty acid composition of backfat from boars in comparison with gilts and barrows. *Animal Production* **31**, 217–219

SUMIDA, D.M., VOGT, D.W., COBB, E.H., IWANGA, I.I. and REIMER, D. (1972). Effect of breed type and feeding regime on fatty acid composition of certain bovine tissues. *Journal of Animal Science* **35**, 1058–1063

THRALL, B.E. and CRAMER, D.A. (1971). Relationships of serum, muscle and subcutaneous lipids to beef carcass traits and flavour. *Journal of Food Science* **36**, 194–198

THOMPSON, E.H., ALLEN, C.E. and MEADE, R.J. (1973). Influence of copper on stearic acid desaturation and fatty acid composition in the pig. *Journal of Animal Science* **36**, 868–873

TICHENOR, D.A., KEMP, J.D., FOX, J.D., MOODY, W.G. and DEWEESE, W. (1970). Effect of slaughter weight and castration on ovine adipose fatty acids. *Journal of Animal Science* **31**, 671–675

TOVE, S.B. and MATRONE, G. (1962). Effect of purified diets on the fatty acid composition of sheep tallow. *Journal of Nutrition* **76**, 271–277

TOVE, S.B. and MOCHRIE, R.D. (1963). Effect of dietary and injected fat on the fatty acid composition of bovine depot fat and milk fat. *Journal of Dairy Science* **46**, 686–689

TRUSCOTT, T.G. (1980). *A Study of the Relationships Between Fat Partition and Metabolism in Hereford and Friesian Steers.* PhD thesis, University of Bristol

TRUSCOTT, T.G., WOOD, J.D. and DENNY, H.R. (1983). Fat deposition in Hereford and Friesian steers. 2. Cellular development of the major fat depots. *Journal of Agricultural Science* **100**, 271–276

TRUSCOTT, T.G., WOOD, J.D. and MacFIE, H.J.H. (1983). Fat deposition in Hereford and Friesian steers. 1. Body composition and partitioning of fat between depots. *Journal of Agricultural Science* **100**, 257–270

TULLIS, J.B. and WHITTEMORE, C.T. (1980). Digestibility of a fully hydrogenated tallow for growing pigs. *Animal Feed Science and Technology* **5**, 87–91

VICKERY, J.R. (1977). *Influence on Carcass Lipids of the Condition of Cows at Slaughter. Division of Food Research and Technology, paper No. 42.* Melbourne, Australia, CSIRO

VILLEGAS, F.J., HEDRICK, H.B., VEUM, T.L., McFATE, K.L. and BAILEY, M.E. (1973). Effect of diet and breed on fatty acid composition of porcine adipose tissue. *Journal of Animal Science* **36**, 663–668

VOLD, E. (1976). Production of raw sausage with backfat from carcasses of genetically fat or genetically lean pigs. *NINF – informasjon 1976* (2), 1–35

WESTERLING, D.B. and HEDRICK, H.B. (1979). Fatty acid composition of bovine lipids as influenced by diet, sex and anatomical location and relationship to sensory characteristics. *Journal of Animal Science* **48**, 1343–1348

WOOD, J.D. and ENSER, M. (1982). Comparison of boars and castrates for bacon production. 2. Composition of muscle and subcutaneous fat, and changes in side weight during curing. *Animal Production* **35**, 65–74

WOOD, J.D., ENSER, M.B. and RESTALL, D.J. (1978). The cellularity of backfat in growing pigs and its relationship with carcass composition. *Animal Production* **27**, 1–10

WOOD, J.D., ENSER, M.B., MacFIE, H.J.H., SMITH, W.C., CHADWICK, J.P., ELLIS, M. and LAIRD, R. (1978). Fatty acid composition of backfat in Large White pigs selected for low backfat thickness. *Meat Science* **2**, 289–300

WOOD, J.D., JONES, R.C.D., BAYNTUN, J.A. and DRANSFIELD, E. (1983a). Fat quality in boars. *Animal Production* **36**, 517 (abstr.)

WOOD, J.D., WHELEHAN, O.P., ELLIS, M., SMITH, W.C. and LAIRD, R. (1983b). Effects of selection for low backfat thickness in pigs on the sites of tissue deposition in the body. *Animal Production* **36**, 389–397

21

FAT DEPOSITION IN BROILERS

C. FISHER
Agricultural and Food Research Council's Poultry Research Centre, Roslin, Midlothian EH25 9PS, Scotland

Introduction

A typical broiler chicken, when killed at 49 days of age, weighs about 2.4 kg if male and 2.0 kg if female. The body of the former will contain about 140 g/kg lipid and the latter 160 g/kg, some 300 g total lipid in each sex. The content of fat tissue will be slightly higher because it has a lipid content of 800–900 g/kg (Evans, 1977). These typical birds will have eaten about 250 g dietary fat providing 13% of their total energy intake. The level of dietary fat used is determined mainly by economic considerations.

In recent years the view has been frequently expressed that broilers have become fatter and are now 'too fat'. This, at first sight, is surprising since over the same period, broilers have been slaughtered at an earlier and earlier stage of maturity. The disadvantages of excessive fat growth in broilers have been reviewed recently by Jensen (1982). These include consumer resistance to fatty foods and to losses in food preparation, losses in processing and consequential changes in the composition of poultry offal meals, increased cleaning costs in factories, pollution problems in waste water disposal, the presence of fat on the skin of birds and concern about taint, taste and keeping quality. As a contribution to liveweight gain, fat growth is energetically inefficient and this may be exacerbated by the fact that birds with higher skin fat contents have fewer pin feathers and more blemishes on the carcass (Quarles *et al.*, 1968). Concern about these issues has led to a remarkable increase in research effort in recent years.

In spite of this concern, however, the poultry industry in the United Kingdom has not, and still does not, use differential payments to influence the quality of the product (other than overt downgrading). Even in countries such as Canada, where more sophisticated grading and payment systems are used, the dual emphasis on both conformation of the carcass and 'finish' tends to encourage as much as it discourages fatness (Moran, 1977). Indeed, there is a great deal of ambivalence about what is the desired degree of fatness and, certainly, with a flat-rate payment based on liveweight, the interests of the producer, processor and consumer are unlikely to concur.

In both research and practice most attention is focused on the fat in the abdominal cavity. This is the largest discrete fat depot, it is readily removed and measured and it constitutes an unambiguous loss to both processor and consumer. The amount of abdominal fat in a growing bird is very variable. Among birds within-strains coefficients of variation of 30–35% are typical (Becker *et al.*, 1979) although non-normal distributions, with an excess of very fat birds, may be found (R.M. Gous, unpublished results). Between flocks of modern broilers grown to 56 days under 'normal' commercial conditions abdominal fat varies from about 3.4 or 4.1% of bodyweight in males and females respectively (Leeson and Summers, 1980) to 1.1 or 1.9% (unpublished data, Poultry Research Centre). Such variation is not unexpected because if depot-fat growth is considered simply to be a function of energy consumption in excess of heat loss and energy deposition in the non-fat body, then any factor, in the bird or in its environment, which modifies food intake, non-fat growth or heat loss, clearly has a potential to modify fat growth. The integration and quantification of all such factors in a satisfactory theory is an essential step in the understanding of this topic but has yet to be achieved. A knowledge of fat metabolism and its control is also useful for describing the response to environmental or nutritional circumstances and for understanding the long-term consequences of genetic selection against fatness.

One of the problems of formulating an integrative theory of fat growth is that it requires a motive or purpose to be attributed to the bird. Superficially, there is no obvious reason why a chicken should consume energy excessively and grow fat. It seems unlikely to be a residue of any seasonal behaviour patterns associated with variations in food supply or migration (Bartov, Jensen and Veltmann, 1980) and is equally unlikely to contribute anything to reproductive fitness. The often-quoted view that fatness has had a selective advantage in modern breeding programmes is discussed below. The adverse effects of excessive fatness on fertility in mature animals have been offset in broiler breeders by the use of restricted feeding techniques during growth and development. If this was not done, then presumably selection for growth rate would have had different consequences for fatness and it is interesting that, in the turkey, where fertility is ensured by the use of artificial insemination, equally successful selection for growth has not led to excessive fatness. There may, of course, be other reasons for this.

The alternative view—that fatness is a consequence of the way that birds are kept, and in particular, fed—has been most clearly stated by Emmans (1981) in defining a set of relationships and conversion ratios which together form the Edinburgh Growth Model. This is based on the idea that birds strive to achieve their potential for non-fat growth and will only fatten, above a minimum level, if the environment obliges them to do so in seeking to reach this objective. Although this formulation of ideas is recent, and entirely original, many of the basic propositions of Emmans' theory have been accepted for many years.

Methods of describing fat growth in the broiler

In reviewing the literature a considerable problem has arisen from the variety of (often) poorly defined methods that are used, especially in dissection, to describe fat growth in broilers. There is no authoritative description or nomenclature for the fat depots and, indeed, these are not discrete or bounded by membranes and therefore such order may not be achievable. Lucas (1979) in the authoritative *Nomina Anatomica Avium* lists only one reference to the *corpora adiposa* (Liebelt and Eastlick, 1954) which names the subcutaneous fat bodies in the chicken embryo. The abdominal or retroperitoneal depot, which is frequently called the abdominal fat pad, lies between the abdominal muscles and the intestines and extends within the ischium to surround the cloaca and bursa of Fabricius. It normally, but not always (e.g. Lilburn *et al.*, 1982a) includes the fat surrounding the gizzard. Other depots have sometimes been identified from their association with a muscle, e.g. the sartorius depot. Confusion can also arise from the comparison of dissected fat tissue and chemically determined lipid. Langslow and Lewis (1974) found that wet tissue from around the gizzard contained 613 and 695 mg/g triglyceride at 4 and 8 weeks of age respectively, rising to over 800 mg/g in the drier tissues from mature birds. On a dry basis, triglyceride accounted for more than 900 mg/kg tissue. The tissue from females was found to contain more triglyceride than that from males. These data were for Rhode Island Red chickens; in the duck, Evans (1969) found up to 920 mg triglyceride/g wet tissue and such higher values may be more appropriate for modern, fatter chickens.

This issue is raised here only in the context of comparing different research reports, but it is clear that more standardized procedures and nomenclature need to be established and that further studies on the triglyceride content of fat tissues from broilers are required. There are also problems with fat extraction. Langslow and Lewis (1972) found that the widely used Folch chloroform–methanol extraction recovered only 88% of triglyceride whereas a comparison of petroleum-ether and chloroform extractions revealed a difference of 21 g/kg lipid (7.3% of the mean) in the analysis of duck carcasses (Clayton *et al.*, 1974). The final source of confusion in comparing different reports is that fat is frequently expressed as a percentage, but variously using live bodyweight, starved bodyweight, plucked weight, 'New York dressed' weight (minus feathers, blood, head and feet) or carcass weight as the reference.

No proper solution to these problems has been found and they have frequently been ignored. In order to facilitate quantitative comparison of experiments in which abdominal fat weight or total body lipid have been reported, the concept of calculated storage lipid has been used. The total body lipid (TL) is considered in two compartments, storage lipid (SL) and non-storage lipid (NSL). NSL is defined as the body lipid content when abdominal fat is zero and it is assumed to represent mainly the essential or structural lipid components of the body, although estimates obtained seem rather higher than this. Abdominal fat is assumed to account for 22.5% of SL while extrapolation of the regression of TL on abdominal fat suggests that in slaughter weight broilers (about 2 kg liveweight) 70 g lipid (NSL) is found at zero abdominal fat (range 50–90 g). This is about 4% of bodyweight.

In declining order of confidence, therefore, storage lipid in broilers weighing about 2 kg may be estimated as

SL = abdominal fat/0.225 (weight or percentage bodyweight)
SL = $TL - 70$ (weight)
SL = $TL - 4$ (percentage bodyweight)

These can be only approximations. Hood (1982) found a curvilinear relationship between abdominal fat and TL and therefore extrapolation of straight lines would give different intercepts in birds of varying fatness. Furthermore, the partition of SL between abdominal fat and other depots may be affected by genetics or nutrition. However, the calculations do allow some otherwise irreconcilable reports to be compared quantitatively and the distinction between storage fat and non-storage fat is useful when trying to predict the effects of treatments on total body fat levels.

Fatness in broilers

Insufficient comparable and chronologically ordered reports have been found to test the argument that fatness in broilers has increased with time. Osbaldiston (1967) grew commercial male broilers to 56 d of age and found 72 g/kg lipid in the whole body. More recently Holsheimer (1975) found 130 and 138 g/kg lipid in males and females at 42 d; Becker and Spencer (unpublished experimental report) found 134 and 151 g/kg lipid in males and females at 55 d; Becker et al. (1979) found 104 and 122 g/kg lipid in males and females at 58 d, and Leeson and Summers (1980) found 211 and 233 g/kg lipid in males and females at 56 d of age. These reports show a general trend but also differ with respect to many factors which will influence fat growth. There has undoubtedly been a transformation in broiler performance over the same period. In 1971, commercial standards in the UK were for 1.81 kg liveweight at 58 d of age with a feed conversion ratio of 2.20: by 1983 the figures were 2.09 kg, 50 d and 2.15 (National Farmers Union, 1983). The relatively small change in feed conversion ratio which has accompanied the large change in growth rate may be seen as indirect evidence of increased fat deposition.

It is a widely held view that the modern broiler is fat because selection for increased juvenile bodyweight has increased 'appetite' and this results in an 'excessive' consumption of energy (Lin, 1982). In a recent, and perceptive, analysis McCarthy and Siegel (1983) have assembled considerable evidence in support of the first part of this hypothesis. Selection for growth does appear to have exploited differences in the rate of food consumption with the capacity of the gastrointestinal tract becoming the factor which limits intake in the broiler, rather than the higher-level hypothalamic control mechanisms operating in smaller strains (Nir et al., 1978; Burkhart et al., 1983). The efficiency of food (energy) utilization has been changed to a much smaller extent, if at all (Siegel and Wisman, 1966; Proudman, Mellen and Anderson, 1970; Owens, Siegel and van Krey, 1971; Pym and Nicholls, 1979). Lilburn et al. (1982a) found that pair-feeding of lean and obese birds accounted entirely for the differences in fatness, thus supporting the view that food intake control is largely involved. Pym and Solvyns (1979), however, found that selection for increased weight gain had very different consequences for body composition when compared with selection for increased food intake.

The direct evidence that selection for growth in chickens increases body fat content is equivocal and most has been collected on strains which are considerably smaller than commercial broilers. In general, however, it does not support the view that fatness is an inevitable consequence of the genetic improvement in growth. Both Proudman, Mellen and Anderson (1970) and Burgener, Cherry and Siegel (1981) report data on White Plymouth Rock lines selected for high and low bodyweight. The former authors found bodyweights of 672 and 345 g at 44 d in their lines and respective lipid contents of 69.4 and 47.9 g/kg. Burgener, Cherry and Siegel (1981) had larger birds, 1363 and 571 g at 63 d with identical carcass lipid contents, 128 and 130 g/kg. After five generations of selection for weight gain, Pym and Solvyns (1979) obtained chickens which weighed 1428 g at 63 d in comparison with 1158 g in an unselected control line. The lipid contents of birds from the same two lines were 105.6 and 104.7 g/kg respectively. When birds from both lines were killed at the same weight, about 1350 g, the lipid contents were also very similar, 105.6 and 104.7 g/kg respectively. In this same experiment Pym and Solvyns (1979) showed that direct selection for high food consumption increased fatness, to 134.0 g/kg lipid at 63 d, whereas selection for improved feed efficiency reduced it to 82.5 g/kg lipid at the same age. Chambers, Gavora and Fortin (1981) found that 14 generations of selection in meat strain chickens increased the plucked and bled carcass weight of males at 47 d from 717 to 1088 g and of females from 635 to 914 g. As a consequence, abdominal fat increased from 1.2 and 1.6% of carcass weight to 1.4 and 1.7% respectively in males and females. Commercial broilers reared at the same time were bigger and fatter. The conclusion to be drawn from the only results reported from successful commercial selection is similar. Gristwood (private communication) has shown that 10 years of selection has considerably increased liveweight at 56 d whereas, in birds of the same weight, abdominal fat has decreased slightly.

From observations such as these, and following the hypothesis elaborated by Hayes and McCarthy (1976) to explain their results obtained with mice, McCarthy and Siegel (1983) have proposed that selection for increased weight in chickens has kept the proportion of fat in the body constant up to the age at which selection is made, with a rapid increase thereafter. This is consistent with most of the evidence reviewed here and it must be concluded that the developmental pattern of fat deposition has been changed little, if at all, by selection for growth rate. The modern broiler chicken is fatter than its forebears at a given age simply because it grows faster, but is not fatter at a given weight or stage of maturity.

Notwithstanding this conclusion, there is evidence of both between-strain (e.g. Griffiths, Leeson and Summers, 1978; van Middelkoop, Kuit and Zegwaard, 1977) and within-strain (Ricard, 1974; Becker, 1978) genetic variation in the fat content of broilers, and selection against abdominal fat should be effective (Becker, 1978), although possibly at the expense of carcass finish. Such selection can be based on directly observed abdominal fat levels (Leclercq, Blum and Boyer, 1980) but is facilitated by methods for determining fatness in the live bird. This can be done in a number of ways, including biopsy of the back skin (Guttridge, 1937), but the most useful are the abdominal caliper (Pym and Thompson, 1980) and the measurement of plasma very-low-density lipoproteins, VLDL (Griffin, Whitehead and Broadbent, 1982). The abdominal caliper measures the relative thickness of the fat pad between the inside of the cloaca and the outside of the body wall. Correlations as high as 0.80 between this measurement and abdominal fat levels have been found (Pym and Thompson, 1980) although there is a danger of misreading due to operator error and some very low correlations have been reported (Gyles, Maeza and Goodwin, 1982).

The correlation between plasma VLDL and total carcass fatness may be as high as 0.70 when a low-fat diet is used and suitable precautions are taken to exclude the effects of starvation on VLDL (Whitehead and Griffin, 1982). A simplified assay for VLDL has been described (Griffin and Whitehead, 1982) to facilitate practical application of this method. The alternative approach is to select indirectly for reduced carcass fat as a correlated response to selection for improved feed conversion efficiency (Pym and Solvyns, 1979) or to selection for weight gain on a fixed and restricted amount of feed (Eitan, Agursky and Soller, 1983).

Three experiments have demonstrated the effectiveness of selection against body or abdominal fat levels. Leclercq, Blum and Boyer (1980) based their selection on the direct measurement of abdominal fat and used sib selection. Lilburn *et al.* (1982a) report on eight generations of selection for abdominal fat size as determined by hand palpation in mature females. At the Poultry Research Centre work is in progress to assess the correlated response in carcass fatness to selection for high and low plasma-VLDL levels (Whitehead and Griffin, 1984). All of this work shows that selection is effective but the longer-term effects on other production characteristics remain to be determined and the metabolic consequences of the various approaches used have yet to be fully worked out.

Lilburn *et al.* (1982a,b) show that lines differentiated by fatness at maturity differ in both growth rate and obesity by 6 weeks of age. Differences in growth rate appeared up to 8 weeks prior to differences in food intake, leading the authors to argue that both hyperphagia and the control of metabolism were modified by selection. The latter was reflected in increased rates of hepatic lipogenesis (*in vitro* incorporation of ^3H_2O) in the obese birds and in an increased total capacity (but not rate) of fatty acid esterification in the adipose tissue (Lilburn *et al.*, 1982b). However this must be an incomplete description of the effects of selection because increased diversion of dietary energy into lipid would lead to smaller liveweight gains in the obese birds and not larger ones as observed; in fact, calculations show that the selected obese birds also had larger fat-free bodies than the lean ones.

In the French experiments divergent selection based on the ratio of abdominal fat weight to total bodyweight in male progeny at 63 d of age produced a marked differentiation between fat (FL) and lean (LL) lines (Leclercq, Blum and Boyer, 1980). Although there was no genetic control, selection for fatness appeared to be more successful than selection for leanness and the divergence between FL and LL in the F2 generation was greater in males than in females. This sex effect had disappeared by the F4 generation (Simon and Leclercq, 1982). In the F4 generation 63-day-old males of the FL line contained 137 g/kg lipid and of the LL line, 94.9 g/kg. Similar differences were found in females, being maximal in both sexes at about 63 d and then remaining constant up to 175 d in females. The differentiation of the two lines appeared to be entirely metabolic in origin, for no excess weight gain, no hyperphagia or decrease in body temperature or maintenance were observed in the FL chickens (Touchburn, Simon and Leclercq, 1981; Leclercq and Saadoun, 1983). Observations on plasma insulin and glucose levels and on glucose tolerance led to the hypothesis of an impairment in glucose–insulin balance in the FL chickens (Simon and Leclercq, 1982) which preferentially diverts intracellular utilization of nutrients towards lipid storage as a result of excessive insulin release. However, the primary mechanism remains unknown because, although differences between the lines in adiposity are seen at 2 weeks of age, a decreased glycaemia in FL chickens is found during the final stages of embryonic development.

Touchburn, Simon and Leclercq (1981) showed that both FL and LL birds responded to changes in protein:energy ratios in the expected way (see below). *Figure 21.1* summarizes these results and shows that linear

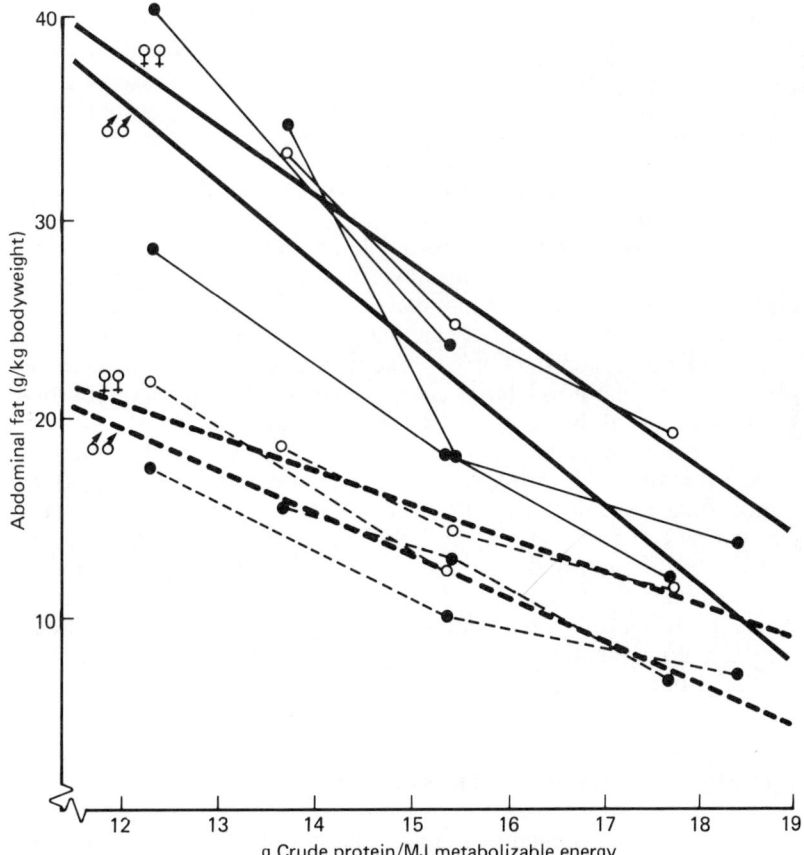

Figure 21.1 The effect of protein:energy ratio (g/MJ) on abdominal fat content (g/kg) of male (●) and female (○) broilers from lines selected for high (———) and low (- - - -) body fat content (Touchburn, Simon and Leclercq, 1981). The linear regressions, fitted within sex and line, intercept the *x*-axis at 21.0, 21.2, 23.3 and 24.4 g CP/MJ ME for fat males, lean males, fat females and lean females respectively

Table 21.1 PRELIMINARY RESULTS OF AN EXPERIMENT[a] INVESTIGATING THE EFFECTS OF CHOICE-FEEDING ON FATNESS OF BROILERS

Treatment[b]	Bodyweight[c] (g)	FCR[d]	Storage fat[e] (g/kg bodyweight)	
			M	F
Single feeds				
1 Control	2230	2.52	73	89
2 Control + protein	2309	2.55	51	75
3 Control + starch	1995	2.56	106	107
Choice feeding				
4 Control starter and balancer	2206	2.56	41	94
5 Control finisher and balancer	2271	2.46	45	81

[a] Commercial broilers reared on litter to 56 d, two pens of 40 male and female birds per treatment. Two-stage feeding programme, 0–28 d, 28–56 d. (R.M. Gous, G.C. Emmans, L. Broadbent and C. Fisher, unpublished results)
[b] Treatments 1–3 show the response, relative to control, of increasing (treatment 2) or decreasing (treatment 3) the protein:energy ratio. Choice-fed birds had free access from 0 to 56 d to both feeds. Balancer diet contained (g/kg) soyabean meal 400, herring meal 594.9, DL-methionine 2.6, vitamin premix 2.5.
[c] Bodyweight at 56 d, mixed sexes.
[d] Feed Conversion Ratio = Feed consumed/bodyweight gain, 0–56 d.
[e] Storage fat estimated as abdominal fat pad/0.225.

extrapolation of the regression lines suggests that zero abdominal fat would be reached in both lines with feeds providing about 22 g crude protein per MJ ME. It would be extremely interesting to see the results of feeding such nutrient levels to these genetically diverse stocks. From *Figure 21.1* a comparison can also be made of the relative merits of genetic selection and dietary manipulation as means of changing carcass fatness.

As the fatness of broilers is so readily modified by environmental variables it is not clear how a bird's characteristic fatness can be established so that genotypes can be compared across environments and across time. Emmans (private communication) has suggested that such a characteristic might be expressed with choice feeding, that is under conditions in which each individual can select both its own nutrient:energy ratios (normally protein:energy) and its own energy intake. The results in *Table 21.1* are taken from an unpublished experiment and suggest that when fatness is responsive to variations in protein:energy ratio, then males are characteristically less fat when choice-fed than when given a single feed. These effects are not seen in the females, which are slightly fatter than males when given a single feed but considerably fatter when choice fed.

The growth and distribution of fats in the broiler

As fat growth is environmentally labile, and because of the problems of technique discussed above, it is difficult to make generalizations about the accumulation and distribution of fat in the broiler body. Possibly the best, and most relevant, data are those of Håkansson, Eriksson and Svensson (1978a,b) and these will be considered here in some detail.

These authors grew broilers on three feeds differing nominally in energy content, and results for the two most extreme of these are considered here.

Diet H (high) provided 13.14 MJ/kg and 249 g/kg crude protein (CP) or 19 g CP/MJ. Diet L (low) provided 10.08 MJ/kg, 224 g/kg CP or 22.2 g CP/MJ. The fat contents of the feeds were 67 and 52 g/kg respectively, providing 15% of the total ME in diet H and 14% in diet L. The data refer to individual birds slaughtered serially over the weight range 0.5–4.0 kg; 20–90 d of age on diet H, 25–104 d on diet L.

The birds given feed H grew slightly faster than those on feed L (*Figure 21.2*), but they were considerably fatter. The difference in fat-free bodyweight is smaller than that in bodyweight. At a given age the total lipid content of a bird is two- to threefold higher on feed H compared with L; this difference is reduced but is still marked when the comparison is made at the same bodyweight (data not shown, but see *Figure 21.4*). There is some indication that the lipid content of the birds on diet H is declining in the older birds but more observations would be required to confirm this. Hood (1982) did not observe any decline in lipid content of broiler chickens reared to 24 weeks of age although stable high levels (*ca.* 300 g/kg in males, 400 g/kg in females) were reached after 18 weeks.

In the study being discussed, six components in the body lipid plus the abdominal fat tissue were separated (*Figure 21.3*). As the birds aged after about 20 days, there is very little change in the proportions of the total fat found in each of these and the figures for 41–60 d can be considered as typical for slaughter-weight broilers. About half of the total lipid is found in the meat and skin; of this about 85%, or 42% of the total, will be found in the skin. The visceral deposits account for almost twice as much of the total lipid on feed H (21.7%) as on feed L (11%). On both feeds, however, the skeletal fat is the second largest depot, accounting for 24% and 34% of the total lipid on feeds H and L respectively. The skeletons of the birds on the two feeds were almost identical in size (401 and 399 g) but those from birds on feed H contained more dry matter (422 and 412 g/kg) and more lipid (276 and 240 g/kg DM) than those on feed L. Total skeletal lipid was 46.7 and 39.5 g on feed H and L respectively. The other components contribute relatively small amounts of lipid although proportionately they are more important on feed L than H.

Fat growth can also be described by its allometric relationship with the growth of the fat-free body (Evans, 1977). In these data (*Figure 21.4*) the coefficient of allometry for feed H shows that fat and non-fat growth occur in constant proportion over the range of weights covered, whereas on feed L, although the birds are less fat, fat growth increased in relative terms with age. This difference can also be seen in *Figure 21.2*.

The allometric coefficients for total fat against fat-free body weights calculated from the data of Leeson and Summers (1980) were 1.302 for males (211 g/kg total lipid at 56 d) and 1.401 for females (233 g/kg total lipid). Values of 1.206 (males, 137 g/kg total lipid at 56 d) and 1.240 (females, 184 g/kg total lipid) were calculated for another flock of commercial birds (Poultry Research Centre, unpublished results). These figures suggest that fat is a relatively late-developing tissue and that it is normal for fatness to increase as growth progresses.

The allometric relationships for the lipid content of different body components on fat-free body weight are shown in *Table 21.2*. There is some uncertainty about the interpretation of allometric coefficients for

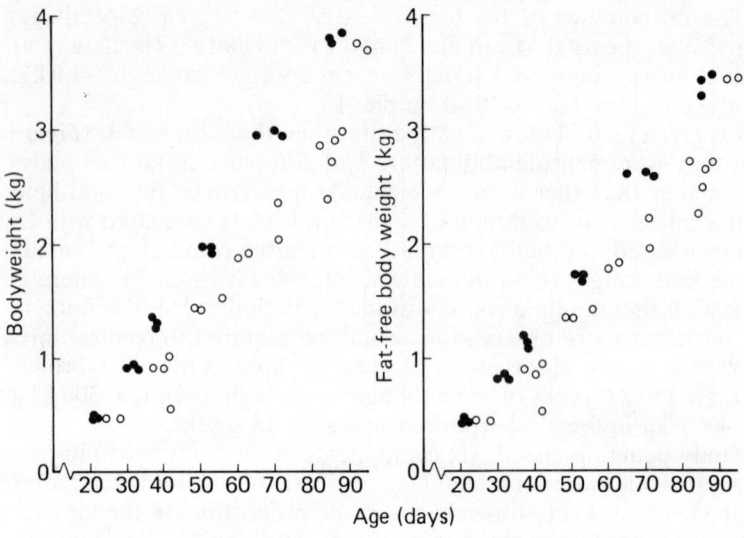
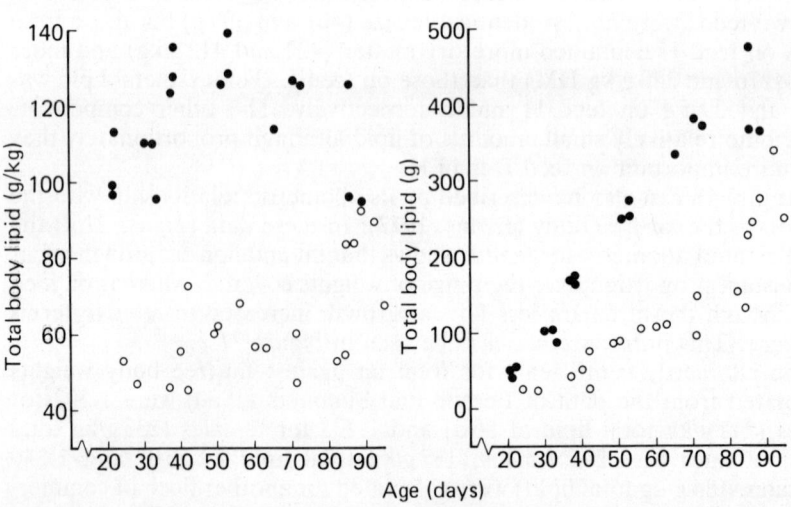

Figure 21.2 Growth and fat growth of male broilers given high (●) and low (○) energy feeds (from Håkansson, Ericksson and Svensson, 1974b)

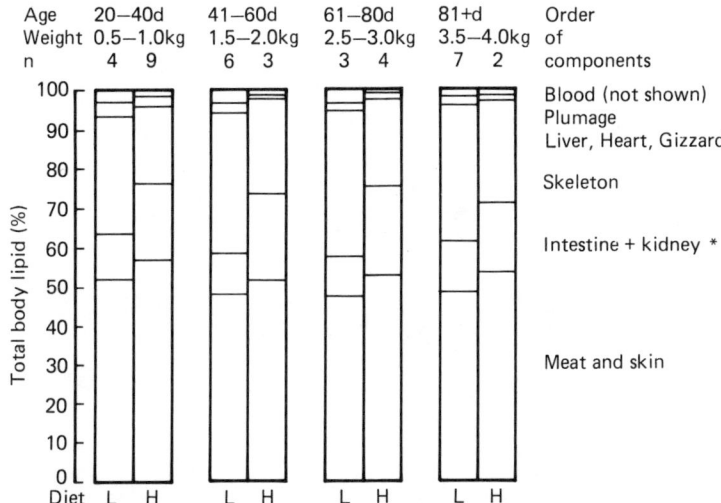

Figure 21.3 The distribution of total body lipid amongst six components in male broilers at different ages (from Håkansson, Ericksson and Svensson, 1974b). The weight ranges given are approximate. For details of low (L) and high (H) energy feeds see text. *This component contains lipid from the neck, lungs, kidneys, gastrointestinal tract, skin from leg, head, internal fat and other minor organs. The blood component, average 0.1% total lipid, cannot be shown

different body components (Parks, 1982, pp. 250–252) but they do provide a summary of trends. On feed H the lipid in the meat and skin grew, over the period 25–90 d, at the same rate as the remaining tissues whereas visceral and skeletal lipid increased slightly as growth advanced. The allometric coefficient for abdominal fat tissue, which forms a major proportion of the visceral lipid, was much higher. The lipid in the other components tended to increase more slowly than the rest of the body. A similar pattern was found on feed L except that the lipid in meat and skin tended to increase as growth advanced.

To some extent, of course, the relationship between fat growth in the different components and total growth are simply a reflection of the relationship between the component weights and the whole body. Therefore the allometric coefficients for component weights are also included in *Table 21.2*. Where the allometric coefficient for lipid in a component is higher than that for the whole, then fat content is increasing during growth and vice versa. The pattern found is much as expected. It can also be seen that there is a wider range of values in the allometric coefficients for fat as opposed to whole component growth. This confirms the labile nature of fat development.

As already noted, the abdominal fat is of most interest and in this experiment constituted 2.1 and 0.33% of bodyweight on feeds H and L respectively. From *Figure 21.3* it can be seen that it is the visceral fat, of which abdominal fat forms the major part, that is mainly affected by the two diets. If it is assumed that the abdominal fat tissue contains 900 g/kg lipid and is the same on the two feeds, then 14.7 and 4.8% of total body

448 Fat deposition in broilers

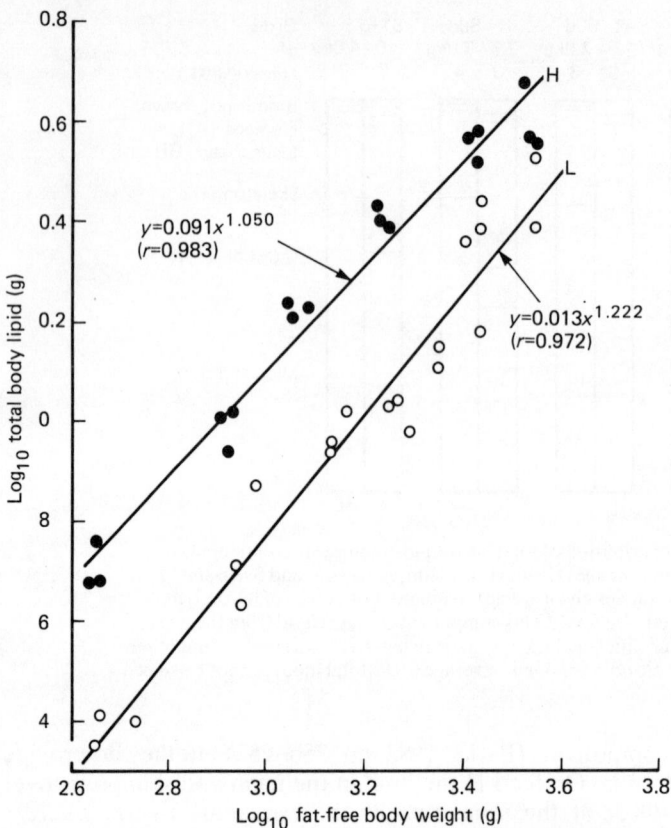

Figure 21.4 Allometric relationship between total body lipid and fat-free bodyweight. Data for male broilers grown on feeds containing high (H) or low (L) energy levels and killed between 20 and 105 days of age (from Håkansson, Ericksson and Svensson, 1974b)

lipid is accounted for by this tissue in birds weighing about 2 kg and given feeds H and L respectively. Similar calculations for modern broiler chickens show that about 15% of total body lipid is found in the abdominal fat tissue with a range from 10% to 20% (Becker *et al.*, 1979; Leeson and Summers, 1980; Chambers, Gavora and Fortin, 1981; Ehinger and Seemann, 1982). After about 40 d of age this proportion increased only slightly with age (Leeson and Summers, 1980; Ehinger and Seemann, 1982) but it is about 1% higher in females than in males. Between modern strains of broiler there are small but significant differences; for example, Ehinger and Seemann (1982) reported values of 15.5, 15.2, 16.1 and 15.7% for four commercial strains. Variations in dietary fat level (Deaton *et al.*, 1981) or protein:energy ratio (Ehinger and Seemann, 1982) cause differences of approximately similar magnitude.

The possibility of redistributing the fat in the body therefore appears to be rather limited, although it is encouraging that dietary means of reducing fatness seem to have a disproportionately large effect on the unwanted

Table 21.2 COEFFICIENTS[a] OF ALLOMETRIC RELATIONSHIPS BETWEEN DIFFERENT BODY COMPONENTS, THE FAT CONTENT OF THOSE COMPONENTS AND FAT-FREE BODYWEIGHT

Body component[b]	Coefficient of allometry			
	Diet H		Diet L	
	Whole component	Fat in component	Whole component	Fat in component
Meat and skin	1.096	1.002	1.101	1.196
Skeleton	0.902	1.177	0.944	1.275
Liver, heart and gizzard	0.737	0.686	0.713	0.856
Intestine + kidneys[c]	0.908	1.093	0.813	1.296
Abdominal fat	—	1.260	—	1.464
Blood	1.069	0.771	1.074	0.718
Feathers	1.201	0.848	1.170	0.933
Total body fat	—	1.050	—	1.222

[a] Calculated from data from Håkansson, Erickssson and Svensson (1978a,b) for individual male broilers killed between 20 and 90 d (diet H) and 25 and 104 d (diet L). For details of diets see text. Correlation coefficients all lie between 0.91 and 0.99.
[b] See *Figure 21.3* for definition of components.
[c] The fraction intestines + kidneys includes the abdominal fat pad.

abdominal tissue. Direct genetic selection against fat in particular depots may of course have larger effects, although the high correlations between abdominal and total fat (Delpech and Ricard, 1965; van Middelkoop, Kuit and Zegwaard, 1976; Becker *et al.*, 1979) suggest that the major effects of such selection will be to reduce fat in all parts of the body.

Although the growth of fat is ordered in birds fed continuously on a single diet, changes from lipogenic to lipolytic diets have very rapid effects on fatness. Khahil, Thomas and Combs (1968) produced lean and obese birds at 9 d of age with 18 and 240 g/kg body lipid by varying the dietary protein:energy ratio and then fed both groups normally. After 9 d these two groups contained 103 and 136 g/kg body lipid respectively. A further illustration of such rapid and reversible effects can be found in Bartov, Bornstein and Lipstein (1974).

The cellularity of fat tissue in chickens has been studied by Pfaff and Austic (1976), March and Hansen (1977), Ballam and March (1979) and Hood (1982). The interpretation of these studies is subject to all the technical qualifications that arise in such work (Kirtland and Gurr, 1979) and the relevance of fat cell number and size in the determination of adiposity remain unclear. Generally, it would be expected that cellularity is of less importance in birds than in animals in which the adipose tissue is the site of lipogenesis. Avian adipocytes are relatively deficient in normal ultrastructural elements and appear to be specialized for their main function of storing fat which has been synthesized elsewhere (Evans, 1977). The overall control of fatness in birds is expected to be found at the metabolic rather than at the tissue level.

An increase in fat cell number was found up to at least 6 weeks of age (March and Hansen, 1977) for the first 12–15 weeks (Pfaff and Austic, 1976) and up to, but not beyond, 14 weeks (Hood, 1982). In older birds, tissue growth is entirely due to cell hypertrophy (Ballam and March, 1979). March and Hansen (1977) found that the retroperitoneal fat in White

Fat deposition in broilers

Table 21.3 CHARACTERISTICS OF GROWTH AND FAT GROWTH IN LINES SELECTED FOR INCREASED BODYWEIGHT (W), FOOD INTAKE (F), FOOD EFFICIENCY (E) AND IN AN UNSELECTED CONTROL LINE (C). FROM HOOD AND PYM, 1982

	Line			
	W	F	E	C
Bodyweight, 63 d (g)	1787	1694	1566	1387
Food intake, relative[a] (35–63 d) (g)	+480	+720	0	0
Abdominal fat (g/kg BW)	21.7	30.9	16.6	23.5
Cell volume (nl)	0.388	0.524	0.310	0.477
No. cells/fat pad × 10^{-6}	100.3	100.1	82.3	68.9
Lipogenesis[b]	43.5	47.9	33.4	44.5
NADF-malate dehydrogenase activity[c]	61.7	70.1	55.4	61.1

[a] Approximate figures given in text of paper.
[b] Nanomoles glucose converted *in vitro* to lipid per hour per g liver.
[c] Nanomoles of substrate converted per mg soluble protein.

Leghorn chickens contained both fewer and smaller cells than tissue from broilers, while both Ballam and March (1979) and Hood (1982) found similar total numbers of cells in the mature abdominal fat pad of broiler chickens (349×10^6 and 270×10^6 respectively).

However, further studies by Hood and Pym (1982) on four selected lines showed that adipose tissue cellularity is only one of several factors involved in genetical differences in fatness. Some of their results are summarized in *Table 21.3*. Comparing lines W and C in this table shows large differences in growth and food intake in lines with similar levels of hepatic lipogenesis, malic enzyme and fatness. The two lines differ, however, in both cell number and cell size. Lines W and F are similar in growth rate but birds in line F eat more food and are fatter. This is reflected in the rate of lipogenesis and level of malic enzyme, but at the tissue level is entirely accounted for by differences in cell volume. In a more recent study of lipid metabolism in lines selected for high and low bodyweight, Calabotta *et al.* (1983) found that the higher fat content of the high-bodyweight birds was primarily the result of decreased lipolysis and not of enhanced lipogenesis. In fact, the low-liveweight lines appeared to have higher levels of both lipogenesis and lipolysis.

Dietary factors influencing fat formation and deposition

It is difficult to describe the effects of diet on fat deposition in broilers and, in particular, the effects of dietary fat, because of the large number of variables involved. Much of this difficulty is associated with food intake, with factors such as dietary bulk, palatability of ingredients and bird preferences all featuring as necessary—but very complex—parts of the observed responses. It is also important to note that nutritional experiments on free-feeding animals can only study the effects of substituting one ingredient, or one nutrient, for another. The direct effects of dietary variables cannot be isolated. This complexity means that at several places in this review exceptional or conflicting results are noted, without any basis for reconciliation being apparent. Some conflicts have been evaded by concentrating on recent studies involving broilers grown to normal slaughter weights so that their empirical interpretation might be relevant to commercial practice.

The general principles underlying the manipulation of carcass composition by dietary means have been understood for a long time and were clearly enunciated by Combs (1962). Since that time they have been demonstrated experimentally many times (for a review see Bartov, 1979) and yet there remains considerable difficulty in quantifying the expected responses. Indeed, if judged from some data collected in Germany in the mid-1970s, this understanding of the principles may not be reflected in uniformity of commercial results. Neupert and Hartfiel (1978) grew broilers to 48 d on eight commercial feeds and obtained carcass lipid contents ranging from 107 to 181 g/kg. These differences were more closely correlated with ME ($r = 0.69$) than with ratio of protein to energy ($r = 0.47$), but further details of the feeds are not available. The correlation amongst feeds, of fatness with growth, was zero but there was a large negative correlation with feed conversion ratio, the leanest birds having the highest FCR. These surprising results, if typical, suggest that there is a lot to be learned about the dietary control of fat growth in commercial broiler production.

Diet and lipogenesis

Several experiments in the chicken, and in other avian species, show that when rates of lipogenesis in the liver and adipose tissue are compared, the former is overwhelmingly the more important site of *de novo* fat synthesis. Fifteen minutes after administration of acetate-1-^{14}C, 69% of fatty acid-^{14}C was recovered from the liver and 16% from the adipose tissue. The remainder was accounted for by plasma fatty acids when only these three tissues were considered (O'Hea and Leveille, 1969). Following glucose-U-^{14}C administration the corresponding figures were 87, 6 and 8% respectively. The importance of extrahepatic lipogenesis is less clear. Yeh and Leveille (1972, 1973) suggest that more than half of the total lipogenesis may occur in tissues other than liver and attribute 5% to the intestine and 7% to the skin. More recently Nir and Lin (1982) have reported that 23% of total lipogenesis may occur in the skeleton. They also found 45% occurring in the liver and 6% and 7% in the intestine and skin respectively, which is in agreement with earlier studies. However, the use of acetate as substrate, as in all these experiments, does not distinguish between *de novo* fatty acid synthesis and chain elongation, and, at least in the rat, appears to underestimate hepatic and overestimate extrahepatic lipogenesis (Rosmos and Leveille, 1974). Calabotta *et al.* (1983) also demonstrated acetate incorporation into bone lipids of mature broiler males but were not able to demonstrate lipogenic enzyme activity in this or in adipose tissue preparations.

The effect of diet on hepatic lipogenesis has been widely investigated but the effects of nutrient levels and nutrient substitutions need to be carefully distinguished. When protein or fat are substituted for carbohydrate hepatic lipogenesis (acetate incorporation), malic enzyme and citrate cleavage enzyme are significantly reduced (Leveille et al., 1975). The authors have argued that these effects are specific because, when the substitutions are expressed as percentage of energy derived from carbohydrate, the rate of decline in lipogenesis caused by protein and fat are quantitatively different. Yeh and Leveille (1970) also showed that the acute administration of corn oil had an effect on lipogenesis within one hour, a rapidity which also encourages interpretation in terms of specific effects. However, Hillard, Lundin and Clarke (1980) have shown that, when carbohydrate intake is held constant, the addition of fat to the diet (with a concomitant increase in energy intake) has no effect on *in vivo* hepatic lipogenesis (^{3}H incorporation).

It is also possible to argue that responses in the rate of lipogenesis to dietary change are simply an effect of, and not a cause of, the dietary effect on fat deposition. Any variation in fatness, mediated in whatever way, must be associated with different rates of lipogenesis. For these reasons the effects of diet on fat growth must normally be determined by empirical experimentation, although the study of causation and of the mechanisms involved will allow such experiments to be interpreted with more confidence.

Nutrient:energy ratios

The dietary variable having the greatest effect on the fat content of broilers is the supply of nutrients per unit of energy. Protein:energy ratios (P:E) have been most widely investigated. Combs (1962) wrote

> "as the protein level is lowered in relation to the energy level in diets fed *ad libitum*, the total amount of food eaten and total energy intake is increased and feed efficiency is reduced where isocaloric rations are used. The positive correlation between C/P ratio [calorie:protein, here expressed as P:E] and body fat indicates that the chick is able to increase considerably its energy intake in an effort to obtain sufficient protein. When the protein deficiency stress can no longer be overcome through increased feed consumption, then differences in growth are noted . . . . In most cases the animal does not completely overcome the need for additional protein through increased food intake but the greater tendency to overconsume in energy is reflected in further increases in carcass fat content as the protein deficiency becomes more severe".

In this succinct description of what is observed, the mechanism of response is seen to be a specific attempt by the bird to overcome a nutrient deficiency by increasing its food consumption. Many others (e.g. Bartov, Bornstein and Lipstein, 1974; Lipstein, Budowski and Bornstein, 1975) have interpreted their data in this way, and Emmans (1981) has built into the Edinburgh Growth Model an assumption that nutrient deficiency, not necessarily protein deficiency, is the only cause of storage fat growth. However, the effects of high-protein diets, if not of all nutrients, may involve other mechanisms. As already noted, hepatic lipogenesis may be directly affected (Yeh and Leveille, 1969, 1972). The high energy cost of uric acid synthesis has also been invoked to explain the reduction in carcass fat on high protein diets. Griffiths, Leeson and Summers (1977a) showed that the addition of a low-quality protein—feather meal—was equally as effective in reducing abdominal fat-pad size as a high-quality protein—methionine-supplemented soyabean meal—under conditions in which neither produced a growth response. This observation has led to the evaluation of non-protein-N sources as lipolytic agents. In studies with free amino acid diets Velu, Scott and Baker (1972) found that glutamic acid, when increased from 100 to 160 g/kg, reduced carcass fat by about two percentage points and was not replaced in this role by essential amino acids.

The elements of the response to P:E are summarized in *Figure 21.5*. With increasing protein supply, growth increases and the food conversion

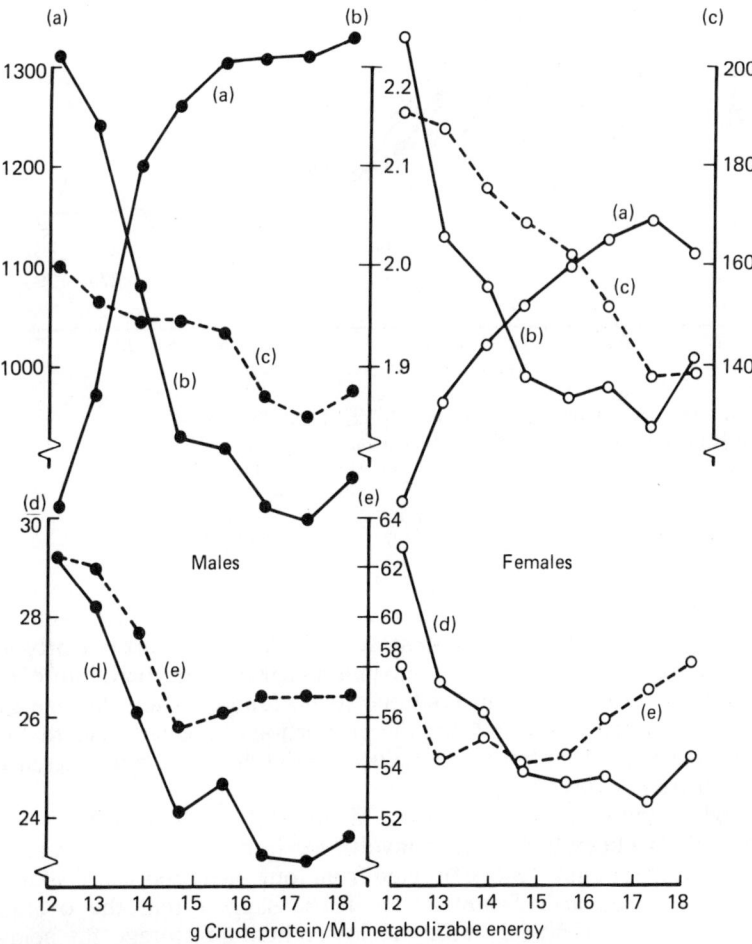

Figure 21.5 The effect of protein:energy ratio on broiler chickens, 0–42 days of age (from Holsheimer, 1975) (a) growth, g/bird; (b) food conversion ratio; (c) total body lipid, g/kg; (d) energy intake per unit bodyweight gain, kJ/g and (e) calculated heat loss, % ME intake

454 *Fat deposition in broilers*

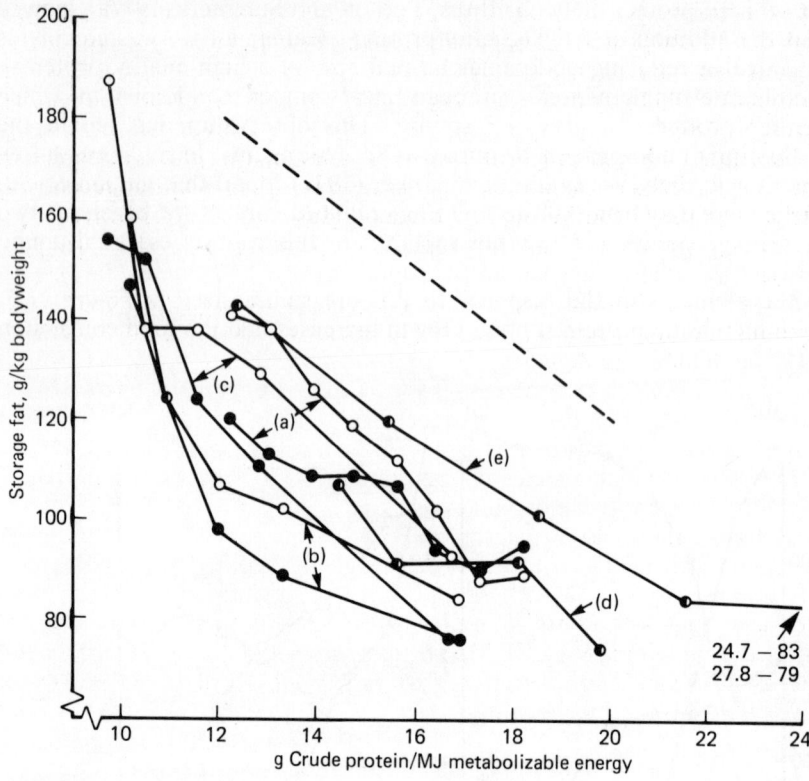

Figure 21.6 The calculated storage fat content (g/kg) of male (●), female (○) and mixed sex (◐) broilers at normal slaughter ages as affected by protein:energy ratio (g/MJ) of the diet. Data from (a) Holsheimer (1975); (b) and (c) Mabray and Waldroup (1981) (ME = 12.4 and 14.3 MJ/kg respectively); (d) Ehinger and Seemann (1982); (e) Jackson, Summers and Leeson (1982). The broken line shows the suggested 'average' rate of linear response for use in predictions

ratio (food/gain) and body-fat content decline. As shown most clearly in the males in this experiment, when maximum growth is reached, improvements in FCR and fat content continue to be found. When the energy balance is calculated it is found that the proportion of ingested ME that is retained increases at first but then declines, so that heat loss must increase at higher protein levels.

Although similar results have been demonstrated many times, few experiments have been found which involve modern broiler strains reared to normal slaughter ages. Data extracted from four such reports are shown in *Figure 21.6* (see also *Figure 21.1*). These suggest that the overall response may be curvilinear, with the lower limit to storage fat being 80–90 g/kg. At very low levels of protein supply, fatness increases at an accelerating rate, presumably because overall growth rate is declining rapidly. However, in spite of these apparent non-linear effects, the data in *Figure 21.6* have been interpreted for practical purposes by an 'average' straight-line response with a slope of -7.5 g/kg SF per g CP/MJ ME between 12 and 20 g/MJ. The results from Holsheimer (1975) in *Figure 21.6* suggest that the response is about twice as great in females as in males, but the results of Mabray and Waldroup (1981) were very consistent across the sexes.

The rather limited practical scope for manipulating the fat content of broilers by means of the protein:energy ratio was demonstrated recently in a trial from the Gleadthorpe Experimental Husbandry Farm (Bray, 1982). In feeds containing 12.4 MJ/kg ME a crude protein level of 181 g/kg (14.6 g/MJ) was found to support maximum growth. Abdominal fat was 3.8 and 4.9% of 'oven-ready' carcass weight in males and females respectively. Feeding additional protein, 18.9 g/MJ, reduced abdominal fat to 3.3 and 4.4% in males and females, the approximate values predicted by the response suggested here being 3.1 and 4.2% respectively. Total storage lipid in the body was therefore reduced by 22 g/kg.

In short-term studies with younger birds, a wider range of response has been reported. In their classic experiments with New Hampshire birds grown to 4 weeks of age, Donaldson, Combs and Romoser (1956) reduced body lipid content to 56 g/kg at an P:E of approximately 22 g/MJ when maximum growth was obtained with 15.7 g/MJ. The body lipid content at this latter ratio was about 87 g/kg. More recently Seaton *et al.* (1978) produced body lipid contents in 21-day-old broilers, ranging from 193 g/kg with a diet providing 8.1 g CP/MJ ME to 61 g/kg with 26.1 g/MJ. Kirchgessner, Roth-Maier and Gerum (1978) reduced body lipid to 90 g/kg in 5-week-old broilers by feeding a diet with 23 g CP/MJ ME from 3 to 5 weeks.

There are few data on the effects of nutrients other than crude protein on body fat growth. Indeed, there is an urgent need to test the extent to which the effects of nutrient supply on body composition are general, and therefore presumably mediated through compensatory intakes of food, or specific, and therefore involving direct metabolic effects. Hughes (1979) has discussed this problem in terms of specific appetites and concludes that these are widespread across nutrients but not apparently universal. The effects of amino-acid level present a confusing picture (see Bartov, 1979 and Boorman, 1979). Velu, Scott and Baker (1972) have tried to present a unifying hypothesis by suggesting that essential amino acids are lipogenic when added to diets with extreme deficiencies, but at higher levels become lipolytic. In their studies with crystalline amino acids the shift between the two states occurred at levels of provision lower than those required for maximum growth. However, Seaton *et al.* (1978), using diets based on intact protein, found that additions of lysine to a deficient basal diet produced small responses in growth but had no effects on body composition. Velu, Scott and Baker (1972) also suggest that individual amino acids have different effects, again when studied in the context of crystalline amino-acid diets.

In general there appears to be no entirely satisfactory theory to explain the effects of amino-acid supply on food intake (Boorman, 1979) and body composition. Over the range of deficiency and imbalance likely to be found in practical feeding, however, variations in essential amino acid:energy ratios can probably be considered in the same way as protein:energy ratios (Lipstein, Budowski and Bornstein, 1975).

Energy level, nutrient density and dietary fat level

These three dietary responses are considered together bcause they involve complex and interacting variations of diet which are difficult to disentangle in a single experiment. The discussion above on nutrient:energy ratios shows that manipulation of energy level *per se*, at constant levels of other nutrients, will produce effects on carcass fat which are predictable from these ratios. The simple substitution of dietary fat for carbohydrate will have this effect and is of little interest (e.g. Kubena *et al.*, 1974).

The manipulation of nutrient density, that is energy level at constant nutrient:energy ratios, and the addition of fat, with or without a change in nutrient density, have complex consequences for dietary composition and experimental findings tend to be inconsistent. For example, the alteration of nutrient density must involve one or more of the following substitutions: organic matter for inorganic matter; digestible for indigestible organic matter; starch for 'fibre'; fat for carbohydrate; digestible fat for less digestible fat. In feeds made with practical ingredients there is often an improvement in protein quality as nutrient density increases and the bulk (physical density) and palatability of the feed may change. Finally, feed-ingredient substitutions may themselves have specific effects. In view of this catalogue of changes it is not surprising that experimental findings are inconsistent and it is questionable whether a single scale, such as nutrient density, is sufficient to describe such a complex situation. The user of experimental findings must either select a part of the available information on the grounds of empirical suitability or use some method to calculate the 'average' rate of response (e.g. Fisher and Wilson, 1974).

Some components of these complex responses to diet can be considered separately. As already noted, the substitution of fat for carbohydrate reduces hepatic fatty acid synthesis and the activity of associated lipogenic enzymes. However, the work of Hillard, Lundin and Clarke (1980) shows that this is due to the reduction in carbohydrate and not to fat *per se*. When saturated fat was substituted into a semi-purified fat-free diet to replace 25% of the glucose energy, fatty acid synthesis (*in vivo* incorporation of ^3H_2O) declined from 3.85 to 2.07 μmol ^3H_2O per minute per g liver and the hepatic enzymes fatty acid synthetase and acetyl CoA carboxylase also declined. When the diet was supplemented with 20% energy as fat, and total energy intake increased, the rate of fatty acid synthesis and lipogenic enzyme levels remained constant.

When food intake was controlled, the isoenergetic (ME basis) substitution of carbohydrate by fat in the diet of 5-week-old broilers increased growth rate and energy retention but had no effect on body composition (Grimbergen, Stappers and Cornelissen, 1982). The 'extra-caloric' effect of fat which had to be invoked to explain these findings varied in two experiments from 25% of the supposed ME in birds producing approximately 100 g fat over 14 days to 11% when fat deposition was only 80 g. These effects were attributed to the direct deposition of fatty acids into the adipose tissue and the reduction in the energy cost of feeding. The 'extra-caloric' effect of fat may also reflect synergisms between fat and other dietary components; other fats due to complementary fatty acid compositions (Young and Garrett, 1963, and many others) and carbohydrates due to the effects of fat on feed passage rate (Mateos and Sell, 1981).

Variations in diet bulk or physical density undoubtedly contribute to inconsistencies between experiments, especially when cellulose is used as a major substituent in the feeds used. Unfortunately it is not possible to quantify this effect at present or to allow for it in the interpretation of experimental findings. From a limited review of the literature Fisher and Wilson (1974) concluded that growth would be adversely affected when ME concentration fell below 6–7 kJ/ml or about 11 MJ/kg in practical diets. However, as they noted, the extrusion of feed would be expected to modify these numbers in a way that remains unknown.

One aspect of diet palatability that should be mentioned is the preference for feeds containing higher levels of fat that was observed by Dale and Fuller (1979). When birds were offered a choice between isoenergetic and isonitrogenous feeds containing either 14.5% or 27.7% energy from fat, they selected approximately 60% of the latter in their total intake, irrespective of whether the feeds were offered as meal or pellets.

It is also a universal finding that growth rate increases with nutrient density. In an analysis of 160 experiments of this type, over 95% showed a positive response (Fisher and Wilson, 1974). For 78 of these experiments three or more energy levels were used and, among these, 29 of the fitted regressions were significant ($P < 0.05$) when analysed individually. The pooled estimate of response was 0.8 g/d per MJ/kg increase in ME at constant nutrient:energy ratios. In most of these experiments dietary fat was varied over at least part of the range of energy levels studied and the growth response obtained may be associated directly with this.

From the results of several experiments it has been concluded that increasing nutrient density by adding fat has no effect on carcass lipid content (Bartov, Bornstein and Lipstein, 1974; Fuller and Rendon, 1977). Others have concluded that nutrient density level does influence fatness (Farrell, 1974; Neupert and Hartfiel, 1978; Mabray and Waldroup, 1981). Bartov (1977) reports a result of each kind but refers to the one showing an increase in carcass fat as the 'exception' and the nil response as 'well-documented'. While this is true of his own work, it is not really an accurate reflection of the weight of evidence in the literature. These contrasts between the conclusions reached by different authors are exacerbated by the very variable nature of fat growth, and differences between diets which may be of practical importance are rejected as statistically non-significant. For example, although Fuller and Rendon (1977) concluded that there was no effect of nutrient density on fatness, in comparison with a basal feed containing 4.5% energy from fat, the addition of 100 or 200 g/kg fat (to provide 30% and 49.5% energy) increased carcass lipid by about 20 g/kg bodyweight. Similar comments apply to some of the observations of Bartov and Bornstein (1976a).

From an analysis of four published experiments, Fisher and Wilson (1974) concluded that carcass dry matter increased by 5.7 g/kg per MJ/kg increase in ME at constant nutrient:energy ratios, the fat content of the carcass dry matter increased by 16.7 g/kg, and the fat content of the carcass increased by 6.7 g/kg. In *Figure 21.7* these estimates of average response are compared with those observed in more recent experiments and, given the variation in the results, they still seem to provide a reasonable prediction for practical purposes. It can be seen in *Figure 21.7* that most of

458 Fat deposition in broilers

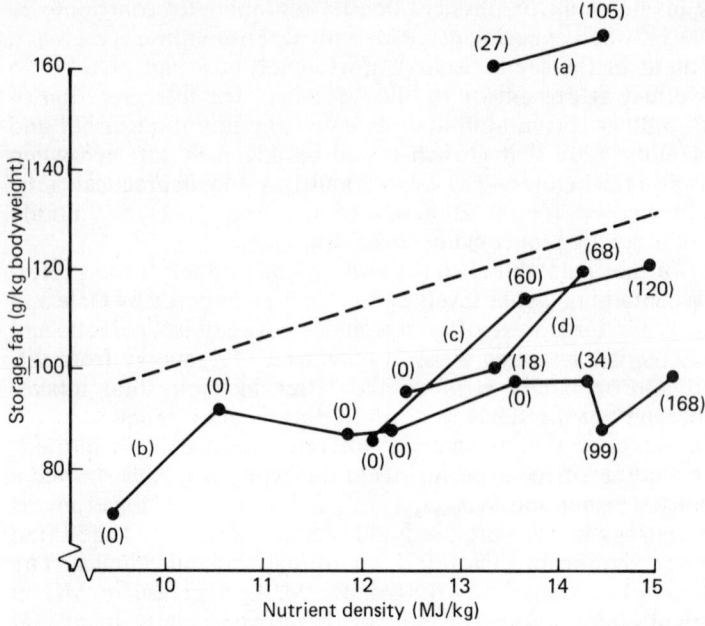

Figure 21.7 The calculated storage fat content of mixed sex broilers (g/kg) as affected by the nutrient density of the diet (MJ/kg). Data from (a) Coon, Becker and Spencer (1981); (b) Farrell (1974); (c) Neupert and Hartfiel (1978); (d) Mabray and Waldroup (1981). The figures in parentheses show the level of added fat, calculated as a weighted average of starter, rearer and finisher feeds in ref (d). The broken line is the predicted response suggested by Fisher and Wilson (1974)

the changes in dietary energy content were achieved by varying the levels of dietary fat, but the experiment reported by Farrell (1974) supports the conclusion that the response is independent of these substitutions and is found among diets containing no added fat.

It is believed that the situation shown in *Figure 21.7* would be reflected in practical feed formulation at this time. However it should be noted that in all of these experiments the diet compositions were calculated in terms of ME. There remains a possibility (Bartov, 1979) that as nutrient density increases the NE:ME ratio also increases when fat is used, and that these observations on carcass fatness can again be interpreted as a response to nutrient:energy ratio. This, on the other hand, would be hard to reconcile with the widespread increases in bodyweight found in these experiments.

A similar variability in experimental findings, and an identical difficulty over their interpretation, both arise when diets calculated to be isoenergetic on an ME basis but with different proportions of carbohydrate and fat are compared. In recent experiments Bartov, Bornstein and Lipstein (1974) and Griffiths, Leeson and Summers (1977b) have found no increase in carcass fatness as dietary fat levels increased at constant ME. In the latter experiment all the birds were usually lean, for reasons which are not apparent. Deaton *et al.* (1981), however, found that both abdominal fat and total body lipid levels increased significantly as dietary tallow increased from 40 to 100 g/kg, abdominal fat changing from 17 to 21 g/kg bodyweight in males and from 20 to 24 g/kg in females.

It is concluded that the trend, in practice, towards higher nutrient density feeds containing more added fat has contributed to the observed increase in body-fat levels. However, whether this is inevitable or a consequence of imperfect diet design, remains to be determined conclusively.

Food restriction and fat growth

The restriction of food or energy intake might facilitate the manipulation of body composition for two reasons: restriction during early growth might reduce fat cell hyperplasia and therefore limit the potential for growth of fat; restriction at the end of the growth period must allow manipulation of fat growth which might, in some circumstances, be economically beneficial. At present neither of these possibilities looks very promising for commercial application although they have been examined only in a rather superficial way.

The restriction of energy intake in the 0–3-week period had no significant effect on abdominal fat when measured at 8 weeks of age (Griffiths, Leeson and Summers, 1977a). Hargis and Creger (1980) compared various feeding programmes which varied in rather complex ways. They found that feeding no supplemental fat from 0 to 7 days of age, which also lowered dietary energy level, consistently reduced fatness in 49-day-old broilers. Considerably more basic information is obviously required before the potential of this sort of dietary manipulation is established. The same is true when restriction is achieved by reducing protein levels (Moran, 1980).

Three experiments have explored the restriction of food and nutrient intake during the final stages of production (Auckland and Fulton, 1973; Simon et al., 1978; Arafa et al., 1983). In the first two of these, intake of a single feed, and hence of all nutrients, was restricted; Arafa et al. (1983) adjusted the concentration of major nutrients, but, because of the procedures used, obtained more severe proportional restrictions than intended. Thus no experiment has been found in which energy intake alone was accurately restricted while maintaining intakes of protein and other nutrients.

From the results of these experiments there is no evidence that a slight restriction of food intake can be used to reduce storage fat growth without affecting the growth of the normal body. However, the proportion of fat in the body is reduced and a rather finely balanced economic analysis would be required to determine whether the procedure is worthwhile in different circumstances. The potential complexity of such an analysis has been shown by Arafa et al. (1983), who found that the level at which the benefit is assessed is of crucial importance. A restriction programme which reduced energy intake to 0.81 ad libitum from 46 to 56 days, reduced carcass lipid from 90 to 76 g/kg bodyweight. Live bodyweight was reduced by 149 g, shrunk bodyweight (12 hours' starvation) by 103 g, 'ready-to-cook' weight by 69 g and the yield of cooked meat by 2 g. Clearly, the

Fat deposition in broilers

interpretation of data of this sort depends entirely on the economic circumstances in which it is to be used.

Environmental temperature and fat growth

Kubena *et al.* (1972) and Bray (1983) have reported that carcass fat increases with environmental temperature. The effect is consistent across dietary variables and sex (*Figure 21.8*). If it is assumed to be linear,

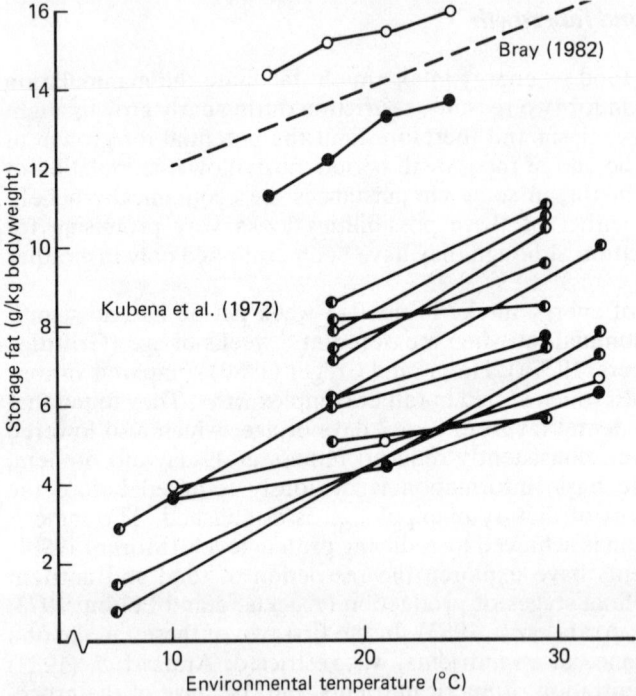

Figure 21.8 Calculated storage fat content of male (●), and female (○) and mixed sex (◐) broilers kept at different environmental temperatures. Data from Bray (1982) (in this case oven-ready bodyweight is the basis on which fat content is defined); Kubena *et al.* (1972) (data from four experiments involving various energy and protein levels). The broken line is the average linear response for use in predictions

estimates of the slope vary from 0.3 to 3.5 g/kg SF per °C, the average being 1.9 g/kg/°C ± 0.24 (SEM). Although this is a wide range the average value is in close agreement with the recent results of Bray (1983), see *Figure 21.8*, and can be used for predictive purposes.

Composition and stability of carcass lipid

The fatty acid composition of the body lipid, and hence its stability against oxidation, is largely determined by the relative importance of hepatic lipogenesis and of dietary fat as sources of deposited fatty acids and by the composition of the dietary fat. Changes in chain length and degree of saturation subsequent to deposition are not very important. The addition of dietary fat will influence both of these and they may act in the same, or in opposite, directions, depending on the composition of the diet.

The data in *Table 21.4* summarize the general pattern of findings. Endogenous fat, as measured on a fat-free diet, is composed overwhelmingly of palmitic and oleic acid with smaller amounts of palmitoleic and stearic acid. Linoleic and linolenic acid are not synthesized and the main effect of dietary fat is to introduce these into the tissues.

Table 21.4 EXAMPLES OF THE EFFECT OF DIET ON THE FATTY ACID COMPOSITION OF BODY LIPIDS

	\multicolumn{10}{c}{Fatty acid}	Index[d]
	12.0	14.0	16.0	16.1	18.0	18.1	18.2	18.3	20.1	22.1	
Endogenous fat[a]		9	251	64	84	581					12
E:P =											
0.536, 2% fat[b]			258	63	56	311	283	27			2.21
0.791, 2% fat			295	88	39	341	211	25			3.23
0.531, 5% fat			196	27	52	242	434	48			1.07
0.791, 5% fat			236	53	43	316	312	38			1.85
Maize soya diet[c]		8	257	74	62	386	201	6			
+ 5% coconut oil[c]	175	113	206	38	46	230	161	8			
+ 5% safflower oil[c]	2	5	144	22	45	206	566	9			

[a] Fatty acid composition (g/kg adipose tissue lipid) of lipid from birds given a fat-free diet (Bottino, Anderson and Reiser, 1970).
[b] Data from Bartov and Bornstein (1976a), expt 3, g/kg total fatty acids, E:P = energy to protein ratio, MJ/kg per protein %, added fat is acidulated soyabean soapstock.
[c] Data from Marion and Woodroof (1966), g/kg total fatty acids, practical (maize – soyabean) basal diet containing 240 g/kg protein.
[d] Ratio of 16:0 + 16:1 + 18:0 + 18:1 to 18:2 + 18:3, index of fatty acid synthesis (see text).

The results from Bartov and Bornstein (1976a) show how a change in protein:energy ratio produces alterations in fatty acid composition which reflect the relative contributions of endogenous and exogenous fat. As the level of protein decreases, the birds are fatter, linoleic acid decreases and palmitic, palmitoleic and oleic acids tend to increase. This pattern led Bartov and Bornstein (1976a) to suggest that the ratio of saturated plus monoenoic fatty acids to polyenoic acids could be used as an index of biosynthesis from carbohydrate when different nutrient:energy ratios were compared. This ratio, calculated for C16 and C18 acids only, is shown in *Table 21.4* and shows clearly the greater importance of biosynthesis on protein-deficient feeds.

The results of Bartov and Bornstein (1976a) also show that this pattern of change in fatty acid composition is not altered by the addition of an unsaturated dietary fat (acidulated soyabean soapstock) which, like the addition of safflower oil in the experiments of Marion and Woodroof (1966), markedly increase the linoleic acid content and the overall degree of unsaturation of the carcass fat.

When unusual dietary fatty acids are given—such as lauric and myristic from coconut oil, erucic and eicosenoic from rapeseed oil and long-chain

PUFAs from marine oils—these will appear in the carcass fat to some degree (*Table 21.4*). One exception to these generalizations is that the stearic acid level in body fat is little changed by the diet, even when high levels of tallow are given (Marion and Woodroof, 1966). This may reflect decreased synthesis or increased desaturation of this fatty acid in response to dietary level.

The practical consequences of differences in fatty acid composition of body lipids are not very apparent. In extreme cases, consumer reaction to soft fat might be important, but this is not widely discussed in the UK industry. The enigmatically named 'Oily Bird Syndrome', which is of particular importance in the USA, does not appear to be a reflection of changes in body lipid composition (Horvat, 1978) and a high environmental temperature is the most clearly reproducible factor in its occurrence (Jensen *et al.*, 1980). One feature of the condition, an easily torn skin, may reflect the influence of nutritional and environmental factors on connective tissue formation, and some studies on skin collagen levels have been reported (Smith *et al.*, 1976). While the aetiology of this condition remains to be firmly established, there is no reason to suppose that its occurrence should limit the use of dietary fats in N. Europe.

The changes in fatty acid composition shown in *Table 21.4* will influence the stability of body lipids against oxidation. The reduced level of saturation that accompanies the increased fatness of birds given narrow protein:energy ratios increases the stability of the lipid (Bartov and Bornstein, 1976b). For a range of oils of plant origin there is a reasonable correlation between the proportion of dienoic and more unsaturated fatty acids in the body lipids and stability, but the instability caused by fish oils is greater than expected from simple measures of unsaturation (Marion and Woodroof, 1966). Supplementation of the diet with tocopherol results in the storage of vitamin E in the body fat and improves stability (Bartov and Bornstein, 1976b). However, maximal effect of vitamin E is found when the fat is rather saturated (Bartov and Bornstein, 1977). Synthetic antioxidants can also have a protective role (Bartov and Bornstein, 1978; Bartov, 1979).

Summary and conclusions

There is little doubt that the fat content of broilers has increased in recent years and that the industry now has a great interest in reducing these levels. An early requirement is that this interest should be reflected in the price structure of the industry. The origins of the increases in fatness are obscure, but it appears that a genetically susceptible animal has been exposed to several nutritional and environmental variables which all act in the direction of tending to increase fatness. Whatever the cause, it is clear that genetic selection against fatness offers an effective way of ameliorating the problem. Until such genetic change has been wrought, nutritional and environmental factors can be manipulated to some extent to modify body composition. However, all of the techniques available to do this involve both gains and losses in productivity which can be resolved only in a cost-benefit analysis peculiar to each set of economic circumstances.

Wherever possible, estimates of the average rates of response have been given to facilitate such an analysis, although in all cases these estimates rest on empirical statistical interpretation of rather variable experimental results. The overwhelming need for further progress in this field is the development of an integrative theory of fat growth which brings together all of the variables affecting this important characteristic.

Although the use of dietary fat *per se* may have contributed to the increase in body lipid levels, it seems unlikely that considerations of body composition will lead to serious limitations on the use of this feed ingredient in N. Europe. The effects of dietary fat composition on the fatty acid make-up of body lipids is briefly reviewed, but again, within the limitations that apply in commercial circumstances, this is not a factor of great economic importance. Fat will continue to be an important source of feed energy in broiler production and the efforts of research should be directed at ensuring its optimum use for that purpose.

Acknowledgements

I am grateful to Mrs Jenny Arnott for help in preparing this paper and to Mr G.C. Emmans for many useful discussions and ideas.

References

ARAFA, A.S., BOONE, M.A., JANKY, D.M., WILSON, M.R., MILES, R.D. and HARMS, R.M. (1983). Energy restriction as a means of reducing fat pads in broilers. *Poultry Science* **62**, 314–320

AUCKLAND, J.N. and FULTON, R.B. (1973). Effect of feeding restricted amounts of a medium and a high protein diet during the finishing period on growth, fat deposition and feed efficiency of male and female broilers. *Journal of the Science of Food and Agriculture* **24**, 709–717

BALLAM, G.C. and MARCH, B.E. (1979). Adipocyte size and number in mature broiler-type female chickens subjected to dietary restriction during the growing period. *Poultry Science* **58**, 940–948

BARTOV, I. (1977). Pro- and antioxidants in the diets of broilers and their effects on carcass quality: copper, selenium and acidulated soybean-oil soapstock. *Poultry Science* **56**, 829–835

BARTOV, I. (1979). Nutritional factors affecting quantity and quality of carcass fat in chickens. *Federation Proceedings* **38**, 2627–2630

BARTOV, I. and BORNSTEIN, S. (1976a). Effects of degree of fatness in broilers on other carcass characteristics: relationship between fatness and the composition of carcass fat. *British Poultry Science* **17**, 17–27

BARTOV, I. and BORNSTEIN, S. (1976b). Effects of degree of fatness in broilers on other carcass characteristics: relationship between fatness and the stability of meat and adipose tissue. *British Poultry Science* **17**, 29–38

BARTOV, I. and BORNSTEIN, S. (1977). Stability of abdominal fat and meat of broilers: relative effects of vitamin E, butylated hydroxytoluene and ethoxyquin. *British Poultry Science* **18**, 59–68

BARTOV, I. and BORNSTEIN, S. (1978). Stability of abdominal fat and meat of broilers: effect of duration of feeding antioxidants. *British Poultry Science* **19**, 129–135

BARTOV, I., BORNSTEIN, S. and LIPSTEIN, B. (1974). Effect of calorie to protein ratio on the degree of fatness in broilers fed on practical diets. *British Poultry Science* **15**, 107–117

BARTOV, I., JENSEN, L.S. and VELTMANN, J.R. (1980). Effect of corticosterone and prolactin on fattening in broiler chicks. *Poultry Science* **59**, 1328–1334

BECKER, W.A. (1978). Genotypic and phenotypic relationships of abdominal fat in chickens. In *Proceedings of the 27th Annual National Breeder's Roundtable, Washington*

BECKER, W.A., SPENCER, J.V., MIROSH, L.W. and VERSTRATE, J.A. (1979). Prediction of fat and fat free live weight in broiler chickens using backskin fat, abdominal fat and live bodyweight. *Poultry Science* **58**, 835–842

BOORMAN, K.N. (1979). Regulation of protein and amino acid intake. In *Food Intake Regulation in Poultry* (K.N. Boorman and B.M. Freeman, Eds), pp. 87–126. Edinburgh, British Poultry Science Ltd

BOTTINO, N.R., ANDERSON, R.E. and REISER, R. (1970). Animal endogenous triglycerides: 2. Rat and chicken adipose tissue. *Lipids* **5**, 165–170

BRAY, T. (1982). Fat deposition in broilers. In *Gleadthorpe Experimental Husbandry Farm, Poultry Booklet No. 9*, pp. 18–23. Ministry of Agriculture, Fisheries and Food

BRAY, T. (1983). Broilers—Why is temperature so important? In *Gleadthorpe Experimental Husbandry Farm, Poultry Booklet No. 10*, pp. 4–9. Ministry of Agriculture, Fisheries and Food

BURGENER, J.A., CHERRY, J.A. and SIEGEL, P.B. (1981). The association between sartorial fat and fat deposition in meat-type chickens. *Poultry Science* **60**, 54–62

BURKHART, C.A., CHERRY, J.A., VAN KREY, H.P. and SIEGEL, P.B. (1983). Genetic selection for growth alters hypothalamic satiety mechanisms. *Behavioural Genetics* **13**, 295–300

CALABOTTA, D.F., CHERRY. J.A., SIEGEL, P.B. and GREGORY, E.M. (1983). Lipogenesis and lipolysis in normal and dwarf chickens from lines selected for high and low body weight. *Poultry Science* **62**, 1830–1837

CHAMBERS, J.R., GAVORA, J.S. and FORTIN, A. (1981). Genetic changes in meat-type chickens in the last twenty years. *Canadian Journal of Animal Science* **61**, 555–563

CLAYTON, G.A., FOXTON, R.N., NOTT, H. and POWELL, J.C. (1974). Estimating carcass composition in the duck (*Anas platyrhynchos*). *British Poultry Science* **15**, 153–158

COMBS, G.F. (1962). The interrelationships of dietary energy and protein in poultry nutrition. In *Nutrition of Pigs and Poultry* (J.T. Morgan and D. Lewis, Eds), pp. 127–147. London, Butterworths

COON, C.N., BECKER, W.A. and SPENCER, J.V. (1981). The effect of feeding high energy diets containing supplemental fat on broiler weight gain, feed efficiency, and carcass composition. *Poultry Science* **60**, 1264–1271

DALE, N.M. and FULLER, H.L. (1979). Effects of low temperature, diet density and pelleting on the preference of broilers for high fat rations. *Poultry Science* **58**, 1337–1339

DEATON, J.W., McNAUGHTON, J.L., REECE, F.N. and LOTT, B.D. (1981). Abdominal fat of broilers as influenced by dietary level of animal fat. *Poultry Science* **60**, 1250–1253

DELPECH, P. and RICARD, F.H. (1965). Relation entre les dépôts adipeux visceraux et les lipides corporels chez le poulet. *Annales de Zootechnie* **14**, 181–189

DONALDSON, W.E., COMBS, G.F. and ROMOSER, G.L. (1956). Studies on energy levels in poultry rations. 1. The effect of calorie-protein ratio of the ration on growth, nutrient utilization and body composition of chicks. *Poultry Science* **35**, 1100–1105

EHINGER, F. and SEEMANN, G. (1982). The influence of feed, age and sex on the growing performance and the carcass quality of broilers from different strains. 2. Fat content. *Archiv für Geflügelkunde* **46**, 177–188

EITAN, Y., AGURSKY, T. and SOLLER, M. (1983). Genetic aspects of feed efficiency under food intake restriction in broiler chickens. In *Proceedings 2nd World Congress on Genetics Applied to Livestock Production* **7**, 417–423

EMMANS, G.C. (1981). A model of the growth and feed intake of *ad libitum* fed animals, particularly poultry. In *Computers in Animal Production. Occasional Publication No. 5.* (G.M. Hillyer, C.T. Whittemore and R.G. Gunn, Eds), pp. 103–110. British Society of Animal Production

EVANS, A.J. (1969). *Fat Deposition during Postembryonic Growth in the Domestic Duck,* Anas platyrhynchos, *with Special Reference to the Action of Some Hormones*. PhD thesis, University of Edinburgh

EVANS, A.J. (1977). The growth of fat. In *Growth and Poultry Meat Production* (K.N. Boorman and B.J. Wilson, Eds), pp. 29–64. Edinburgh, British Poultry Science Ltd

FARRELL, D.J. (1974). Effects of dietary energy concentration on utilisation of energy by broiler chickens and on body composition determined by carcass analysis and predicted using tritium. *British Poultry Science* **15**, 25–41

FISHER, C. and WILSON, B.J. (1974). Response to dietary energy concentration by growing chickens. In *Energy Requirements of Poultry* (T.R. Morris and B.M. Freeman, Eds), pp. 151–184. Edinburgh, British Poultry Science Ltd

FULLER, M.L. and RENDON, M. (1977). Energetic efficiency of different dietary fats for growth of young chicks. *Poultry Science* **56**, 549–557

GRIFFIN, H.D. and WHITEHEAD, C.C. (1982). Plasma lipoprotein concentration as an indicator of fatness in broilers: development and use of a simple assay for plasma very low density lipoproteins. *British Poultry Science* **23**, 307–313

GRIFFIN, H.D., WHITEHEAD, C.C. and BROADBENT, L.A. (1982). The relationship between plasma triglyceride concentrations and body fat content in male and female broilers—a basis for selection. *British Poultry Science* **23**, 15–23

GRIFFITHS, L., LEESON, S. and SUMMERS, J.D. (1977a). Fat deposition in broilers: effect of dietary energy to protein balance, and early life caloric restriction on productive performance and abdominal fat pad size. *Poultry Science* **56**, 538–646

GRIFFITHS, L., LEESON, S. and SUMMERS, J.D. (1977b). Influence of energy system and level of various fat sources on performance and carcass composition of broilers. *Poultry Science* **56**, 1018–1026

GRIFFITHS, L., LEESON, S. and SUMMERS, J.D. (1978). Studies on abdominal fat with four commercial strains of male broiler chicken. *Poultry Science* **57**, 1198–1203

GRIMBERGEN, A.H.M., STAPPERS, H.P. and CORNELISSEN, J.P. (1982). The influence of an isocaloric substitution of soyabean oil for carbohydrates and of the nutrient density of the feed on growth and efficiency of energy utilisation in broiler chickens. *Netherlands Journal of Agricultural Science* **30**, 115–125

GUTTRIDGE, H.S. (1937). Methods and rations for fattening poultry. 2. Experimental technique and comparative value of fattening rations. *Scientific Agriculture* **18**, 198–206

GYLES, N.R., MAEZA, A. and GOODWIN, T.L. (1982). Regression of abdominal fat in broilers on abdominal fat in spent parents. *Poultry Science* **61**, 1809–1814

HÅKANSSON, J., ERIKSSON, S. and SVENSSON, S.A. (1978a). *The Influence of Feed Energy Level on Feed Consumption, Growth and Development of Different Organs of Chicks. Report No. 57.* Swedish University of Agricultural Science, Department of Animal Husbandry

HÅKANSSON, J., ERIKSSON, S. and SVENSSON, S.A. (1978b). *The Influence of Feed Energy Level on Chemical Composition of Tissues and on the Energy and Protein Utilisation by Broiler Chicks. Report No. 59.* Swedish University of Agricultural Science, Department of Animal Husbandry

HARGIS, P.H. and CREGER, C.R. (1980). Effects of varying dietary protein and energy levels on growth rate and body fat of broilers. *Poultry Science* **59**, 1499–1504

HAYES, J.F. and McCARTHY, J.C. (1976). The effects of selection at different ages for high and low body weight on the pattern of fat deposition in mice. *Genetical Research* **27**, 389–433

HILLARD, B.L., LUNDIN, PAULA and CLARKE, S.D. (1980). Essentiality of dietary carbohydrate for maintenance of liver lipogenesis in the chick. *Journal of Nutrition* **110**, 1533–1542

HOLSHEIMER, J.P. (1975). The effect of changing energy-protein ratios on carcass composition of broilers. *Spelderholt Mededeling* **241**. Spelderholt Institute of Poultry Science, The Netherlands

HOOD, R.L. (1982). The cellular basis for growth of the abdominal fat pad in broiler-type chickens. *Poultry Science* **61**, 117–121

HOOD, R.L. and PYM, R.A.E. (1982). Correlated responses for lipogenesis and adipose tissue cellularity in chickens selected for body weight gain, food consumption and food conversion efficiency. *Poultry Science* **61**, 122–127

HORVAT, R.J. (1978). Oily bird skin lipids. *Poultry Science* **57**, 1187

HUGHES, B.O. (1979). Appetite for specific nutrients. In *Food Intake Regulation in Poultry* (K.N. Boorman and B.M. Freeman, Eds), pp. 141–169. Edinburgh, British Poultry Science Ltd

JACKSON, S., SUMMERS, J.D. and LEESON, S. (1982). Effect of dietary protein and energy on broiler carcass composition and efficiency of nutrient utilisation. *Poultry Science* **61**, 2224–2231

JENSEN, J.F. (1982). Quality of poultry meat—an issue of growing importance. *World's Poultry Science Journal* **38**, 105–111
JENSEN, L.S., BARTOV, I., BEIRNE, M.J., VELTMANN, J.R. and FLETCHER, D.L. (1980). Reproduction of the oily bird syndrome in broilers. *Poultry Science* **59**, 2256–2266
KHALIL, A.A., THOMAS, O.P. and COMBS, G.F. (1968). Influence of body composition, methionine deficiency or toxicity and ambient temperature on feed intake in the chick. *Journal of Nutrition* **96**, 337–341
KIRCHGESSNER, M., ROTH-MAIER, D. and GERUM, J. (1978). Körperzusammensetzung und Nährstoffansatz 3-5 Wochen alter Broiler bei unterschiedlicher Energie–und Eiweissversorgung. *Archiv für Geflügelkunde* **42**, 62–69
KIRTLAND, J. and GURR, M.I. (1979). Adipose tissue cellularity: A review. 2. The relationship between cellularity and obesity. *International Journal of Obesity* **3**, 15–55
KUBENA, L.F., LOLT, B.D., DEATON, J.W., REECE, F.N. and MAY, J.D. (1972). Body composition of chicks as influenced by environmental temperature and selected dietary factors. *Poultry Science* **51**, 517–522
KUBENA, L.F., CHEN, T.C., DEATON, J.W. and REECE, F.N. (1974). Factors influencing the quantity of abdominal fat in broilers. 3. Dietary energy levels. *Poultry Science* **53**, 974–978
LANGSLOW, D.R. and LEWIS, R.J. (1972). The compositional development of adipose tissue in *Gallus domesticus*. *Comparative Biochemistry and Physiology* **43b**, 681–688
LANGSLOW, D.R. and LEWIS, R.J. (1974). Alterations with age in composition and lipolytic activity of adipose tissue from male and female chickens. *British Poultry Science* **15**, 267–273
LECLERCQ, B., BLUM, J.C. and BOYER, J.P. (1980). Selecting broilers for low or high abdominal fat: initial observations. *British Poultry Science* **21**, 107–113
LECLERCQ, B. and SAADOUN, A. (1983). Selecting broilers for low or high abdominal fat: comparison of energy metabolism of the lean and fat lines. *Poultry Science* **61**, 1799–1803
LEESON, S. and SUMMERS, J.D. (1980). Production and carcass characteristics of the broiler chicken. *Poultry Science* **59**, 786–798
LEVEILLE, G.A., ROMSOS, D.R., YEH, Y.Y. and O'HEA, E.K. (1975). Lipid biosynthesis in the chick. A consideration of site of synthesis, influence of diet and possible regulatory mechanisms. *Poultry Science* **54**, 1075–1093
LIEBELT, R.A. and EASTLICK, H.L. (1954). The organlike nature of the subcutaneous fat bodies in the chicken. *Poultry Science* **33**, 169–179
LILBURN, M.S., LEACH, R.M., BUSS, E.G. and MARTIN, R.J. (1982a). The developmental characteristics of two strains of chickens selected for differences in mature abdominal fat pad size. *Growth* **46**, 171–181
LILBURN, M.S., MORROW, F.D., LEACH, R.M., BUSS, E.G. and MARTIN, R.J. (1982b). A comparison of the *in vitro* lipogenic rates and other physiologic parameters in two strains of lean and obese chickens. *Growth* **46**, 163–170
LIN, C.Y. (1982). Fatness: a result of selection for fast growth. *Poultry International*, May 1982, pp. 62–64

LIPSTEIN, B., BUDOWSKI, P. and BORNSTEIN, S. (1975). The replacement of some of the soybean meal by the first-limiting amino acids in practical broiler diets. 3. Effect of protein concentrates and amino acid supplementation in broiler finisher diets on fat deposition in the carcass. *British Poultry Science* **16**, 627–635

LUCAS, A.M. (1979). Integumentum commune. In *Nomina Anatomica Avium* (J.J. Baumel, Ed.), pp. 19–51. London, Academic Press

McCARTHY, J.C. and SIEGEL, P.B. (1983). A review of genetical and physiological effects of selection in meat-type poultry. *Animal Breeding Abstracts* **51**, 87–94

MABRAY, C.J. and WALDROUP, P.W. (1981). The influence of dietary energy and amino acid levels on abdominal fat pad development of the broiler chicken. *Poultry Science* **60**, 151–159

MARCH, B.E. and HANSEN, G. (1977). Lipid accumulation and cell multiplication in adipose bodies in White Leghorn and broiler-type chicks. *Poultry Science* **56**, 886–894

MARION, J.E. and WOODROOF, J.G. (1966). Composition and stability of broiler carcasses as affected by dietary protein and fat. *Poultry Science* **45**, 241–247

MATEOS, G.G. and SELL, J.L. (1981). Influence of fat and carbohydrate source on rate of food passage of semi-purified diets for laying hens. *Poultry Science* **60**, 2114–2119

MORAN, E.T. (1977). Growth and meat yield in poultry. In *Growth and Poultry Meat Production* (K.N. Boorman and B.J. Wilson, Eds), pp. 145–173. Edinburgh, British Poultry Science Ltd

MORAN, E.T. (1980). Early protein restriction of the broiler chicken and carcass quality upon later marketing. *Poultry Science* **59**, 378–382

NATIONAL FARMERS UNION (1983). *Quarterly Broiler Bulletin No. 45. Notes on Standard Production Data and Costs* (1st July 1983) (J. Holton, Ed.), p. 3. Spalding, UK, The National Farmers Union

NEUPERT, B. and HARTFIEL, W. (1978). Untersuchungen zur Mastleistung und Schlachtkörperzusammen-setzung von Broilern in Abhängigkeit von Herkunft und Futterzusammensetzung. *Archiv für Geflügelkunde* **42**, 150–158

NIR, I. and LIN, H. (1982). The skeleton, an important site of lipogenesis in the chick. *Annals of Nutrition and Metabolism* **26**, 100

NIR, I., NITSAN, Z., DROR, Y. and SHAPIRA, N. (1978). Influence of overfeeding on growth, obesity and intestinal tract in young chicks of light and heavy breeds. *British Journal of Nutrition* **39**, 27–35

O'HEA, E.K. and LEVEILLE, G.A. (1969). Lipid biosynthesis and transport in the domestic chick (*Gallus domesticus*). *Comparative Biochemistry and Physiology* **30**, 149–159

OSBALDISTON, G.W. (1967). Chemical analysis of the chicken carcass. *The Poultry Review*, April 1967, pp. 10–13

OWENS, C.A., SIEGEL, P.B. and VAN KREY, H.P. (1971). Selection for body-weight at eight weeks of age. 8. Growth and metabolism in two light environments. *Poultry Science* **50**, 548–553

PARKS, J.R. (1982). *A Theory of Feeding and Growth of Animals. Advanced Series in Agricultural Sciences, II.* Berlin, Heidelberg, New York, Springer-Verlag

PFAFF, F.E. and AUSTIC, R.E. (1976). Influence of diet on development of the abdominal fat pad in the pullet. *Journal of Nutrition* **106**, 443–450

PROUDMAN, J.A., MELLEN, W.J. and ANDERSON, D.L. (1970). Utilisation of feed in fast- and slow-growing lines of chickens. *Poultry Science* **49**, 961–972

PYM, R.A.E. and NICHOLLS, P.J. (1979). Selection for food conversion in broilers: direct and correlated responses to selection for bodyweight gain, food consumption and food conversion ratio. *British Poultry Science* **20**, 73–86

PYM, R.A.E. and SOLVYNS, A.J. (1979). Selection for food conversion in broilers: Body composition of birds selected for increased body-weight gain, food consumption and food conversion ratio. *British Poultry Science* **20**, 87–97

PYM, R.A.E. and THOMPSON, J.M. (1980). A simple caliper technique for the estimation of abdominal fat in live broilers. *British Poultry Science* **21**, 281–286

QUARLES, C.L., BURR, T.W., MacNEIL, J.H. and BRESSLER, G.O. (1968). The effects of varying levels of hydrolysed animal and vegetable fat upon growth and carcass characteristics of broilers. *Poultry Science* **47**, 1764

RICARD, F.H. (1974). Etude de la variabilité génétique de quelques caractéristiques de carcasses en vue de sélectionner un poulet de qualité. In *Proceedings, 1st World's Congress on Genetics Applied to Livestock Production, Madrid* **1**, pp. 931–940

ROSMOS, D.R. and LEVEILLE, G.A. (1974). Effect of dietary fructose on *in vitro* and *in vivo* fatty acid synthesis in the rat. *Biochemica et biophysica acta* **360**, 1–11

SEATON, K.W., THOMAS, O.P., GOUS, R.M. and BOSSARD, E.H. (1978). Effect of diet on liver glycogen and body composition in the chick. *Poultry Science* **57**, 692–698

SIEGEL, P.B. and WISMAN, E.L. (1966). Selection for bodyweight at eight weeks of age. 6. Changes in appetite and feed utilisation. *Poultry Science* **49**, 1341–1345

SIMON, J. and LECLERCQ, B. (1982). Longitudinal study of adiposity in chickens selected for high or low abdominal fat content: further evidence of a glucose-insulin imbalance in the fat liver. *Journal of Nutrition* **112**, 1961–1973

SIMON, J., ZYBKO, A., GUILLAUME, J. and BLUM, J.C. (1978). Recherche d'une limitation de l'engraissement du poulet de chair pour un eger rationement alimentaire entre 6 et 8 semaines. *Archiv für Geflügelkunde* **42**, 6–9

SMITH, T.W., COUCH, J.R., CREGER, C.R. and GARRETT, R.L. (1976). The relationship of sex, dietary energy and meat protein with collagen in broiler skin tissue. *Poultry Science* **55**, 2093–2094

TOUCHBURN, S., SIMON, J. and LECLERCQ, B. (1981). Evidence of a glucose-insulin imbalance and effect of dietary protein and energy level in chickens selected for high abdominal fat content. *Journal of Nutrition* **111**, 325–335

VAN MIDDELKOOP, J.H., KUIT, A.R. and ZEGWAARD, A. (1976). In *Study of Fat Deposition in broilers*, pp. 37–38. Annual Report 1975, Netherlands Institute for Poultry Research 'Het Spelderholt', Beekbergen (In Flemish)

VAN MIDDELKOOP, J.H., KUIT, A.R. and ZEGWAARD, A. (1977). Genetic factors in broiler fat deposition. In *Growth and Poultry Meat Production* (K.N. Boorman and B.J. Wilson, Eds), pp. 131–143. Edinburgh, British Poultry Science Ltd

VELU, J.G., SCOTT, H.M. and BAKER, D.H. (1972). Body composition and nutrient utilisation of chicks fed amino acid diets containing graded amounts of either isoleucine or lysine. *Journal of Nutrition* **102**, 741–748

WHITEHEAD, C.C. and GRIFFIN, H.D. (1982). Plasma lipoprotein concentration as an indicator of fatness in broilers: effect of age and diet. *British Poultry Science* **23**, 299–305

WHITEHEAD, C.C. and GRIFFIN, H.D. (1984). Direct and correlated response to selection for decreased body fat in broilers. In *Poultry Genetics and Breeding* (D. Hewitt, W.G. Hill and J. Manson, Eds). Edinburgh, British Poultry Science Ltd (in press)

YEH, Y.Y. and LEVEILLE, G.A. (1969). Effect of dietary protein on hepatic lipogenesis in the growing chick. *Journal of Nutrition* **98**, 356–366

YEH, Y.Y. and LEVEILLE, G.A. (1970). Hepatic fatty acid synthesis and plasma free fatty acid levels in chicks subjected to short periods of food restriction and refeeding. *Journal of Nutrition* **100**, 1389–1398

YEH, S.J.C. and LEVEILLE, G.A. (1972). Cholesterol and fatty acid synthesis in chicks fed different levels of protein. *Journal of Nutrition* **102**, 349–358

YEH, S.J.C. and LEVEILLE, G.A. (1973). Significance of skin as a site of fatty acid and cholesterol synthesis in the chick. *Proceedings of the Society of Experimental Biology and Medicine* **142**, 115–119

YOUNG, R.J. and GARRETT, R.L. (1963). Effect of oleic and linoleic acids on the absorption of saturated fatty acids in the chick. *Journal of Nutrition* **81**, 321–329

22

NUTRIENT PARTITIONING IN DOMESTICATED AND NON-DOMESTICATED ANIMALS

M.A. CRAWFORD, W.R. HARE and D.B. WHITEHOUSE
Institute of Zoology, Regent's Park, London, NW1 4RY, UK

Meat fats and human health

In response to consumer demand and recommendations from Expert Committees (DHSS, 1974; FAO/WHO, 1978; WHO, 1982) on nutrition and health, there has been a move towards leaner animal carcasses. Despite this, the Royal College of Physicians (1983), in their current report on obesity and related diseases, again emphasize the need for a reduction in the intake of fats and sugars. In particular, the need for leaner animals is stressed. They state: 'Steps could be taken to encourage the breeding of cattle, sheep and pigs with a lower fat content', and that 'The present fat content of meat products is far too high'.

It is relevant to note that animal meats are rich sources of essential fatty acids (EFA) and their long-chain derivatives. The latter can supply human EFA requirements particularly efficiently. However, their value is suppressed by the high content of saturated fatty acids in fatty meats.

There can be little doubt that these issues are of current concern to those involved in animal production and to those interested in human health.

Relevance to modern domesticated breeds of meat animals

Our ancestral breeds of domestic livestock were derived from wildlife. The process of selecting desired traits in the development of our modern breeds may have unintentionally introduced undesirable qualities into these animals. For example, a faster rate of weight gain may be associated with a genetic tendency towards obesity, consumer perception of meat flavour with the accumulation of fat within muscle, and higher fat 'quality' with a greater proportion of saturated fats.

In the meat industry, 'quality' refers to firmness and whiteness (properties of the saturated fats). However, in the nutritional context, the quality of meat is associated with its content of essential nutrients. For fat, this implies a high proportion of the essential fatty acids, effectively the opposite of the commercial definition.

In addition to possibly possessing characteristics for meat quality (i.e. its nutritional value) which now require modification, we may find that our domestic breeds of meat animal, which were rigorously selected for traits considered desirable at the time of their development, may have lost the genetic potential necessary for the changes which we may now seek to make in their meat quality. Irrespective of whether it is preferable to introduce characteristics into established breeds or to attempt to develop new breeds from non-domesticated animals (the approach varies with the species; the introduction of breeds with a high performance for a given characteristic is still the most important means of improvement for sheep and goats for example), it is important to understand why and how differences in the partitioning of dietary nutrients and energy occur.

Until recently, the practical approaches to these questions required multiple carcass analyses. However, in the same way that it has become possible, for example, to assess the tendency for 'pale, soft, exudative' meat (PSE) production in pigs by its association with a sensitivity to halothane, the technology and background now exist to monitor the partitioning of energy and nutrients relevant to meat quality without the requirement for expensive slaughter programmes. Moreover, these measurements may be performed in the same animal at different times during its growth, greatly reducing the problems of carrying out such studies.

Contrasts in meat fat content and composition

There is little information on the carcass composition of breeds not commonly reared for meat production or on the quality of their fat; there is, however, information on the carcass and meat composition of wild and domestic species. These provide contrasts which indicate the direction and priorities that could be applied to the selection of domestic breeds. The information on the wild species is of added interest because it represents the biology of meat which was applicable to the period of human evolution and which, moreover, is likely to be closely representative of the meat component of human diets prior to urbanization and the intensification of animal production.

Gross carcass composition of wild and domestic species

In 16 different African ruminants, the mean carcass content of lean tissue was about 75% and of fat 3%; in African domesticated cattle, carcasses contained less lean and significantly more fat tissue (up to 33%; Ledger, 1968). In addition, Crawford *et al.* (1970a) have demonstrated substantially higher fat contents in the lean tissues of domestic animals compared with those of wild animals. These contrasts were not a consequence of differences in maturity.

Even within one domestic species, differences in growth characteristics can be large. For example, the growing Large White pig will consume 40 g food/kg body weight/day and gain about 40% more weight than the

Pietrain pig which consumes about 30 g/kg/day. Faster growth is usually considered to be associated with an increase in the feed utilization efficiency of the animal and usually (as in the case of the Large White pig) also with an increase in the fatness of the carcass.

Generally, animals will store nutrients in lean tissue to the limit permitted by diet and/or metabolism, will burn energy at a rate determined partly by metabolic characteristics, and will store excess energy (the intake of which reflects appetite) as fat.

The differences described are substantial and, if they are not a function of maturity, can be attributable only to the effects of differences in nutrition, genetic selection and exercise.

Differences in the types of meat fat

It is generally considered that dietary fats are reflected in the tissue fatty acids. However, in ruminants, this relation is obscured by the hydrogenation of most of the dietary unsaturated fatty acids in the rumen; the fatty acids which predominate in the adipose tissue of ruminant animals are therefore saturated or monoenoic. Despite this, examination of the meat fatty acid content of wild and domestic bovids showed that the balance of non-EFAs to essential polyunsaturated fatty acids was approximately 50:1 in domestic species but 3:1 in the wild (Crawford, 1968). This striking difference is caused by the combination of alterations in the types of lipid present in the meat tissue and in the fatty acid content of those lipids. There are several reasons for these alterations. A large effect is produced by the usually substantial accumulation of adipose fat in meat tissues in the domestic ruminant. Consequently, the types of fatty acids present in triglycerides (the predominant lipid of adipose tissue) dominate the fatty acid profile of the meat. In contrast, the meat from the wild species contains relatively little adipose fat and the fatty acid profile reflects the cellular structural lipids, i.e. the phosphoglycerides as opposed to the triglycerides.

However, this is not the whole explanation of the differences in the fat balance of domestic and wild animal meats: examination of the phosphoglycerides (asociated with membranes) of lean tissue demonstrates that the lipids of the wild ruminants are consistently richer in the EFA and the long-chain EFA derivatives than those of their domestic counterparts (*Tables 22.1a,b*, Crawford *et al.*, 1970a; Crawford, Gale and Woodford, 1970b).

Similarly, comparison of wild and domestic non-ruminants such as the pig (in non-ruminants, the diet fatty acid profile is not modified prior to absorption) shows that, whereas the adipose tissue of the domestic pig may be expected to contain about 6–12% linoleic acid and less than 1% α-linolenic acid, that of the wild wart-hog contained 17% linoleic and 17% α-linolenic acids. The phosphoglycerides of wild pig meat were also found to be significantly richer in EFAs compared with those of the domestic pig (*Table 22.1b*, Crawford *et al.*, 1970b).

Together, this information strongly supports the existence of fundamental differences in both diet and metabolism, relevant to fat quality, and operating between the wild and domestic situations.

Table 22.1a THE MEAT FAT CONTENTS OF WILD AND DOMESTIC BOVIDS

Typical source of meat	Meat fat content[a] (g/100 g fresh weight)	Percentage of the meat fat present as	
		Triglyceride	Phosphoglyceride
Wild species (Buffalo; *Syncerus caffer*)	2.5	4.1	46
Domestic (*Bos taurus*)	10	81	11
Domestic (*Bos taurus*)	15	87	7.5

[a] As the fat content rises in meat, so the amount of phosphoglyceride diminishes, both as a proportion of the total fat and as the amount present on a wet weight of tissue basis. Effectively, the cellular material of the muscle is diluted by infiltration of triglycerides.

Table 22.1b FATTY ACID COMPOSITION (% WT) OF MEAT TRIGLYCERIDES AND PHOSPHOGLYCERIDES FROM WILD SPECIES AND DOMESTIC BREEDS

	16:0	18:0	18:1	9,12-18:2 (n-6)	9,12,15-18:3 (n-3)	Long-chain derivatives	
						(n-6)	(n-3)
Phosphoglycerides							
Wild ruminant[a]	15	18	21	18	5.5	10	4.7
Domestic ruminant	20	19	29	11	0.9	8	3.8
Triglycerides							
Wild ruminant	21	18	38	2.6	2.7	<1	<1
Domestic ruminant	22	15	40	1.2	1.0	<1	<1
Phosphoglycerides							
Wild pig[b]	15	9	9	35	4.1	14	8
Domestic pig	22	14	31	14	0.5	12	6
Triglycerides							
Wild pig	20	11	20	17	17	<1	<1
Domestic pig	26	14	39	12	0.5	<1	<1

[a] Buffalo (*Syncerus caffer*).
[b] Warthog (*Phacochoerus aethiopicus*).

The compartmentation of different fatty acids into triglycerides and phosphoglycerides is illustrated above. In all cases the phosphoglycerides of meat are a rich source of EFA. Triglycerides in ruminants contain little EFA which explains why infiltration of fat (largely triglycerides) into muscle will dilute the EFA components. In addition, the phosphoglycerides and triglycerides of wild species tend to contain a higher proportion of EFA.

Selection of n-6 EFA

The principal EFA of seeds is linoleic acid, the parent n-6 EFA, whereas that of leaves is linolenic acid, the parent n-3 EFA. The n-3 EFA are the preferred substrate of many of the enzymes of EFA metabolism. However, the balance of the two families of EFA in the lean tissues of grazing ruminant and non-ruminant herbivores (n-6:n-3 > 5:1) is the reverse of that in their diet (n-6:n-3 < 1:5). Grazing animals appear to possess particularly effective mechanisms for the conservation of linoleic acid. The nature of the conservation and particularly the n-6 vs. n-3 EFA discrimination processes are unknown. In wild zebra, for example, the diet would be expected to be n-3 rich and indeed it has been found that some 42% of the

Table 22.2 SELECTION OF ESSENTIAL FATTY ACIDS EXHIBITED BY THE ZEBRA[a]

Source	Fatty acid composition (% wt) of total lipid			
	9,12-18:2 (n-6)	9,12,15-18:3 (n-3)	Long-chain derivatives (n-6)	(n-3)
Muscle	45	1.5	6.0	3.2
Adipose tissue	16	42	<1	<1
Stomach contents	24	41	0	0

[a] Zebra (*Equus burchelli*)
The selective compartmentation of fatty acids to produce a different balance of fatty acids in muscle tissue compared to that available in food is illustrated in the case of the zebra. As a non-ruminant grazing animal, the fatty acid profile of zebra's stomach contents is dominated by 18:3 (n-3); the same pattern is found in the adipose tissue. By contrast, the muscle tissue lipid exhibits selection for the (n-6) fatty acids.

Table 22.3 DIFFERENCES IN FATTY ACID COMPOSITION OF TOTAL LIPIDS BETWEEN BROWSING AND GRAZING SPECIES

Animal[a]	Fatty acid composition (% wt) of total lipid						
	16:0	18:0	9-18:1	9,12-18:2 (n-6)	9,12,15-18:3 (n-3)	Long-chain derivatives (n-6)	(n-3)
Woodland buffalo	16	20	21	16	5.0	9.0	4.2
Grassland buffalo	19	22	28	8.5	2.3	6.0	3.0
Kob (grazing)	19	22	23	16	3.0	6.0	3.0
Eland (browsing)	17	19	16	20	4.3	9.6	6.0
Giraffe (browsing)	13	17	15	28	3.4	11.0	6.6

[a] Buffalo (*Syncerus caffer*), kob (*Adenata kob*), eland (*Taurotragus oryx*), giraffe (*Giraffa camelopardalis*).
The differences between the tissue contents of EFA in browsing and grazing species are illustrated both in the same species (buffalo) living in different environments, and by the comparisons between the kob (a grazing species), the eland (which will obtain about 50% of its food from browse material) and the giraffe (which will obtain 95% of its food by browsing).

fatty acids in the adipose tissue of the zebra was accounted for by α-linolenic acid. In contrast, the choline and ethanolamine phosphoglycerides contained more than 40% of their fatty acids as linoleic acid and its derivatives (*Table 22.2*).

In previous studies of wild life, it was observed that there was a marked contrast between the tissue-lipid fatty acid profile in animals of the same species which were living in open grassland compared with those in bush and woodland where they made use of browse as a food source. Similarly a higher tissue content of EFAs was found in browsing compared with grazing species (Crawford and Woodford, 1971; *Table 22.3*).

Although wild ruminants may consume (perhaps select) food sources which are relatively rich in unsaturated fatty acids, for these fatty acids to be utilized they must escape ruminal hydrogenation. It is possible that plant structures themselves may give unsaturated fatty acids protection from ruminal hydrogenation. For example, the disproportionate tissue contents of n-6 EFA may be partly due to the fast transit of unbroken seeds which retain a natural protective coat. Seeds can constitute a significant proportion of the diet of wild ruminants (M.A. Crawford, unpublished observation), and it is known that small particles have a faster passage rate through the rumen (Balch, 1950), thus reducing the time for the occurrence of ruminal biohydrogenation.

Inter- and intramolecular partitioning of fatty acids

Mechanisms exist by which different fatty acids are selected for different triglyceride or phospholipid pools (intermolecular partitioning). Additionally, intramolecular segregation (e.g. between the 1- and 2- positions of a phospholipid) allows the dual functions of the close regulation of the synthesis of highly active prostaglandins from EFA derivatives, and the production of phosphoglycerides with the particular physicochemical characteristics required for membranes.

For example, a specific requirement for fatty acid composition appears to be shown by the major phospholipid of lung tissue which is dipalmitoylphosphatidylcholine in both ruminant and non-ruminant species (Montfoort, van Golde and Van Deenen, 1971). In contrast, the major phospholipid of biliary secretions is 1-palmitoyl,2-linoleoyl-phosphatidylcholine in non-ruminant species, but 1-palmitoyl,2-oleoyl-phosphatidylcholine in ruminants (Christie, 1973), an alteration which would assist the conservation of n-6 EFA by ruminants, but which indicates no functional requirement for a particular fatty acid in this situation.

Genetic considerations

It is known that animal products are subject to substantial variation of nutrient composition, and that many of these differences are genetically determined to a significant extent (Warwick, 1980). For instance, the distribution of nutrients in milk shows considerable variations between species such as reindeer and cattle and is an example of compartmentation on a gross scale. Again, an individual component such as energy can be compartmented into fat or lean muscle to different degrees in different species or animals within a species. The importance of genetic determination on this compartmentation will depend largely on the relative importance of heredity and environment.

Variation between species and breeds in any character suggests a degree of genetic determination, hence the differential compartmentation of fatty acids found in domesticated and wild species is likely to have an element of genetic control. The capacity to compartment differentially appears to be a quantitative character and therefore the inherited component is likely to be the product of gene action at a number of loci. If it were possible to remove all environmental influences, a reduction of the observed differences in fats between domesticated and wild species would probably occur, leaving the residual genetic difference in evidence. In general, the measurement of the degree to which a character is genetically determined demands that the two species or breeds are subjected to identical environmental influences such as food, management and age at slaughter.

To derive a precise estimate of genetic determination further requires that inbred lines be produced, which is clearly not feasible in species such as cattle, pigs and sheep that have long generation times. However, precise knowledge of the degree of genetic determination alone is of little practical value as it cannot be used to predict the rapidity by which the character (in this case, the compartmentation of fatty acids) can be altered by selection (Falconer, 1963). Of most use to animal breeders is an estimate of the heritability of the phenotypic character being considered. Heritability is the proportion of the observed variation in a character that is attributable to additive genetic differences and, in practical terms, the response to selection may be predicted from an estimate of the heritability of the character. Thus, if the differential compartmentation of fatty acids, which is already known to be a variable trait, is also found to have a high heritability, the probability of modifying compartmentation ratios by selective breeding is high.

Biochemical studies combined with fine-structure genetic analyses are potentially able to increase our understanding of the mechanisms of compartmentation. For instance, some recent studies suggest that the percentages of protein and fat and even the amino acid composition of protein in cows' milk are strongly associated with certain combinations of genotypes at just four gene loci, α-s1-casein, β-casein, κ-casein and β-lactoglobulin (Jebrovoski, Mithioutka and Babonkov, 1982; McLean *et al.*, 1982). Clearly, biochemical and genetic monitoring may offer considerable rewards with regard to the understanding of fatty acid, fat and lean muscle compartmentation.

Over the past twenty years, selection for reduced fat levels in pigs has been notably successful. On the other hand, attempts to alter characters such as the levels of proteins, amino acids and vitamins have been less rewarding; this, in part, has been due to the lack of variability in the control of these parameters. In contrast, there are substantial differences in the quantity and quality of fats both within and between breeds and species, for example, the differences in body fat and lean body mass of the Aberdeen Angus and the Charolais. Experience in humans demonstrates that there are significant heritable differences in the handling of dietary fats, for example the individual differences in cholesterol metabolism and the hyperlipidaemias. Together, the available evidence indicates that there is considerable variability in compartmentation, and we suggest that it would be well worth while to establish the degree of heritability of this characteristic with a view to modifying compartmentation ratios in domestic food species.

Future breeding directions

Some European countries, e.g. Norway, have already taken the health of their populations into account in determining their meat-production policies. Trends in the consumption of pig, beef and sheep meats in the United Kingdom (and in Europe generally) suggest that little significant downward movement is likely to occur before the year 2000. In view of this, the quality of the meat consumed is likely to be of major importance both in terms of its effect on consumer demand and as part of national health policy.

Concluding remarks

We have emphasized fats because of their dominant role in energy storage and metabolism. The contrasts described indicate that there are differences between breeds or species in:

1. The partitioning of energy between requirements for structure, metabolism and storage.
2. The partitioning of the two families of EFA into selected lipids (phosphoglycerides and triglycerides).
3. For both (1) and (2), the sites of deposition, i.e. lean or fat tissues.
4. Inter- and intramolecular partitioning of fatty acids, which is relevant to meat quality as expressed by essential nutrient density.

Examination of the above details could involve slaughter and carcass dissection. However, as our data on wild and domestic species indicate, much of the difference lies in the compartmentation between phosphoglycerides (structural) and triglycerides (storage fats). As with the example of the use of halothane in the context of PSE, it may be possible to use the compartmentation between blood phosphoglycerides and triglycerides for genetic selection without slaughter.

The differences in carcass and tissue composition between wild animals and their domestic counterparts indicate that they are caused by fundamental differences in the utilization of nutrients as well as by environmental differences. We have not discussed the role of exercise although this represents a further difference relevant to the products (i.e. meat) of the wild and intensive systems. It is of interest that, at this symposium, concern has been voiced about the high carcass fats now being found in broiler chickens; the concepts discussed here may be relevant in this situation also. It is clear that we do not as yet understand the degree to which the differences are a product of food or genetics, but this is an important question to answer if we are to return to producing leaner animals and are to become concerned with the issue of 'fat quality' as well as 'fat quantity' in its context of human nutrition.

References

BALCH, C.C. (1950). factors affecting the utilization of food by dairy cows.
 1. The rate of passage of food through the digestive tract. *British Journal of Nutrition* **4**, 361–388
CHRISTIE, W.W. (1973). The structures of bile phosphatidylcholines. *Biochimica et biophysica acta* **316**, 204–211
CRAWFORD, M.A. (1968). Fatty acid ratios in free living and domestic animals: possible implications for atheroma. *Lancet* **1**, 1329–1333

CRAWFORD, M.A., GALE, M.M., WOODFORD, M.H. and CASPERD, N.M. (1970a). Comparative studies of fatty acid composition of wild and domestic meats. *International Journal of Biochemistry* **1**, 295–305

CRAWFORD, M.A., GALE, M.M. and WOODFORD, M.H. (1970b). Muscle and adipose tissue lipids of the wart hog. *Phacochoerus aethiopicus. International Journal of Biochemistry* **1**, 654–658

CRAWFORD, M.A. and WOODFORD, M.H. (1971). Fatty acid composition in liver, aorta, skeletal and heart muscle of two free-living ruminants. *International Journal of Biochemistry* **2**, 493–496

DHSS (1974). *Diet and Coronary Heart Disease, Report on Health and Social Subjects 7*. London, HMSO

FALCONER, D.S. (1963). In *Methodology in Mammalian Genetics* (W.J. Burdette, Ed.), pp. 193–216. San Francisco, Holden-Day Inc.

FAO/WHO (1978). *The Role of Dietary Fats and Oils in Human Nutrition. Nutrition Report No. 3*. Rome, FAO

JEBROVSKI, L.S., MITHIOUKA, V.E. and BABONKOV, A.V. (1982). Biochemical genetic factors influencing milk production of black and white cattle. In *21st International Dairy Congress* Vol 1, Book 1, p. 47. Moscow, USSR, Mir Publishers

LEDGER, H.P. (1968). Body composition as a basis for a comparative study of some East African mammals. In *Symposia of the Zoological Society of London* **21**, 289–310

McLEAN, D.M., GRAHAM, E.R.B., PONZONI, R.W. and McKENZIE, H.A. (1982). In *21st International Dairy Congress* Vol 1, Book 1, pp. 54–55. Moscow, USSR, Mir Publishers

MONTFOORT, A., VAN GOLDE, L.M.G. and VAN DEENEN, L.L.M. (1971). Molecular species of lecithins from various animal tissues. *Biochimica et biophysica acta* **231**, 335–342

ROYAL COLLEGE OF PHYSICIANS OF LONDON (1983). Obesity. *Journal of the Royal College of Physicians* **17**, 5–65

WARWICK, E.J. (1980). *Animal Breeding Abstracts* **48**(12), 843–858

WHO (1982). *The Prevention of Coronary Heart Disease. Technical Report Series 678*. Geneva, WHO

VII

Practicalities of Fat Utilization

23

BLEND SOURCES AND QUALITY CONTROL

A.J. HOWARD
Procter & Gamble Limited, Newcastle upon Tyne, UK

Introduction

Up to 200 000 tonnes of fats were added to feed compounds in the UK in 1982. This represents a considerable increase over the past five years and demonstrates the increasing importance that the compounding industry places upon fat as an ingredient.

A significant proportion of these fats are blended. This is done to meet the needs of the market at optimum cost per unit of energy.

It is not the purpose of this chapter to review in depth the nutritional qualities of the component fatty acids: this has already been done in previous chapters. Nor will this chapter review detailed methods of fat handling, as this will be discussed subsequently. However, it is necessary to state the principles upon which fat blends are formulated as a prelude to describing blend sources. These principles are:

1. The components must be safe.
2. The components must contribute to maximum energy benefit per unit cost and other nutritional requirements.
3. The product must be handled easily by the feed mill.
4. The product must satisfy certain non-nutritional requirements, e.g. carcass quality.

Safety

This will be reviewed more fully under the heading of Quality Control. However, it is important to recognize that safety is of paramount importance and that there is the potential for toxic contaminants to enter into fatty materials by accident.

Energy benefit

Energy benefit is largely concerned with digestibility. Freeman (1974) listed practical factors which influenced digestibility, as follows:

1. Degree of unsaturation
2. Fatty acid chain length
3. Free fatty acid (FFA):glyceride ratio
4. Level of inclusion
5. Age of animal/bird
6. Environment

Of these, the first three can be influenced by the fat blender although it must be conceded that the 'basal' fat in the rest of the ration can also influence digestibility of the added fat, which could affect the formulation of the fat added to the best 'value for money' blend.

Degree of unsaturation

The degree of unsaturation is important, as shown in the work of Lewis and Payne (1966). Here a synergistic relationship between beef tallow and soyabean oil was clearly demonstrated (level of fat addition was 10%) (*Figure 23.1*).

Thus for poultry and pig feeding, manufacturers aim to blend unsaturated and saturated components to achieve high digestibility at least cost. Ruminant nutrition does not require the same unsaturated components—in fact it has been suggested that too much unsaturation can cause problems in the rumen.

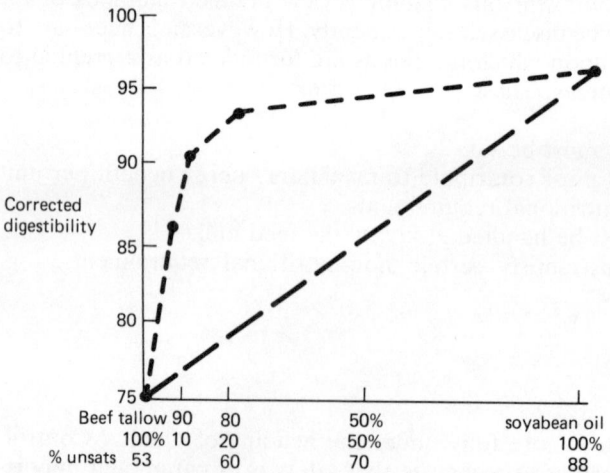

Figure 23.1 Synergistic effect of soyabean oil upon tallow digestibility. From Lewis and Payne (1966)

Fatty acid chain length

The saturated fats of shorter chain-length (e.g. those present in coconut oil) are highly digestible although there is some evidence of lack of palatability of short-chain free fatty acids. Longer chain-length (C20 and above) materials are in nature mainly unsaturated and thus also well digested—although there is justifiable concern over the potential transmission of 'fish' flavour into chicken meat, eggs and bacon if these materials are not limited. Most blend materials are derived predominantly from C16 and C18 chain length so the most important factor in their formulation is degree of unsaturation. However, restrictions may be necessary on C8/C10 and C20 and over for the above reasons.

Free fatty acid:glyceride ratio

The free fatty acid (FFA) content of a fat influences digestibility. Freeman (1976) considered the available data and suggested a fall of about 15% in digestibility from 0 to 100% free fatty acid. Lewis and Wiseman (1977), however, suggested that there is no significant fall in digestibility until the free fatty acid reaches 50%. For this reason FFA levels should be limited to a maximum of 50%. In addition, when water is present, its corrosive potential is increased by higher FFA levels.

The presence of mono- and diglyceride species is important to enhance the formation of micelles which are then absorbed in the intestine. Because economic blend materials tend to be derived from materials containing FFA it is important to understand the influence of the FFA:glyceride ratio.

Handling in the feed mill

It is important that the fat is stable not only through its storage period before compounding, but also in the feed compound. Antioxidants are added to help preserve fat stability. Fat application is considered in the next chapter but it is worth mentioning the practical observation that many mills have found that fat blends containing appreciable levels of FFA are more easily absorbed into pellets from fat spray-on operations than are materials containing low levels of FFA.

Non-nutritional requirements

It is accepted that, as the level of dietary fat is increased, the deposition of body fat chemically resembling the dietary fat is increased (Salmon and O'Neil, 1973). The composition of body fat has a major influence on both the physical quality of the carcass and its taste. Soft fat can cause excessive losses at the processing plant and poor subsequent visual appeal. The effect of nutrition on carcass quality was reviewed by Hunton (1977). The taste of the chicken meat is said by at least one major retailer to be largely determined by the type of interstitial fat deposited. It has been suggested that the fat characteristics for broiler feeding should include a *soft* fat blend starter, a *medium* fat blend grower, and a *hard* fat finisher ration. For pig and ruminant feeding other characteristics are necessary.

486 *Blend sources and quality control*

It is possible to meet all of these needs by blending edible soft oils and high-grade tallows. However, the current cost would be of the order of 30% higher than the present market, and, in addition, a multiplicity of tanks would be required by the feed compounder. When considering the second principle outlined above—that the formulation must deliver the maximum energy benefit per unit cost—the most economical sources of fats are found to be from by-product operations which, by their very nature, do not obtain the quality control shown for the principal products of the operation. It is these economical sources which will now be considered.

Blend component sources

Tallow rendering

Tallow rendering is a by-product operation of the meat industry. In this country the main constituents are beef fat and sheep fat; pork fat or lard is usually processed separately. The raw materials include butcher's shop scrap, bones, offal, fleshing grease from hides, dead stock and chicken grease. All of this material is handled in the same way through one of three process variants—dry rendering, wet rendering and solvent extraction—as shown in *Figure 23.2*.

In dry rendering the raw material is roasted to split open the fat cells. The result is a dry fat material which does not undergo subsequent hydrolysis. The danger, however, is that the product can be spoilt by overheating.

The wet rendering process employs water mixed with the raw material. This reduces the risk of overheating. The fat and water are then normally separated by centrifuge. However, this leaves tallow containing some water which can result in hydrolysis to FFA.

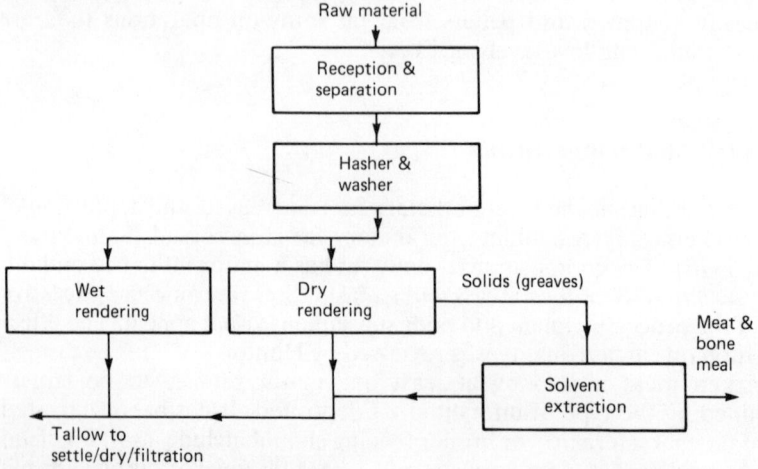

Figure 23.2 Tallow rendering processes

The solvent extraction process is usually applied to the extracted solids (or greaves) from either dry or wet rendering; solvents used include hexane or heptane. The product of this operation tends to be of lower quality than those from the other operations because quantities of fine solids may be left suspended in the oil/solvent mixture.

The quality of tallow produced is therefore dependent on (1) the raw material, (2) the process used. Tallows ranging from Human Food Grade, and Toilet Soap Grade to Industrial Soap and Fatty Acids Grades are all produced. The top-quality grades are produced from suet, butchers' scrap and bones; the lower grades come from offal processing. Ideally, the manure should be removed from offal before cooking but frequently this is not done; needless to say, this has a major impact on the quality of tallow produced. Feed Grade Fat lies within this lower-quality grouping.

By-products from edible-oil refining

Acid oils from edible-oil refining are a second source of blend components. The traditional edible-oil refining step uses caustic soda to neutralize FFA present in the crude oil, as shown in *Figure 23.3*. The soap stock formed emulsifies approximately its own weight of neutral oil: therefore, for crude oil containing 2% FFA, a 4% refining loss can be expected. When this material is acidulated the result is an approximate 50:50 mixture of neutral oil and FFA. Acid oils also contain many minor components and impurities, such as phospholipids and some water. Of course if the acidulation is incorrectly done there can be free mineral acid which has a detrimental effect upon mild steel tanks.

A second technique for refining is the physical refining process. It was originally applied in the 1930s but has become much used in the last 10 years with the development of the Malaysian palm oil industry. The process is shown in *Figure 23.4*. Because no alkali is used there is a relatively small loss of neutral oil. This means that the fatty acid distillate, especially that from palm oil, is 80–90% FFA. This product is also

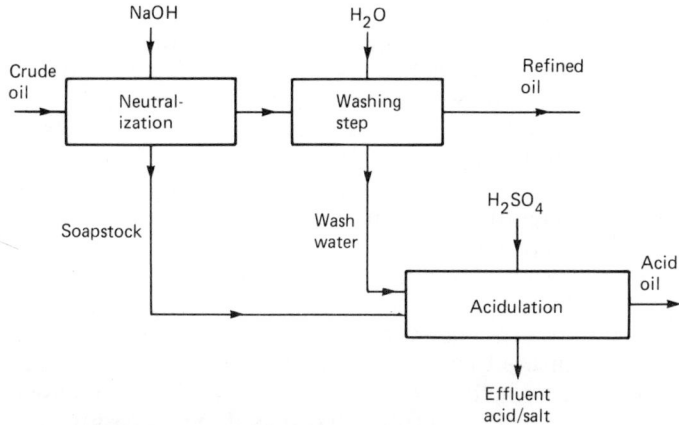

Figure 23.3 Chemical refining of edible oils

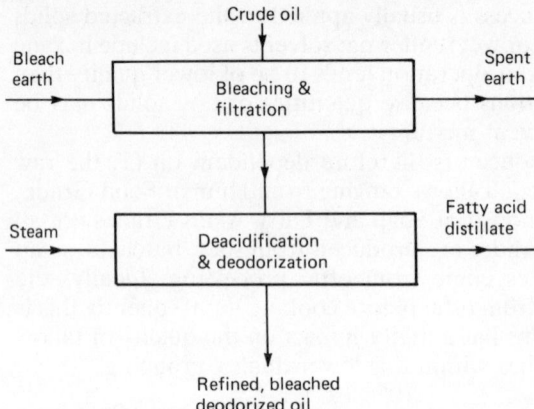

Figure 23.4 Physical refining of edible oils

characterized by a much higher level of unsaponifiable material (up to 3.0%).

Recovered vegetable oils

Recovered vegetable oils (RVO) are a third source of blend components. Despite this collective term, it is unusual for the materials to be wholly vegetable. They comprise a collection of triglyceride oils which have been used previously in various processes ranging from potato-crisp frying and other manufacturing procedures, to spent oils from the fish-and-chip shops. Because of this wide range of sources and lack of control by the collecting trade their variation in quality is not surprising. It is worth noting that RVO is sometimes blended and sold as mixed soft acid oil.

The fatty chemical processing industry

As shown in *Figure 23.5*, fats are chemically split into fatty acids and glycerol. Purification of the fatty acids is by distillation at low absolute pressure, e.g. 4 mmHg. In this process any unsplit fat remains in the bottom of the still and in a way analogous to chemical refining some free fatty acid is retained with it. It is possible for these materials to be recycled to the splitter but an alternative is their use in feed fats. Fully recycled materials have at times been offered to the feed industry but these have low nutritional benefit and should not be used.

This completes the review of the sources of the components of feed fat blends. They are all by-product materials which have been used or abused to varying degrees. This means that careful analysis and control by the blender is necessary to ensure that all the raw materials are of the right nutritional and safety quality before the blending process. This introduces the second consideration—quality control—an essential tool to ensure that the components and resultant blends meet the principles outlined earlier.

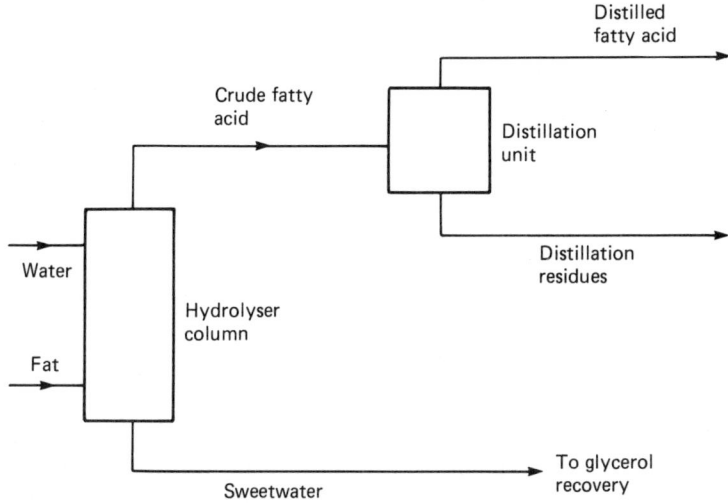

Figure 23.5 Hydrolysis of fats

Quality control

Quality control analyses can be grouped under three headings:

1. Nutritional value
2. Ease of processing
3. Contaminants

Nutritional value

There are many chemical quality control analyses which are important guides to nutritional value. Fatty acid chain-length distribution is ascertained by qualitative gas–liquid chromatography (GLC) to ensure that the correct balance is obtained between unsaturated and saturated components. It is also used to ensure that acids with very short chains are kept within limits, as are those with long chains (C20 and above) which might subsequently impart a fishy taste, and also erucic acid, which may have harmful clinical effects.

As many of the potential raw materials may be high in FFA it is necessary to measure their FFA content and ensure that the blend meets the nutritional requirements outlined earlier.

Although iodine value analysis has been largely superseded by GLC techniques it is still a useful and rapid indicator of unsaturation.

Peroxide value has been used as an indicator of incipient rancidity in edible oils. It is of doubtful value for feed grade materials as it is possible that the peroxide value has risen and then fallen as oxidation of the fat has progressed. However, it is included as an 'indicator' of the stability of the fat blend.

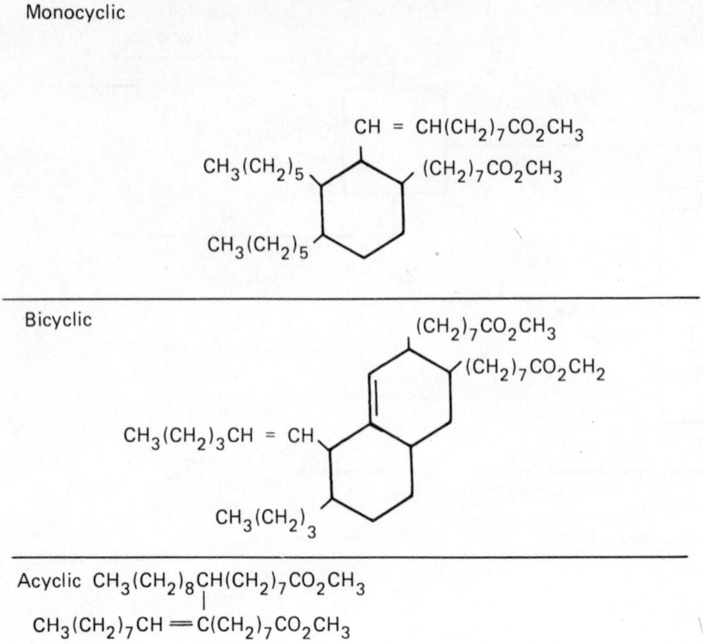

Monocyclic

[structure: cyclohexane ring with substituents CH=CH(CH$_2$)$_7$CO$_2$CH$_3$, CH$_3$(CH$_2$)$_5$, (CH$_2$)$_7$CO$_2$CH$_3$, CH$_3$(CH$_2$)$_5$]

Bicyclic

[structure: decalin-type bicyclic with substituents (CH$_2$)$_7$CO$_2$CH$_3$, (CH$_2$)$_7$CO$_2$CH$_2$, CH$_3$(CH$_2$)$_3$CH=CH, CH$_3$(CH$_2$)$_3$]

Acyclic CH$_3$(CH$_2$)$_8$CH(CH$_2$)$_7$CO$_2$CH$_3$
         |
CH$_3$(CH$_2$)$_7$CH=C(CH$_2$)$_7$CO$_2$CH$_3$

Figure 23.6 Some possible dimer acid (methyl ester) structures

The unsaponifiable content represents non-metabolizable material in the component or blend. As such it must be strictly monitored in the components, which have to be then adjusted in the formulation to meet the blend unsaponifiable limit.

Recent developments include assessments of the non-metabolizable content of fats. Many of the raw materials for fat blends may suffer abuse during by-product processing, giving rise to oxidized and polymerized fatty acids: these may have structures similar to those shown in *Figure 23.6* (Leonard, 1979). When such compounds were fed to rats, ring structures were excreted rather than absorbed (Leonard, 1979). This work supports that of many reports on animal tests to check the effects of used frying oil for human consumption.

Although oxidized and polymerized oils may not be actually harmful, they certainly are not beneficial. Billek, Guhr and Waibel (1978) have correlated the levels of oxidized and polymerized materials in used frying fats by the methods of gel permeation chromatography (GPC), liquid chromatography (LC) and column chromatography (CC) comparing results with those for oxidized fatty acids insoluble in petroleum ether. A simpler system for gel chromatography has recently been proposed by Schulte (1982).

Finally, a quantitative GLC method adopted by the Association of Official Analytical Chemists (AOAC) (1980) shows some promise as a detector of materials which have no metabolic value. One or other of these methods may be of use in helping to predict the nutritional value of by-product fats. The following relationship is proposed:

Metabolizable energy is proportional to $(D \times E)$

where D = a function of fatty acid chain-length distribution, i.e. ratio of unsaturates/saturates
E = percentage of fat eluted by quantitative GLC.

The fat blender should take into account these factors when evaluating potential blend components for maximum energy benefit per unit cost.

Ease of processing

From the description of raw materials it can be seen that moisture and insoluble impurities may be associated with many blend components. As well as being a waste of money, moisture can cause further hydrolysis and also tank corrosion; insoluble impurities can cause blockages in feed mill equipment; both are unwanted and must be controlled to meet blend specifications. Fortunately, this is an area where, with a properly designed process and careful control, moisture and insoluble impurities can be reduced to an insignificant level.

The relatively innocuous but troublesome material polyethylene has plagued the rendering industry since the growth in plastic packaging in the 1960s. Polyethylene package still enters the tallow rendering process despite industry efforts to prevent it. Once in the process it melts below rendering temperatures and passes through with the tallow. It starts to solidify again below 80°C and can then form lumps which will block fat spray nozzles. Blenders are aware of the problems and should either carefully prevent ingress of polyethylene in their blend components or operate a process to remove it.

Table 23.1 EEC (1977) PROPOSED LIMITS FOR PESTICIDE RESIDUES ETC. IN FEED FATS/COMPOUNDS

Contaminant	Maximum content proposed by EEC (ppm)
Aldrin	0.2
Dieldrin ⎫ Individually or	0.05
Chlordane ⎭ in combination	
DDT, DDE and DDE (TDE) individually or in combination	0.5
Endosulfan	0.1
Endrin	0.2
Heptachlor (including heptachlor epoxide)	0.2
Hexachlorobenzene	0.3
Hexachlorocyclohexane (BHC)	
α-isomer	0.5
β-isomer	0.1
γ-isomer (Lindane)	1.0
Methoxychlor	0.5
Toxaphene	0.5
Chlorinated Hydrocarbons ($C_xH_yCl_z$) not otherwise specified (single or combined)	0.5

Contaminants

This area of concern includes that of chlorinated compounds (pesticides, their residues and the stable polychlorinated biphenyls—PCBs). There is justifiable concern over their build-up in the human food chain and the EEC has proposed legislation limiting pesticides and their residues in animal feedingstuffs (*Table 23.1*). PCBs of the empirical formula $C_x H_y Cl_z$, are particularly tightly controlled. These materials are being phased out but are still in use in closed systems such as industrial electricity transformers. An incident in late 1979 in which it is believed transformer

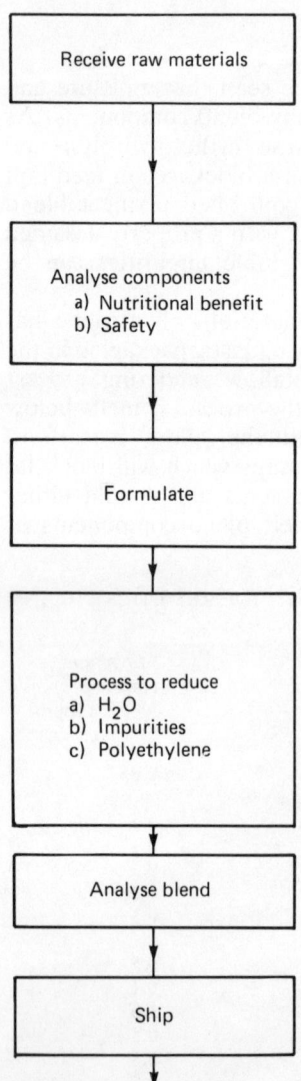

Figure 23.7 Summary of blender operation

fluid may have been dumped into RVO caused a level of about 1000 ppm PCB in a considerable tonnage of oil. This was detected by Procter and Gamble routine analysis of all raw materials and the delivery was rejected. Investigation with the supplier failed to discover the source of the PCB. However it was made clear to the supplier that his material could cause serious problems and that particular RVO should not go into animal feed. A second possible contaminant is the family of chemicals known as chlorodioxins. These are extremely toxic and were implicated in producing chick oedema factor in 1957. Both of these types of chlorinated compound have to be detected at very low levels (10^{-6} (ppm) for pesticides; 10^{-9} (ppb) for dioxins); this requires sophisticated and expensive electron capture GLC equipment.

The purpose of discussing these potential contaminants is to show that, in ensuring their absence, the fat blender needs to use sophisticated methodology and expensive equipment to give feed compounders fat blends which are safe and nutritious. It may be suggested that such operations are best left to the fat blender to give the quality assurance the feed industry needs. *Figure 23.7* summarizes the various operations discussed which are essential to ensure optimum blend quality.

References

AOAC (1980). *Polymers and Oxidation Products of Heated Vegetable Oils: Gas Chromatographic Method for Non-Elution Materials. AOAC Method 28.070*

BILLEK, G., GUHR, G. and WAIBEL, J. (1978). Quality assessment of used frying fats. *Journal of the American Oil Chemists Society* **55**, 728–733

EEC (1977). *Official Journal*, No C197/21-24 (18 August 1977)

FREEMAN, C.P. (1974). Fats as dietary energy source in pig and poultry rations. In *Proceedings, International Symposium on Energy Management in Mixed Feeds, Luxembourg*, pp. 83–88. Brussels, National Renderers Association

FREEMAN, C.P. (1976). Digestion and absorption of fat. In *Digestion in the Fowl* (K.N. Boorman and B.M. Freeman, Eds), pp. 117–142. Edinburgh, British Poultry Science

HUNTON, P. (1977). Nutrition and carcass quality in broilers and turkeys. In *Recent Advances in Animal Nutrition—1977* (W. Haresign and D. Lewis, Eds), pp. 149–157. London, Butterworths

LEONARD, E.C. (1979). Polymerisation—dimer acids. *Journal of the American Oil Chemists Society* **56**, 782A–785A

LEWIS, D. and PAYNE, C.G. (1966). Fat and amino acids in broiler rations. 6. Synergistic relationships in fatty acid utilisation. *British Poultry Science* **7**, 209–218

LEWIS, D. and WISEMAN, J. (1977). The use of added fats in rations for growing pigs. In *Proceedings, International Symposium on Animal Fats in Pig Feeding*, pp. 101–110. Dubrovnik, National Renderers Association

SALMON, R.E. and O'NEIL, J.B. (1973). The effect of the level and source of a change of source of dietary fat on the fatty acid composition of the depot fat and the thigh and breast meat of turkeys as related to age. *Poultry Science* **52**, 302

SCHULTE, E. (1982). Gelchromatographische Bestimmung polymerisierter Triglyceride. *Fette Seifen Anstrichmittel* **84**, 178–180

24

APPLICATION OF FATS IN THE MILL

R.E. ATKINSON
NRA Consultant, 10 Gwentlands Close, Chepstow, Gwent, UK

Introduction

The economics of feed formulations, not only for high-density feeds although especially so in their case, indicate that significant savings are available if it is possible to include liquid fat in the options offered. It has to be added that in the case of the highest energy levels it may be impossible to manufacture the diet without added fat. Whilst fat is available from a number of sources in dry form, and they are valuable ingredients, they inevitably carry a heavy cost penalty and in only a few cases has the carrier any significant energy value. Thus the benefits of including fat as an excellent source of energy which takes up little formulation space may be lost.

In these circumstances it is surprising that there has been so much reluctance to set up effective standard procedures to handle liquid fat.

The addition of fat to feed should not present problems if appropriate guidelines are followed, the most important of which are that the fat must be clean and that it must be very hot. The primary place at which cleanliness of fat is controlled is the point of manufacture or distribution, and adequate filtration must take place at these points. However, further filtration is essential at the feed plant and normally takes place as the fat is drawn from storage. Additionally, a large filter on the intake line can prevent poor fat ever being taken in.

Another factor to be considered is the corrosive potential of the fat, ranging from a relatively inert series of animal fats to something like palm acid oil, the use of which precludes the employment of metals other than good-quality stainless steel in fat-handling equipment, if machinery is to last. The use of copper or any copper-containing alloys in such machinery is precluded because not only are these metals attacked but they act as catalysts in speeding up the oxidation of the fat and the development of rancidity. For similar reasons, moisture must be excluded from fat: an increase of moisture from 1% to 2% can double the rate of corrosion caused by a fat and speed up oxidation.

Storage and distribution

From the above, a number of criteria to be observed in designing a fat-handling and addition system become apparent:

1. If mild steel is used for tanks and pipelines, as may be economically desirable, it must be of adequate thickness and, in the case of pipes, these should preferably be solid drawn, not seamed. Joints should be flanged or welded sockets: the use of screwed joints is unsatisfactory because the amount of metal removed in cutting the threads produces weak spots.
2. Pump, valve and meter parts in contact with fat should be stainless steel, synthetic rubber or inert plastic.
3. The ingress of moisture must be strenuously guarded against. This means completely enclosed tanks, carefully vented to avoid condensation and preferably without internal steam coils which could leak. Steam heating is, in any case, less efficient than is normally the case because it is unsafe to return condensate to the hot well in case it may have picked up fat.

With consideration of an actual storage and distribution installation, it is necessary first to consider the size of the load that it is economic to receive

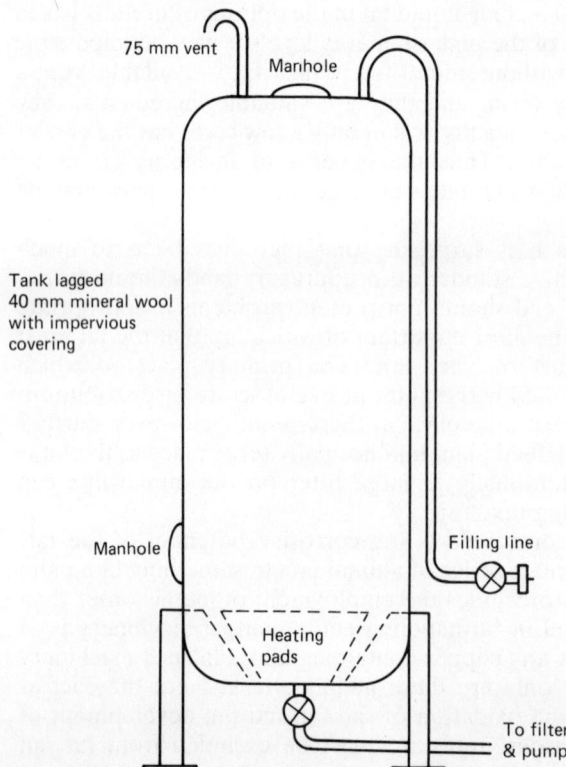

Figure 24.1 General description of a fat-storage tank

and currently this can be expected to be not less than 25 tonnes. Thus a tank of 30 tonnes storage capacity becomes the minimum. It is desirable to have a pair of tanks to allow for their periodic cleaning although, provided that adequate precautions are taken, this need not be frequent. The second tank need not necessarily take a full load if the rate of take-off is such that a smaller tank will cover an adequate period, but it is preferable to have a pair of similar units. To avoid accumulation of sludge, the tank bottom should slope to the outlet, which may be either centrally positioned (requiring a dished or concave bottom) or located at one side (or corner if square) with a sloping bottom. It is desirable that the bottom of the tank should be heated to allow convection full scope to assist in sludge prevention. Tanks are normally circular and supported on four or more short legs to give clearance of about 1 metre underneath (*Figure 24.1*).

Heating of the tank may be accomplished by an internal steam coil but is much better achieved by an external coil, some part of which is in contact with the bottom as well as the lower side walls, or by the use of electric pad heaters. The latter are available in a flexible form, in units about 1 metre square and of 1 kW rating. Provided that a 30 tonne tank is installed indoors, in a situation free from draughts and adequately lagged, then 4–5 kW will provide adequate heat. Thermostatic control should be fitted to keep the fat at a storage temperature of 50°C which will also avoid waste of electricity. If it is appreciated that a wind of 20 mph (9/m/s) can increase heat loss by a factor of four, the importance of a sheltered situation becomes apparent. Whatever source of heat is used, it is important that temperature should be maintained at all times, to discourage the sludge formation that occurs on lowering the temperature to the point where the constituents with highest melting point tend to separate.

Whether fat is to be added to meal in a special high-speed blender, or to the meal mixer, or sprayed on to pellets, it is essential (as indicated earlier) that further heating and filtering take place.

In order adequately to protect meters and, where applicable, spraying jets, it is necessary to filter out foreign matter and a 40 (2 mm) mesh is recommended in a large duplex unit; however, practical considerations usually dictate a 20 mesh (1 mm). The duplex unit allows for cleaning of filters without closing down the line. There have been considerable difficulties with polythene in fat deliveries at various times and filtration at a storage temperature of 50°C is necessary to eliminate this: if the filtration temperature is too low then filters will block, with large quantities of fat adhering to the polythene; if too high, then polythene will pass as liquid and be deposited on any cold spot later.

Additional heat is required to increase absorption by maintaining the fat in a liquid form for as long as possible, even when it touches the relatively cold surfaces of the meal or pellets to which it is being added. Traditionally, this is done by pumping fat from the main storage tank to a header or day tank holding about 1 tonne and fitted with a steam coil for heating. This system is liable to overflow, exposes a heated fat surface to air, and requires a second pump. Additionally, such a system may produce problems associated with siting and steam coil leakage. Therefore the use of a ring main and a line heater, as is normal when firing steam boilers with heavy fuel oil, have been recommended for some time. *Figure 24.2* shows

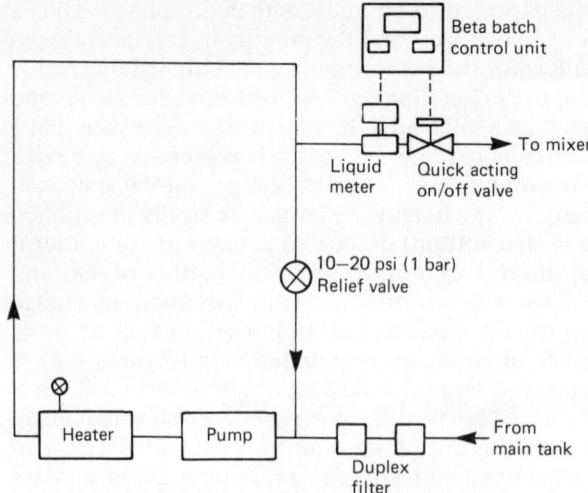

Figure 24.2 Fat-distribution system (feeding batch mixer)

the system as applied to mixer addition and fitted with a relief valve to maintain pressure in the line at 10–20 psi (1 bar). From the main tank, fat is filtered by the duplex unit and then pumped by a mono type pump with suitable stator and rotor or a jacketed gear pump through a steam-fed line heater with thermostatic control, round the circuit and back to the pump inlet. All fat lines are traced, preferably with electric heating tape which can be left on permanently, under thermostatic control, to prevent damage to plant by fat solidifying in the line at weekends. Pipe and heating tape are lagged together; valves and meters etc. should be traced and lagged, as well as the pipe run.

Recent experience has shown that steam trace heating can be used successfully even in cases of weekend boiler shut-down if instead of a single 6 mm copper trace, two 7.5 mm pipes are used. These will provide sufficient heat to free in a relatively short time, say a couple of hours, any pipes which may have solidified. However, thermostatic control is essential. In order to avoid water-logging the system will have to be split into several sections: the simplest control is therefore an electric thermostat in a stainless steel pocket in the line, or a surface thermostat, or thermostats controlling good-quality solenoid valves in the steam line.

For many years Albany gear pumps or the Mono pump have been recommended. The latter is well known to feed manufacturers and, provided that it is not allowed to run dry, it is a very effective unit. Despite this, a number of problems have arisen on units recently and the source of the trouble proved elusive, especially as the positioning of the filters before the pump had been implicated. However, even when these were moved, the problem of poor stator life and high-speed pressure gauge fluctuation continued. It now seems certain that pump speed is the culprit. It may be concluded that for fat a Mono should run at 400 rev/min with an absolute maximum of 500 rev/min.

Batch mixing

The simplest and most accurate method of adding fat to a dry meal mix is to add it in the batch mixer. While successful results can be obtained in both vertical and horizontal mixers, the latter is capable of achieving a good blend in a much shorter time and will accept higher fat levels (*Figure 24.3*).

In practice, meal mixtures rarely require more than 2–2.5% of fat added on nutritional grounds but feed for pelleting, especially high-energy pig and broiler feed, may require up to 6% or 7%. While it would be practicable to add this to the mixer, the resultant blend would not produce a satisfactory pellet and it is normal to add only up to 2–2.5% to blends for pelleting, the remainder being sprayed on to the pellet after manufacture.

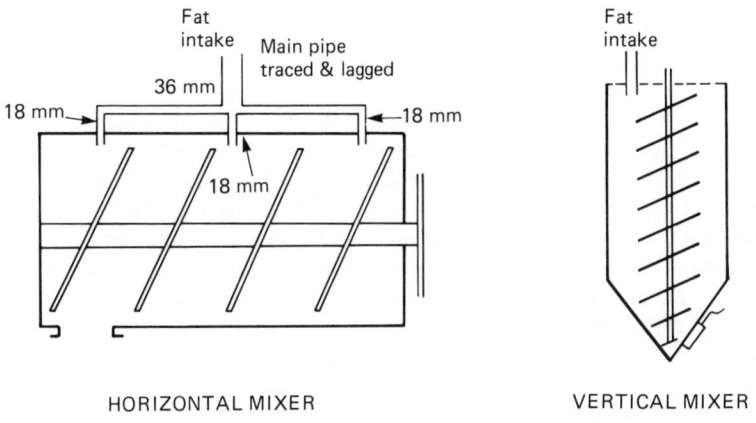

Figure 24.3 Mixer types indicating position of fat addition

It has been found that it is not beneficial to spray fat on to the meal in a mixer: for best results the fat should be heated to 85°C and applied in a steady stream. Spraying serves to reduce the fat temperature before it hits the meal, whereas the requirement is to maintain temperature and therefore liquidity long enough for the fat to be absorbed. The standard installation should therefore be a horizontal, lagged and traced main pipe across the top of the mixer with three down-pipes to deliver the fat on to the meal on the rising side of the mixer in such a position as to avoid impinging on the mixer side or mixer scroll. Fat addition does not commence until the mixer is full and then is completed over the first minute and a half of mixing time.

The measurement of the quantity of fat to be added may be achieved by:

1. Filling a measuring tank to the required level and pumping the contents into the mixer.
2. Measuring the rate of flow of a pump and then running the pump for a timed period to achieve the required addition.
3. Feeding the fat to the mixer through a meter and quick-acting cut-off valve. The required amount is pre-set on the meter and, when that has

passed, the valve is automatically closed. Because of the high rates of flow involved the valve is normally of electro-pneumatic type to achieve a fast enough closure to avoid over-running. The meter will also provide a total of fat used and may be manually set to the required quantity or may accept a signal from the blending computer.

The tank system is messy and very dependent on human supervision; the timed pump can give rise to major inaccuracies; the meter system is therefore to be recommended for all installations.

Long-term conditioning

It was suggested above that not more than 2–2.5% of fat could be added prior to pelleting because of the effect on the extrusion process—the lubrication provided by the fat lowering the resistance of the die to the point where effective operation is impossible at the maximum practicable die thickness.

It has always been known that much higher levels of fat could be pelleted when in bound form and, in the last few years, work started in Germany has resulted in a variation of the pelleting process: the dry matter is conditioned in the normal way by the addition of steam and specified liquids and then retained in a hot moist condition for 15–30 minutes to allow 'ripening' of the mixture; it is then possible to pellet much more satisfactorily. In many ways this is a reversion to the type of pelleting common in the past when a pellet machine of 2–3 t/h would have a 1 t conditioning kettle in which the steam and liquids were added. Today the difference is that the steam and liquids are added before the material enters the kettle, but otherwise both vessel and system are very similar.

Fat spraying

The alternative to long-term conditioning as a means of producing high-energy pellets is to coat the outside of the pellet with fat to be absorbed into the outer layers. This has the added benefit of increasing the effective durability of the pellet, especially by tending to bind the broken ends, and significantly reduces the amount of meal rubbed off.

This reduction of pellet degradation is of considerable importance in the European feed industry where the two important consumers of high-energy feeds, the broiler grower and dairy farmer, both demand clean pellets. In the case of the broiler grower this is directly contrary to American experience where up to 10% of fines seems to be the requirement.

In the Hayes & Stolz machine, pellets are delivered to a surge hopper of appropriate size to allow pellet output to vary without too much interruption to the fat coater—the feeder running from the time that the top level is covered until the bottom level is uncovered (*Figure 24.4*). Pellets are extracted from the bin by a two-deck chain conveyor with adjustable gate giving a wide even feed rate. The gate should be set to give an off-take

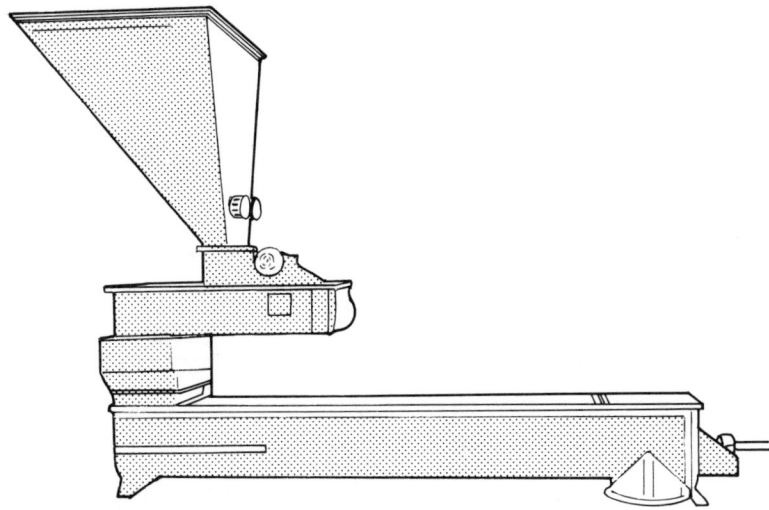

Figure 24.4 Hayes and Stolz fat coater

slightly in excess of the intake rate to the hopper. The pellets fall from the end of the lower deck in a continuous vertical column through a heated spray chamber where two horizontal fan type jets spray fat from each side. The maximum amount of fat is thus sprayed on to the pellets leaving the minimum to run down the chamber into the worm casing where a ribbon scroll and lifter bars encourage the pellets to roll on one another, thereby further spreading an even coating.

The use of fan-type jets allows the maintenance of spray pattern over a reasonable range of flow rates, as opposed to the cone type of nozzle, the pattern of which tends to vary with pressure and flow rate. This, with the ability to vary pellet feed rate and still maintain an even spread, allows maximum flexibility. *Figure 24.4* shows the machine complete with the lower section of the pellet hopper—a capacity of 2 tonnes is desirable to minimize stoppages.

The spray is adjusted by a motorized ball valve, controlled electronically from the Simplitrol which uses a rate-of-flow signal from the meter in the line to modulate the flow of fat through opening and closing the valve, thereby maintaining accurately the predetermined fat flow rate. The three-way valve after the control valve, is actuated by the pellet flow; when pellets are flowing, the valve ports over to the sprayer; when the pellet flow is interrupted, the valve ports over to the return leg. In order to avoid hunting by the control mechanism on change to and from by-pass, a 20 psi (1.5 bar) back pressure is maintained in the return to roughly equate with the back pressure of the spray nozzle. The return is fed to the pump inlet rather than back to the tank, to maintain as low a volume as possible of high-temperature fat. The pressure gauges before and after the control valve indicate line and spray pressure; they must be of the diaphragm type for accuracy and prevention of damage.

The control meter has three needles: a black one, indicating rate of flow, and two red ones for high- and low-level cut-off, moved by the round

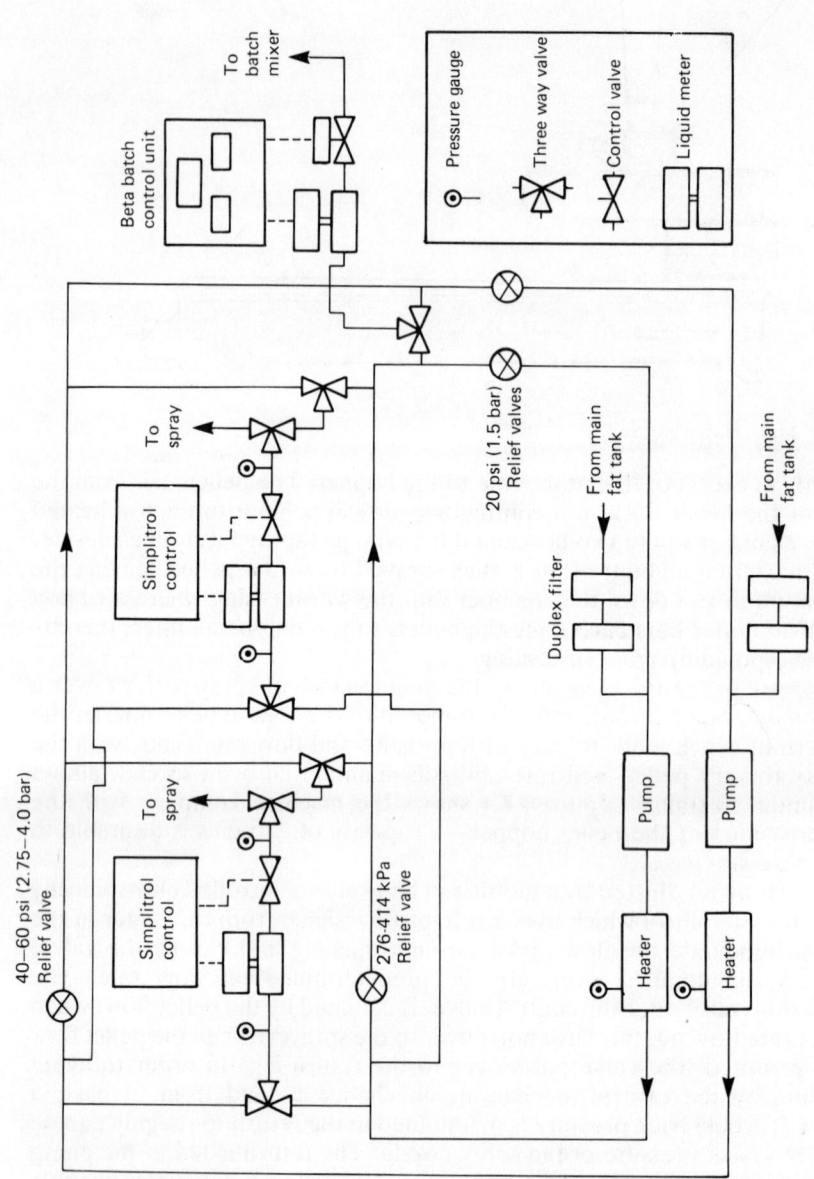

Figure 24.5 Distribution system for two types of fat

knobs. When the flow deviates outside the limits of either needle, the valve is opened or closed on a pulsed basis to bring flow back into line. The counter, top centre, only operates when the unit is spraying, not when on by-pass, and therefore acts as a check on consumption as well as giving a convenient method of setting up the rate-of-flow meter.

The liquid meter and control valve are mounted as a unit as close to the spray head as practicable. The concept has simplicity and the unit performs well and consistently if adequately looked after. Although a wide range of application rates is available, a rate of under 1% has not been found to be capable of covering the pellets effectively and, excluding special formulations, a rate of over 3% to 5% may not allow fat to be absorbed adequately.

A recent development adds to the unit a Driflo Sensor, which actually weighs the pellets between the feeder and spray chamber, so that a slightly more sophisticated control unit can not only register the quantity of dry feed passed but can match the fat exactly to the flow; this gives an even more accurate result and prevents the errors which can arise from miscalculation of the required flow rate or unauthorized alterations to the feed gate. In one unit installed in Eastern England the rate of fat to be sprayed is taken from the blending computer.

The Buhler unit comprises a rotating drum which is fed with pellets from an overhead bin of about 2 tonne capacity via the control unit which maintains a constant pre-set flow rate. The pellets are lifted by inverted angle-iron protuberances in the drum and are sprayed from multiple nozzles as they roll around during their passage down the drum. The rate of fat addition is controlled by the number of jets used, as well as the pressure behind them.

The control of fat pressure and therefore rate of flow on the Buhler is somewhat different, in that a return valve to the tank or pump is opened or closed to bypass sufficient fat to reduce the pressure to the required figure. This has the marginal advantage of using a wider valve opening for control at lower spray rates but means that a separate pump and ring main must be used for each spray unit. The bypass system operates as in the Beta unit.

The Buhler unit has an automatic control based on a flap balanced by air pressure which automatically controls fat addition but does not currently produce rate of flow or totals as a performance check.

Multiple fat installations

In the simplest approach to the use of more than one fat in the mill, the method has been to use one type of fat at the Batch Mixer and to spray a second type on to the pellets. A more sophisticated version, as shown in *Figure 24.5*, will allow the use of either fat in either mixer or fat coaters under totally automatic control from the blending panel. When increasing the number of types beyond two, then it would be normal practice to retain two types for general use, but to provide extra circuits to feed the mixer only, either from ring mains or from day tanks by the use of additional cut-off valves to a weighed process tank above the mixer, or through individual meters, or a common meter, to the mixer.

LIST OF PARTICIPANTS

Aeschbacher, Miss G.	Swiss Federal Research Station for Animal Production, Grangeneuve, 1725 Posieux/FR, Switzerland
Alderman, Mr G.	MAFF, Great Westminster House, Horseferry Road, London SW1P 2AE, UK
Annison, Prof. E.F.	Department of Animal Husbandry, University of Sydney, Werombi Road, Camden, NSW 2570, Australia
Atkinson, Mr R.E.	10 Gwentlands Close, Chepstow, Gwent NP6 5JH, UK
Bath, Dr I.H.	Faculty of Veterinary Medicine, University College, Dublin, Ireland
Bell, Dr B.	Feedex 9 Feeds Ltd, Daisy Hill, Burstwick, Hull, North Humberside HU12 9HE, UK
Boak, Mr W.	Carrs Farm Foods Ltd, Stanwix, Carlisle, Cumbria CA3 9BA, UK
Bogaert, Mr H.	Aan-en verkoopvennootschap van de Belgische Boerenbond NV, Eugeen Meeusstraat 6, 2060 Merksam (Belgium)
Boorman, Dr K.N.	Department of Applied Biochemistry and Food Science, University of Nottingham School of Agriculture, Sutton Bonington, Loughborough, Leics. LE12 5RD, UK
Boixel, Mr J.L.	Union des Fabricants D'Aliments Composés, 95450 Vigny, France
Bosi, Dr P.	Instituto di Allevamenti Zootechnici, Facolta di Agraria, Universita di Bologna, Reggio Emilia, Italy
Brett, Dr P.A.	North Western Farmers Ltd, The Mill, Wardle, Nantwich, Cheshire CW5 6BP, UK
Brindley, Dr D.N.	Department of Biochemistry, University Hospital and Medical School, Clifton Boulevard, Nottingham NG7 2UH, UK
Brooks, Dr P.H.	Seale-Hayne College, Newton Abbot, Devon TQ12 6NQ, UK
Brown, Mr G.	Colborn-Dawes Nutrition Ltd, Heanor Gate, Heanor, Derbyshire, DE7 7SG, UK
Burrows, Mr C.C.	Marfleet Refining PLC, Heddon Road, Hull, North Humberside, UK

List of participants

Chandler, Mr N.J.	Vitafoods Northern Ltd, 5–14 Cotton Exchange Building, Old Hall Street, Liverpool, UK
Campbell, Dr R.G.	Animal Research Institute, Werribee, Victoria 3030, Australia
Christie, Dr W.W.	Hannah Research Institute, Ayr, Scotland KA6 5HL, UK
Clapperton, Dr J.L.	Hannah Research Institute, Ayr, Scotland KA6 5HL, UK
Cole, Dr D.J.A.	Department of Agriculture and Horticulture, University of Nottingham School of Agriculture, Sutton Bonington, Loughborough, Leics. LE12 5RD, UK
Cooke, Dr B.C.	Dalgety Spillers Feed Ltd, Dalgety House, The Promenade, Clifton, Bristol, UK
Corbett, Mr M.A.	Cranswick Mill Ltd, Cranswick, Driffield, North Humberside, UK
Crawford, Prof. M.A.	Institute of Zoology, The Zoological Society of London, Regent's Park, London NW1 4RY, UK
Crehan, Mr M.P.	Nutec Ltd, Eastern Avenue, Lichfield, Staffs. WS13 7SE, UK
Cunnane, Dr S.C.	Efamol Research Institute, Kentville, Nova Scotia, Canada B4N 4H8
Curry, Mr H.A.	Curry, Morrison & Co, Ltd, Northern Road, Belfast Harbour Estate, Belfast BT3 9AL, UK
Dawkins, Mr C.W.C.	ADAS, Welsh Office, Agriculture Department, Trawscoed, Aberystwyth ST23 4HT, UK
Deeley, Miss S.M.	Butterworth & Co (Publishers) Ltd, Borough Green, Sevenoaks, Kent TN15 8PH, UK
De Mulder, Mr A.J.	Prosper de Mulder Ltd, Ings Road, Doncaster, DN5 9SW, UK
Easden, Mr S.	Pauls Agriculture Ltd, Unit 141, Walton Summit Industrial Estate, Bamber Bridge, Preston, Lancs, UK
Edmunds, Dr B.K.	R & A Department, Pauls Agriculture Ltd, New Cut West, Ipswich, Suffolk, UK
Enser, Dr M.B.	Agricultural Research Council, Meat Research Institute, Langford, Bristol BS18 7DY, UK
Fairbairn, Dr C.A.	Ministry of Agriculture, Fisheries and Food, Block C, Brooklands Avenue, Cambridge, UK
Fernandez, Mr J.A.	National Institute of Animal Science Research in Pigs & Horses, 25 Rolighedsvej, DK-1958 Copenhagen, Denmark
Ferns, Mr H.R.	Agricultural Chemistry Division, West of Scotland Agricultural College, Auchincruive, Ayr, Scotland, UK
Filmer, Mr D.G.	BOCM Silcock Ltd, Basing View, Basingstoke, Hants. RG21 2EQ, UK
Fisher, Dr C.	Agricultural Research Council, Poultry Research Centre, Roslin, Midlothian, EH25 9PS, UK
Fitt, Dr T.J.	Colborn-Dawes Ltd, Barton Mills, Canterbury, Kent, UK
Flanagan, Mr P.	C. Czarnikow (Liverpool) Ltd, Norwich House, Water Street, Liverpool, L2 8TA, UK
Foxcroft, Mr P.D.	Prosper de Mulder Ltd, Ings Road, Doncaster DN5 9SW, UK

List of participants

Freeman, Dr C.P.	Unilever Research Ltd, Colworth House, Sharnbrook, Bedford, UK
Fulleylove, Miss T.J.	Burgess Feeds Ltd, Thornton Dale, Pickering, N. Yorks, UK
Garnsworthy, Dr P.C.	Department of Agriculture and Horticulture, University of Nottingham School of Agriculture, Sutton Bonington, Loughborough, Leics. LE12 5RD, UK
Garton, Dr A.G.	Rowett Research Institute, Bucksburn, Aberdeen AB2 9SB, UK
Gibson, Mr J.E.	Farm Feed Formulators, Darlington Road, Northallerton, North Yorkshire DL6 2NW, UK
Gill, Mr G.A.	Vitafoods Northern Ltd, 5–14 Cotton Exchange Buildings, Old Hall Street, Liverpool, L3 9LH, UK
Gurr, Dr M.I.	Nutrition Department, National Institute for Research in Dairying, Shinfield, Reading, RG2 9AT, UK
Hannagan, Mr M.J.	Dalgety Spillers Feed Limited, Dalgety House, The Promenade, Clifton, Bristol, UK
Haresign, Dr W.	Department of Agriculture and Horticulture, University of Nottingham School of Agriculture, Sutton Bonington, Loughborough, Leics. LE12 5RD, UK
Harrington, Mr G.	Meat and Livestock Commission, PO Box 44, Queensway House, Bletchley, Milton Keynes MK2 2EF, UK
Harris, Mr C.I.	ADAS, Ministry of Agriculture, Fisheries and Food, Block A, Coley Park, Reading RG1 6DT, UK
Hazzledine, Mr M.	RHM Agriculture, Deans Grove, Colehill, Wimborne, Dorset, UK
Hermansen, Mr J.E.	National Institute of Animal Science Research in Pigs and Horses, 25 Rolighedsvej, DK-1958 Copenhagen, Denmark
Hewson, Mr R.	Vitrition Ltd, Ryhall Road, Stamford, Lincs, UK
Holmes, Mr W.B.	Favor Parker Ltd, The Hall, Stoke Ferry, Kings Lynn, Norfolk, UK
Homan, Dr G.W.	Hendrix' voeders b.v., Veerstraat 38. 5831 JN Boxmeer, The Netherlands
Howard, Mr A.J.	Procter & Gamble Ltd, Whiteley Road, Longbenton, Newcastle upon Tyne, NE12 9TS, UK
Hudson, Mr K.A.	Vitamealo, Broadmead Lane, Keynsham, Bristol, UK
Ivins, Prof. J.D.	Department of Agriculture and Horticulture, University of Nottingham School of Agriculture, Sutton Bonington, Loughborough, Leics. LE12 5RD, UK
Jones, Miss C.M.	Pauls Agriculture Ltd, Unit 141, Brierley Road, Walton Summit Industrial Estate, Bamber Bridge, Preston, Lancs, UK
Jones, Mr W.T.	J. Bibby Agriculture Ltd, Adderbury, Banbury, Oxon OX17 3HL, UK
Kaminska, Dr B.Z.	Institute of Animal Husbandry, Sarego 2, Krakow 31-047, Poland
Kelly, Mr P.	ADAS, Ministry of Agriculture, Fisheries and Food, Olantigh Road, Wye, Ashford, Kent, UK

List of participants

Kendall, Dr P.T.	Animal Studies Centre, Freeby Lane, Waltham-on-the-Wolds, Leics, UK
Koorman, Mr A.W.	Hoffman-La Roche & Co. Ltd, 4002 Basle, Switzerland
Kronaner, Mr M.	Institut für Tierproduktion, Gruppe Ernährung, ETH Zentrum, CH-8092, Zurich, Switzerland
Lake, Mr P.W.G.	Vitamealo, Broadmead Lane, Keynsham, Bristol, UK
Leat, Dr W.M.F.	ARC Institute of Animal Physiology, Babraham, Cambridge CB2 4AT, UK
Lebzien, Dr P.	Federal Research Center of Agriculture, Institute of Animal Nutrition, Bundesallee 50, D-3300 Braunschweig
Lee, Dr P.A.	National Institute for Research in Dairying, Shinfield, Reading, RG2 9AT, UK
Leeson, Dr S.	Department of Animal and Poultry Science, University of Guelph, Guelph, Ontario, Canada
Lewis, Prof. D	Department of Applied Biochemistry and Food Science, University of Nottingham School of Agriculture, Sutton Bonington, Loughborough, Leics. LE12 5RD, UK
Lewis, Mr T.	Vitamealo, Broadmead Lane, Keynsham, Bristol, UK
Lindeman, Mr M.A.	BOCM Silcock Ltd, Basing View, Basingstoke, Hants. RG21 2EQ, UK
Lindsay, Prof. D.B.	Institute of Animal Physiology, Babraham, Cambridge CB2 4AT, UK
Major, Mr R.	The Major Pig Consultancy, 48, Kenilworth Avenue, Reading RG3 3DN, UK
Marangos, Dr A.	Peter Hand (GB) Ltd, 15–19 Church Road, Stanmore, HA7 4AR, UK
McFarquhar, Dr A.M.	Bernard Matthews PLC, Great Witchingham Hall, Norwich, Norfolk, UK
Mead, Mr S.J.	Pauls Agriculture Ltd, Unit 141, Walton Summit Industrial Estate, Bamber Bridge, Preston, Lancs, UK
Meggison, Dr P.A.	BP Nutrition (UK) Ltd, Wincham, Northwich, CW9 6DF, UK
Miller, Dr E.L.	Department of Applied Biology, University of Cambridge, Pembroke Street, Cambridge CB2 3DX, UK
Moore, Prof. J.H.	Department of Biochemistry, Physiology & Soil Science, Wye College (University of London), Ashford, Kent TN25 5AH, UK
Mortensen, Mr H.P.	National Institute of Animal Science Research in Pigs & Horses, 25 Rolighedsvej, DK-1958 Copenhagen, Denmark
Morton, Dr P.	Procter & Gamble Ltd,. Industrial Chemicals Division, PO Box 9, Hayesgate House, Hayes, Middlesex, UK
Noble, Dr R.C.	The Hannah Research Institute, Ayr, KA6 5HL, Scotland, UK
Ojala, Mr J.	Vaasamills Ltd, Kolmas linja 22, 00530 Helsinki 53, Finland
Opstvedt, Dr J.	Norwegian Herring Oil and Meal Industry Research Institute, 5033 Fyllingsdalen, Bergen, Norway
Orr, Dr R.M.	Seale-Hayne Agricultural College, Newton Abbot, Exeter, Devon, UK

List of participants 509

Owers, Dr M.J.	Pauls Agriculture Ltd, New Cut West, Ipswich, Suffolk, UK
Palmer, Mr F.G.	ADAS, Ministry of Agriculture, Fisheries and Food, Lawnswood, Leeds, LS16 5PY, UK
Palmquist, Dr D.L.	Department of Dairy Science, Ohio Agricultural Research & Development Centre, Wooster, OH 44691, USA
Patterson, Dr D.	Agricultural Research Institute of Northern Ireland, Hillsborough, Co. Down, BT26 6DP, Northern Ireland
Pearson, Dr J.	Peter Hand (GB) Ltd, 15–19 Church Road, Stanmore, HA7 4AR, UK
Peers, Dr D.G.	ADAS, Ministry of Agriculture, Fisheries and Food, Burghill Road, Westbury-on-Trym, Bristol, UK
Perry, Mr F.G.	BP Nutrition (UK) Ltd, 1 Stepfield, Witham, Essex, UK
Pettigrew, Dr J.E.	Department of Animal Science, University of Minnesota, St Paul, MN 55108, USA
Phillips, Mr G.	Cleeve Hall, Bishops Cleeve, Cheltenham, Glos., UK
Phoya, Mr R.K.D.	ARC Meat Research Institute, Langford, Bristol BS18 7DY, UK
Plonka, Dr S.S.	Institute of Animal Production, Sarego 2, Krakow 31-047, Poland
Poutiainen, Prof. E.K.	Department of Animal Husbandry, University of Helsinki, 00710 Helsinki 71, Finland
Prescott, Miss N.J.	ARC Meat Research Institute, Langford, Bristol BS18 7OY, UK
Raddison, Dr	Director of Research, Ralston Purina, 391 Avenue Louise, Brussels, Belgium
Raine, Dr H.	RHM Agriculture Ltd, Dean Grove, Colehill, Wimborne, Dorset, UK
Rice, Dr R.D.	Marfleet Refining PLC, Hedon Road, Hull, North Humberside, UK
Rypkemo, Dr Y.	Institute for Livestock, Feeding and Nutrition Research, (IVVO) Runderweg 2, Postbox 160, 8200 AD, Lelystad, The Netherlands
Sabine, Dr J.R.	Department of Animal Sciences, Waite Agricultural Research Institute, University of Adelaide, Adelaide, Australia
Sambrook, Dr I.E.	National Institute for Research in Dairying, Shinfield, Reading RG2 9AT, Berks, UK
Scaife, Dr J.R.	Department of Agriculture, University of Aberdeen, Aberdeen, Scotland, UK
Seerley, Prof. R.W.	Animal & Dairy Science Dept, University of Georgia, Athens, GA 30602, USA
Shipston, Mr A.H.	RHM Agriculture, Deans Grove, Colehill, Wimborne, Dorset, UK
Skuse, Mr J.	Wandalup Farms, PO Box 642, Mandurah, W.A. 6210, Australia
Speight, Mr D.	Nitrovit Ltd, Nitrovit House, Dalton, Thirsk, North Yorkshire YO7 3JE, UK

List of participants

Speight, Dr J.	Procter and Gamble Ltd, Industrial Chemicals Division, PO Box 9, Hayesgate House, Hayes, Middlesex, UK
Spencer, Mr P.G.	Bernard Matthews PLC, Great Witchingham Hall, Norwich, Norfolk, UK
Spreeuwenberg, Dr W.M.M.	Cehave NV, Pater van de Elsenlaan 4, 5462 GG Veghel, The Netherlands
Stahly, Prof. T.	University of Kentucky College of Agriculture, Lexington, Kentucky 40546-0215, USA
Steele, Dr W.	Hannah Research Institute, Ayr, Scotland KA6 5HL, UK
Summers, Dr J.D.	Department of Animal and Poultry Science, University of Guelph, Guelph, Ontario, Canada N1G 2W1
Tame, Dr M.J.	J. Bibby Agriculture Ltd, Adderbury, Banbury, Oxon, UK
Thompson, Mr R.J.	Preston Farmers Ltd, Kinross, New Hall Lane, Preston, UK
Tomkins, Dr T.	Volac Ltd, Orwell, Royston, Herts. SG8 5QX, UK
Tonks, Mr W.P.	Park Tonks Ltd, 104, High Street, Gt. Abington, Cambs, UK
Turner, Miss R.E.	Pauls Agriculture Ltd, Mill Road, Radstock, Nr. Bath, Avon, UK
Twigge, Mr J.R.	BP Nutrition (UK) Ltd, Wincham, Northwich, Cheshire, CW9 6DF, UK
Van Aelten, Mr G.	Aan-en verkoopvennootschap van de Belgische Boerenbond NV, Eugeen Meeusstraat 6, 2060 Merksem (Belgium)
Van Eenoo, Dr W.	NV Dossche, Vaartoever, 9800 Deinze, Belgium
Van Gils, Dr L.G.M.	Trouw & Co, B.V. (International), R & D Department, Postbus 50, 3880 AB Putten, The Netherlands
van der Honing, Dr Y.	Institute for Livestock Feeding and Nutrition Research, (IVVO) Runderweg 2, Postbox 160, 8200 AD, Lelystad, The Netherlands
Van Hoecke, Mr P.	NV Radar, Dorpsstraat 4, 9800 Deinze (Belgium)
Veen, Dr W.A.G.	C.L.O.-Institut voor de Veeveoding, "de Schothorst", Meerkoetenweg 26, 8218 NA Lelystad, The Netherlands
Vernon, Dr B.G.	Dalgety Spillers Feed Ltd, Dalgety House, The Promenade, Clifton, Bristol, UK
Vernon, Dr R.G.	Hannah Research Institute, Ayr, Scotland KA6 5HL, UK
Wahle, Dr K.W.J.	Department of Lipid Biochemistry, Rowett Research Institute, Bucksburn, Aberdeen AB2 9SB, UK
Wakelam, Mr J.A.	George A. Palmer Ltd, Oxney Road, Peterborough, UK
Walker, Dr T.	BP Nutrition Ltd, Britannic House, Moor Lane, London EC2Y 9BU, UK
Ward, Mr J.H.	Nitrovit Ltd, Nitrovit House, Dalton, Thirsk, North Yorkshire, UK
Weir, Mr J.	Chemistry Division, West of Scotland Agricultural College, Auchincruive, Ayr, KA6 5HW, Scotland, UK
Whitehead, Dr C.	ARC Poultry Research Centre, Roslin, Midlothian EH25 9PS, UK

List of participants

Wilson, Prof. P.N.	BOCM Silcock Ltd, Basing View, Basingstoke, Hants. RG21 2EQ, UK
Wilson, Dr R.H.	Wandalup Farms, PO Box 642, Mandurah, W.A. 6210, Australia
Wiseman, Dr J.	Department of Agriculture and Horticulture, University of Nottingham School of Agriculture, Sutton Bonington, Loughborough, Leics. LE12 5RD, UK
Witt, Mr G.T.	Gro-Well (Pig Feeds) Ltd, Bowerhill Trading Estate, Melksham, Wilts, UK
Wood, Dr J.D.	ARC Meat Research Institute, Langford, Bristol BS18 7DY, UK
Wyatt, Mr D.H.	John Wyatt Ltd, Braithwaite Street, Holbeck Lane, Leeds LS11 7PJ, UK

INDEX

Abdominal fat pad,
    effect of dietary manipulation, 448–449
    effect of protein:energy ratio, 453–456
    in broilers, 438–440
    selection against, 441–444
Absorption,
    cellular mechanism, 93–95
    effect of age species and breed, 117, 280–283
    effect of dietary protein, 287–288
    effect of feeding regime, 117–118
    effect of gut microflora, 117
    effect of poorly absorbed carbohydrate, 316–317
    effect of soaps in the pig, 287
    efficiency in the ruminant, 133
    in non-ruminants, 107–113
    in ruminant small intestine, 128–133, 188
    into enterocytes, 89
    malabsorption, 95, 390
    non-micellar, 110
    of bile salts, 89, 112–113
    of cholesterol, 107
    of glycerol, 113
    of medium chain fatty acid, 13, 86
    of micelles into mucosal cells, 110–112
    of plant phospholipids, 13
Acetate,
    cost of ATP generation, 364
    energy source in muscle, 235–236
    incorporation into PUFA, 128
    metabolism in ruminant liver, 230
    oxidation in mammary tissue, 241–242
Activity,
    of essential fatty acid, 154–156
Amphipiles,
    classification, 85
    in absorption of non-polar lipid, 116
    in fat digestion, 87
Antioxidants, 13, 28, 202–203
Antiperistalsis, 112

Apoprotein,
    in ruminant VLDL, 137
    incorporation into lipoprotein, 135
    transport of fat across enterocytes, 94–95
Arachidonic acid,
    EFA activity, 154–156
    in cat tissue, 393–394
    in sow milk, 339
    in young ruminants, 36, 158–160, 192
    indicator of EFA status, 36, 157
    interconversion with other EFA, 169
    position in EFA pathways, 14
    role of vitamin E in metabolism, 203–204
Arterio-venous difference,
    for study of lipid metabolism, 227
Autocatalytic autoxidation, 28

Bacterial lipid,
    composition, 124
    contribution to ruminant diet, 259–261
    proportions in rumen, 187
Bile,
    absorption and recycling, 89, 112–113
    action in fat digestion, 87
    as detergents, 108
    cholecystolimin in release, 113
    composition of ruminant bile, 129–130
    effect on micelles, 88
    importance in ruminant, 132
Binding protein,
    in jejunal absorption, 89
    in ruminant mucosal cells, 134
Biohydrogenation,
    in domestic ruminants, 125–127, 186–187, 252, 360
    in wild ruminants, 477–478
Body fat composition,
    effect of diet in pigs, 421–425
    effect of dietary copper, 176
    effect of forage:concentrate ratio, 417
    effect of level of intake, 425
    effect of unsaturation in ruminants, 418

Body fat composition (*cont.*)
  in the broiler, 461–463
  influence of dietary fat, 15, 44, 414–416, 487
Branched chain fatty acid,
  effect on fat quality, 418
  melting point, 25
Broiler,
  abdominal fat, 438
  composition and stability of carcass lipid, 461–463
  effect of feed restriction on fatness, 460–461
  effect of EFA deficiency, 156–158
  EFA requirements, 158–162
  fat deposition, 438–440, 457–460
  fat distribution, 444–453
  lipogenesis, 452–453
  protein:energy ratio, 453–456
  selection against abdominal fat, 441–444
By-product fats, 489–490

Calcium,
  effect of fat on ruminant requirement, 358
  in blood coagulation, 210
  interaction with zinc, 172
  maintenance of homeostasis by vitamin D, 208–209
Calcium soaps,
  as protected fat, 371–372
  in rumen, 26, 251, 358–360
  reduction of added fat availability, 287
Carbohydrate,
  effect of source on energy utilization, 271–272
  effect on fat digestibility, 316
  metabolism in young pigs, 334–336
Carbon, 14
  in measurement of tissue balance, 228
Carcass taint,
  effect of fish oil, 72
β-carotene, 207
Cat,
  energy value of fats, 385–389
  EFA metabolism, 383
  EFA requirement, 392–395
  fat soluble vitamins, 383, 396–397
  palatability of cat diets, 397–399
  tolerance to dietary fat level, 389–390
Cetoleic acid, 45–46
Chain length,
  effect on melting point, 25
  effect on micelle formation, 316
Cholecystokinin,
  in release of bile, 113
Cholesterol,
  absorption, 92, 107
  synthesis in ruminant enterocyte, 134–135
Chylomicron,
  formation of, 92–95
  metabolism of, 95–98

Chylomicron (*cont.*)
  structure and function in ruminants, 135–139
Coconut oil, 7, 13, 282
Cofactors,
  in EFA metabolism, 168, 177
Colipase, 87, 106–107
Compartation,
  of rumen microbial population, 250–251, 254–255
Competition,
  for desaturase and elongase, 36, 153, 174
Composition,
  effect of environment on plant glycerides, 10–11
  effect on fat assimilation, 114–116
  EFA composition of young ruminants, 191–192
  of broiler carcass lipid, 461–463
  of fish oil, 60
  of lipids in rumen, 124
  of ruminant bile, 129–130
  of wild and domestic tissue, 474–475
Copper,
  effect on body fat composition, 176
  effect of deficiency in pigs, 173
  effect on fat quality, 426
  effect of other minerals on pig requirement, 173–174
  effect of zinc:copper on blood cell stability, 170
  in fatty acid synthesis, 37
  in oxidation of EFA, 177
  interaction in EFA metabolism, 175–176
Critical micellar concentration, 107
Cyclopropene fatty acids, 15, 45
Cycloxygenase,
  in prostaglandin formation, 170

Deficiency,
  effect of EFA deficiency on lipoprotein, 157
  effect of EFA deficiency on phospholipid, 157
  general effects of EFA deficiency, 45, 170
  of EFA in poultry, 156–158
  of EFA in pigs, 170–171
  of vitamin A, 206
  of vitamin E, 164, 201–202
  of zinc in pigs, 171–172
  role of PUFA in zinc deficiency, 164
Deposition of fat
  in broilers, 438–440
  in cold and warm blooded animals, 65
  in pigs, cattle and sheep, 408–410
  influence of dietary factors in broilers, 451–461
Δb-desaturase, 36, 174–175
  in the cat, 383, 393
Desaturation, 14, 34, 36, 153, 174–175, 176

Diet formulation,
  for ruminants, 370–372
  with added fat, 326–327
Digestible energy,
  effect of fat structure, 283–284
  methods of determination, 277–280
Digestibility,
  effect of chain length, 61
  effect of emulsification and hydrogenation, 278
  effect of fat in ruminant diet, 360–362
  effect of fat structure, 283–284, 314, 316
  effect of inclusion rate of fat, 288–293
  effect of poorly absorbed carbohydrate, 316
  effect of unsaturation, 61, 270, 486
  factors affecting digestibility, 486–487
  of fat, for measuring energy value, 278
  of fat in cat and dog diets, 385–389
  of fish oil, 61
Diluents,
  effect on digestible energy, 317
Distribution,
  of fat in the mill, 498–500
Dog,
  energy value of fat, 385–389
  EFA metabolism, 383
  EFA requirements, 391–392
  fat malabsorption, 390
  fat soluble vitamins, 396–397
  tolerance to dietary fat levels, 389–390

Efficiency,
  of fat absorption in ruminant, 133
  of transfer of dietary fat to milk, 364
  of utilization of fat, 317–318
Egg formation,
  effect of EFA supplementation, 160
  role of vitamin D, 209
Eicosatrienoic acid, 36, 157
  in cats, 393
  in growing birds, 158–160
  in young ruminants, 192
Elongation, 34, 134, 153, 174
Emulsification,
  effect on fat digestibility values, 278
  of fat in rumen, 250
Endogenous fat,
  secretion of, 278
Energy metabolism,
  circulating lipid as energy reserves, 225–226
  effect of underfeeding, 233
  energy sources in the ruminant, 226
  influence of dietary fat in ruminants, 362–366
  of mammary tissue, 240–242
  VFA as energy sources, 229–230
Energy value,
  effect of age, species and breed, 280–283
  effect of fat structure, 283–284
  effect of inclusion level, 288–293

Energy value (cont.)
  methods of determination, 277–280
  of fat, 324–325
  of fat in cat and dog diets, 385–389
  of fish oil, 63
  prediction of, 293–294
Erucic acid,
  effect of temperature on content of rapeseed, 11
  toxic effects, 16, 45–46
  low erucic acid rapeseed, 17
Essential fatty acid,
  biochemical role, 156, 170
  composition of young ruminants, 191–192
  effects of deficiency, 45, 156–158, 170–171
  effect on egg size, 160
  excess, 164
  for cell growth, 167–168
  in sow's milk, 339
  inhibition by saturated fat, 175–176
  interaction with copper, 174–175
  interaction with zinc, 168
  metabolism in the cat, 383
  metabolism in the dog, 383
  metabolism in the rumen, 186–187
  mineral cofactors, 168, 177
  pathway in animal tissues, 14, 34, 36
  placental supply in ruminants, 192–193
  relative EFA activities, 154–156
  requirement for dog and cat, 390–395
  requirement for growing pig, 313–314
  requirement for poultry, 158–162
  requirements of a dietary source, 154
  reserves in adult birds, 158, 160
  sources of EFA, 162–164
  status in newborn ruminants, 193–195
  stepwise metabolism, 169
Extension, 34, 134, 153, 174

Fat distribution,
  in broilers, 444–451
  redistribution by dietary manipulation, 448–449
Fat quality,
  consequences for human nutrition, 473–474
  defined, 407
  effect of copper, 426
  effect of diet in pigs, 421–426
  effect of diet in ruminants, 417–419
  factors affecting fat quality, 410–416
  genetic effects in pigs, 426–428
  genetic effects in ruminants, 419–420
  in cattle and sheep, 416–421
  in pigs, 421–429
  sex effects in pigs, 428–429
  sex effects in ruminants, 420–421
Fat soluble vitamins,
  in plant fats, 13
  requirements of cats and dogs, 383, 396–397

Fat tissue,
  composition, 412–416
  effect of age on texture, 410
  effect of breed on total fat, 419–420
  effect of dietary fat, 414–416
  growth differences between depots, 411–412
  in broilers, 438–440
Fatness,
  effect of altering nutrient density, 457–460
  effect of environment, 461
  effect of genetic improvement, 440–441
  effect of protein:energy ratio, 443–444
  effect of restriction, 460–461
  in broilers, 440–441
Fatty acid,
  composition of animal lipid, 412–416
  composition of milk, 23
  effect of chain length on digestibility, 486–487
  effect of position on energy digestibility, 283–284
  effect on rumen microbial activity, 255–259
  lipoprotein lipase in removal from chylomicrons, 97
  metabolism in muscle, 234
  partitioning into micellar phase, 86–89
  regulation of synthesis, 36–38
  sites of synthesis, 226
  synthesis, 11–12, 32–36
  synthesis in rumen, 127–128
  turnover in adipose tissue, 43–44
Fibre,
  effect of dietary fat in ruminants, 302
  reduction of nutrient density, 357–358, 362
Filtration,
  of added fat, 497, 499
Fish,
  characteristics of lipids, 53
  composition of oil, 58–61
  digestibility of oil, 61
  effect of oil on lipoprotein lipase activity, 369
  effect of oil on rumen fermentation, 68–71
  energy value, 63
  production of off flavour by oil, 72
  source of EFA, 126, 162–163
  storage lipid, 55
  structure of fat, 53
  toxicity, 45–46, 71–73
Forage plants,
  typical lipids, 123
Free fatty acid,
  as energy source in muscle, 234, 236
  as indicator of nutritional value, 491
  effect of sow intake on piglet FFA, 341
  effect of stress, 227
  FFA:glyceride ratio, 487
  influence on micelle formation, 314, 316

Free fatty acid (*cont.*)
  metabolism of plasma FFA, 230–233
  plasma FFA in neonatal piglet, 336–337

Gastric lipase, 85–86, 105
Glucorticoids,
  maintenance of lipoprotein lipase activity, 98
Gluconeogenesis,
  in the newborn pig, 335–336
  stimulation by fatty acid, 337
Glucose,
  energy source in mammary tissue, 241
  fatty acid precursor, 226
  interaction with dietary fat, 271
  metabolism in muscle, 237
Glutathione peroxidase, 202
Glycerides,
  composition in plants, 8, 10–11
  major types in pig backfat, 30
  structure in animals, 28–30
Glycerol,
  absorption, 113
Glycogen,
  in the newborn piglet, 335, 340
Growth hormone,
  fat mobilization in ruminants, 365

Handling,
  fat in the milk, 372
Health,
  fat and human health, 190, 473
Heating,
  of fat storage tanks, 499
Hydrogenation,
  effect on fat digestibility in the pig, 278
  in wild ruminants, 477–478
  of lipids in the rumen, 125–127, 186–187, 252, 360
  of PUFA, 42
Hydrolysis,
  of esters, 26
  of phospholipid in the duodenum, 88, 131–132
Hyperketonaemia, 233
Hypoglycaemia,
  in the neonatal piglet, 334–335

Ideal protein, 303–304
Inclusion rate,
  effect on fat digestibility, 288–293
Interaction,
  of fat and other dietary components, 249–250, 284–288
Intramuscular fat, 409
Insulin,
  effect on lipoprotein lipase activity, 42, 98
  fat mobilization in the ruminant, 365

Ketone bodies,
  as an energy source, 236

Index 517

Ketone bodies (*cont.*)
  effect of starvation, 237
  effect of supplementary fat in ruminants, 365

Lecithin,
  in absorption, 108, 109
Linoleic acid,
  biohydrogenation, 126, 186
  changes in ruminant small intestine, 187–188
  effect on melting point of pig fat, 412
  EFA activity, 154–156
  in fish lipid, 58–60
  in forage, 123
  in sows' milk, 339
  incorporation into ruminant tissue, 189–192
  interconversion with other EFA, 169
  oxidative breakdown during cooking, 418
  requirement in pig diets, 45, 313–314
  requirements of poultry, 158–162
  selection by ruminant, 476–478
  transport and metabolism in ruminant, 188–189
Linolenic acid,
  biohydrogenation, 126
  EFA activity, 154–156, 162–163
  in fish lipid, 58–60, 66–68
  in forage, 123
  interconversion with other EFA, 169
  nutritional importance, 45
  off flavours in meat, 72
  selection by ruminants, 476–478
Linseed oil, 256–258
Lipase,
  gastric, 85–86, 105
  in herbage, 186
  pancreatic, 87, 107, 114, 130–131
  selectivity, 316
Lipid metabolism,
  in the neonatal piglet, 336–337
Lipogenesis,
  effect of dietary fat, 37
  effect of protected fat, 190
  in broilers, 452–453
  site of lipogenesis, 226
Lipolysis,
  in non-ruminants, 106–107
  in the neonatal piglet, 336
  in the rumen, 125, 186–187, 251
Lipoprotein,
  effect of EFA deficiency, 157
  formation in the ruminant enterocyte, 135
  metabolism in ruminant blood, 137–139
Lipoprotein lipase,
  activity in rat at parturition, 226
  effect of fish oil, 369
  effect of insulin, 42
  hormonal control of activity, 98
  in milk formation, 138
  regulation of plasma triglyceride distribution, 42

Lipoprotein lipase (*cont.*)
  removal of fatty acid from chylomicrons, 97
Liver,
  in fatty acid metabolism, 33, 230–233
  lipogenic activity in broilers, 452
  metabolism of acetate in ruminants, 230
  vitamin A in cat, 396

Malabsorption, 95
  in dogs, 390
Mammary gland,
  energy metabolism, 240
  fatty acid synthesis, 33
  oxidation of acetate, 241–242
Medium chain fatty acid,
  absorption, 13, 86
  energy value of, 13
  in treatment of malabsorption, 95
  occurrence, 7
Melting point,
  effect of fatty acid conformation, 25, 416
  of fish oil, 60
  of triglycerides, 30
Metabolic faecal fat loss,
  in cats and dogs, 389
Metabolism,
  energy metabolism in mammary gland, 240
  of acetate in ruminant liver, 230
  of chylomicrons, 95–98
  of digested fat in growing pigs, 317–320
  of fatty acids in the rumen, 42
  of linoleic acid in the ruminant, 188–189
  of lipid in neonatal piglet, 336–337
  of lipid in ruminant small intestine, 131–133
  of lipoprotein in ruminant, 135–139
  of short chain fatty acid, 229–230
Metabolizable energy,
  effect of added fat, 321–322
  effect of age, species and breed, 280–283
  effect of carbohydrate, 271–272
  effect of fat structure, 283–284
  effect of fat type, 270
  effect of inclusion level, 288–293
  effect of protein, 267
  effect of rate of passage, 272–273
  methods of determination, 277–280
  TME and AME, 277, 292–293
Micelle,
  absorption into mucosal cells, 110–112
  effect of chain length, 316
  effect on water insoluble material, 108
  factors governing entry of lipid, 116
  formation in the growing pig, 314
  influence of FFA on formation, 314, 316
  non-micellar absorption, 110
  partitioning of fatty acid in the gut, 88–89
  structure, 108–109
Microbial lipid,
  formation in rumen, 124, 187, 259–261

Micro-organisms,
    effect on fat assimilation in
        non-ruminants, 117
Milk,
    digestion of fat by suckling animal, 106
    effect of dietary fat, 338–340, 357, 363–369
    effect of polyethylenic fatty acids on fat
        content, 68–71
    fatty acid composition, 23
    lipoprotein lipase in formation, 138
Mixing,
    of fat supplements, 501–502
Modification,
    of fat in the rumen, 25
Mortality,
    in the neonatal piglet, 333–334
Muscle
    acetate as an energy source, 235–236
    effect of added fat, 322–324
    fatty acid metabolism, 234
    FFA as energy source, 234, 236
    substrate utilization, 234–240

NADPH,
    in mammary tissue, 240–241
Non-micellar absorption, 110
Nutrient balance,
    in relation to feed allowances, 302–305
Nutrient density,
    effect of altering nutrient density, 304, 305
    effect on broiler fat deposition, 457–460
    modification of, 302

Odour, 26
Oestrogen,
    effect on lipoprotein lipase activity, 98
Oily bird syndrome, 463
Osteocalcin, 211
β oxidation, 65–66
    of FFA, 231
Oxidative rancidity, 16–17, 26–28
    effect of copper storage vessels, 497
    in stored meat, 425
    mineral cofactors in EFA peroxidation,
        168, 177
    oxidation of EFA, 164
    role of vitamin E and selenium, 202–203
Oxidized fats,
    effects on pigs and poultry, 28
    oxidized fish oil, 71
    toxic effects in rats, 17

Palatability,
    of cat and dog diets, 397–399
Palmitic acid,
    in liver and adipose tissue, 33
Pancreatic lipase,
    action, 87
    in ruminants, 130–131
    positional specificity, 107, 114
Parakeratosis, 171–172, 175

Partitioning,
    of fatty acid, 478
    of long chain fatty acid in ruminants, 365
Peroxide value, 16, 17, 491
Peroxisome, 65–66
Pesticide residues, 494–495
Phosphatidyl choline, 92
Phospholipids,
    absorption of plant phospholipid, 13
    effects of EFA deficiency on composition,
        157
    hydrolysis in duodenum, 87–88, 131–132
    in animal tissues, 30–32
    in plant structural lipids, 3
    in ruminant intestinal lymph, 188
    incorporation of PUFA, 156
    synthesis in ruminant enterocyte, 13
Pig,
    absorption in, 107–113
    carbohydrate metabolism, 334–336
    copper metabolism, 173–174
    effect of diet on body composition,
        421–425
    effect of EFA deficiency, 170–171
    effect of hydrogenation on fat digestibility,
        278
    effect of oxidized fat, 28
    effect of temperature on energy intake,
        318–320
    EFA requirement of growing pigs,
        313–314
    fat deposition, 408–410
    fat metabolism, 317–320
    fat quality, 421–429
    hypoglycaemia, 334–335
    micelle formation, 314
    parakeratosis, 171–172, 175
    soap formation, 287
    value of supplemental fat, 320–326
    zinc requirement, 172–173
Piglet,
    effect of sow FFA intake, 341
    gluconeogenesis in neonate, 335–336
    interaction of EFA and zinc in neonate,
        177–178
    lipid metabolism of neonate, 336–337
    mortality, 333–334
    post partum glycogen, 335, 340
    post partum plasma FFA, 336–337
    supplementary fat, 354
Plant,
    absorption of plant phospholipid, 13
    composition of glycerides, 8, 10–11
    membrane lipid, 6
Polyethylene,
    contaminant in tallow rendering, 493
Polyethylenic fatty acid,
    effect on milk fat, 68–71
Polyunsaturated fatty acid,
    as prostaglandin precursors, 14
    effect on ruminant methane production, 256

Polyunsaturated fatty acid (*cont.*)
  hydrogenation, 42
  in regulation of immunity, 15
  in wild and domestic ruminants, 475
  in zinc deficiency, 164
  incorporation into phospholipid, 156
  interaction with age on absorbability, 283
  metabolism in the cat, 383, 392–395
  metabolism in the dog, 391–392
  regulation of formation of long chain PUFA, 153
Prediction,
  of dietary energy values, 293–294
Prostaglandin,
  effect of EFA deficiency, 157
  effect of oleic acid, 176
  EFA as precursors, 14, 156, 170
Protected fat,
  effect on forage intake, 363
  effect on meat quality, 418–419
  effect on milk fat composition, 363–364
  effect on VLDL, 93
  in ruminant diets, 123, 133, 190–191
  practical considerations, 371–372
Protein,
  effect of added fat on milk protein, 369
  effect on dietary protein, 172
  effect on response to added fat, 267, 287–288
  ideal protein, 303–304
  in broilers (protein:energy ratio), 453–356
Protozoal lipid,
  composition, 124
  contribution to ruminant diet, 259–261
  proportions in rumen, 187

Quality control,
  of fat supplements, 326, 491–495

Rancidity,
  through ester hydrolysis, 26
  through oxidation, 16–17, 26–28
Rapeseed,
  breeding for low erucic acid, 17
  effect of temperature on erucic acid content, 11
  toxic effects of oil in rats, 16
Rate of passage,
  effect of added fat, 286, 290
  effect on fat digestibility, 272–273
Recovered vegetable oils, 490
Reducing agents,
  in mammary tissue, 241
Requirements,
  for EFA in cats, 392–395
  for EFA in dogs, 391–392
  for EFA in growing pigs, 45, 313–314
  for EFA in poultry, 158–162
  for zinc in pigs, 172–173
Reserves,
  of EFA in poultry, 157–158, 160

Resynthesis,
  of absorbed lipid, 90–92, 113
Retinoids, 207
Rumen,
  biohydrogenation, 123–127, 186–187, 252, 360, 477–478
  compartmentation of microbial population, 250–251
  composition of lipids in rumen, 124
  effect of fat on microbial metabolism, 255–259
  effect of fish oil on fermentation, 68–70
  effect of polyethylenic fatty acids on metabolism, 68–70
  effect of PUFA on methane production, 256
  effect on fatty acids presented to duodenum, 185
  emulsification, 250
  fate of dietary fat in, 250–255
  influence of dietary fat on fermentation, 357–360
  lipolysis in, 125, 186–187, 251
  metabolism of fatty acid, 42
  modification of fat, 25
  soap formation, 26, 251
  synthesis of fatty acid, 125–128
Rumen microflora,
  enzymes, 125
  inhibition by long chain fatty acid, 358
Ruminant,
  absorption in small intestine, 128–133, 188
  binding protein in mucosal cells, 134
  biohydrogenation, 125–127, 186–187, 252, 360, 477–478
  calcium requirement, 358
  composition of bile, 129–130
  contribution of microbial lipid to diet, 259–261
  diet formulation, 370–372
  dietary fat and energy metabolism, 362–366
  effect of PUFA on methane production, 256
  efficiency of fat absorption, 133
  EFA composition of young ruminant, 191–192
  EFA metabolism, 186–187
  EFA status in neonate, 193–195
  Energy sources, 226
  fat quality, 416–420
  formation of lipoprotein, 135
  linoleic acid changes in small intestine, 187–188
  lipid metabolism in small intestine, 131–133
  metabolism of acetate, 230
  metabolism of blood lipoprotein, 137
  pancreatic lipase, 130–131
  partitioning of long chain fatty acid, 365
  phospholipid in intestinal lymph, 188

Ruminant (*cont.*)
  prolected fat, 123, 133, 190–191
  structure of chylomicron, 135–136
  synthesis of cholesterol, 134–135
  synthesis of triacyl glycerol, 134
  triene:tetraene ratio of young ruminants, 192

Saturated fatty acid,
  inhibition of EFA metabolism, 168
  potential for micelle formation in pigs, 314
  synergism with unsaturated fat, 270–271, 285–286
Selenium,
  in cat diets, 397
  in vitamin E antioxidant activity, 202–203
Shortening of fatty acid, 34
  role of peroxisomal oxidation, 66
Slip point, 30
Soap formation,
  in rumen, 26, 251
  reduction of added fat availability, 287
Source,
  of added fat, 488–490
Soybean oil equivalent, 279
Spraying,
  fat in milk, 502–504
Stearic acid,
  formation in the rumen, 126, 186
Storage,
  of fat in tanks, 497–500
  of oil in plant cells, 12
Storage lipids,
  in broilers, 439–440
  in fish, 55
  in plants, 6
Structure,
  effect on energy utilization, 283–284
  of animal fats, 23–26
  of chylomicron in ruminants, 135–136
  of fish fats, 53
  of plant fats, 3–8
  of micelles, 108
Structural lipid,
  in plants, 3
Supplemental fat,
  direct to piglet, 354
  effect on circulating ketones, 366
  effect on dietary ME content, 268–270, 321–322
  effect on medium chain fatty acid in sow milk, 354
  effect on milk production, 366
  effect on milk protein, 369
  effect on piglet composition, 340–345, 356
  effect on piglet survival, 337–338
  effect on ruminant energy metabolism, 362–366
  effect on ruminant fibre digestion, 357–358, 362
  effect on ruminant VFI, 362–363

Supplemental fat (*cont.*)
  in modification of nutrient density, 302
  mixing, 501–502
  nutritional and economic value for pigs, 320–326
  pre farrowing, 353
  quality control, 326, 491–495
  vitamin C in supplements for sows, 353
Synergism,
  of added fat with dietary fat, 288
  of saturated and unsaturated fat, 270–271, 285–286
Synthesis,
  of cholesterol in ruminant enterocyte, 134–135
  of fatty acid in animals, 32–36
  of fatty acid in plants, 11–12
  of fatty acid in the rumen, 127–128
  of glycerolipid in animals, 38–44
  of phospholipid in ruminant enterocyte, 13
  of prostaglandin, 14, 156–157, 170, 176
  resynthesis of absorbed lipid, 90, 113

Tallow,
  contamination with polyethylene, 493
  effect on sow milk composition, 339
  rendering, 488–489
  solvent extraction, 489
Tank storage,
  of fat, 497–500
Thermoregulation,
  in piglets, 334
TBA (thiobarbituric acid),
  test for rancidity, 28
Trans isomer fatty acid, 25, 155–156
Transport,
  by chylomicrons, 92–98
  in the fowl, 114, 135–139
  of fat across enterocytes, 93
  of linoleic acid in ruminants, 188–189
  of triacylglycerol in ruminants, 136–139
  of unesterified fat, 95
Triacylglycerol,
  content of broiler tissue, 439
  digestibility, 284, 314, 316
  hormonal control of tissue uptake, 98
  melting point, 30
  regulation of plasma distribution by lipoprotein lipase, 42
  synthesis in enterocyte, 90, 134
  transport in ruminants, 136–139
Triene-tetraene ratio, 36
  in poultry nutrition, 158–160
  in young ruminants, 192
Tri-stearin,
  digestibility by cats, 385
Tolerance,
  of dietary fat level by cats and dogs, 389–390

Index

Toxicity,
  of lipids, 15–17
  of vitamin A in cats, 396

Unsaturated fatty acids,
  effect on body fat composition of ruminants, 418
  effect on digestibility, 61, 270, 486
  effect on micelle formation, 314
  synergism with saturated fat, 270–271, 285–286
  synthesis in animal, 34–36

Very low density lipoprotein, 93
  effect of feeding prolected fat, 190
  effect of starvation, 42
  plasma VLDL (index of fatness), 441–442
  structure and function in ruminants, 114, 135–139

Vibrio fibrisolvens, 252

Vitamin A,
  deficiency symptoms, 206
  interaction with other vitamins, 206
  requirements of cats and dogs, 396
  toxicity to cats, 396

Vitamin C,
  in fat supplements for sows, 353

Vitamin D,
  hormonal activity, 207–208
  maintenance of Ca and P homeostasis, 208–209
  requirements of cats and dogs, 396–397

Vitamin E,
  deficiency symptoms in animals, 164, 201–202
  inhibition of free radical oxidation, 13, 28
  interaction with other vitamins, 204
  prolective functions with selenium, 202–203

Vitamin E (*cont.*)
  requirements of cats and dogs, 397
  role in arachidonic acid metabolism, 203–204

Vitamin K,
  in blood coagulation, 210
  pharmacological properties, 211–212
  requirements of cats and dogs, 397

Volatile fatty acid,
  as energy source in ruminants, 229–230

Voluntary feed intake,
  compensatory feed intake, 307
  effect of restriction on broiler fatness, 460–461
  effect of temperature on pig energy intake, 318–320
  factors affecting VFI, 305–311
  in ruminants fed fat supplements, 362–363
  response to individual amino acid levels, 310

Waxes,
  definition, 3
  digstion of plant waxes, 13

Zinc,
  effects of deficiency, 171–172
  effect of PUFA, 164
  effect of zinc:copper on blood cell stability, 169
  in diets for early weaned pigs, 177–178
  in lipid peroxidation, 177
  induction of secondary EFA deficiency, 168
  interaction on EFA metabolism, 174–175
  interaction with calcium, 172
  requirement in pig diets, 172–173